ENCYCLOPEDIA OF
ENVIRONMENTAL ISSUES

ENCYCLOPEDIA OF ENVIRONMENTAL ISSUES

REVISED EDITION

Volume 4

Shantytowns—Zoos

Appendixes

Indexes

Editor

Craig W. Allin

Cornell College

SALEM PRESS

Pasadena, California   Hackensack, New Jersey

Editor in Chief: Dawn P. Dawson

Editorial Director: Christina J. Moose   *Photo Editor:* Cynthia Breslin Beres

Development Editor: R. Kent Rasmussen   *Production Editor:* Joyce I. Buchea

Project Editor: Judy Selhorst   *Graphics and Design:* James Hutson

Acquisitions Manager: Mark Rehn   *Layout:* William Zimmerman

Research Supervisor: Jeffry Jensen

Cover photo: ©George Hammerstein/CORBIS

Library of Congress Cataloging-in-Publication Data

Encyclopedia of environmental issues / editor, Craig W. Allin. — Rev. ed.
     Planned in 4 v.
     Includes bibliographical references and index.
     ISBN 978-1-58765-735-1 (set : alk. paper) — ISBN 978-1-58765-736-8 (vol. 1 : alk. paper) — ISBN 978-1-58765-737-5 (vol. 2 : alk. paper) — ISBN 978-1-58765-738-2 (vol. 3 : alk. paper) — ISBN 978-1-58765-739-9 (vol. 4 : alk. paper)
     1. Environmental sciences—Encyclopedias. 2. Pollution—Encyclopedias.
I. Allin, Craig W. (Craig Willard)
     GE10.E523 2011
     363.7003—dc22

2011004176

PRINTED IN THE UNITED STATES OF AMERICA

Contents

Complete List of Contents

Volume 1

Volume 2

Volume 3

Volume 4

Contents lxxix
Complete List of Contents lxxxi

ENCYCLOPEDIA OF
ENVIRONMENTAL ISSUES

Shantytowns

CATEGORY: Urban environments
DEFINITION: Communities of poor people housed in small, poorly built dwellings
SIGNIFICANCE: The environmental hazards faced by the residents of shantytowns include disease caused by polluted water and air, exposure to industrial waste, and vulnerability to natural disasters.

Millions of people throughout the world cannot afford housing that is safe, sturdy, and reasonably spacious. While inhabitants of poor housing in sparsely settled farm or forest regions run certain risks, the environmental dangers attached to living in poor housing multiply when many such dwellings are huddled closely together in urban areas.

The most acute housing problems are found in the world's developing nations. Since the mid-twentieth century, millions of migrants from rural areas have crowded into these nations' cities. The jobs the cities provide are a powerful magnet for citizens whose customary farming life no longer sustains them. Indeed, in many cases the farmlands have been ruined by destructive processes such as deforestation or soil erosion. While the cities seem to promise a better life, most migrants arrive unable to afford the most basic housing with amenities. They are left to build their own shelter from whatever materials they can scavenge.

Statistics show the importance of shantytowns in housing these poorer residents. In Bogotá, Colombia, 59 percent of the population lives in self-built housing, and 60 percent of the people in Dar es Salaam, Tanzania, live in such dwellings. In Addis Ababa, Ethiopia, the proportion is close to 85 percent. It has been estimated that in most of the cities of developing nations, 70 to 95 percent of all new housing is illegally built and held.

The flimsy houses that make up shantytowns usually occupy land that is unsuitable for other purposes. Shantytowns spring up on floodplains, steep hillsides, lots adjacent to contaminated industrial sites, and rail or highway rights-of-way. In Manila, Philippines, some 30,000 squatters have lived for more than forty years on and around Smokey Mountain, a municipal dump that the squatters scavenge for materials to sell. In Cairo, Egypt, some 500,000 people live in a squatter community built amid ancient mausoleums.

Their shacks are constructed of corrugated tin, packing crates, mud bricks made on-site, and similar materials.

ENVIRONMENTAL HAZARDS

The greatest environmental dangers in these communities stem from the lack of sufficient drinking water and the lack of infrastructure to remove human and household wastes. Few residents have water piped into their houses, so they must buy water from street vendors or manually carry it from a common spigot shared by many people. With either source, the expense or effort of obtaining water means that households seldom have enough for healthy day-to-day living. In some cases residents may use water from surface sources or shallow wells; both are likely to be contaminated with biological or industrial wastes.

Without sewage treatment and disposal systems, disease vectors get cycled back into the immediate environment. On-site methods such as septic tanks or pit latrines, which are adequate for low-density conditions, break down in crowded settlements. Waterborne diseases such as dysentery, cholera, and infant diarrhea are common in shantytowns. Standing water and waste also shelter mosquitoes, which spread malaria. Dumped garbage attracts rats.

Industrial pollutants are another hazard. Many shantytowns are built next to factories at which the inhabitants work, and most such sites generate contaminants. This is doubly true of factories in developing nations, where public awareness and legal regulation of environmental toxins are much weaker than in the United States. The chemical disaster in Bhopal, India, in 1984 killed mostly low-income people who lived near the Union Carbide pesticide plant that leaked toxins into the air. Shantytown dwellers who live beside highways are subject to a constant barrage of fumes from motor traffic.

Many added environmental stresses arise in these communities. Families of four, five, or more people living in a single room readily exchange airborne infections. Some 60 percent of slum children in Kanpur, India, are estimated to have tuberculosis. The most dramatic threat, however, comes from natural disasters. Squatter settlements on bare hillsides, and houses built of mud or flimsy materials, are vulnerable to hurricane, flood, or earthquake damage.

Although their impacts on surrounding regions are hard to separate from more general urban pollution, shantytown living conditions do affect wider ar-

A shantytown on the outskirts of Cape Town, South Africa. (AP/Wide World Photos)

eas. Fecal material and garbage from urban slums travel downstream in rivers for hundreds of miles. Yards of packed dirt and lack of drainage systems slow the absorption of rainwater into the soil. The costs of residents' excess illnesses and deaths, let alone time wasted in daily tasks such as hauling water from a far-away spigot, are further drains on developing countries' limited resources.

Shantytowns are a rational response to housing needs by people who cannot afford standard houses. Merely bulldozing the shanties or forcing their residents out seldom works; such actions simply compound problems. Providing squatter villagers with a basic infrastructure of piped water and sewage disposal immediately upgrades their living conditions. It is also much cheaper, liter by liter, than the hauled water that slum dwellers buy. When such systems are provided, shantytown residents may further improve their homes with materials at hand. Economic, political, and cultural pressures on local governments of-

ten prevent such plans from going forward, however. It helps if shantytown residents first gain legal title to their dwellings and a modicum of political power. Given the urban growth rates and economic problems in developing nations, however, these problems may not be solved for many years.

Shantytowns are also found in developed countries, but in such nations they constitute a much smaller proportion of the total housing stock than they do in developing nations. Because the settlements are smaller and attention to public health issues is greater, the environmental hazards that the residents of these shantytowns face are less overwhelming, although they are still significant.

Emily Alward

FURTHER READING

Auyero, Javier, and Débora Alejandra Swistun. *Flammable: Environmental Suffering in an Argentine Shantytown.* New York: Oxford University Press, 2009.

Benton-Short, Lisa, and John R. Short. "Contemporary Urbanization and Environmental Dynamics." In *Cities and Nature.* New York: Routledge, 2008.

Hardoy, Jorge E., and David Satterthwaite. *Squatter Citizen: Life in the Urban Third World.* 1989. Reprint. London: Earthscan, 1995.

McNeill, J. R. "More People, Bigger Cities." In *Something New Under the Sun: An Environmental History of the Twentieth-Century World.* New York: W. W. Norton, 2000.

SEE ALSO: Bhopal disaster; Environmental health; Environmental illnesses; Environmental justice and environmental racism; Urban sprawl; Water quality.

Shark finning

CATEGORY: Animals and endangered species

DEFINITION: Practice of cutting off the fins of a captured shark and discarding the shark, dead or alive

SIGNIFICANCE: Shark finning has been criticized internationally for several reasons. It is considered wasteful, as it utilizes only about 5 percent of a shark's body, and cruel when the shark is not killed after finning. It also contributes to potential overfishing of a top predator, which could have severe harmful effects on species survival and marine ecosystems.

Historically, sharks have rarely been fished commercially because of the poor quality of their meat. Shark fins, however, are the prized ingredient in shark fin soup, a traditional delicacy of imperial China that is still served around the world, although primarily in China. Shark fin soup is valued as the expensive end product of an elaborate, days-long cooking process. It is built around the special texture of the needlelike collagenous fibers inside the fins, which are tasteless but are given flavor by additional ingredients. In 2010 it was reported that shark fins were selling for as much as $700 per kilogram (about $318 per pound) on the Hong Kong market.

Since the 1980's, as increasing numbers of Chinese have been able to afford this special dish, demand for shark fins has risen consistently. Traditionally, commercial fishers caught sharks merely as unwanted bycatch, particularly when they were using long-line or purse-seine fishing gear. When a shark was caught, the fishers would remove the shark's fins and discard the shark back into the ocean to save storage space on their boats, as the rest of the shark's body was of so little value.

In the well-documented practice that has arisen since demand for shark fins became high, fishers capture sharks intentionally to take their fins. Typically, a shark's trademark first dorsal (back) fin is carefully sliced off so that it is not contaminated with unwanted matter at the base of the fin. The same is done to the

An Australian Customs Service official displays drying shark fins found on a suspected illegal fishing boat in the waters off Cape Wessel in northern Australia. The practice of shark finning is widely criticized as cruel and wasteful. (AP/Wide World Photos)

two pectoral (front) fins and the lower half of the caudal (tail) fin. These four parts are most valuable and are typically sold as a set. Less valuable are the so-called chips, the lower dorsal and three pelvic fins, which are also taken.

Generally, shark finners do not bother to kill a shark after finning it. They simply discard it into the ocean, where, as it can no longer swim, the shark is doomed to a slow death by suffocation, by bleeding, or as prey to other fish. It was especially this latter practice that became the subject of international outrage beginning in the late 1990's, combined with environmentalists' concerns that increased shark fishing overall could lead to unsustainable overfishing of shark species.

Scientists have found it very difficult to gather precise data on the numbers and species of sharks that are killed, whether for their fins or for other reasons. Because sharks were considered only unwanted by-catch for many years, few fisheries reported on their numbers. A study in 2006 estimated that between 26 and 73 million sharks were being killed for their fins annually, with a median of 38 million. That number was at least three times as much as previously had been estimated by the United Nations.

Growing concerns about shark populations and outrage against shark finning have led international agencies such as the United Nations Food and Agriculture Organization (FAO) to address the issue. In 2000 FAO published *International Plan of Action for the Conservation and Management of Sharks*, a document discouraging the practice of shark finning and promoting the full utilization of any caught sharks. When FAO reviewed the plan in 2005, it found that global implementation of its voluntary guidelines was still seriously lacking. By 2010, however, a number of individual governments, including those of the United States, Canada, Japan, the Republic of China (Taiwan), and the European Community, had adopted legal guidelines that generally ban the practice of shark finning in their waters or by their fishing fleets.

R. C. Lutz

FURTHER READING

Abel, Daniel C., and Robert L. McConnell. "The Demise of Sharks." In *Environmental Oceanography: Topics and Analysis.* Sudbury, Mass.: Jones and Bartlett, 2010.

Carrier, Jeffrey C., John A. Musick, and Michael R. Heithaus, eds. *Sharks and Their Relatives II: Biodiver-*

sity, Adaptive Physiology, and Conservation. Boca Raton, Fla.: CRC Press, 2010.

Musick, John A. "Shark Utilization." In *Management Techniques for Elasmobranch Fisheries,* edited by John A. Musick and Ramon Bonfil. Rome: United Nations Food and Agriculture Organization, 2005.

Skomal, Greg, and Nick Caloyianis. *The Shark Handbook: The Essential Guide for Understanding the Sharks of the World.* Kennebunkport, Maine: Cider Mill Press, 2008.

SEE ALSO: Commercial fishing; Extinctions and species loss; Fisheries; Food and Agriculture Organization; Food chains; Gill nets and drift nets; Public opinion and the environment; Sea Shepherd Conservation Society; Turtle excluder devices.

Sick building syndrome

CATEGORY: Human health and the environment
DEFINITION: Illness effects among building occupants that are linked to time spent in the building and for which no other apparent causes can be identified
SIGNIFICANCE: Awareness of the health problems caused by indoor air pollution and the recognition of sick building syndrome led governments in some industrialized nations to establish guidelines for acceptable levels of gaseous indoor air pollutants and to develop recommendations regarding procedures for detoxifying indoor spaces.

The illness known as sick building syndrome (SBS) is often a problem in office buildings but can be evident in homes, schools, nurseries, and libraries. In industrialized nations, the average person spends 80 to 90 percent of the day indoors. The longer an individual spends in an affected environment, the more severe the symptoms of SBS can become. Since the 1970's, buildings have become more airtight in response to concerns about conservation of energy. Many heating, ventilating, and air-conditioning systems are designed to recirculate indoor air rather than to draw in fresh, filtered air from outside. Also, systems that bring in polluted outdoor air may contribute to building-related illnesses. Lack of ventilation is one primary cause of the trapping of natural and anthropogenic (human-caused) indoor pollutants and the onset of SBS.

The symptoms of SBS appear to increase in severity with time spent in an affected building and decrease or disappear with time away from the building. Symptoms include respiratory problems such as coughing; increased allergic reactions and sneezing; irritation of the eyes, throat, nose, and skin; and headaches. Some people experience severe fatigue and flulike symptoms. Not all individuals exposed to the same indoor contaminants experience similar symptoms, and some may feel no ill effects. Microenvironments may exist within a building; for example, inhabitants of offices sharing a common wall might experience unique symptoms. Many symptoms of SBS are similar to those of common allergies or illnesses, and often sufferers do not seek help or may be misdiagnosed. The rise in SBS cases seen in the late twentieth century may be related to several factors, such as increased usage of synthetic building materials, higher stress and greater regimentation in workplaces, and a general increase in number of workers employed in office settings.

There is no single cause of SBS, but many sources of indoor pollution may contribute to SBS, including secondhand tobacco smoke, ozone and heat from photocopiers and computers, off-gassing of volatile organic compounds (compounds composed exclusively of hydrogen and carbon) from new carpet and furniture, asbestos, lead in paint, formaldehyde, microbials, respirable and inhalable particulates, dust mites, and gases such as carbon monoxide. Biological contaminants that are found where temperature and moisture levels are high are bacteria, pollen, mildew, and mold, all of which can increase SBS symptoms when inhaled. The presence of computers seems to be a significant contributor to the onset of SBS because of their heat production, which often results in the installation of air-conditioning systems that can exacerbate the problem.

Sick Building Syndrome and Building-Related Illness

The U.S. Environmental Protection Agency differentiates between sick building syndrome and building-related illness:

The term "sick building syndrome" (SBS) is used to describe situations in which building occupants experience acute health and comfort effects that appear to be linked to time spent in a building, but no specific illness or cause can be identified. The complaints may be localized in a particular room or zone, or may be widespread throughout the building. In contrast, the term "building related illness" (BRI) is used when symptoms of diagnosable illness are identified and can be attributed directly to airborne building contaminants. . . .

Indicators of SBS include:

- Building occupants complain of symptoms associated with acute discomfort, e.g., headache; eye, nose, or throat irritation; dry cough; dry or itchy skin; dizziness and nausea; difficulty in concentrating; fatigue; and sensitivity to odors.

- The cause of the symptoms is not known.

- Most of the complainants report relief soon after leaving the building.

Indicators of BRI include:

- Building occupants complain of symptoms such as cough; chest tightness; fever, chills; and muscle aches.

- The symptoms can be clinically defined and have clearly identifiable causes.

- Complainants may require prolonged recovery times after leaving the building.

Source: Environmental Protection Agency, Office of Air and Radiation, "Indoor Air Facts No. 4 (Revised): Sick Building Syndrome" (2006).

SBS is generally distinguished from building-related illnesses that are caused by exposure to specific indoor contaminants that can be definitively diagnosed. Alveolitis, bronchospasm, rhinitis, and conjunctivitis can be caused by the lodging of airborne allergens in the alveoli of the lungs, the bronchi of the lungs, the nose, or the eyes, respectively. Legionnaires' disease is associated with bacteria commonly transmitted through contaminated water sources, such as evaporative condensers or potable water distribution systems. SBS is rarely attributed to one specific exposure and is therefore more difficult to detect and treat.

Public health officials have advised that seemingly minor complaints by office workers should be accepted as serious, and, if SBS is suspected, investigation should begin to determine if the building's indoor air quality is acceptable and to identify the cause of the suspected outbreak. If no obvious breakdown of ventilation systems or major pollution sources are identified, additional precautionary measures should be taken. Filters should be cleaned, ventilation systems checked, humidity conditions lowered, and new carpets and paints allowed to off-gas and settle before any workers are expected to spend time in newly renovated areas. Increasing the rate at which the ventilation system supplies outdoor air to the interior has also been found to be helpful.

Alternative solutions include the addition of common houseplants that extract contaminants in the air through their leaves, particularly formaldehyde, benzol, phenol, and nicotine. A study conducted at the Botanical Institute at the University of Cologne in Germany showed that certain hydroculture plants (plants grown in a nonsoil medium) are efficient at absorbing pollutants and transforming 90 percent of these chemical substances into sugars, oxygen, and new plant material. These plants provide moisture to the air without contributing fungus spores.

Diane Stanitski-Martin

FURTHER READING

Godish, Thad. *Sick Buildings: Definition, Diagnosis, and Mitigation.* Boca Raton, Fla.: Lewis, 1995.
Murphy, Michelle. *Sick Building Syndrome and the Problem of Uncertainty: Environmental Politics, Technoscience, and Women Workers.* Durham, N.C.: Duke University Press, 2006.
Redlich, Carrie A., Judy Sparer, and Mark R. Cullen. "Sick-Building Syndrome." *The Lancet* 349 (April 5, 1997): 1013-1016.
Vallero, Daniel. "Indoor Air Quality." In *Fundamentals of Air Pollution.* 4th ed. Boston: Elsevier, 2008.

SEE ALSO: Air pollution; Asbestos; Carbon dioxide; Carbon monoxide; Indoor air pollution; Radon; Secondhand smoke.

Sierra Club

CATEGORIES: Organizations and agencies; activism and advocacy; preservation and wilderness issues
IDENTIFICATION: American environmental organization
DATE: Established in 1892
SIGNIFICANCE: The Sierra Club is one of the largest, oldest, and most influential conservation organizations in the world. Unique among large environmental groups because of its reliance on volunteer activists and its democratic structure, the Sierra Club pioneered many grassroots political techniques in its efforts to preserve wilderness and protect parks and other natural areas.

John Muir, amateur naturalist and writer, discovered for himself the wonders of Yosemite Valley in California's Sierra Nevada in 1868. He soon realized that such areas needed to be protected from development and resource extraction, and he used his writings to influence his friends and others to lobby for the creation of Yosemite National Park.

Muir conceived of the idea of an organization that would ensure the protection of Yosemite and the surrounding wildlands, and in 1892 the Sierra Club was born, with Muir as its first president. The club's stated purpose was

> to explore, enjoy, and render accessible the mountain regions of the Pacific Coast . . . and to enlist the support and cooperation of the people and government in preserving the forests and other natural features of the Sierra Nevada.

This statement set the tone for an organization that went on to combine the recreational goals of a hiking and climbing club with political savvy and influence. Muir's idea was to build a constituency for nature by getting people out into the mountains, showing them areas that needed to be saved, and explaining how those areas were endangered, a technique that the Sierra Club has now used successfully for more than a century.

After the relatively easy success of the creation of Yosemite National Park, the Sierra Club's next major campaign was to be more difficult and ultimately unsuccessful. In 1907 a dam and reservoir to supply water to San Francisco were proposed for Hetch Hetchy Valley, an area within Yosemite National Park that many considered to be nearly equal to Yosemite Valley in scenic grandeur. A long campaign to prevent the building of the dam, and thus the flooding of the valley, revealed dissent within the Sierra Club and made clear the difficulties inherent in the organization's strictly democratic structure. The battle was finally lost, and, through an act of Congress, the dam was built. Out of this failure, however, the Sierra Club gained experience in managing its own growing organization and building national support for its views by using media publicity and connections with a network of other conservation organizations.

Muir died in 1914, one year after the Hetch Hetchy defeat, but the Sierra Club continued to grow in membership and influence. The outings program introduced people to the wonders of the Sierra Nevada and other wild areas as far afield as Montana and the Canadian Rocky Mountains. Club members developed mountaineering and rock-climbing techniques, as well as low-impact camping ethics. On the conservation front, the club was involved in campaigns to create Kings Canyon and Olympic national parks and to prevent the damming of rivers in Yellowstone and Glacier parks.

In 1951 another proposed dam project pushed the Sierra Club to national prominence and influence. Another federally protected area, in this case Utah's Dinosaur National Monument, was to be the site for the Echo Park Dam and another dam as part of the Colorado River Project. Remembering the loss of Hetch Hetchy, the club vowed to fight harder this time and hired David Brower as its first executive director. Brower led the successful campaign against these dams with float trips on the river for influential politicians and members of the media, articles in national magazines, and a film titled *Wilderness River Trail.* As part of the compromise that saved the canyons at Dinosaur National Monument, the Sierra Club agreed not to challenge other dams in the project. One of those, the Glen Canyon Dam in Arizona, flooded a canyon that turned out to be a magnificent slickrock gorge that Brower and others believed, in

Naturalist John Muir, first president of the Sierra Club. (Library of Congress)

retrospect, should have been saved as well. This was another hard lesson learned by the Sierra Club, and one that made club members and leaders suspicious of compromises.

Brower led the Sierra Club for sixteen years, through a period of major expansion. He opened the membership rolls and actively sought new members from the public. He developed a publications program, putting the Sierra Club name on a series of calendars and coffee-table books that combined beautiful photography with messages of preservation for specific areas. To prevent yet another dam, this time in the Grand Canyon, Brower used full-page advertisements in *The New York Times* to enlist the general public in a letter-writing campaign to dissuade Congress from authorizing the project.

The Sierra Club was also instrumental in creating the field of environmental law. For many years, conservation organizations had a difficult time pursuing court cases regarding land use because they were deemed to have no "standing"—that is, no financial stake—in the decisions being made. In a case in New York State in the late 1960's, attorneys representing local conservation organizations and the Sierra Club won the right to claim standing in land-use cases based on recreational, conservational, and aesthetic interests. In 1971 the Sierra Club created the Sierra Club Legal Defense Fund (renamed Earthjustice in 1997), a legally and financially distinct organization that would represent the club, other conservation organizations, and individuals in environmental litigation.

By the beginning of the twenty-first century, its policies driven by the interests of its more than one million members, the Sierra Club had taken on issues of water and air pollution, recycling, nuclear energy, population, and global warming. Through public education, grassroots letter-writing campaigns, lobbying, and litigation, the Sierra Club continues to affect environmental opinion and policy in the United States and, increasingly, the world.

Joseph W. Hinton

FURTHER READING

Bevington, Douglas. *The Rebirth of Environmentalism: Grassroots Activism from the Spotted Owl to the Polar Bear.* Washington, D.C.: Island Press, 2009.
Cohen, Michael P. *The History of the Sierra Club, 1892-1970.* San Francisco: Sierra Club Books, 1988.
Jones, Holway R. *John Muir and the Sierra Club: The Battle for Yosemite.* San Francisco: Sierra Club Books, 1965.
McCloskey, J. Michael. *In the Thick of It: My Life in the Sierra Club.* Washington, D.C.: Island Press, 2005.
Miller, Norman. *Environmental Politics: Stakeholders, Interests, and Policymaking.* 2d ed. New York: Routledge, 2009.
Turner, Tom. *Sierra Club: One Hundred Years of Protecting Nature.* San Francisco: Sierra Club Books, 1991.

SEE ALSO: Brower, David; Echo Park Dam opposition; Glen Canyon Dam; Hetch Hetchy Dam; Muir, John; National parks; *Sierra Club v. Morton*; Yosemite Valley.

Sierra Club v. Morton

CATEGORIES: Treaties, laws, and court cases; preservation and wilderness issues
THE CASE: U.S. Supreme Court decision regarding the principle of standing in U.S. environmental law
DATE: Decided on April 19, 1972
SIGNIFICANCE: *Sierra Club v. Morton* opened the federal courts to a wealth of environmental litigation because it settled the question about whether an injury to a noneconomic interest could provide the basis for a challenge to a federal agency decision.

During the early 1970's the Sierra Club sued the U.S. Forest Service to prevent the agency from approving permits that would allow Walt Disney Enterprises to construct a $35 million complex of motels, restaurants, swimming pools, and ski trails in Mineral King Valley, a quasi-wilderness area located in the Sierra Nevada of California. Up to fourteen thousand visitors per day were expected to gain access to the resort by using a 32-kilometer (20-mile) highway to be built, in part, through Sequoia National Park. The federal district court granted an injunction, but the federal court of appeals reversed it. The Supreme Court did not consider whether the building of the proposed development would violate federal law; rather, its focus was on whether the Sierra Club, as an organization with a special interest in the preservation of national parks and forests, had standing to challenge a federal agency's decision to issue the permits.

The Supreme Court had addressed the standing issue in *Association of Data Processing Service Organizations*

Justice Douglas's Dissent

In his dissenting opinion in the case of Sierra Club v. Morton, *Justice William O. Douglas argued for the legal standing of environmental objects:*

The critical question of "standing" would be simplified and also put neatly in focus if we fashioned a federal rule that allowed environmental issues to be litigated before federal agencies or federal courts in the name of the inanimate object about to be despoiled, defaced, or invaded by roads and bulldozers and where injury is the subject of public outrage. Contemporary public concern for protecting nature's ecological equilibrium should lead to the conferral of standing upon environmental objects to sue for their own preservation. . . . This suit would therefore be more properly labeled as *Mineral King v. Morton.*

Inanimate objects are sometimes parties in litigation. A ship has a legal personality, a fiction found useful for maritime purposes. The corporation sole—a creature of ecclesiastical law—is an acceptable adversary and large fortunes ride on its cases. The ordinary corporation is a "person" for purposes of the adjudicatory processes, whether it represents proprietary, spiritual, aesthetic, or charitable causes.

So it should be as respects valleys, alpine meadows, rivers, lakes, estuaries, beaches, ridges, groves of trees, swampland, or even air that feels the destructive pressures of modern technology and modern life. The river,

for example, is the living symbol of all the life it sustains or nourishes—fish, aquatic insects, water ouzels, otter, fisher, deer, elk, bear, and all other animals, including man, who are dependent on it or who enjoy it for its sight, its sound, or its life. The river as plaintiff speaks for the ecological unit of life that is part of it. Those people who have a meaningful relation to that body of water—whether it be a fisherman, a canoeist, a zoologist, or a logger—must be able to speak for the values which the river represents and which are threatened with destruction. . . .

The voice of the inanimate object, therefore, should not be stilled. That does not mean that the judiciary takes over the managerial functions from the federal agency. It merely means that before these priceless bits of Americana (such as a valley, an alpine meadow, a river, or a lake) are forever lost or are so transformed as to be reduced to the eventual rubble of our urban environment, the voice of the existing beneficiaries of these environmental wonders should be heard.

Perhaps they will not win. Perhaps the bulldozers of "progress" will plow under all the aesthetic wonders of this beautiful land. That is not the present question. The sole question is, who has standing to be heard?

v. Camp (1970), in which it had held that people who seek judicial review of a federal agency's action under Section 10 of the Administrative Procedure Act have to claim that the agency caused them injury in fact and that the injury was a harm within the zone of interests protected or regulated by statutes the agency was said to have violated. In *Sierra Club v. Morton,* Justice Potter Stewart's opinion for the Court addressed only the "injury in fact" element of the *Data Processing* test. The Court accepted the Sierra Club's claim that noneconomic injury constitutes injury in fact, that the "change in the aesthetics and ecology" caused by Mineral King's development "would destroy or otherwise adversely affect the scenery, natural and historic objects and wildlife of the park and would impair the enjoyment of the park for future generations." However, the Court rejected the Sierra Club's argument that it did not have to claim that its members would be adversely affected by the Mineral King development because the club's long-standing concern for and expertise in environmental matters gave it standing as a "representative of the public." In denying standing,

the Court held that an organization's sincere interest in an environmental problem, even if the interest is of long duration and the organization is highly qualified to speak on behalf of the public, is not enough to satisfy the "injury in fact" requirement. If it were, the Court feared, there would be "no objective basis on which to disallow a suit by any other bona-fide organization no matter how small or short-lived."

Justice William O. Douglas, in an eloquent dissent, argued that the case should have been entitled *Mineral King v. Morton* and that the Court should have designed a standing rule that would have permitted the Sierra Club to litigate the Forest Service use permit on behalf of the valley. Drawing upon and citing Christopher Stone's law review article "Should Trees Have Standing?," Douglas proceeded to sketch the broad outlines of an imaginative redefinition of standing. The law, he said, indulges a fiction that inanimate objects such as ships and corporations are people and may, therefore, be parties to litigation. So should it be with valleys, such as Mineral King, and with lakes, rivers, and forests. Who should speak for these inanimate objects and defend their rights? Congress, he argued, is "too remote . . . and too ponderous." Federal agencies, including the Forest Service, "are notoriously under the control of powerful interests." Only those who have an intimate relationship with valleys, lakes, rivers, and forests, because they hike, fish, or "merely sit in solitude and wonderment," may speak for the values that these natural objects represent.

Justice Harry Blackmun's dissent was much more direct in its criticism of the Court's "practical" decision. The Court's "somewhat modernized" conception of standing, he argued, was not adequate to deal with the novel issues raised by the deteriorating state of the environment. He suggested two alternatives: Either approve the district court's decision on the condition that the Sierra Club amend its complaint to comply with the Court's standing rule or redefine standing, as Justice Douglas had, to permit any bona fide environmental organization, such as the Sierra Club, to litigate on behalf of Mineral King. He did not fear, as the majority on the Court did, that an expanded definition of standing would open a Pandora's box of litigation, and he had much greater faith that appropriate restraints could be imposed on an "imaginative expansion" of standing.

Sierra Club v. Morton opened the federal courts to a wealth of environmental litigation because it settled the question left open by the *Data Processing* case about whether an injury to a noneconomic interest could provide the basis for a challenge to a federal agency decision. The Court further broadened its standing test in *United States v. Students Challenging Regulatory Agency Procedures* (1973) and *Duke Power v. Carolina Environmental Study Group* (1978) by allowing environmental groups to gain standing based on tenuous claims of causation between a proposed federal agency action and fairly speculative injuries. In *Lujan v. National Wildlife Federation* (1990), the Court tightened up standing and made it more difficult for environmental groups to gain access to federal courts and challenge federal programs by requiring them to allege that the specific lands involved were actually used by their members.

William Crawford Green

FURTHER READING

Buck, Susan J. *Understanding Environmental Administration and Law.* 3d ed. Washington, D.C.: Island Press, 2006.

Cox, Robert. "Public Participation in Environmental Decisions." In *Environmental Communication and the Public Sphere.* 2d ed. Thousand Oaks, Calif.: Sage, 2010.

Findley, Roger W., and Daniel A. Farber. *Environmental Law in a Nutshell.* 7th ed. St. Paul, Minn.: Thomson/West, 2008.

Hoban, Thomas, and Richard Brooks. *Green Justice: The Environment and the Courts.* 2d ed. Boulder, Colo.: Westview Press, 1996.

Stone, Christopher D. *Should Trees Have Standing? Law, Morality, and the Environment.* 3d ed. New York: Oxford University Press, 2010.

SEE ALSO: Kings Canyon and Sequoia national parks; National parks; Nature preservation policy; Preservation; Sierra Club.

Silent Spring

CATEGORIES: Activism and advocacy; pollutants and toxins

IDENTIFICATION: Book by Rachel Carson that presents an account of the dangers of toxic substances in the environment

DATE: Published in 1962

SIGNIFICANCE: Often credited with helping to launch the modern environmental movement, *Silent Spring* exposed the toxic effects of chemical pesticides on the natural environment. The national debate sparked by the book led to an investigation of pesticide use by the U.S. Congress.

Silent Spring is the best-known work of marine biologist and ecologist Rachel Carson, who is widely considered to be the founder of the research field of environmental ethics. *Silent Spring*, like Carson's many pamphlets and other books on conservation and natural resources, frames an environmental ethics around the holistic thesis that human beings are but a single part of the whole of nature, distinguished primarily by their power to alter the natural world, which all too often they do for the worse and irreversibly.

Carson was motivated to write *Silent Spring* by a report from friends regarding the broad lethal effects of the aerial spraying of the insecticide dichlorodiphenyl-trichloroethane (DDT) on the wildlife at their bird sanctuary. The book was more than four years in preparation; Carson investigated the effects of chemical pesticides on the chain of natural life, poring over the work of a wealth of researchers, and secured damning evidence of the heedless pesticide poisoning of American air, rivers, and soils. The title of the book derives from its apocalyptic opening chapter, which pictures how the world would look and sound if indiscriminate spraying of pesticides were to continue. With *Silent Spring*, Carson challenged the U.S. government, agricultural scientists, and chemical pesticide producers with evidence of ecological and societal carelessness in the unrestricted use of chemical toxins. She warned the public that pesticides that had been in widespread use since World War II had not been tested to determine their long-term effects.

Silent Spring launches its environmental ethics from the grounding assumption that because human beings are an integral part of the natural world, the largely confrontational approach that humans, and in particular the scientific community, have traditionally assumed toward nature is an inappropriate one, governed by notions of control and manipulation. Since this orientation places human beings in a state of war against nature, it simultaneously sets them at war against themselves, as earthly beings. Carson contends that humankind's integral position and remarkable scientific and technological power places on human beings a moral requirement: to approach their relationship with nature as a calling, a moral duty to stewardship.

Responses

When it first appeared in 1962, *Silent Spring* immediately sparked a highly publicized national debate between opponents and proponents of the use of synthetic chemical pesticides. The public was shocked to learn the extent of the peril associated with the widespread use of pesticides across the country. The response from the chemical industry and the government was to name Carson an alarmist; she was accused of being excessively emotional and lacking in rational objectivity—in short, unscientific. The Monsanto Company, one of the nation's largest chemical companies, launched a counterattack to *Silent Spring* that included commissioning articles that adopted Carson's poetic style but presented a different apocalyptic vision, one in which insects and other pests strip the countryside, leaving the planet uninhabitable by humans.

Despite the virulent counterattacks, Carson's credibility was affirmed in 1963 when she was called as an expert to testify before the U.S. Congress, which was investigating the dangers of pesticide use. In her testimony she repeated the main theses of *Silent Spring*, revealing that since the 1940's the number of chemicals created to control insects, rodents, weeds, and other organisms had proliferated to more than two hundred, and that poisonous sprays, dusts, and aerosols, under thousands of brand names, were being applied universally on farms, in forests, in homes, and in gardens. These chemical toxins are so dangerous for people and the environment, Carson argued, that they should not be called insecticides but biocides, for their ability to poison all earthly life. Carson called on Congress to take decisive action to protect human health and the environment against the toxins.

Even before its publication date, *Silent Spring* had become a best seller, and the following year (1963), when it was published in England, it again reached best-seller status. One of this work's lasting effects has been to bring the topic of environmental ethics into the public sphere. *Silent Spring* gave pause to a generation that had previously trusted the future of the natural and human world to science, the government, and corporate interests. Carson died in 1964, after a long battle against cancer, but the fight she began during the 1950's to preserve the beauty and integrity of earthly life continues to inspire new generations. The fierce debate touched off by *Silent Spring* continues in the twenty-first century among scientists and philosophers as much as among laypersons: whether the purpose of science is to dominate and alter nature to serve human purposes or to preserve nature, study its mysteries, and find the place of humankind within it.

Wendy C. Hamblet

Further Reading

Carson, Rachel. *Silent Spring*. 40th anniversary ed. Boston: Houghton Mifflin, 2002.

Dunlap, Thomas R., ed. *DDT, "Silent Spring," and the Rise of Environmentalism: Classic Texts*. Seattle: University of Washington Press, 2008.

Lytle, Mark Hamilton. *The Gentle Subversive: Rachel Carson, "Silent Spring," and the Rise of the Environmental Movement*. New York: Oxford University Press, 2007.

Murphy, Priscilla Coit. *What a Book Can Do: The Publication and Reception of "Silent Spring."* Amherst: University of Massachusetts Press, 2005.

SEE ALSO: Agricultural chemicals; Air pollution; Anthropocentrism; Biocentrism; Carcinogens; Carson, Rachel; Dichloro-diphenyl-trichloroethane; Dioxin; Ecocentrism; Environmental ethics; Extinctions and species loss; Habitat destruction; Land pollution; Pesticides and herbicides; Soil contamination.

Silicosis

CATEGORY: Human health and the environment
DEFINITION: Disabling condition of the lungs caused by the inhalation of crystalline silica
SIGNIFICANCE: The recognition of silicosis as a distinct occupational illness led industrialized nations to create regulations aimed at protecting workers from inhaling crystalline silica. Such regulations have not been adopted in many less developed nations, however.

Silica, or silicon dioxide, often referred to as quartz, is one of the most common minerals on earth, occurring in a wide variety of natural and industrial settings. It is found in almost all rock and sand; sandstone, for example, is composed primarily of silica. Silicosis thus occurs most often as an occupational illness among workers in what are known as the dusty trades: foundry work, construction, and mining.

Ancient Greek and Roman physicians recognized the high risk of lung diseases among quarrymen and miners, which is one reason criminals were often sentenced to work at those occupations. Still, researchers did not recognize silicosis as a distinct disorder until almost the twentieth century. Indeed, following the discovery of the tuberculin bacillus during the nineteenth century, some physicians disputed the existence of environmental disorders such as silicosis. Rather than accepting that a specific environmental factor, such as high levels of quartz dust, caused disabling lung conditions, many physicians blamed workers' ill health on a combination of unsanitary living conditions and bacterial infections. By the mid-twentieth century, however, health professionals recognized silicosis as a distinct occupational illness.

Silicosis is considered to be an environmental illness because the disorder is caused by exposure to silica rather than transmitted as an infectious disease. Although silica is not a toxic substance in itself, crystalline silica can scar sensitive lung tissue and lead to the development of fibrotic nodules in the lungs. As these nodules grow in size and number, the lungs stiffen, and breathing becomes more difficult. Depending on the level of exposure to silica, the onset of disabling and even fatal silicosis can occur within only a few weeks or months. More commonly, however, silicosis takes many years to develop. Sufferers of silicosis may eventually become disabled from diminished lung capacity and may be at higher risk for developing other diseases such as tuberculosis and lung cancer. In some cases, the decreased flow of blood to the lungs caused by silicosis can cause the heart to enlarge. Although silicosis is becoming increasingly rare in industrialized nations because of improvements in industrial hygiene, more than two hundred people still die in the United States each year from the effects of the disease.

In most industrialized nations, workers whose jobs expose them to high levels of crystalline silica are now required to wear protective masks and respirators. Some environmental ethicists are concerned that more stringent occupational health and safety standards in countries such as the United States and Canada could cause industries to move dusty, high-risk manufacturing operations to less developed nations. Rather than improving working conditions at existing foundries in the American Midwest, for example, some corporations have chosen to close those facilities and build plants in countries where regulations are perceived as less burdensome, workers are less well educated and therefore unlikely to be aware of the risks associated with silica, and no system of workers' compensation insurance for occupational diseases exists.

Nancy Farm Männikkö

FURTHER READING

Blanc, Paul D. *How Everyday Products Make People Sick: Toxins at Home and in the Workplace.* Berkeley: University of California Press, 2007.

Greenberg, Michael I., et al., eds. *Occupational, Industrial, and Environmental Toxicology.* 2d ed. St Louis, Mo.: C. V. Mosby, 2003.

Waite, Donald E. *Environmental Health Hazards: Recognition and Avoidance.* Rev. ed. Columbus, Ohio: Environmental Health Consultants, 2002.

SEE ALSO: Asbestos; Environmental illnesses; Environmental justice and environmental racism; Hawk's Nest Tunnel disaster.

Silkwood, Karen

CATEGORY: Nuclear power and radiation
IDENTIFICATION: American nuclear industry worker
    and union activist
BORN: February 19, 1946; Longview, Texas
DIED: November 13, 1974; near Crescent,
    Oklahoma
SIGNIFICANCE: Until her death in an automobile
    crash Silkwood was not widely known, but after the
    accident, which occurred under mysterious cir-
    cumstances, many antinuclear activists saw her as a
    martyr to their cause.

Karen Silkwood, the daughter of Bill and Merle
Silkwood, grew up in the oil and gas fields around
Nederland, Texas. She eloped with Bill Meadows after
one year of study at Lamar College, where she was en-
rolled in a course of study leading to a degree in medical technology. The couple separated in 1972, and
their three children remained with their father.
Silkwood moved to Oklahoma to be near her parents
and found work as a laboratory technician with the
Kerr-Magee Corporation at its Cimarron plant in the
town of Crescent.

Silkwood joined the Oil, Chemical, and Atomic
Workers Union and was soon involved in actions protesting Kerr-Magee's lax health and safety procedures. One allegation she made was that the company
had falsified records to cover up the fact that it was
missing some nuclear material. Her concern in this
area led to her being called to testify before the
Atomic Energy Commission in Washington, D.C., in
the early part of 1974. At that time Silkwood apparently agreed to clandestinely obtain film evidence of
poor workmanship in Kerr-Magee's manufacture of
nuclear reactor fuel rods.

In early November of 1974, monitoring equipment
in the Kerr-Magee plant detected that Silkwood had
become contaminated with a radioisotope. Further
testing showed that the contamination was also present in her apartment. Silkwood was taken to the Los
Alamos National Laboratory, where more refined
testing was performed. It was determined that the
level of the contamination was not serious and did not
constitute a threat to her health.

On the night of her death, Silkwood left a union
meeting supposedly carrying an envelope that contained proof of wrongdoing by Kerr-Magee. She was
on her way to a meeting with Drew Stephens, a *New
York Times* reporter and union representative, when
her car crashed into a culvert alongside a dry, straight
section of Oklahoma Highway 74. Silkwood was killed
in the crash, and the envelope was never recovered.
The Oklahoma Highway Patrol concluded that she
had fallen asleep while driving, but a private investiga-
tor concluded that her car had been forced off the
road.

The resulting controversy, charges of a cover-up,
and lawsuits provided a focal point for those concerned with safety in the nuclear power industry. Congressional hearings brought forth intriguing and bizarre stories but led to no definite conclusions. The
Atomic Energy Commission confirmed that safety violations had occurred at the Cimarron plant, and it was
eventually closed. After years in court, most of the
questions surrounding Silkwood's death remained
unanswered, but her father did win a large settlement
from Kerr-Magee on behalf of her children. Her
death and the events leading up to it are portrayed in
the award-winning motion picture *Silkwood* (1983).

Kenneth H. Brown

FURTHER READING
Lief, Michael S., H. Mitchell Caldwell, and Benjamin
    Bycel. "Death by Plutonium: Fallout from Karen
    Silkwood's Death Brings the Nuclear Industry to
    Its Knees." In *Ladies and Gentlemen of the Jury: Great-
    est Closing Arguments in Modern Law.* 1998. Reprint.
    New York: Charles Scribner's Sons, 2008.
Rashke, Richard. *The Killing of Karen Silkwood: The
    Story Behind the Kerr-McGee Plutonium Case.* 2d ed.
    Ithaca, N.Y.: Cornell University Press, 2000.

SEE ALSO: Antinuclear movement; Hazardous and
toxic substance regulation; Nuclear accidents; Nuclear regulatory policy; Power plants; Radioactive pollution and fallout.

Silver Bay, Minnesota, asbestos releases

CATEGORY: Human health and the environment
THE EVENT: The dumping of mining wastes into Lake Superior by the Reserve Mining Company
DATE: Begun in 1947
SIGNIFICANCE: Mining wastes deposited directly into Lake Superior were found to be the source of asbestos fibers in the drinking-water supply of Duluth, Minnesota, and other nearby communities.

Taconite is a low-grade iron ore used in the making of steel products. Rocks are crushed and the ore is magnetically removed; the residual materials, or tailings, are industrial waste. In 1947 the state of Minnesota, in an attempt to revive mining in the region known as the Iron Range, granted the Reserve Mining Company of Silver Bay, Minnesota, permission to dump the tailings from its taconite processing plant directly into Lake Superior. The company began depositing the tailings into a large chasm in the lake at the rate of 67,000 tons per day.

On June 14, 1973, the U.S. Environmental Protection Agency (EPA) announced that high concentrations of asbestos fibers had been found in the drinking water of Duluth, Minnesota. The fibers were identified by Irving Selikoff, director of an environmental sciences laboratory in New York, as amosite, the same asbestos fibers that, when inhaled, are known to cause lung, stomach, and colon cancers after an incubation period of twenty to thirty years. Duluth's drinking water was coming directly from Lake Superior, which was thought to have the purest lake water in the world.

State and U.S. government experts charged that the Reserve Mining Company's tailings were the source of these fibers and that lake currents had brought them into the water supply of Duluth and surrounding communities. At the time of the EPA warning, court action was under way to force Reserve to cease dumping its slurry into Lake Superior. Company executives protested that the fibers had come into the lake naturally from eroding rocks located in tributary streams. They also contended that the "water scare" was a ploy to influence the court's decision.

With Reserve threatening to close down its mining operation if it were to be forced to find another dump site, Silver Bay residents, nearly all of whom owed their jobs to Reserve, strongly defended the company. Duluth's residents, while wanting Reserve to stop its dumping, generally reacted with equanimity when confronted with the EPA report, many noting that they had been drinking Lake Superior water for more than thirty years without ill effects. As no filtration system existed that could remove the asbestos fibers from the water, most Duluthians had no option except to continue to drink the water. Moreover, it was unclear whether ingesting the fibers might have the same effect as inhaling them.

Numerous mortality studies were initiated by Selikoff and other scientists in 1973 to evaluate the impact of the fibers on Duluth's population. These studies failed to confirm that the water posed any serious threat to those who drank it. By 1975 the water scare had dissipated, and soon thereafter the fibers in the water ceased to be an issue. After years of litigation, Reserve agreed to stop dumping tailings in the lake and instead dispose of them on land. The company constructed a basin about 11 kilometers (7 miles) inland from Silver Bay to contain the tailings. In 1990 Reserve announced that it was bankrupt, and the company closed. Subsequently, another mining company reopened the Silver Bay taconite plant and continued to use the tailings basin built by Reserve.

Ronald K. Huch

FURTHER READING

Bartrip, Peter. *Beyond the Factory Gates: Asbestos and Health in Twentieth Century America.* New York: Continuum, 2006.

Farber, Daniel A. "Economics Versus Politics." In *Ecopragmatism: Making Sensible Environmental Decisions in an Uncertain World.* Chicago: University of Chicago Press, 1999.

SEE ALSO: Asbestos; Asbestosis; Clean Water Act and amendments; Drinking water; Mine reclamation; Strip and surface mining; Water pollution.

Singer, Peter

CATEGORIES: Activism and advocacy; animals and endangered species

IDENTIFICATION: Australian philosopher and bioethicist

BORN: July 6, 1946; Melbourne, Victoria, Australia

SIGNIFICANCE: Singer, author of the 1975 book *Animal Liberation* and numerous other works in applied ethics, is considered by many to have launched the modern animal liberation movement.

Peter Singer is the son of Jewish parents who fled Vienna in 1938, after the Nazi occupation of Austria. His father was a coffee salesman and his mother one of the first female graduates of the University of Vienna. They settled in Melbourne, Australia. As a young student, Singer excelled in math and science, but preferred the humanities and eventually studied history, philosophy, and law. He received undergraduate and master's degrees from the University of Melbourne and went on scholarship to University College, Oxford, where he received a bachelor of philosophy degree. He held various positions at Monash University in Melbourne, where he helped establish the Centre for Human Bioethics and served as codirector of the Institute for Ethics and Public Policy. In 1999 he took the position of Ira W. DeCamp Professor of Bioethics at Princeton University. He is also Laureate Professor at the University of Melbourne's Centre for Applied Philosophy and Public Ethics.

The widespread popularity of his 1975 book *Animal Liberation* cemented Singer's reputation as one of the most influential moral philosophers of his generation. He is a utilitarian, which means that for him the rightness of an action is determined solely by its consequences. In assessing consequences, Singer argues that one must give equal consideration to the interests of all beings affected. All sentient beings, he argues, share certain interests, such as experiencing pleasure and avoiding pain. In determining the correct course of action, one must weigh interests according to their strength rather than according to one's affinity for the being that has the interest. According to Singer, humankind's current treatment of nonhuman animals does not reflect an equal consideration of the animals' interests. In fact, it barely considers them at all. Singer contends that humans' poor treatment of nonhuman animals is simply the result of speciesist attitudes that are no more justifiable than racism or sexism. Just as racists violate the principle of equality by favoring members of their own race, speciesists violate the principle of equality by allowing the interests of their own species to outweigh the greater interests of members of other species.

Singer has also discussed the connection between a human-centered ethic and environmental values. Essentially, he argues that environmental policy decisions must give equal consideration to all those affected by an action. Cattle farms may stimulate economic growth for humans but cause suffering and death for cattle, and perhaps loss of habitat for other species. The destruction of rain forests, the building of dams, and other human actions may have consequences that affect members of other species with whom humans share the planet. Singer asserts that the interests of these other species cannot be ignored; he argues that appropriate environmental values give equal consideration to the interests of all sentient beings. Moreover, these interests include the consequences of animal production, such as that required

Philosopher and bioethicist Peter Singer. (©Najiah Feanny/CORBIS)

for the meat and egg industries. In addition to the suffering of the animals involved, animal production creates massive amounts of waste, increases the use of fossil fuels, pollutes air and water, and releases greenhouse gases into the atmosphere. According to Singer, intensive animal production is a disaster, and he advocates a vegan diet and lifestyle.

Jonelle DePetro

FURTHER READING

Beers, Diane L. *For the Prevention of Cruelty: The History and Legacy of Animal Rights Activism in the United States.* Athens: Swallow Press/Ohio University Press, 2006.

Schaler, Jeffrey A., ed. *Peter Singer Under Fire.* Chicago : Open Court, 2009.

Singer, Peter. *Animal Liberation.* 1975. Reprint. New York: HarperPerennial, 2009.

_____. *Writings on an Ethical Life.* New York: Ecco Press, 2000.

SEE ALSO: Animal rights; Animal rights movement; Anthropocentrism; Deep ecology; Environmental ethics; People for the Ethical Treatment of Animals; Speciesism; Vegetarianism.

SL-1 reactor accident

CATEGORIES: Disasters; nuclear power and radiation

THE EVENT: Destruction of the SL-1 nuclear reactor, a test reactor at the Idaho National Engineering Laboratory

DATE: January 3, 1961

SIGNIFICANCE: The accident involving the SL-1 reactor was the first nuclear reactor accident in the United States that resulted in fatalities. As a result of this incident, new design criteria were developed to ensure that a complete reactor shutdown could be accomplished safely.

In February, 1954, the U.S. secretary of defense authorized the U.S. Army to develop small nuclear reactors to provide electrical power and heat for military facilities in remote locations, such as Distant Early Warning (DEW) Line radar sites in Alaska and Greenland. A prototype, the Stationary Low-Power Reactor Number One (SL-1), began operation on August 11, 1958, at the National Reactor Testing Station in the desert west of Idaho Falls, Idaho. This small reactor was designed to operate at a maximum power of 3 megawatts, to produce electricity and steam heat for the crews and equipment at DEW Line radar sites or other military installations.

Combustion Engineering, the contractor that operated SL-1 for the Army, shut the reactor down on December 23, 1960, for the Christmas holiday. As part of the shutdown procedure, the control rod, which determined the rate of reaction in the core, was fully inserted and disconnected from its drive mechanism. A three-man crew arrived at the reactor site on the evening of January 3, 1961, to reconnect the drive mechanism for the main control rod in preparation for restarting the reactor. This required them to move the control rod out about 10 centimeters (4 inches). For reasons that were never determined, the crew instead moved the rod out about 51 centimeters (20 inches), resulting in an extreme power surge, with the reactor producing an estimated 20,000 megawatts. Within a few milliseconds the core overheated and the resulting steam explosion propelled the control rod upward and disrupted the reactor core. All three of the crew, the only people near the reactor, were killed.

Because the Army was intending to operate reactors like SL-1 in remote locations, the reactor had been built to be small and lightweight; it had no containment structure like the ones on commercial power reactors. The reactor was housed in a steel cylinder with walls 0.63 centimeter (0.25 inch) thick, which trapped most of the radioactivity released in the explosion. Some radioactive iodine escaped, and monitoring downwind indicated that its radioactivity reached fifty times background levels near the plant. Because of the remote location of the SL-1 reactor no humans outside the test site were adversely affected.

Radioactivity inside the reactor structure was so high that each member of the rescue team, wearing whole-body protective clothing, was permitted to enter the site only once for a period of one minute to assess the situation and recover the victims. Because of its high radioactivity the reactor was buried on-site.

An investigation of the accident found that the direct cause was the improper withdrawal of the main control rod, but the investigators also concluded that the design of the SL-1 reactor was poor, because a single rod controlled 80 percent of the activity in the core. As a result, new design criteria were put in place requiring that a complete reactor shutdown could be accomplished with the most reactive control rod in its full out position. Nonetheless, the SL-1 accident demonstrated the general safety of the water-moderated reactor design, since even this severe accident resulted in dispersal of the nuclear core, shutting down the reaction, rather than a more serious nuclear meltdown.

George J. Flynn

FURTHER READING

McKeown, William. *Idaho Falls: The Untold Story of America's First Nuclear Accident.* Toronto: ECW Press, 2003.

Stacy, Susan M. *Proving the Principle: A History of the Idaho National Engineering and Environmental Laboratory, 1949-1999.* Washington, D.C.: Government Printing Office, 2000.

Tucker, Todd. *Atomic America: How a Deadly Explosion and a Feared Admiral Changed the Course of Nuclear History.* New York: Free Press, 2009.

SEE ALSO: Chalk River nuclear reactor explosion; Chernobyl nuclear accident; Nuclear accidents; Nuclear power; Three Mile Island nuclear accident.

Slash-and-burn agriculture

CATEGORY: Agriculture and food

DEFINITION: Practice in which forestland is cleared and burned for use in crop and livestock production

SIGNIFICANCE: Among the negative environmental impacts of the widespread use of slash-and-burn agriculture are habitat fragmentation, air pollution, and soil erosion.

Slash-and-burn agriculture has been practiced for many centuries among people living in tropical rain forests. Initially, this farming system involved small populations. Therefore, land could be allowed to lie fallow for many years, leading to the full regeneration of secondary forests and hence restoration of the ecosystems. During the second half of the twentieth century, however, several factors led to drastically reduced fallow periods. In some places the use of such fallow systems ended, resulting in the transformation of forests into shrub and grasslands, negative effects on agricultural productivity for small farmers, and disastrous consequences for the environment.

Among the factors responsible for reduced or nonexistent fallow periods during the 1980's and 1990's were increased population in the Tropics, increased demand for wood-based energy, and, perhaps most important, increased worldwide demand for tropical commodities, especially products such as palm oil and natural rubber. These factors helped industrialize slash-and-burn agriculture, which had been practiced for centuries by small farmers. Ordinarily, when small farmers engage in slash-and-burn farming, they are able to control their fires, which might be compared in size to small forest fires triggered by lightning in the northwestern or southeastern United States. However, the continued reduction in fallow periods, coupled with increased burning by subsistence farmers and large agribusiness, especially in Asia and Latin America, resulted in increased negative impacts on the environment.

Although slash-and-burn agriculture seldom takes place in temperate regions, some agricultural burning occurs in the Pacific Northwest of the United States, where it has been estimated that three thousand to five thousand agricultural fires are set each year in Washington State alone. These fires also create problems for human health and the environment.

HABITAT FRAGMENTATION

One of the most easily recognizable results of slash-and-burn agriculture is habitat fragmentation, which leads to significant loss of the vegetation needed to maintain effective gaseous exchange in tropical regions and throughout the world. For every hectare of land lost to slash-and-burn agriculture, ten to fifteen times that amount of land is fragmented, resulting in the loss of habitat for wildlife, plant species, and innumerable macro- and microorganisms yet to be identified. This also creates problems for management and wildlife conservation efforts in parts of the world with few or no resources to feed their large populations. The recognition of the problems caused by habitat fragmentation led to intensive discussions on global warming. While slash-and-burn agriculture by itself is not completely responsible for global warming, the industrialization of the process could make it a significant component of the problem, as more and more vegetation is fragmented.

HUMAN HEALTH

The impacts of slash-and-burn agriculture on human health and the environment are best exemplified by the impacts related to the 1997 Asian fires that resulted from such practices. Monsoon rains normally extinguish the fires set by farmers, but a strong El Niño weather phenomenon delayed the expected rains in 1997, and the fires burned out of control for months. Thick smoke caused severe health problems. It has been estimated that more than twenty million people in Indonesia alone were treated for asthma, bronchitis, emphysema, and eye, skin, and cardiovascular problems as a result of the fires. Similar problems have been reported for smaller agricultural fires.

Three major problems are associated with air pollution: particulate matter, pollutant gases, and volatile organic compounds. Particulate compounds of 10 microns or smaller that are inhaled become attached to the alveoli of the lungs and can result in severe illness. Research conducted by the U.S. Environmental Protection Agency (EPA) and the University of Washington found that death rates associated with respiratory illnesses increase when fine-particulate air pollution increases. Pollutant gases such as carbon monoxide, nitric oxide, nitrogen dioxide, and sulfur dioxide become respiratory irritants when they combine with vapor to form acid rain or fog. Until the Asian fires, air pollutants stemming from the small fires of slash-and-burn agriculture that occur every planting season often went unnoticed. Thus it is likely that millions of people in the Tropics experience environmental health problems because of slash-and-burn agriculture that are never reported.

SOIL AND WATER QUALITY

The loss of vegetation that follows slash-and-burn agriculture causes an increased level of soil erosion. The soils of the humid tropics create a hard pan underneath a thick layer of organic matter. Therefore, upon the removal of vegetation cover, huge areas of land become exposed to the torrential rainfalls that occur in these regions. The result is severe soil erosion. As evidenced by the impact of Hurricane Mitch on Honduras during 1998, these exposed lands can give rise to large mudslides that can lead to significant loss

Workers in Rondonia, Brazil, burn a portion of the Amazon rain forest to clear land for agriculture. (Getty Images)

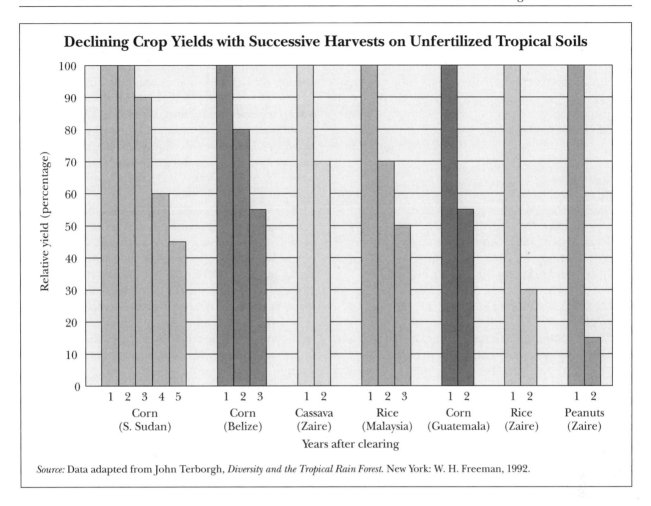

Declining Crop Yields with Successive Harvests on Unfertilized Tropical Soils

Source: Data adapted from John Terborgh, *Diversity and the Tropical Rain Forest.* New York: W. H. Freeman, 1992.

of life. While slash-and-burn agriculture may not be the ultimate cause of sudden mudslides, it contributes to the problem by predisposing the land to erosion.

Associated with erosion is the impact of slash-and-burn agriculture on water quality. As erosion continues, sedimentation of streams increases. This sedimentation affects stream flow and freshwater discharge for catchment-area populations. Mixed with the sediment are minerals such as phosphorus and nitrogen-related compounds that enhance algal growth in streams and estuaries, which depletes the supply of oxygen that aquatic organisms require to survive. Although fertility is initially increased on noneroded soils, nutrient deposition and migration into drinking-water supplies continues to increase.

CONTROLLING SLASH-AND-BURN AGRICULTURE

Given the fact that slash-and-burn agriculture has significant effects on the environment, not only in regions where it is the mainstay of the agricultural sys-

tem but also in other regions of the world, environmental activists and scientists have encouraged the exploration of different approaches to agriculture in the regions using the slash-and-burn system. Because slash-and-burn agriculture has evolved to be a socio-culturally accepted way of making a livelihood, to be successfully implemented any recommended changes to this practice must be consistent with the ways of life of peoples who have minimal resources for extensive agricultural systems.

Among the alternatives are new agroecosystems, such as agroforestry systems and sustainable agriculture systems, that do not rely so much on the slashing and burning of forestlands. These systems allow for the cultivation of agronomic crops and livestock within forest ecosystems, thus protecting soils from being eroded. Another approach involves the education of small rural farmers, absentee landlords, and big agribusiness concerns in developing countries so that they understand the environmental impacts of slash-and-burn agriculture. Small rural farmers do not have the resources to renovate forestlands that have been slashed and burned for farming, but the big businesses that profit from such farming can organize restoration of the original ecosystems; this kind of restoration has been undertaken in many of the world's developed nations.

Oghenekome U. Onokpise

FURTHER READING

Dean, Warren. *With Broadax and Firebrand: The Destruction of the Brazilian Atlantic Forest.* Berkeley: University of California Press, 1997.

Mazoyer, Marcel, and Laurence Roudart. *A History of World Agriculture: From the Neolithic Age to the Current Crisis.* New York: Monthly Review Press, 2006.

Palm, Cheryl A., et al., eds. *Slash-and-Burn Agriculture: The Search for Alternatives.* New York: Columbia University Press, 2005.

Terborgh, J. *Diversity and the Tropical Rain Forest.* New York: Scientific American Library, 1992.

Vandermeer, John H. *The Ecology of Agroecosystems.* Sudbury, Mass.: Jones and Bartlett, 2011.

SEE ALSO: Deforestation; Rain forests; Sustainable agriculture.

Sludge treatment and disposal

CATEGORIES: Waste and waste management; water and water pollution

DEFINITION: Processing and disposal of the residue that is left after water is removed from sewage and industrial waste

SIGNIFICANCE: The proper treatment and disposal of sewage sludge is a critical element in environmental planning because improper disposal or inadequate treatment can result in the contamination of groundwater and of drinking-water supplies.

Most wastewater, whether from industrial discharge, storm drains, or sewage systems, goes through a process that separates the solids from the water. This process takes place in up to three stages: primary, secondary, and tertiary treatment. Although it is desirable for wastewater to undergo tertiary treatment before it is discharged into lakes or oceans, in the United States secondary treatment meets the minimum requirement for such sewage. Tertiary treatment involves polishing, or further treating, the liquid—or effluent—that is removed from wastewater during the dewatering process.

Primary treatment involves collecting the wastewater in a sedimentation lagoon or clarifier. The water is allowed to settle so that the solids and liquids separate. The sediment that is left after the liquid is pumped out is called sludge. Sludge usually contains about 95 percent water. It is filtered to remove more water in a process known as dewatering.

The open-lagoon method filters sludge through sand beds, allowing it to air dry. This may take several months. To accelerate the process and reduce offensive conditions, most treatment facilities use clarifying tanks; in such a tank, rotating mechanical rakes move the settled solids to the center of the tank, where they are drawn off. This sludge is further dewatered before it is disposed of in a landfill or through incineration. Sometimes lime or other chemicals are added to increase the amount of solids that settle on the tank's bottom.

Secondary treatment involves using bacteria to break down the organic contaminants in the sludge. Oxygen must be supplied to the sludge so that the microorganisms present can do their work; this is accomplished through the use of aerated lagoons, digesting tanks, trickling filters, or oxidation ponds. The methods differ in the ways in which they supply oxygen.

Aerated lagoons and oxidation ponds both use large, shallow, open pits, where air and sunlight encourage bacteria and algae to grow. These organisms work together to break down the organic matter: The bacteria consume the organic matter, and the algae "feeds" on the sun and provides further oxygen, allowing the bacteria to thrive. Sometimes oxygen is supplied mechanically, enabling relatively small ponds to process larger amounts of sludge than they could otherwise process. Sludge deposits are removed from the ponds on a regular basis through dredging.

Trickling filters can take different forms: Some are large tanks filled with stones, whereas others are large plastic tanks. Settled sewage is sprayed on top of the stones or on the top and walls of the plastic tank. Water from the sediment trickles to the bottom of the tank, where it is collected, removed, and treated. Bacteria attack and metabolize the sediment that clings to the rocks or the tank walls. Sealed digesting tanks use anaerobic bacteria, which work best without oxygen, to break down the organic matter in sludge.

The most common method of sludge treatment is the activated sludge process, which is an aerobic (with oxygen) biological system. Microbes that rely on air are used to help metabolize, or break down, organic waste. During the first stage of this treatment, sludge is mixed with settled sewage in an aeration tank. Large amounts of microorganisms are collected in the aeration tank and mixed with the semisolid slurry. Oxygen is added so that the microorganisms will feed on the organic matter in the wastewater. The organisms consume tiny particles of waste, and large particles are broken down. After about twelve hours, the slurry in the aeration tank is pumped to a sedimentation tank. The sediment settles, and the organisms are returned to the aeration tank where they consume more organic matter. Any remaining liquid is treated to remove nitrogen and phosphorus, which may cause excessive growth of plants or algae, before it is discharged into lakes or the ocean.

About 30 percent of the sludge is pumped back to the aeration tank to repeat the process. Recirculating the sludge and mixing it with fresh sewage sediment in the aeration tank is a key part of the activated sludge treatment process. Remaining sludge is dewatered using centrifuges, which spin the sludge to draw out water. Filter presses use belts or plates to squeeze out excess water. They create drier sludge cake.

The method used to dewater sludge depends on the disposal method. The more water that is removed from sludge, the less volume of sludge cake remains. Sludge cake needs to be very dry if it is to be incinerated. If the sludge cake is to be used as fertilizer, added to composting facilities, or disposed of in landfills, the water content can be higher. Chemicals are sometimes added to the sludge to encourage particles to clump together, which speeds the removal of water.

Separation of sludge from wastewater does not prevent toxins or other pollutants from remaining in the sludge, which means that toxins can enter landfills or crops. Bioremediation remedies this problem; in this process, microorganisms known to consume inorganic particles, oil, and other toxins are added to the sludge. Some agencies responsible for waste treatment in the United States, including the Palm Beach County Solid Waste Authority in Florida, have had great success using bioremedial treatment. Following bioremediation, sludge and solid waste can be composted into usable soil conditioners.

Lisa A. Wroble

Further Reading

Hill, Marquita K. "Water Pollution." In *Understanding Environmental Pollution*. 3d ed. New York: Cambridge University Press, 2010.

Lester, J., and D. Edge. "Sewage and Sewage Sludge Treatment." In *Pollution: Causes, Effects, and Control*, edited by Roy M. Harrison. 4th ed. Cambridge, England: Royal Society of Chemistry, 2001.

Miller, G. Tyler, Jr., and Scott Spoolman. "Water Pollution." In *Living in the Environment: Principles, Connections, and Solutions*. 16th ed. Belmont, Calif.: Brooks/Cole, 2009.

Qasim, Syed A. *Wastewater Treatment Plants: Planning, Design, and Operation*. 2d ed. Lancaster, Pa.: Technomic, 1999.

Roseland, Mark. "Water and Sewage." In *Toward Sustainable Communities: Resources for Citizens and Their Governments*. Rev. ed. Gabriola Island, B.C.: New Society, 2005.

Snyder, James D. "Off-the-Shelf Bugs Hungrily Gobble Our Nastiest Pollutants." *Smithsonian*, April, 1993, 67-70, 72, 74, 76.

See also: Bioremediation; Composting; Septic systems; Sewage treatment and disposal; Wastewater management; Water treatment.

Smallpox eradication

Category: Human health and the environment

The Event: Medical science's elimination of the infectious disease smallpox worldwide

Date: Announced on October 26, 1979

Significance: The eradication of smallpox by scientists working under the auspices of the World Health Organization represented the first time medical science was able to eliminate an infectious disease. Concern remains that eradication of a viral species may create an environmental niche into which other agents may enter.

The precise origins of smallpox are unknown, but it clearly was a disease of antiquity. Pharaoh Ramses V of Egypt (twelfth century B.C.E.) reportedly died of a disease resembling smallpox. The disease was prevalent in China and India for at least fifteen centuries before its appearance in the Middle East in the sixth century C.E. Crusaders returning from the

Middle East spread smallpox throughout Europe during the twelfth and thirteenth centuries, and the disease was introduced to the Western Hemisphere by the Spanish under Hernán Cortés (1520). Reportedly, more than three million Aztecs died from smallpox within two years, opening Mexico to Spanish invaders. It is estimated that the annual death rate from smallpox in Europe during this period approached 400,000; those who survived the disease were often disfigured by scarring.

The first attempts at immunization against the disease were practiced by the Chinese. In a process called variolation, dried powder from smallpox crusts (the scabs that formed over the lesions caused by the disease) was inhaled. The procedure was taken to the Middle East during the seventeenth or eighteenth centuries, probably by Arab traders, where Lady Mary Wortley Montagu, wife of the British ambassador to the Ottoman Empire, became aware of the practice. Lady Montagu, herself a scarred survivor of the disease, had her young son successfully variolated in 1718. The British royal family soon heard of the practice and introduced it into England several years later.

While variolation was useful in producing immunization against smallpox, it remained a difficult and dangerous procedure. An alternative had long been practiced by British dairy farmers: the inoculation of people with material taken from the lesions that developed on the udders of cows infected with cowpox. In the 1790's, Edward Jenner, an English country physician, became aware of the practice and tested the procedure on himself and several volunteers. With the publication of his work *An Inquiry into the Causes and Effects of the Variolae Vaccinae* (1798), the use of vaccination quickly spread through both Europe and the Americas. Though smallpox was not eliminated, its incidence declined steadily over the next century. Owing to widespread, systematic vaccination, it was eliminated from the United States by 1949.

The key to global eradication of smallpox was based on a characteristic of the virus that made it unlike most viral diseases: Humans represent the only reservoir for smallpox. Because both vaccination and recovery from infection result in lifelong immunity,

Children in Mali in the late 1960's read a poster notifying the public of the country's campaign to eradicate smallpox. (CDC)

once the chain of infection was disrupted, smallpox would cease to be an epidemic disease.

In 1950 a program was developed to eradicate smallpox in the Western Hemisphere through widespread immunization among susceptible populations. By 1958 the disease had been eradicated in most of the Americas, lending credibility to a Soviet proposal for the global elimination of the disease. Beginning in 1965 a program was developed to meet that goal. The program was based on the realization that a goal of immunizing 100 percent of the world's population was unrealistic. Rather, the goal had to be to detect and contain local outbreaks of disease, breaking any chain of transmission. Once the disease could no longer spread beyond local borders, any outbreak would die out.

In 1967 the incidence of smallpox approached an estimated ten to fifteen million cases in forty-six countries. Within ten years, however, the disease virtually ceased to exist; the last reported natural case was that of a young Somali in 1977. In 1978 a single fatal laboratory-associated infection occurred at the Birmingham University Medical School in Great Britain, in which a medical photographer was infected by a virus being studied in an adjacent laboratory. On October 26, 1979, the World Health Organization (WHO) announced the global eradication of the disease.

Following the eradication of smallpox, the issue was raised as to whether all existing laboratory stocks of the virus should be destroyed, which would mean the deliberate destruction of a species. By the mid-1990's, the only remaining virus stocks known to exist were stored at the Centers for Disease Control and Prevention in Atlanta, Georgia, and the Institute for Viral Preparations in Moscow, Russia (these were later moved to the State Research Center of Virology and Biotechnology VECTOR in Koltsovo, Novosibirsk, Russia). The deoxyribonucleic acid (DNA) genomes of several species of the virus were sequenced, and many argued that the remaining stocks of smallpox should be destroyed. Others offered arguments to the contrary, however, and after years of debate, WHO in 2002 took the stance that the remaining stocks should not be destroyed. Scientists continue to disagree on this topic.

Some have expressed concern that with the elimination of a viral species, an environmental niche may be created that could provide a means for other viruses to enter the human population. Widespread infection by smallpox, in addition to general use of vaccination, created a population resistant to infection by most other forms of poxviruses. Whereas smallpox was species-specific in only infecting humans, other viruses such as monkeypox and the whitepox viruses can infect a variety of primates, including humans. These viruses are neither as disfiguring nor as deadly as smallpox, and they are not transmitted as readily as smallpox, but the combination of poverty and social upheaval in the world's developing countries could conceivably facilitate the spread of such diseases among human populations.

Richard Adler

FURTHER READING

Behbehani, Abbas M. *The Smallpox Story: In Words and Pictures.* Kansas City: University of Kansas Medical Center, 1988.

Fenner, Frank, et al. *Smallpox and Its Eradication.* Geneva: World Health Organization, 1988.

Koplow, David. *Smallpox: The Fight to Eradicate a Global Scourge.* Berkeley: University of California Press, 2003.

Tucker, Jonathan B. *Scourge: The Once and Future Threat of Smallpox.* New York: Grove Press, 2001.

SEE ALSO: Bacterial resistance; Centers for Disease Control and Prevention; Genetically altered bacteria; World Health Organization.

Smart grids

CATEGORY: Energy and energy use

DEFINITION: Power grids that efficiently connect users with electrical power produced by a variety of sources

SIGNIFICANCE: Rebuilding the current power grid with smart-grid technology would provide the opportunity to make the delivery of electricity to consumers more efficient. Smart grids are also versatile enough to use different types of power sources, so their proliferation would encourage growth in the development of renewable energy sources such as solar and wind power.

The basic design of the current power grid in the United States and the design of many of its components go back to decisions made during the 1890's and the early twentieth century. The grid was originally constructed with power lines radiating outward from power plants to consumers and with interconnections so that if a generator went offline another could pick up the load. The U.S. grid has become a complex of some 482,800 kilometers (300,000 miles) of transmission lines connecting more than 9,200 power generators producing 1 million megawatts of power for American factories and homes. It is estimated that 9 percent of the energy fed into the grid is lost in transmission. A smart grid is more efficient and communicates with customers; it has the ability to shift load to off-peak hours and is versatile enough to accept power from a range of sources.

Generators may produce direct current (DC), which flows in one direction only, or alternating current (AC), which alternates its direction sixty times each second (in the United States). Almost from the first, it was decided to use AC because it can easily be transformed to high voltage and low current, and this reduces transmission losses. If power from another

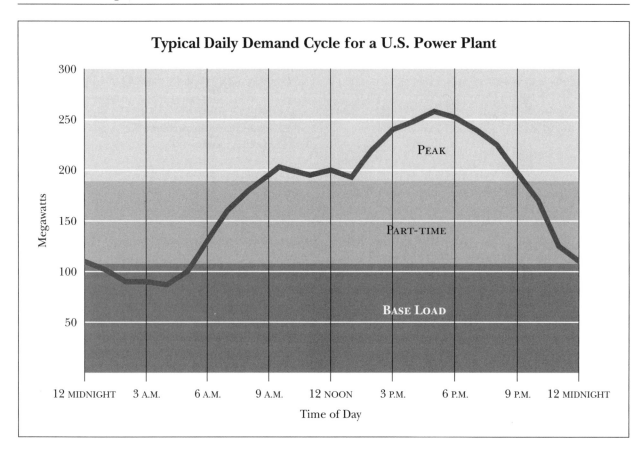

Typical Daily Demand Cycle for a U.S. Power Plant

AC generator is to be added to the grid, it must have exactly the same frequency (60 hertz or cycles per second) and the same phase. It must rise and fall in exact lockstep with the power already on the grid; otherwise, power from the generator will cancel some of the power already on the grid, and it will be wasted. A smart grid has interfaces that exactly match the power from generators to the power already on the grid. As increasing numbers of electricity users install their own small windmill generators and solar cells, some have begun to produce sufficient power that they can sell what they do not use to the grid. Solar cells require an interface that converts DC to AC of the proper frequency and phase so that it can be added to the grid.

The amount of power used over the course of a day is not constant; rather, there are times of peak load, when more users—such as factories or home owners and businesses running air conditioners—are drawing power from the grid. Newer generators are usually more efficient and less polluting than older generators, so these generators are always in use. Power companies run their least efficient generators only during peak load periods, and this electricity costs more to produce. Smart power meters inform consumers of changing power costs so that they may shift their usage to off-peak hours if they wish. This not only saves the consumers money but also allows the power companies to use only their more efficient generators if enough load is shifted.

High-voltage DC (HVDC) lines can transmit power with less loss than AC lines, so for transmission lines over 400 miles (650 kilometers) long, DC may be used. Alternating current is first transformed up to as much as 800 kilovolts and then converted to DC. At the destination power station the DC is converted back to AC, and the voltage is reduced to typical substation voltage (2.4 to 33 kilovolts). Superconducting wire would be even more efficient, and trial smart grid projects involving such wire have been undertaken, such as Long Island's Project Hydra.

Charles W. Rogers

FURTHER READING

Blume, Stephen W. *Electric Power System Basics for the Nonelectrical Professional.* Hoboken, N.J.: Wiley-IEEE, 2007.

Bowman, Ron. *The Green Guide to Power: Thinking Outside the Grid.* Charleston, S.C.: BookSurge, 2008.

Fox-Penner, Peter. *Smart Power: Climate Change, the Smart Grid, and the Future of Electric Utilities.* Washington, D.C.: Island Press, 2010.

SEE ALSO: Alternative energy sources; Alternative fuels; Biomass conversion; Coal-fired power plants; Fossil fuels; Geothermal energy; Hydroelectricity; Power plants; Solar energy; Tidal energy; Wind energy.

Smog

CATEGORY: Atmosphere and air pollution

DEFINITION: Air pollution resulting from the combination of smoke with fog or from sunlight acting on unburned hydrocarbons emitted from automobiles

SIGNIFICANCE: Severe smog episodes have been responsible for many deaths and widespread illness in cities around the world. Growing recognition of the detrimental health effects of smog have led many governments to pass laws designed to reduce chemical pollutants in the air.

Originally a blend of the words "smoke" and "fog," the term "smog" was coined to describe the severe air pollution that results when smoke from factories combines with fog during a temperature inversion. As one ascends upward from the earth's surface, the air temperature drops by about 3 degrees Celsius (5.5 degrees Fahrenheit) every 300 meters (1,000 feet). Temperature inversions occur when this normal condition is reversed so that a blanket of warm air is sandwiched between two cooler layers. A temperature inversion restricts the normal rise of surface air to the cooler upper layers, in effect placing a lid over a region. When the air above a city cannot rise, the air currents that carry pollutants away from their sources stagnate, causing pollution levels to increase drastically. A combination of severe air pollution, prolonged temperature inversion, and moisture-laden air may result in what has been termed "killer fog."

KILLER FOGS

Several acute episodes of killer fog occurred during the twentieth century. One was in the Meuse Valley of Belgium. During the first week of December, 1930, a thick fog and stagnant air from a temperature inversion concentrated pollutants spewing forth from a variety of factories in this heavily industrialized river valley. After three days of such abnormal conditions, thousands of residents became ill with nausea, shortness of breath, and coughing. Approximately sixty people died, primarily elderly people and persons with chronic heart and lung diseases. The detrimental effects on health were later attributed to sulfur oxide gases emitted by combusting fossil fuels; the gases were concentrated to lethal levels by the abnormal weather. The presence of coal soot, combined with moisture from the fog, exacerbated the effect.

A second episode occurred in Donora, Pennsylvania, during the last week of October, 1948. Donora is situated in a highly industrialized river valley south of Pittsburgh. A five-day temperature inversion with fog concentrated the gaseous effluents from steel mills with the sulfur oxides released by burning fossil fuels. Severe respiratory tract infections began to occur, especially in the elderly, and 50 percent of the population became ill. Twenty people died, a tenfold increase in the normal death rate.

A third major episode occurred in London, England, in early December, 1952. At that time, many residents burned soft coal in open grates to heat their homes. When a strong temperature inversion and fog enveloped the city for five consecutive days, Londoners began complaining of respiratory ailments. By the time the inversion had lifted, four thousand excess deaths had been recorded. In this case it was not only the elderly who were affected—deaths occurred in all age categories. During the next decade London experienced two additional episodes: one in 1956, which claimed the lives of one thousand people, and one in 1962, which caused seven hundred deaths. The decline in mortality rates resulted from the restriction of the use of soft coal, with its high sulfur content, as a source of fuel. Sulfur oxide compounds are responsible for causing lung problems during such episodes; therefore, the term "killer fog" has come to be replaced by the more accurate "sulfurous smog."

PHOTOCHEMICAL SMOG

Photochemical smog, first noticed in the Los Angeles basin in the late 1940's, has been an increasingly serious problem in cities around the world. Moisture is not part of the equation in this type of air pollution, and smoke-belching factories dumping tons of sulfur oxide compounds into the atmosphere are not required. Rather, photochemical smog results when unburned hydrocarbon fuel, emitted in automobile exhaust, is acted upon by sunlight. The Los Angeles basin, hemmed in by mountains to the east and ocean to the west, has a high density of automotive traffic and plenty of sunshine. Varying driving conditions mean that gasoline is never completely consumed by automobile engines; instead, it is often changed into other highly reactive substances. Sunlight acts as an energy catalyst that changes these compounds into the variety of powerful oxidizing agents that constitute photochemical smog. This type of smog has a faint bluish-brown tint and typically contains several powerful eye irritants. The chemical reactions also produce aldehydes, a class of organic chemical best typified by an unpleasant odor.

The complicated chemistry of photochemical smog also produces ozone, which is extremely reactive; it damages plants and irritates human lungs. Because ozone production is stimulated by sunlight and high temperatures, it becomes a particularly pernicious problem during the summer, especially during morning rush hours. Under temperature inversion conditions, the ozone created in photochemical smog can increase to dangerous levels. Ozone is highly toxic. It irritates the eyes, causes chest irritation and coughing, exacerbates asthma, and damages the lungs.

Photochemical smog and ozone are now common ingredients in urban air. Although acute episodes of ozone-induced mortality are rare, concerns have grown about the detrimental long-term consequences of the brief but repetitive exposures to ozone consistently inflicted on commuters. It appears as though no curtailment of the problem will be possible in urban areas in the United States without significant changes in transportation systems, strict limits on growth, and radical alterations in lifestyle, including automobile use.

George R. Plitnik

The city of Los Angeles is well known for its photochemical smog. (©David Mcshane/Dreamstime.com)

FURTHER READING

Elsom, Derek M. *Smog Alert: Managing Urban Air Quality.* London: Earthscan Publications, 1996.

Grant, Wyn. *Autos, Smog, and Pollution Control: The Politics of Air Quality Management in California.* Brookfield, Vt.: Edward Elgar, 1995.

Hinrichs, Roger A., and Merlin Kleinbach. *Energy: Its Use and the Environment.* 4th ed. Belmont, Calif.: Thomson Brooks/Cole, 2006.

Jacobs, Chip, and William J. Kelly. *Smogtown: The Lung-Burning History of Pollution in Los Angeles.* Woodstock, N.Y.: Overlook Press, 2008.

Vallero, Daniel. *Fundamentals of Air Pollution.* 4th ed. Boston: Elsevier, 2008.

SEE ALSO: Air pollution; Air-pollution policy; Atmospheric inversions; Automobile emissions; Black Wednesday; Catalytic converters; Clean Air Act and amendments; Donora, Pennsylvania, temperature inversion; London smog disaster.

Snail darter

CATEGORY: Animals and endangered species
DEFINITION: Small fish native to the Little Tennessee River
SIGNIFICANCE: The fight by environmentalists and others to prevent the building of the Tellico Dam, which would destroy the snail darter's critical habitat, was a test of both the National Environmental Policy Act of 1969 and the Endangered Species Act of 1973.

The Tennessee Valley Authority (TVA) began building dams in the Tennessee Valley watershed in 1936. In 1960 the TVA had more than sixty dams in the region and focused on building the Tellico Dam on the Little Tennessee River. The Little Tennessee River and its valley were sacred to the Cherokee and used by more than three hundred farm families; in addition, the region's only remaining stretch of natural river was enjoyed by hundreds of canoeists and fishing enthusiasts. In response to the proposed Tellico Dam, a citizens' coalition was formed in 1964. Although the local citizens were unable to stop the TVA from beginning the Tellico project, enactment of the National Environmental Policy Act (NEPA) in 1970 gave them a way to stop it. A court order stopped the construction of the dam until 1973, when the TVA produced a legally sufficient environmental impact statement on the project.

Around the same time, a University of Tennessee professor discovered an endangered species of perch, the snail darter (*Percina tanasi*), living in the Tellico project area. The Endangered Species Act (ESA) prohibits federal actions that jeopardize the existence of endangered species or modify the critical habitats of such species. Citizens opposed to the building of the dam began administrative and court proceedings based on the ESA to stop the project. The Tellico citizens' group presented the argument in court that the TVA did not properly value the river as a major recreational resource, the valley for its rich agricultural lands, or the area's historic resources. The citizens pointed to the snail darter as an indicator species, a barometer of endangered human and economic values the disappearance of which would signal devastating environmental degradation.

In 1978 the U.S. Supreme Court decided that construction of the Tellico Dam violated the ESA; the decision permanently ended attempts to complete the project. However, members of Congress held meetings to consider the extreme nature of the law's application: a $100 million dam stopped because of a fish. The conclusion drawn as the result of these meetings was that the Court's decision was rational. Congress later created the Endangered Species Committee to authorize the extinction of certain species in compelling cases. The committee reviewed the case of Tennessee's snail darter and unanimously upheld protection for the endangered fish.

The legal victory for the snail darter was spoiled by political maneuvers, however. A senator and a congressman from Tennessee inserted language into an appropriations bill that overrode the Supreme Court's Tellico decision. Despite talk of a veto, President Jimmy Carter signed the bill into law. Consequently, the TVA finished the Tellico Dam and closed its floodgates in November, 1979. Although none of the estimated twenty thousand snail darters in the Little Tennessee River survived, small populations of the fish had been transplanted to another river, and others were later discovered in several downstream sites. In 1984 the snail darter's status under the ESA was reduced from "endangered" to "threatened."

Michael D. Kaplowitz

A snail darter. (USFWS)

FURTHER READING

Chiras, Daniel D. "Preserving Biological Diversity." In *Environmental Science.* 8th ed. Sudbury, Mass.: Jones and Bartlett, 2010.

Murchison, Kenneth M. *The Snail Darter Case: TVA Versus the Endangered Species Act.* Lawrence: University Press of Kansas, 2007.

SEE ALSO: Dams and reservoirs; Endangered Species Act; Endangered species and species protection policy; Tellico Dam; *Tennessee Valley Authority v. Hill.*

Snyder, Gary

CATEGORIES: Activism and advocacy; preservation and wilderness issues

IDENTIFICATION: American poet, essayist, and environmental activist

BORN: May 8, 1930; San Francisco, California

SIGNIFICANCE: Snyder was one of the first writers to base his poetry, ethics, and spirituality in environmental ideas and values. He is one of the most influential figures in American nature writing.

Gary Snyder grew up in a rural area outside Seattle, Washington, and started hiking in the mountains early in his life. The radical labor politics of the region in which he lived laid the foundation for his critique of the dominant ideology and social structure of Western society, and the presence of American Indian and Asian culture in the Pacific Northwest attracted him to other traditions as constructive alternatives. In college he studied anthropology, followed by graduate study in Asian languages. He was a logger, mountain lookout, and merchant seaman; a love for work and an emphasis on the physical are key themes in his writings.

In the 1950's Snyder moved to the San Francisco area, where he was associated with (but not a member of) the Beat movement. His early poetry was particularly influenced by Native American myth and Chinese nature poetry, which he translated into English. He studied Zen Buddhism in Japan during the late 1950's and early 1960's, developing a rich understanding of both the philosophy and practice of that tradition. Following his return to the United States in the late 1960's, Snyder began emphasizing Native American wisdom along with Buddhism in his view of nature and the place of humans within it.

Poet and environmental activist Gary Snyder. (Courtesy, author)

Snyder writes of the sacredness of natural processes, including predation and decomposition. He argues for the intrinsic value of all of nature and the importance of biodiversity, celebrating "the preciousness of mice and weeds." He thus goes beyond the traditional romantic view of nature. He is known not only for his praise of wilderness but also for his exploration of its connection to the rich wilderness in the human mind. His perspective on the self and nature exhibits the Buddhist view of radical interrelationship, which combines a holistic vision of nature with an affirmation of the reality and value of individuals.

Snyder's poetry is suffused with a sensuous and mystical intimacy with nature, and Snyder has proposed a shamanistic view of the poet's role as one who heals by bringing people into a close relationship with the natural world. Yet he also insists on attention to the practical details of living as a full member of one's bioregion. He sees the interplay of culture and nature as central to the development of a deeply rooted sense of place. Snyder is considered a major voice of the deep ecology movement, and his critique of Western society and ideology relates him to social ecology and ecofeminism as well. Among the honors Snyder has received for his work are the Pulitzer Prize in poetry for his collection *Turtle Island* (1974) and the 2008 Ruth Lilly Poetry Prize, which is presented for lifetime achievement.

David Landis Barnhill

Further Reading

Hart, George. "Gary Snyder, *Turtle Island* (1974)." In *Literature and the Environment*, edited by George Hart and Scott Slovic. Westport, Conn.: Greenwood Press, 2004.

Snyder, Gary. *A Place in Space: Ethics, Aesthetics, and Watersheds.* 1995. Reprint. Berkeley, Calif.: Counterpoint, 2008.

See also: Biodiversity; Deep ecology; Ecofeminism; Environmental ethics; Social ecology.

Social ecology

Category: Philosophy and ethics

Definition: Philosophical movement based in the belief that the domination of nature by humans is derived from the domination of human society by the capitalist mode of production

Significance: The philosophy and set of organizing principles known as social ecology has had some political success with the Green movement and provides trenchant criticism of modern society; nevertheless, social ecology is often treated with suspicion by many mainstream environmentalists.

Social ecologists try, in many ways, to combine the environmental concerns of American preservationist John Muir with the economic concerns of German political philosopher Karl Marx. They argue that the domination of both disadvantaged peoples and the environment is derived from modern capitalist society and that both forms of domination must be corrected together. All social ecologists look to the views of Marx and Friedrich Engels for guidance, although they may differ in the policies they advocate.

There are two main theoretical approaches to social ecology: anarchist and socialist. Murray Bookchin was one of the leading proponents of anarchist social ecology; his writings emphasize the interdependence of humans and nonhuman nature. Anarchist social ecologists believe that hierarchy in society leads to the domination of some people by others and to the domination of nature by humankind. They are suspicious of the state as an agent of domination. Ideally, they would do away with all hierarchy in society, which they contend would remove the hierarchy of exploitation of nature by humans. Social ecologists should thus work to eliminate the domination of nature by the material world by first taking into account social issues in order to address environmental problems. Those who hold this view of social ecology are critical of mainstream environmentalism, which they see as a mechanistic approach that treats nature as a resource for humans to use. They are also critical of some deep ecologists for being insensitive to social issues. Successful social ecology in this perspective emphasizes the achievement of small-scale communities in which local groups live in harmony with one another and the environment.

Socialist ecology is also rooted in the Marxian tradition, but instead of an emphasis on hierarchy and domination and the achievement of a utopian society modeled on nature, this school of thought calls for an economic transformation of society into an ecological socialist system. This variant is often the motivating force for much of the Green social and political movement. Particularly in the work of James O'Connor, socialist ecology starts from a Marxian framework of society but then incorporates the concepts of the autonomy of nature and ecological science. In an ecological socialist society, nature would be recognized as autonomous rather than as humanized and capitalized. The interrelatedness of all living organisms would be recognized, and socially and environmentally harmful means of production would be curtailed. Socialist ecology often emphasizes the importance of central planning and organizing the working class to achieve its aims of remaking first the state and then the human-nature relationship. An example of socialist ecology at work is integrated pest management, which is based on using biological rather than chemical means of controlling pests, a process that, compared with other methods, is less exploitative of nature and society.

Social ecology, no matter what its perspective, has much to offer anyone concerned about the environment. It also has many critics from a variety of perspectives. Deep ecologists, for example, criticize social ecology for its essentially human-centered perspective. Some ecofeminists contend that social ecology does not do enough regarding the domination of women. Social ecologists respond to the first criticism by maintaining that humans are part of the environment and that domination of humans is just as wrong as the domination of nature. Moreover, they argue that because the domination of nature is centered in the capitalist mode of production—a socioeconomic phenomenon—protecting nature must deal with capitalist, industrial society in a realistic fashion. Social ecologists also contend that dealing with the exploitation of humans includes dealing with the exploitation of women.

A more telling criticism is that some socialist ecologists are overly caught up in Marxist rhetoric and are unable go beyond doctrine to offer workable alternatives to modern capitalist society. The deep ecologist argument that social ecology does not pay enough attention to environmental ethics would also appear to have some validity. Anarchist social ecology has a tendency to become utopian rather than practical in orientation, and socialist ecology is often doctrinaire and rigid. Social ecologists in both groups often expend a good deal of energy criticizing those in the other group.

A far-ranging attack on social ecology comes from a market-based perspective. This approach, which social ecologists would argue is a mechanistic one that is ultimately exploitative of people and nature, argues that small-scale communal efforts at dealing with environmental issues are ineffective. Because modern society is based on property rights, environmental solutions must take property rights into account. Such critics argue that social ecology is too utopian and unable to deal with the problems facing modern society. One social ecologist response would be that this may be true, but that is why capitalist, industrial society needs to be remade rather than taking for granted a perspective that is, at base, exploitative of humans and nature.

Social ecology is both a philosophy and a set of organizing principles for social reformers who are concerned about the environment. It aims to achieve an ecosystem in which humans and the rest of the natural world live in harmony in a nonexploitative setting. It has had some political success with the Green movement and provides trenchant criticism of modern society. Both variants of social ecology, however, are treated with suspicion by many mainstream environmentalists.

John M. Theilmann

FURTHER READING
Barry, John. "Murray Bookchin, 1921-   ." In *Fifty Key Thinkers on the Environment*, edited by Joy A. Palmer. New York: Routledge, 2001.
Bookchin, Murray. *The Ecology of Freedom: The Emergence and Dissolution of Hierarchy*. 1982. Reprint. Oakland, Calif.: AK Press, 2005.
_____. "What Is Social Ecology?" In *Earth Ethics: Introductory Readings on Animal Rights and Environmental Ethics*, edited by James P. Sterba. 2d ed. Upper Saddle River, N.J.: Prentice Hall, 2000.
Eckersley, Robyn. *Environmentalism and Political Theory: Toward an Ecocentric Approach*. 1992. Reprint. New York: Routledge, 2003.
Pepper, David. *Eco-Socialism: From Deep Ecology to Social Justice*. New York: Routledge, 1993.

SEE ALSO: Bookchin, Murray; Deep ecology; Ecofeminism; Environmental ethics; Green movement and Green parties.

Soil conservation

CATEGORIES: Agriculture and food; resources and resource management
DEFINITION: Use of agricultural and other cultivation practices aimed at maintaining soil quality and reducing erosion
SIGNIFICANCE: By using agricultural methods that protect soil from degradation, farmers can reduce erosion, prevent soil particles from contributing to air and water pollution, and improve crop production.

According to the United Nations Environment Programme, approximately 17 percent of the earth's vegetated land is degraded, a situation that poses a threat to agricultural production around the world. The introduction of minerals, metals, nutrients, fertilizers, pesticides, bacteria, and pathogens suspended in topsoil runoff into waterways is a significant source of water pollution and is a threat to fisheries, wildlife habitats, and drinking-water supplies. The introduction of soil particles into the air through wind erosion is a significant source of air pollution. Soil conservation is the effort by farmers and other land users to prevent the loss of topsoil from wind erosion, water erosion, desertification, and chemical deterioration such as the buildup of salts and fertilizer acids.

The Industrial Revolution of the nineteenth century and the population explosion of the twentieth century encouraged people to till new land, cut down forests, and disturb soil for the expansion of towns and cities. The newly exposed topsoil quickly succumbed to erosion from rainfall, floods, wind, ice, and snow. The Dust Bowl, which occurred in the Great Plains in the United States during the 1930's, is one example of the devastating effects of wind erosion.

Hugh Hammond Bennett, often called the father of soil conservation, lobbied for congressional establishment of the Soil Erosion Service, which was formed in the U.S. Department of the Interior in 1933, and the establishment of voluntary Soil Conservation Districts in each state. Bennett was named the first chief of the renamed Soil Conservation Service, now part of the Department of Agriculture, in 1937 (in 1994 the name of the agency changed again, to the Natural Resources Conservation Service). On August 4, 1937, the Brown Creek Conservation District in Bennett's home county, Anson County, North Carolina, became the first Soil Conservation District in the United States. Local landowners voted to establish the district by three hundred to one, proving that farmers were concerned about soil conservation. A reporter for the *Charlotte Observer* newspaper sought out the one negative voter, and after the program was explained to him, he changed his opinion. By 1948 more than 2,100 districts had been established nationwide; this number grew to 3,000 by the early years of the twenty-first century. The districts were eventually renamed Soil and Water Conservation Districts.

The Food Security Act of 1985 authorized the Conservation Reserve Program to take out of production any land deemed to be highly susceptible to erosion; it also required farmers to develop soil conservation plans for any remaining susceptible land. The Natural Resources Conservation Service has estimated that with such soil conservation measures, the loss of top-

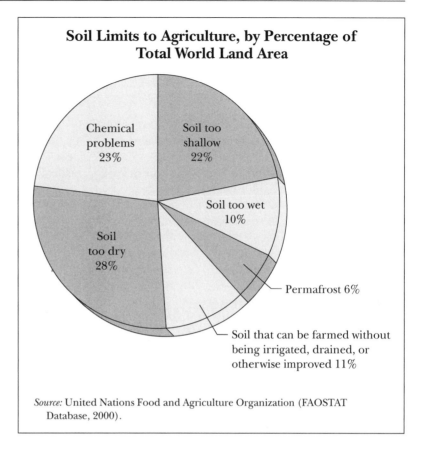

Soil Limits to Agriculture, by Percentage of Total World Land Area

Chemical problems 23%

Soil too shallow 22%

Soil too wet 10%

Soil too dry 28%

Permafrost 6%

Soil that can be farmed without being irrigated, drained, or otherwise improved 11%

Source: United Nations Food and Agriculture Organization (FAOSTAT Database, 2000).

soil in the United States was cut nearly in half, reduced from 1.6 billion tons per year to 0.9 billion tons. The European Community and Australia also adopted soil conservation measures during the 1990's.

Soil conservation practices include covering the soil with vegetation, reducing soil exposure on tilled land, creating wind and water barriers, and installing buffers. Vegetative cover slows the wind at ground level, slows water runoff, protects soil particles from being detached, and traps blowing or floating soil particles, chemicals, and nutrients. Because the greatest wind and water erosion damage often occurs during seasons in which no crops are growing or natural vegetation is dormant, soil conservation often depends on permitting the dead residues and standing stubble of the previous crop to remain in place until the next planting time. In forested areas, annual tree foliage loss serves as a natural ground mulch. Farmers can also reduce erosion by planting grass or legume cover crops until the next planting season for their primary crops or as part of a crop rotation cycle or no-till planting system.

Modern no-till and mulch-till planting systems reduce soil exposure to wind and rain. No-till systems leave the soil cover undisturbed before planting; crop seeds are inserted into the ground through narrow slots in the soil. In mulch-till planting, a high percentage of the dead residues of previous crops are retained on the soil's surface when a new crop is planted.

The ways in which crops are planted can also help to reduce erosion. Row crops can be planted at right angles to the prevailing winds and to the slope of the land in order to absorb wind and rainwater runoff energy and trap moving soil particles. Crops may be planted in small fields to prevent the avalanching caused by an increase in the amount of soil particles transported by wind or water as the distance across bare soil increases. As the amount of soil moved by wind or water increases, the erosive effects of the wind and water also increase. Smaller fields reduce the length and width of unprotected areas of soil.

Wind and water barriers include tree plantings and crosswind strips of perennial shrubs and tall grasses, which act as windbreaks, slowing wind speeds at the surface of the soil. The areas protected by such windbreaks extend for ten times the height of the barriers. In alley cropping, which is used in areas of sustained high winds, crops are planted between rows of larger mature trees. Contour strip farming on slopes, planting grass waterways in areas where rainwater runoff concentrates, and planting grass field borders 3 meters (10 feet) wide on all edges of cultivated or disturbed soil are additional methods for reducing wind speed and rainwater runoff and trapping soil particles, chemicals, and nutrients.

Soil conservation buffers work to filter agricultural runoff to remove sediments and chemicals. Riparian buffers are waterside plantings of trees, shrubs, and grasses, usually 6 meters (20 feet) in width. Riparian buffers planted only in grass are called filter strips. Grassed waterways, field borders, water containment ponds, and contour grass strips are other types of soil conservation buffers.

Gordon Neal Diem

FURTHER READING

Blanco, Humberto, and Rattan Lal. *Principles of Soil Conservation and Management.* New York: Springer, 2008.

Field, Harry L., and John B. Solie. "Erosion and Erosion Control." In *Introduction to Agricultural Engineering Technology: A Problem Solving Approach.* 3d ed. New York: Springer, 2007.

Plaster, Edward. *Soil Science and Management.* 5th ed. Clifton Park, N.Y.: Delmar Cengage Learning, 2008.

Schwab, Glen, et al. *Soil and Water Conservation Engineering.* 5th ed. Clifton Park, N.Y.: Delmar Cengage Learning, 2005.

SEE ALSO: Agricultural chemicals; Air pollution; Clean Water Act and amendments; Deforestation; Desertification; Dust Bowl; Erosion and erosion control; Groundwater pollution; Runoff, agricultural; Sedimentation; Soil salinization; Strip farming; Water pollution; Watershed management.

Soil contamination

CATEGORY: Human health and the environment
DEFINITION: Presence of hazardous substances in soil
SIGNIFICANCE: Soils contaminated with high concentrations of hazardous substances pose potential risks to human health and the earth's thin layer of productive soil.

To be productive, soil depends on bacteria, fungi, and other microbes to break down wastes and release and cycle nutrients that are essential to plants. Healthy soil is essential for humankind's ability to grow enough food for the world's increasing population. Soil also serves as both a filter and a buffer between human activities and natural water resources, which ultimately serve as the primary source of drinking water. Soil that is contaminated may cause water pollution through the leaching of contaminants into groundwater and through runoff into surface waters such as lakes, rivers, and streams.

The U.S. government has tried to address the problem of soil contamination by passing two landmark legislative acts. The Resource Conservation and Recovery Act (RCRA) of 1976 regulates hazardous and toxic wastes from the point of generation to disposal. The Comprehensive Environmental Response, Compensation, and Liability Act (CERCLA) of 1980, also known as Superfund, identifies past contaminated sites and implements remedial action.

Soils can become contaminated by many human activities, including fertilizer or pesticide application, direct discharge of pollutants at the soil surface, leaking of underground storage tanks or pipes, leaching from landfills, and atmospheric deposition. Additionally, soil contamination may be of natural origin. For example, soils with high concentrations of heavy metals can occur naturally because of their close proximity to metal ore deposits. Common contaminants include inorganic compounds such as nitrate and heavy metals (for example, lead, mercury, cadmium, arsenic, and chromium); volatile hydrocarbons found in fuels, such as benzene, toluene, ethylene, and xylene (BTEX) compounds; and chlorinated organic compounds such as polychlorinated biphenyls (PCBs) and pentachlorophenol (PCP).

Contaminants may also include substances that occur naturally when these appear in concentrations that are elevated above normal levels. Examples are substances such as nitrogen- and phosphorus-containing compounds, which are often added to agricultural lands as fertilizers. Since nitrogen and phosphorus are typically the limiting nutrients for plant and microbial growth, their accumulation in the soil is usually not a concern, but the leaching and runoff of these nutrients into nearby water sources is a problem, as it may lead to oxygen depletion of lakes. Furthermore, nitrate is a concern in drinking water because it poses a direct risk to human infants (blue-baby syndrome).

Contaminants may reside in the solid, liquid, and gaseous phases of the soil. Most will occupy all three phases but will favor one phase over the others. The physical and chemical properties of the contaminant and the soil will determine which phase the contaminant favors. The substance may preferentially adsorb to the solid phase. This may include either the inorganic (mineral) or the organic (organic matter) fraction of the soil. The attraction to the solid phase may be weak or strong. The contaminant may also volatilize into the gaseous phase of the soil. If the contaminant is soluble in water, it will dwell mainly in the liquid-filled pores of the soil.

Contaminants may remain in soils for years or wind up in the atmosphere or nearby water sources. Additionally, contaminating compounds may be broken down or taken up by the biological component of the soil. This may include plants, bacteria, fungi, and other soil-dwelling microbes. The volatile compounds may slowly move from the gaseous phase of the soil into the atmosphere. The contaminants that are bound to the solid phase may remain intact or be carried off in runoff attached to soil particles and flow into surface waters. Compounds that favor the liquid phase, such as nitrate, will either wind up in surface waters or leach down into the groundwater.

Metals display a range of behaviors. Some bind strongly to the solid phase of the soil, while others easily dissolve and wind up in surface water or groundwater. PCBs and similar compounds bind strongly to the solid surface and remain in the soil for years. These compounds can still pose a threat to waterways because, over long periods of time, they slowly dissolve from the solid phase into the water at trace quantities. Fuel components favor the gaseous phase but will bind to the solid phase and dissolve at trace quantities into the water. However, even trace quantities of some compounds can pose serious ecological and health risks. When a contaminant causes a harmful effect, it is classified as a pollutant.

One of two general approaches is used in cleaning up a contaminated soil site: treatment of the soil in place (in situ) or removal of the contaminated soil to another location for treatment (non-in situ). In situ methods, which have the advantage of minimizing exposure pathways, include biodegradation, volatilization, leaching, vitrification (glassification), and isolation or containment. Non-in situ methods generate additional concerns about exposure during the process of transporting the contaminated soil. Non-in situ options include thermal treatment (incineration), land treatment, chemical extraction, solidification or stabilization, excavation, and asphalt incorporation. The choice of method depends on the quantity and type of contaminants and the nature of the soil.

John P. DiVincenzo

FURTHER READING

Connell, Des W. "Soil Contamination." In *Basic Concepts of Environmental Chemistry.* 2d ed. Boca Raton, Fla.: CRC Press, 2005.

Pierzynski, Gary M., J. Thomas Sims, and George F. Vance. *Soils and Environmental Quality.* 3d ed. Boca Raton, Fla.: CRC Press, 2005.

Sparks, Donald L. "Soil Decontamination." In *Handbook of Hazardous Materials*, edited by Morton Corn. San Diego, Calif.: Academic Press, 1993.

Testa, Stephen M. *The Reuse and Recycling of Contaminated Soil.* Boca Raton, Fla.: Lewis, 1997.

SEE ALSO: Agricultural chemicals; Bioremediation; Landfills; Pesticides and herbicides; Polychlorinated biphenyls; Restoration ecology; Soil conservation; Water pollution.

Soil salinization

CATEGORIES: Agriculture and food; water and water pollution

DEFINITION: Process in which water-soluble salts build up in soil within the root zone of plants

SIGNIFICANCE: The salinization of soil on agricultural lands can result in poor plant growth and reduced crop yields, and attempts to reduce salinity through increased irrigation can lead to contamination of water supplies.

In the process of soil salinization, water-soluble salts build up in the part of the soil known as the root zone, where the soil comes into contact with the roots of plants, blocking the movement of water and nutrients into plant tissues. Soil salinization rarely occurs naturally. Rainwater is virtually free of dissolved solids, but surface waters and groundwater contain significant quantities of dissolved solids, ultimately produced by the weathering of rocks. Evaporation of water at the land surface results in an increase in dissolved solids in soil that may adversely affect the ability of plant roots to absorb water and nutrients.

In arid regions, evaporation of soil water potentially exceeds rainfall. Shallow wetting of the soil followed by surface evaporation lifts the available dissolved solids to near the surface of the soil. The near-surface soil therefore becomes richer in soluble salts. In natural arid areas, soluble salts in the subsurface are limited in quantity because rock weathering is an extremely slow process, and degrees of soil salinization detrimental to plants are uncommon.

The irrigation of arid-climate soils with surface water or groundwater provides a constant new supply of soluble salt. As the irrigation water evaporates and moves through plants to the atmosphere, the dissolved solid content of the soil water increases. Eventually, the increase in soil salt will inhibit or stop plant growth. It is therefore necessary to apply much more water to fields in arid climates than the amount required for plant growth, so that the water flushes salts away from the plant root zone. If the excess water drains easily to the groundwater zone, however, the groundwater becomes enriched in dissolved solids, which may be detrimental.

If the groundwater table is near the surface, or if impermeable soil zones are close to the surface, overirrigation will not alleviate the problem of soil salinization. Alleviation of this condition requires the installation of subsurface drains to carry the excess soil water and salts to a surface outlet. The problem with this method is that disposing of the salty drain water is difficult. If the drain water is released into surface streams, it degrades the quality of the stream water, adversely affecting downstream users. If the water is discharged into evaporation ponds, it has the potential to seep into the groundwater zone or produce a dangerously contaminated body of surface water, as occurred at the Kesterson National Wildlife Refuge in California, where concentrations of the trace element selenium rose to levels that interfered with the reproduction of resident birds.

Robert E. Carver

FURTHER READING

Blanco, Humberto, and Rattan Lal. *Principles of Soil Conservation and Management.* New York: Springer, 2008.
Vengosh, A. "Salinization and Saline Environments." In *Environmental Geochemistry,* edited by Barbara Sherwood Lollar. Oxford: Elsevier, 2005.

SEE ALSO: Heavy metals; Irrigation; Kesterson Reservoir; Runoff, agricultural; Soil contamination.

Solar automobile racing

CATEGORY: Energy and energy use

DEFINITION: Competitions in which drivers travel specified distances in vehicles powered by the sun's energy

SIGNIFICANCE: Races featuring solar automobiles demonstrate the potential of solar energy to provide sustainable transportation. Solar car races motivate engineers and scientists to refine vehicle and energy technologies utilizing solar power. Several manufacturers have appropriated innovative features and designs developed for those competitions to use in the vehicles they produce.

In 1982 Hans Tholstrup departed from Sydney, Australia, in a solar-energy-powered automobile. He traveled west, covering 4,052 kilometers (2,518 miles) in twenty days, averaging 23 kilometers (14.3 miles) per hour, to reach Perth. Inspired by that experience, Tholstrup envisioned a transcontinental competition in Australia to promote solar vehicle technology research. The detrimental impact of fossil-fuel emissions on the environment concerned Tholstrup, who emphasized that solar cars do not produce such damaging emissions. He established the World Solar Challenge (WSC), with the debut race occurring in 1987.

The first WSC started in Darwin in northern Australia, following the Stuart Highway 3,000 kilometers (1,864 miles) south to Adelaide. Tholstrup stressed that participants' vehicles had to rely completely on solar power. According to the rules, the cars could race from 8:00 A.M. to 5:00 P.M. daily, and the first vehicle to reach the finish line would be named the winner. The drivers endured extremely hot temperatures, windstorms, and other difficulties. A solar automobile developed by General Motors, the Sunraycer, won the race, averaging 66.9 kilometers per hour (41.6 miles per hour).

Solar cars raced in the WSC in 1990, 1993, and 1996 along the same route, before the South Australian government purchased the competition from Tholstrup; since that time the race has been held every two years. The WSC remains the most important solar automobile racing contest in the world, but solar automobile racing has gone on to attract diverse par-

ticipants and sponsors, and other races are held worldwide, including the North American Solar Challenge. In 2009 the WSC became part of the Global Green Challenge, an event that encompasses both the WSC competition and the Eco Challenge, a competition for "environmentally friendly production and experimental vehicles."

A solar car is typically constructed with light carbon fiber composite materials, such as Kevlar, that form a wide surface area, approximately 6 square meters (65 square feet), on which are mounted solar cells, typically made from silicon, which convert solar energy into electricity; this energy is stored in lithium batteries, which power electric motors that power the vehicle. Competitors in solar races strive to reduce the weight and drag of their vehicles and improve the vehicles' aerodynamics. Many invest in relatively costly materials, such as gallium arsenide, for their solar cells to maximize the amount of solar energy converted into electricity; for example, solar cells made from gallium arsenide can convert 26 percent of solar energy into electricity, compared with 20 percent for solar cells made from silicon. Solar cars can move very quickly; racers in the WSC often exceed South Australia's speed limit of 110 kilometers (68 miles) per hour.

Solar-powered cars at the start of an eight-hour endurance race in Suzuka, Japan, in July, 2010. (AP/Wide World Photos)

The strategies used by solar car racing teams include computer modeling to simulate races. Some teams use weather balloons to assess cloudiness on race routes. Racers often position mirrors on their solar cars to direct sunlight to the solar cells. Competitors also place microprocessors on batteries and utilize wireless communications between the solar cars and their support teams so the teams can remotely set the solar cars' speeds after interpreting data.

Many automobile makers have been impressed by the technologies developed by solar automobile racing competitors, some of which may be employed to enhance the energy efficiency and reduce the carbon emissions of other vehicles. As a result of advances made in solar automobile racing, for example, some manufacturers have incorporated tires with reduced rolling resistance, batteries that store more energy for greater durations, and improved aerodynamics in mainstream vehicles.

Elizabeth D. Schafer

FURTHER READING

Barry, Courtney. "Racing: Here They Come, as Fast as the Sun Will Carry Them." *The New York Times*, October 22, 2003, G34.

Excell, Jon. "Lighting the Way." *The Engineer*, April 6, 2009, 22-25.

Roche, David M., et al. *Speed of Light: The 1996 World Solar Challenge*. Sydney: Photovoltaics Special Research Centre, University of New South Wales, 1997.

Thacher, Eric F. *A Solar Car Primer*. Hauppauge, N.Y.: Nova Science, 2005.

Upson, Sandra. "Across the Outback on Photons Alone." *IEEE Spectrum* 45, no. 2 (February, 2008): 50-55, 59-60.

Zorpette, Glenn. "Sun Kings Cross the Outback." *IEEE Spectrum* 39, no. 2 (February, 2002): 41-46.

SEE ALSO: Alternative energy sources; Alternative fuels; Alternatively fueled vehicles; Automobile emissions; Electric vehicles; Fossil fuels; Hybrid vehicles; Photovoltaic cells; Renewable energy; Solar energy.

Solar energy

CATEGORY: Energy and energy use

DEFINITION: Conversion of energy from the sun into thermal or electrical energy

SIGNIFICANCE: Diminishing supplies of fossil fuels and the pollution problems that result from their combustion have made renewable resources such as solar energy increasingly important. While creating solar cells and solar power facilities is energy- and resource-intensive, once in operation solar power plants provide low-carbon, low-pollution energy.

The sun was essentially humankind's only source of energy before the Industrial Revolution. In its broadest definition, solar energy is remarkably diverse and takes many forms. Hydropower, which is derived from the kinetic energy of moving water, is possible because water evaporated from the oceans by solar radiation subsequently falls in mountainous areas and flows to lower elevations. Wind energy results from the uneven solar heating of the planet's surface. Biomass contains solar energy that has been stored by photosynthesis, as do fossil fuels. Defined more narrowly, viable solar energy technologies include active and passive systems that make use of the sun's heat energy and photovoltaic cells that convert the sun's energy into electricity.

According to the National Renewable Energy Laboratory (NREL), each day 46,700 quadrillion British thermal units (quads) of energy fall on the coterminous United States alone—substantially more than the 101.527 quads the nation consumed in peak year 2007, and more than one hundred times the total world consumption for the year 2006. This energy is available to anyone who is able to collect, transform, and store it.

Between 2000 and 2009 solar energy generation in the United States nearly quadrupled, rapidly increasing from 909 million kilowatt-hours at the beginning of the decade to almost 3.6 billion kilowatt-hours by its end. Between 2001 and 2009 U.S. venture capital and private equity investment in solar technology companies surged from $5 million to more than $1 billion. While solar energy has become an increasingly important source of power in the United States, by 2009 solar accounted for only 0.1 percent of the nation's total energy production and consumption, trailing behind biomass, geothermal, and wind.

By the end of the first decade of the twenty-first century, solar and wind energy had become the world's fastest-growing energy technologies. The total solar energy installed capacities in 2009 among the world's top users of the technology were 9,677 megawatts in Germany, 3,595 megawatts in Spain, 2,628 megawatts in Japan, 2,108 megawatts in the United States, and 1,158 megawatts in Italy.

Solar energy is both a centralized and a decentralized technology. There are many different methods for collecting and converting solar energy, from small-scale home installations to massive commercial power facilities. Research into solar power technologies is carried out by major corporations and universities and by small start-up companies. The NREL coordinates much of this research. Solar energy technologies can be divided into three categories: solar thermal energy, concentrating solar power, and photovoltaic solar power.

SOLAR THERMAL ENERGY

The use of solar energy for heating and cooling is based on the simple fact that when an object absorbs sunlight it gets hot. The heat energy may be used in several ways: to provide space heating and cooling, to drive engines, or to heat water or other fluids. If this is accomplished by nothing other than appropriately designed and situated buildings and without moving parts, it is termed passive solar technology. Passive solar architecture was introduced about two thousand years ago by the Greeks and is a common feature in traditional Islamic architecture.

The simplest passive method of solar heating used in the Northern Hemisphere involves situating a building so that its windows face south and its long axis runs east-west (in the Southern Hemisphere, the windows must face north). During the winter the sun is low in the sky and provides heating to the windows. In the summer, when the sun is high in the sky, most of

Mitch McCullough, manager of site operations for the SkyFuel company, looks over a parabolic-shaped solar collector in Arvada, Colorado. The solar collector harnesses solar radiation to produce steam for electricity generation and industrial applications. (Rick Wilking/ Reuters/CORBIS)

Comparison of Two Types of Solar Energy Collectors

SOLAR PHOTOVOLTAIC COLLECTOR	SOLAR THERMAL COLLECTOR
Converts solar energy directly into electricity for immediate use.	Collects heat from solar energy for conversion into electricity.
Electricity can be converted into heat for thermal use.	Heat is used directly.
Solar radiation of only a very small range of energy can be utilized.	Radiation of a wide range of energy can be used.
Requires additional storage devices that are costly and inefficient.	Some have built-in storage devices that are relatively inexpensive and efficient.
Ideal for micropower and small appliances.	Unsuitable for micropower and small appliances.

the radiation falls on the roof. The building's windows and walls are constructed to minimize heat transfer by conduction, convection, and radiation. Sufficient interior heat capacity, generally in the form of concrete walls or floors, keeps the building from overheating during the day while storing excess heat for release during the night. Double-paned windows with a layer of air between the sheets of glass are effective for preventing conduction and convection. Glass is transparent to visible and infrared radiation, so sunlight enters and infrared radiation exits through the windows. During daylight hours more energy comes in than goes out, and at night drapes or blinds can be closed to reduce energy loss. Specialized coatings can be applied to the windows to enhance their ability to reflect infrared radiation back into the interior.

Some buildings are designed so that the sun's heat creates convection currents, which draw cool air into the buildings through ducts that are underground or north-facing (south-facing for the Southern Hemisphere) while hot air is vented to the outside. Traditional Islamic architecture frequently uses chimneys to vent this hot air. The Trombe wall, invented by French designer Félix Trombe during the 1970's, is a modern version of this feature. Well-designed passive solar buildings are economical to heat and cool because little or no additional thermal regulation is needed.

The term "passive solar" is also sometimes used to refer to solar water-heating systems that have no moving parts and operate without the use of electricity. In such a system, cold water flows under local water pressure into a collector, where it is heated by solar energy. The heated water rises and flows from there into a storage tank.

One small-scale passive application of solar heat that has the potential to have significant environmental impacts is the solar cooker. Low-cost passive solar cookers made from cardboard and reflective foil have been distributed in developing countries, particularly among refugee populations living in temporary settlements, as an alternative to cooking over open fires. Widely used, solar cooking can save tons of firewood every year, thereby reducing deforestation and desertification. The use of solar cookers also cuts back on the number of open cooking fires, which generate unhealthful levels of smoke and particulate matter when they burn within poorly ventilated structures. Solar cookers also offer the added advantage of costing nothing to operate.

Active solar heating involves heating a working substance, usually a liquid, and pumping it to where the heat is needed. Depending on the system used, the temperature increases of the fluid may be a few degrees or as much as a few thousand degrees. Such a system generally has tubes that are painted black or are in thermal contact with a metal sheet that is painted black. Fluid that is circulated through the tubes is heated, then sent to a heat exchanger, where the heat energy is used directly or stored for future use. The technology is fairly simple, but the initial installation costs of such systems continue to be a stumbling block for most home owners in the United States—particularly given that the average American changes residence every five to seven years, a time span too short to allow recoupment of the investment made on many residential solar installations. An advantage to active solar heating systems is that they can be installed in existing buildings.

Concentrating Solar Power

Concentrating solar power (CSP) technology uses direct solar radiation to produce electricity on a commercial scale. It is best suited to locations with direct sunlight, clear skies, and dry air, which is why CSP facilities tend to be located in desert areas. In a CSP system, an array of reflectors or lenses concentrates sunlight onto a collector, producing temperatures as high as 3,000 degrees Celsius (5,432 degrees Fahrenheit). There the heat may be retained in a storage system containing molten salts or some other fluid, or it may be converted into mechanical energy. Sun-tracking systems orient the reflectors or lenses to take best advantage of the available sunlight.

The most common solar concentrator design, and the one that has enjoyed the most commercial success, is the parabolic trough, a long array of curved mirrors that concentrates sunlight on receiver tubes that parallel each mirror along its focal line. Inside the tubes is a liquid medium, which is heated by the sunlight. The liquid is conveyed to a central collector, where it heats water to produce steam that drives an electric turbine. Another design, the linear Fresnel reflector system, involves a receiver tube positioned above several mirrors. This allows the mirrors to track the sun more effectively. Yet another design is the parabolic dish, which concentrates the sun's energy on a central thermal receiver. Such a system is often combined with a Stirling engine. Fluid heated in the receiver is transferred to the engine, where the heat moves pistons to create mechanical power. This in turn powers a generator or alternator to produce electricity. A fourth design, the central receiver, involves a large field of flat mirrors that focuses sunlight on a central location, typically a tower with a generator at its base. Heated fluid within the tower produces steam, which drives the generator below.

CSP systems can also be used in conjunction with photovoltaic cells, a technology known as concentrating photovoltaic (CPV). CPV systems used lenses or mirrors to focus sunlight onto high-efficiency solar cells.

In 2009 the United States led the world in CSP installed capacity, followed by Spain. The capacity of U.S. CSP plants reached 431 megawatts that year. States leading the nation in CSP installed capacity were California, Nevada, Hawaii, and Arizona. Nine commercial parabolic trough facilities have been operating in California's Mojave Desert since the 1980's. World CSP capacity began a rapid increase in 2007, but by 2009 CSP still accounted for less than 0.1 percent of the world's total installed capacity.

Photovoltaic Solar Power

Photovoltaic solar cells, first developed in 1954 by scientists at the Bell Telephone Laboratories, are solid-state devices that convert sunlight into electricity. The earliest solar cells were made using single-crystal silicon wafers. Silicon is an important type of element known as a semiconductor, which has properties between those of conductors and insulators.

Electric conduction in silicon results from the movement of negative charges (electrons) and positive charges (holes). One way to cause this movement is to add arsenic or phosphorus atoms—which have five outer-shell electrons—to the pure silicon, creating a semiconductor that has excess negative charge (n type). The addition of boron atoms with three outer-shell electrons creates a p-type semiconductor. Electric conduction occurs when p- and n-type slices are placed in close contact.

The simplest solar cell is a p-n junction sandwiched between two conductors. One of the tricks to fabricating efficient solar cells centers on the design of the top conductor. It should be large enough to capture the electrons but not so large that it blocks sunlight from passing into the center of the sandwich. Sunlight enters the top of the cell and excites electrons in the n-type layer, causing them to jump to the p-type layer, where they are captured and carried away to do work in the external circuit.

The earliest solar cells converted sunlight to electricity with about 1 percent efficiency and were expensive to produce, but since the 1950's photovoltaic technology has become substantially more efficient as well as less costly. Improvements have been made in crystalline silicon solar cells, and amorphous silicon (a-Si) solar cells have emerged. Although a-Si solar cells are less efficient than crystalline silicon cells, thin-film a-Si cells are comparatively inexpensive to manufacture, and their efficiency is boosted when they are stacked atop one another to create a multilayered structure. While silicon continues to be important for solar cell production, use of other materials such as cadmium telluride, cadmium sulfide, gallium arsenide, gallium phosphide, and copper indium gallium selenide has grown.

The efficiency of silicon cells depends, to a great extent, on the purity of the silicon. Working in vacuum and zero gravity, scientists can prepare exceptionally pure silicon. Transporting the materials into orbit and the finished product back down to earth, however, would substantially increase production costs. An alternative proposal would leave the finished product in space, where the sun shines every day of the year. Although it is technically possible to build enormous solar cells in space and beam the energy down to the planet's surface in the form of microwave or laser energy, the concept is a controversial one. An orbiting solar power station would be very expensive to construct and maintain, and the technical challenges of successfully beaming energy from earth orbit to the planet's surface are considerable. The U.S. National Aeronautics and Space Administration (NASA) initiated the Space Solar Power Exploratory Research and Technology program in 1999 to investigate the possibility of space-based solar power, but two years later it abandoned the program. Japan's space agency, however, has announced plans to develop and deploy such a system and have it in operation by 2030.

Photovoltaics have been used extensively in space to power almost all satellites, and the range of earthbound consumer products with built-in photovoltaic cells is continually expanding. Such cells are used to power radios, calculators, battery rechargers, outdoor lights, and electric fences. One of the advantages of photovoltaics is that they are modular: Individual units can be combined to produce outputs covering a wide voltage range. The technology is also transportable: Units can be set up wherever there is enough sunlight. Photovoltaics are used to power traffic signal controls, roadside signage, scientific instrumentation, anticorrosion systems on bridges, and experimental cars and aircraft.

In 2009 the capacity of photovoltaic power plants in the United States reached 1,677 megawatts. CSP accounted for most of the nation's solar capacity until 2005, when photovoltaics surpassed it. Between 2008 and 2009 cumulative solar photovoltaic capacity in the United States grew nearly 52 percent. In 2009 California led the nation in photovoltaic installed capacity, followed by New Jersey, Colorado, Arizona, Florida, Nevada, New York, Hawaii, Connecticut, and Massachusetts.

The solar cell manufacturing industry experienced major growth in the early twenty-first century, and in 2008 world photovoltaic capacity began to rise. In 2009 photovoltaics accounted for 0.4 percent of the world's total installed capacity and 0.1 percent of total worldwide renewable electricity generation.

Germany was the world leader in cumulative photovoltaic installed capacity that year, followed by Spain, Japan, the United States, and Italy.

Grace A. Banks
Updated by Karen N. Kähler

FURTHER READING

Cassedy, Edward S. "Solar Energy Sources." In *Prospects for Sustainable Energy: A Critical Assessment*. New York: Cambridge University Press, 2000.

Foster, Robert, Majid Ghassemi, and Alma Cota. *Solar Energy: Renewable Energy and the Environment*. Boca Raton, Fla.: CRC Press, 2010.

Kalogirou, Soteris. *Solar Energy Engineering: Processes and Systems*. Burlington, Mass.: Academic Press, 2009.

Naff, Clay Farris, ed. *Solar Power*. Detroit: Greenhaven Press, 2007.

National Renewable Energy Laboratory. *2009 Renewable Energy Data Book*. Golden, Colo.: U.S. Dept. of Energy, Office of Energy Efficiency and Renewable Energy, 2010.

National Research Council. *Electricity from Renewable Resources: Status, Prospects, and Impediments*. Washington, D.C.: National Academies Press, 2010.

O'Keefe, Philip, et al. *The Future of Energy Use*. 2d ed. Sterling, Va.: Earthscan, 2010.

Pimentel, David, ed. *Biofuels, Solar, and Wind as Renewable Energy Systems: Benefits and Risks*. New York: Springer, 2008.

SEE ALSO: Alternative energy sources; Energy policy; Photovoltaic cells; Renewable resources; Solar automobile racing; Solar One; Solar water heaters; Sun Day.

Solar One

CATEGORY: Energy and energy use
IDENTIFICATION: The first solar power tower system
DATES: Operated from 1982 to 1988
SIGNIFICANCE: The Solar One facility proved that power towers work efficiently to produce utility-scale power from sunlight.

Solar One, a pilot plant built as a large-scale test of the tower system of generating power using solar energy, was built in the early 1980's near Barstow, Cali-

Aerial view of Solar Two, the solar power tower facility converted from Solar One's water and steam system to a molten salt system. (George Steinmetz/CORBIS)

fornia, by Southern California Edison, with the support of Sandia Labs, the U.S. Department of Energy, the Los Angeles Department of Water and Power, and the California Energy Commission. Rated at ten megawatts of power, Solar One could efficiently and cost-effectively store energy, making it unique among solar technologies. Solar One operated successfully from 1982 to 1988.

A solar power tower operates by focusing a field of thousands of mirrors (heliostats) onto a receiver located at the top of a centrally located tower. The receiver collects the sun's heat in a heat-transfer liquid. This liquid is used to generate steam for a conventional steam turbine, which then produces electricity at the base of the tower. The mirrors of Solar One reflected and focused sunlight onto a central tower that was nearly 100 meters (328 feet) tall. The tower's absorber panels, which were painted black, absorbed 88 to 96 percent of the incident light. In order to capture

sunlight from the south, the field of mirrors was oriented mostly toward the north. Solar energy was focused onto six panels to preheat the water that traveled to eighteen superheat panels.

After leaving the superheat panels, the water was at 510 degrees Celsius (950 degrees Fahrenheit). The hot water was sent either to turbines, where it generated electricity at 35 percent efficiency, or to a heat exchanger, where it heated oil that was sent to a thermal storage tank and circulated through crushed granite. The stored heat could be drawn back from the tank through the heat exchanger to produce steam for the turbine. In addition, the thermal storage allowed a buffer system for periods of cloudiness so that the plant could keep operating through changes in weather conditions.

Solar One proved that the heat-transfer liquid cycle is reliable, that the system could meet expectations, and that thermal storage is cost-effective. Furthermore, the power tower system with energy storage showed a unique advantage over other solar power systems because it could supply power to the local electrical utility company during peak periods. In Southern California, these periods occur on hot, sunny afternoons and into the evenings during the summer, when needs are high for air-conditioning of homes and workplaces and power production is most valuable to the power company.

Based on what they learned through the operation of Solar One, which used water and steam as the heat-transfer liquid, solar engineers determined that power towers operate more efficiently using molten salt. The salt also has the further advantage of providing a direct, practical way to store heat. The concept of storing energy in molten salt and decoupling solar energy collection from electricity production formed the basis for Solar Two, which operated from 1996 to 1999. The construction of Solar Two involved the conversion of Solar One from its water and steam system to a molten salt system. After Solar Two was decommissioned, the tower remained as a local landmark in the California desert until it was torn down in 2009. Construction on a third solar power tower plant, named Solar Tres, began in 2007 near Seville, Spain; the plant became operational in 2009.

Alvin K. Benson

FURTHER READING

Cassedy, Edward S. "Solar Energy Sources." In *Prospects for Sustainable Energy: A Critical Assessment.* New York: Cambridge University Press, 2000.

Gordon, Jeffrey, ed. *Solar Energy: The State of the Art.* London: James & James, 2001.

Miller, G. Tyler, Jr., and Scott Spoolman. "Energy Efficiency and Renewable Energy." In *Living in the Environment: Principles, Connections, and Solutions.* 16th ed. Belmont, Calif.: Brooks/Cole, 2009.

Naff, Clay Farris, ed. *Solar Power.* Detroit: Greenhaven Press, 2007.

SEE ALSO: Alternative energy sources; Renewable energy; Solar automobile racing; Solar energy; Solar water heaters; Sun Day.

Solar radiation management

CATEGORY: Weather and climate

DEFINITION: Subfield of geoengineering concerned with the alteration of the earth's climate through changes in the interaction between sunlight and the planet

SIGNIFICANCE: A small but growing number of scientists have suggested that strategies to limit global warming through reduction of greenhouse gas emissions may not work, either because of lack of willingness to limit emissions or because global warming may have already progressed to the point that it cannot easily be reversed. Solar radiation management has been proposed as a possible countermeasure in the event that limiting greenhouse gas emissions fails to halt global warming.

Most strategies for combating global warming involve reducing the amount of greenhouse gases in the earth's atmosphere. The strategy of solar radiation management, in contrast, involves either limiting the amount of sunlight reaching the earth's surface (solar irradiation mitigation) or reflecting some of the sunlight that does reach the surface back into space (albedo modification). Both approaches would result in less absorption of sunlight and thus less solar heating of the planet. Theoretically, less solar heating could counter the increased atmospheric heating linked with greenhouse gases.

ALBEDO MODIFICATION

One way of limiting how much sunlight is absorbed to heat the earth is simply to reflect the light back into space. The fraction of sunlight striking a planet that is reflected is called the albedo of the planet. The higher the albedo, the more light is reflected. Variations in albedo occur naturally on local scales. On cloudy days sunlight is reflected into space by clouds, causing the temperature to be lower on the ground. One suggested approach to solar radiation management is to increase the earth's albedo artificially to lower the planet's temperature by reflecting sunlight into space so that it cannot be absorbed.

Several strategies for making modifications to earth's albedo have been suggested. One idea is to mimic the effect of cloudy days by making more clouds or making existing clouds thicker. Cloud seeding can be accomplished through the injection of particulate matter into the atmosphere to form nucleation sites for water droplets. Other suggestions include using ships with powerful pumps to inject a seawater mist into clouds, making them thicker and more reflective.

Very large volcanic eruptions can send sulfur dioxide and other sulfur contaminants high into the stratosphere. These sulfur contaminants can be reflective enough to increase the earth's albedo by several percent. Since the sunlight is reflected in the stratosphere, it does not reach the ground or the lower atmosphere, and the surface temperature of the earth drops. Historically, massive volcanic eruptions have often been followed by years of cooler weather. One idea for increasing the earth's albedo involves injecting large amounts of similar sulfur contaminants into the stratosphere to cause conditions that would mimic those that lead to the temperature reductions that sometimes follow massive volcanic eruptions.

Albedo modifications can be made on land, too. Some scientists have suggested that simply painting roads white and making the roofs of buildings reflective would have a local effect on weather, particularly near large urban areas. It is unclear whether making human construction artifacts more reflective would have a sufficient impact on global climate to compensate for increased greenhouse gases if that technique is not used in conjunction with other measures to limit global warming. More radical proposals have included the suggestion that reflective paint could be applied to deserts, to mimic the increase in albedo of snowfall. Painting the ground white near where ice has melted is one possible strategy to limit the positive feedback of melting polar ice caps.

SOLAR IRRADIATION MITIGATION

Some volcanic eruptions inject more light-absorbing ash than reflective particles into the stratosphere. This ash, by absorbing light, reduces the sunlight that reaches the earth's surface, in much the same way as reflecting light would do, causing a cooling of the planet's surface. Some scientists have suggested that the injection of soot or even dirty airplane exhaust into the stratosphere could act to cool the earth's surface. This strategy has many drawbacks, however. Particles absorbing solar energy in the stratosphere, although reducing the absorption of solar energy at the ground, still result in solar energy being absorbed somewhere in the atmosphere, possibly resulting in

an eventual change in atmospheric structure. This could eventually lead to temporary reduction in the temperature on the ground at the expense of much larger climate disruption years later.

One very ambitious proposal calls for deploying a cloud of mirrors in space between the earth and the sun. Through the deployment of enough objects to reflect less than 2 percent of the sunlight reaching the earth, a thermal balance could be reached, offsetting the effect of increased atmospheric greenhouse gases. Once launched, such a space-based system would require far less effort to maintain than other solar radiation management plans. Reflectors in space are passive systems, so they continue to operate even if they are not actively maintained. The only maintenance required would be to make sure that they stay in stable orbits. This would be achievable if they could be placed in high orbit around the earth, but there they would interfere with spacecraft and the activities of communication satellites. Another possible location for a space-based solar shield would be at the point between the earth and the sun known as the L1 Lagrange point. A gravitational balance between the earth and sun causes objects at the L1 point to tend to stay near that point.

A space-based shield would not raise the concerns associated with the injection of pollutants into the atmosphere to alter the earth's albedo. Furthermore, because such a shield would make no modifications to the atmosphere or to land on the earth, it would not have the possible negative environmental impacts that other approaches to solar radiation management would have. A major difficulty with implementing any plan to create a space-based shield, however, is that it is extremely expensive to launch objects into space.

CRITICISMS

Numerous criticisms have been leveled against the suggestions that have been made by scientists examining the possibility of solar radiation management. Critics point out that any approach to solar radiation management on a global scale would be exceedingly expensive, would require a great deal of international cooperation, and would require significant investments of time and effort. Even proponents of geoengineering admit that implementation of any of these plans on a global scale would be an extremely difficult and expensive undertaking; they assert, however, that solar radiation management may be the only way to check otherwise unstoppable climate change. There is no agreement among climate scientists on how much climate change is unstoppable without such drastic measures.

Solar radiation management is often seen as only a temporary measure—it may have an effect opposite to that of greenhouse gases, but it does not address the root problem of excessive concentrations of atmospheric greenhouse gases. If solar radiation management programs were to be deployed successfully and then ever stopped for some reason—for example, because of war or the economic collapse of the nations supporting the programs—then the high levels of greenhouse gases in the atmosphere could potentially result in a catastrophic rise in global temperatures. Even proponents of solar radiation management thus often suggest that it be only one part of a multipronged approach to controlling climate change.

A further criticism of solar radiation management is rooted in the fact that solar heating drives many aspects of the earth's ecology. Interfering with natural solar radiation could potentially result in such unintended consequences as changes in the character and frequency of storms, which could result in drought and flooding in diverse areas of the earth. Altering the climate in such a way may also change the acidity of the oceans, resulting in wide-scale extinctions.

Environmentalists often oppose atmospheric modifications such as injecting materials into the stratosphere or into clouds because, they argue, these actions amount to intentional pollution of the planet; the materials that scientists have proposed injecting into the atmosphere could have deleterious effects on the health of humans and other species. Proponents of these plans argue that despite the risks of possible negative health effects, these effects would ultimately be less damaging than the effects of unchecked global warming. Space-based solar radiation management plans would have the fewest potential negative impacts on the earth's environment, but they would also be among the most expensive to implement.

Raymond D. Benge, Jr.

FURTHER READING

Angel, Roger. "Feasibility of Cooling the Earth with a Cloud of Small Spacecraft Near the Inner Lagrange Point (L1)." *Proceedings of the National Academy of Sciences* 103, no. 46 (2006): 17184-17189.

Bala, G. "Problems with Geoengineering Schemes to Combat Climate Change." *Current Science* 96, no. 1 (2009): 41-48.

Hoyt, Douglas V., and Kenneth H. Schatten. *The Role of the Sun in Climate Change*. New York: Oxford University Press, 1997.

Kerr, Richard A. "Pollute the Planet for Climate's Sake?" *Science* 314, no. 5798 (2006): 401-402.

Launder, Brian, and J. Michael T. Thompson, eds. *Geo-engineering Climate Change: Environmental Necessity or Pandora's Box?* New York: Cambridge University Press, 2010.

Levi, Barbara Goss. "Will Desperate Climates Call for Desperate Geoengineering Measures?" *Physics Today*, August, 2008, 26-28.

Morton, Oliver. "Is This What It Takes to Save the World?" *Nature* 447, no. 7141 (2007): 132-136.

SEE ALSO: Geoengineering; Krakatoa eruption; Mount Tambora eruption; Positive feedback and tipping points; Volcanoes and weather; Weather modification.

Solar water heaters

CATEGORY: Energy and energy use
DEFINITION: Devices that heat water by using absorption of solar energy either exclusively or primarily
SIGNIFICANCE: Heating water by absorption of solar energy does not require any electricity or burning of fossil fuels, thus solar water heaters have extremely low environmental impacts. Even solar water heaters that use electric pumps to circulate water use far less electricity than water heaters that heat water using electricity.

Heating with sunlight is the easiest and most cost-effective method of utilizing solar energy, and humans have heated water using the sun since ancient times. The biggest drawback to solar water heating, however, has been that the sun heats most effectively during the middle of the day, and hot water is often needed in the evenings. A further difficulty is that if the heated water is not properly insulated from the environment, it will cool off on cold days. These problems, however, are fairly easily overcome by well-designed solar water heaters. Well-insulated systems can keep water warmed during the day and hot through the night and can permit water to be heated even when the outdoor temperature is quite cold.

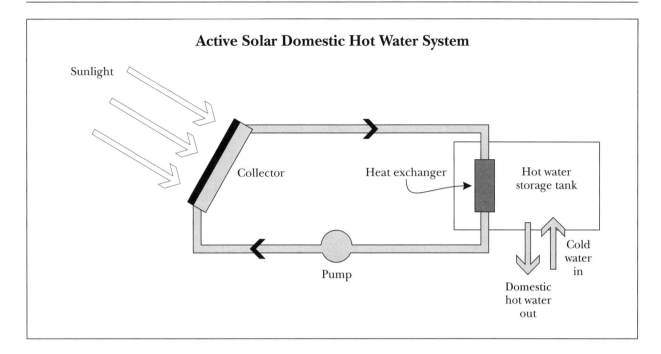

A more difficult problem for solar water heaters to overcome is that they do not work well on cloudy days. They also do not work as effectively at very high latitudes in the winter, when there are few hours of sunlight during the day. Even in summer at high latitudes, the sun is at a low angle in the sky, reducing the effectiveness of solar water heaters. Many solar water-heating systems thus include gas- or electricity-powered backups.

TYPES

The simplest form of solar water heater is the integral collector system (ICS), which is basically a tank, often painted black, placed where the sun can shine on it during the day. This is the least effective system and is often used as a preheater for a conventional gas or electric water heater.

The most commonly installed type of solar water heater is the flat-plate collector, which consists of a series of pipes with fluid flowing through a large flat box, often insulated and with a transparent window through which the sun can heat the interior of the box and the pipes. The biggest drawback to simple ICS systems is that the warming tanks have a large volume compared with the surface area being heated. Flat-plate collectors overcome this difficulty by creating surface areas to be heated that are far larger than those of the storage tanks. In warm climates where temperatures almost never reach freezing, water can be circulated directly through the collector to be heated and then collected in an insulated storage tank that operates much like a conventional water heater. In locations where there is a risk for a freeze, a fluid resistant to freezing (such as water mixed with some sort of antifreeze) circulates in the collector and then through a set of pipes either inside the storage tank or in another tank near the storage tank. The second set of pipes in this indirect system acts as a heat exchanger to warm the water. In many systems, the storage tank has heating elements and serves as a backup heater if necessary.

In very cold climates an evacuated-tube solar collector may be used. The evacuated-tube collector is very similar to the flat-plate collector, except that the pipes in the collector are replaced by double-walled glass tubes. The space between the tubes is a vacuum, thus the loss of heat through conduction in the tubes is limited, much as heat loss is limited in a vacuum Thermos bottle. Evacuated-tube systems can be either direct or indirect.

Both flat-plate and evacuated-tube collectors can be constructed as either active or passive systems. An active system uses a pump to circulate the fluid through the collectors; it is thus not a zero-energy-use system, as it requires a small amount of electricity to run the circulation pump. A passive system, in contrast, does not require any electricity. Instead, the heated fluid rises to a heat exchanger in a storage tank located above the collector. In most solar water heater installations, however, the collector is placed on a building's roof, so placement of the storage tank at a greater height than the collector is impractical. Active systems are thus more common than passive systems.

COSTS OF INSTALLATION AND OPERATION

Because solar water heaters involve more plumbing than conventional systems, they are more expensive to install than conventional water heaters. The biggest cost of a conventional water heater is the tank, and a solar water heater also requires a tank in addition to other expensive components. Solar water heaters thus cost considerably more than conventional systems in terms of both parts and installation. A further reason for the high cost of solar water heaters is their comparative rarity compared with conventional water heaters. Relatively few plumbers and builders have had much experience with solar water heaters, so specialists who charge more for their services are often needed to install the systems.

ICS heaters are the least expensive solar water heaters, but they are also the least efficient. Evacuated-tube heaters are often so expensive that they are not practical except in commercial installations. Thus the flat-plate collector is the type of solar water heater most commonly used in residential systems.

Solar water heaters are much less expensive to operate than conventional water heaters, but it can still take many years for a system to save a home owner enough to cover the cost of the initial investment in the equipment. The high up-front cost is a significant deterrent to many people considering installing such systems.

Raymond D. Benge, Jr.

FURTHER READING

Galloway, Terry R. *Solar House: A Guide for the Solar Designer.* Burlington, Mass.: Architectural Press, 2004.

Laughton, Chris. *Solar Domestic Water Heating.* Sterling, Va.: Earthscan, 2010.

National Renewable Energy Laboratory. *Heat Your Water with the Sun.* Washington, D.C.: Government Printing Office, 2003.

Ramlow, Bob. *Solar Water Heating.* Gabriola Island, B.C.: New Society, 2006.

SEE ALSO: Energy conservation; Green buildings; Renewable energy; Solar energy.

Solid waste management policy

CATEGORY: Waste and waste management
DEFINITION: Procedures and regulations put in place by communities for dealing with the creation, accumulation, utilization, and eventual deposition of solid waste materials
SIGNIFICANCE: When the disposal of solid wastes is mismanaged or inadequately addressed, the negative results for the environment can include air, water, and soil pollution. The policies that governments put in place for managing solid waste can thus have wide-ranging impacts.

No internationally recognized definition has been established regarding what constitutes solid waste. The Organization for Economic Cooperation and Development, a multinational body, employs a definition that excludes radioactive waste but includes hazardous waste. The federal government of the United States has adopted a definition that excludes most hazardous waste, except that included within municipal waste. Despite these definitional differences, the public policies adopted in regard to waste management have been reasonably similar among most economically advanced nations. The purpose behind the definitions has been to identify levels of danger posed by waste materials so that adequate management regimes can be constructed. Materials defined as radioactive, hazardous, or toxic require stricter controls than do those defined as solid waste.

HISTORY

Although it has been said that in nature there is no waste, it would appear that there has always been waste in human communities. For many centuries, however, the amount of garbage generated and the dispersal of such materials were not recognized as problems. Like many environmental issues, solid waste became problematic largely as a result of the increase in human population on the planet. As people began to live in more densely populated areas, they could no longer ignore the accumulation of solid waste. Evidence indicates that in the ancient cities of Africa and the Roman Empire people threw their solid waste on the floors of their dwellings. They lived in the midst of their waste, then built new streets and housing on top of the resultant mess. Eventually urban dwellers developed a variety of ways of removing their waste products from their living areas: Wastes were gathered in cesspools, directed through drainage systems and sewage systems, redistributed by scavengers, and dumped in designated places outside densely populated areas.

Urbanization and industrialization led to greater accumulations of waste and to collective decisions regarding how to deal with those accumulations. In addition to the problems generated by high concentrations of people on relatively small tracts of land, industrialization led to greater affluence, which correlates positively with the generation of solid waste. Municipal governments began to be expected to develop policies regulating waste disposal and to provide services to assist urban dwellers in dealing with their waste. In England, for example, the Poor Law Commission found in 1842 that a filthy environment promotes the spread of disease. This finding led to an increase in municipal sanitation services.

Sanitation services initially tended to operate on an "out of sight, out of mind" basis and focused on removing solid waste from densely populated areas. Collectors and scavengers gathered some of the waste and put it to other uses. Some waste was burned in incinerators, and some of the steam or electrical energy thus generated was put to use. Most of the solid waste, however, was dumped either into bodies of water or onto land.

CONTEMPORARY TRENDS

The United States leads the world in the generation of solid waste; its per-capita production is about twice that of other economically and industrially advanced nations. The People's Republic of China, the most populous nation in the world, generates less than one-third the total amount of solid waste generated in the United States.

In 1970 William E. Small brought attention to solid waste as an environmental health problem with his book *Third Pollution: The National Problem of Solid Waste Disposal*, which discussed the problem of waste as a form of pollution on a par with air and water pollution. It has now become clear, however, that solid waste is a multimedia environmental problem. In one sense there is no such thing as "disposal" of solid waste. Once solid waste is generated, it must be physically located in at least one of the three key environmental media: land, water, or air. When it is disposed of, solid waste remains in one of those three media. Thus all solid waste management policies must include

guidelines as to how much waste will be allowed to be generated, how much will be redirected toward continued utility, how much will deposited into water, how much will be emitted into the air, and how much will be deposited on or into the land.

Although developed nations around the world are involved in ongoing efforts to reduce the volume of materials that are deposited in landfills, land dumping is still the most common answer to the garbage problem. Approximately 70 percent of the solid waste in economically advanced countries is deposited in landfills. Even when other solid waste management techniques are utilized, landfilling is often the final answer.

A wide variety of landfill technologies are in use. The first landfills were nothing more than places to dump garbage, but approaches to landfilling have become more sophisticated over the years. During the twentieth century, many town dumps were replaced by municipal sanitary landfills; in some advanced

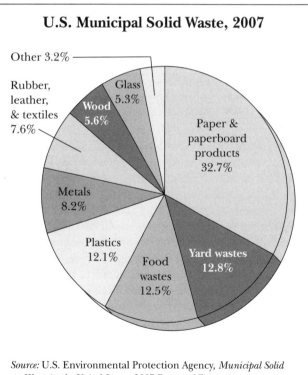

U.S. Municipal Solid Waste, 2007

Other 3.2%

Rubber, leather, & textiles 7.6%

Glass 5.3%

Wood 5.6%

Paper & paperboard products 32.7%

Metals 8.2%

Plastics 12.1%

Food wastes 12.5%

Yard wastes 12.8%

Source: U.S. Environmental Protection Agency, *Municipal Solid Waste in the United States: 2007 Facts and Figures.*

Note: Total U.S. municipal solid waste generated in 2007 was about 230 million metric tons, or 2.1 kilograms per person per day. Not included in these figures are mining, agriculture, industrial, and construction wastes; junked automobiles and equipment; or sewage.

countries, sanitary landfills have been replaced by technologically sophisticated and carefully lined regional megalandfills. By the end of the twentieth century, the U.S. Environmental Protection Agency was predicting the closure of three-fourths of the landfills then in existence. This development was fueled by communities' resistance to the locating of solid waste management facilities in their vicinity (the "not in my backyard," or NIMBY, syndrome) and by the increasingly stringent technical requirements placed on landfills. The trend is toward fewer landfills, but the typical size of newer landfills is much larger, and the technology is more complex than that used to operate landfills in the past; consequently, the costs are also much higher.

At the beginning of the twentieth century, deposition of solid waste into bodies of water was seen as a good method of disposal, but this is no longer a popular option. Numerous laws have been put in place to prohibit the dumping of solid wastes into the oceans, such as the Marine Protection, Research, and Sanctuaries Act of 1972 (also known as the Ocean Dumping Act), but illegal dumping continues. Moreover, ocean dumping remains a legal option for many cities and for most commercial vessels. However, even if the practice of dumping solid waste into bodies of water were to stop today, the problem of groundwater pollution caused by such dumping would remain.

INCINERATION, RECYCLING, AND SOURCE REDUCTION

Incinerating solid waste reduces the volume of material that needs to be placed in a landfill, but the ash it produces must still be landfilled. Incineration also emits air pollutants, including sulfur dioxide, nitrogen dioxide, carbon monoxide, hydrocarbons, polychlorinated biphenyls (PCBs), and dioxins. Incineration has been an institutionalized practice since at least 1865, but it is still not a management technique used on a large percentage of solid waste. A small resurgence of incineration technology in the United States during the 1980's quickly subsided because of problems with the technology, the economics of the approach, and citizen resistance. Incineration is used more widely in Europe and Japan than in the United States.

If waste materials are not to be deposited in water or on land, they must be reused. Resource recovery and recycling are politically popular aspects of solid waste management policy, but their contributions are limited in three ways. First, not all solid waste materials appear to have recycling potential. Second, the costs of recovery may be prohibitive. Third, even recovered and recycled products may eventually be deposited in a landfill.

One approach to recycling is the conversion of materials into other products. Cardboard and newspaper are sometimes recycled into other paper products. Aluminum cans may be recycled into other aluminum products or into new aluminum cans. Some plastic products can be recycled into other plastic products—for example, milk cartons into fibers for apparel or foam cups into plastic lawn furniture. Yard waste (grass, leaves, and other organic matter) may be composted, but commercial compost facilities often experience difficulties with odor control. Solid waste may also be seen as an alternative energy source, given that a wide variety of materials may be burned in waste-to-energy facilities to generate steam or electricity; incineration, however, has its own drawbacks, as noted above.

Source reduction is the most fundamental answer to the solid waste problem, but it is also the most difficult to establish as public policy. It stands to reason that reducing the generation of waste will reduce the need to manage it, but the economic impacts of regulating waste generation have thus far prevented waste reduction from becoming a significant part of solid waste management policy. In the United States the Comprehensive Environmental Response, Compensation, and Liability Act (CERCLA) of 1980, also known as Superfund, created a "cradle-to-grave" legal regime that attempts to hold the original generators of hazardous wastes responsible for the costs of managing those wastes, even after they are deposited in legally approved sites.

POLICY MAKING

It has been traditional in economically advanced, democratic nations to leave policy making regarding waste management to local governments. When the problems associated with solid waste were seen as primarily issues of disposal and local health effects, it made sense to leave policy making at that level, but after solid waste was identified as an environmental problem, policy makers became more inclined to recognize that, like other environmental problems, the impacts often extend beyond local political jurisdictions. Moreover, as the costs of managing solid waste increased, the need for funding assistance from larger units of government increased.

In the United States, for example, the federal government now plays a major role in developing solid waste management policy. Congress has created legislation, the Environmental Protection Agency and other regulatory agencies have promulgated regulations, and federal courts have set parameters within which state and local governments must operate. President Lyndon B. Johnson was a leader in bringing the federal government into solid waste policy making, initiating action that resulted in passage of the Solid Waste Disposal Act in 1965 as an amendment to the Clean Air Act. In 1976 Congress passed the Resource Conservation and Recovery Act (RCRA), Subtitle D of which created the first national waste management program. In combination with other environmental laws, RCRA changed the way solid waste was managed in the United States.

Federal court decisions have further diminished local governments' control over solid waste management policy making. In its decision in *Philadelphia v. New Jersey* (1978), the U.S. Supreme Court said that garbage is to be treated the same as any other commercial commodity under the commerce clause of the U.S. Constitution. This means that state laws and local government ordinances that interfere with interstate commerce in solid waste are unconstitutional. Consequently, state and local governments are not allowed to restrict solid waste that was generated outside their jurisdictions from entering into their jurisdictions. Moreover, the Supreme Court's decision in *C. & A. Carbonne, Inc., v. Town of Clarkstown, New York* (1994) held that local governments cannot restrict solid waste that was generated within their jurisdictions from leaving. Without the ability to control solid waste's entry into or exit from their jurisdictions, state and local governments are severely restricted in their ability to set solid waste management policy.

Another development in solid waste management policy has been the trend toward privatization. Privatization is not a new approach to solid waste management, but at the end of the twentieth century it saw a resurgence in popularity. Legislative antipathy toward command-and-control regulatory approaches, combined with the legal system's treatment of solid waste management as a commercial activity rather than an issue of health and welfare, strengthened the role of private commercial enterprises in solid waste management.

Larry S. Luton

Further Reading

Blumberg, Louis, and Robert Gottlieb. *War on Waste: Can America Win Its Battle with Garbage?* Washington, D.C.: Island Press, 1989.

Hickman, H. Lanier, Jr. *American Alchemy: The History of Solid Waste Management in the United States.* Santa Barbara, Calif.: Forester Communications, 2003.

Luton, Larry S. *The Politics of Garbage: A Community Perspective on Solid Waste Policy Making.* Pittsburgh: University of Pittsburgh Press, 1996.

Melosi, Martin V. *Garbage in the Cities: Refuse, Reform, and the Environment.* Rev. ed. Pittsburgh: University of Pittsburgh Press, 2005.

Rogers, Heather. *Gone Tomorrow: The Hidden Life of Garbage.* New York: New Press, 2005.

Twardowska, Irena, et al. *Solid Waste: Assessment, Monitoring, and Remediation.* San Diego, Calif.: Elsevier, 2004.

See also: Hazardous and toxic substance regulation; Landfills; Recycling; Superfund; Waste management.

South America

Categories: Places; ecology and ecosystems; forests and plants; resources and resource management

Significance: The subregions of South America, the fourth-largest continent in the world, contain the most extensive diversity of flora and fauna on the planet. South America is known for its massive mountain ranges and very long rivers, especially the Amazon, and for its rain forests, which are threatened by pressures toward clearance of the land for agriculture.

The major geographical and geological features of the South American continent determine in large measure the differences in regional environmental conditions. Geologically the most spectacular feature of the continent is the mountain range of the Andes, which stretches some 7,000 kilometers (4,350 miles) from Venezuela in the north to Argentina in the south. The mountains pass through Colombia, Ecuador, Peru, and Chile and reach average heights of 4,000 meters (about 13,000 feet). The highest peak (7,000 meters, or nearly 23,000 feet) is Aconcagua in western Argentina near the Chilean border. The Andes, which form two main ranges in South America (the Western and Eastern Cordilleras), are essentially the continuation of the tectonic uplifting that formed the North American Rocky Mountains, with the accompanying phenomenon of major volcanic activity in Mexico and Central America.

After the Andes, much of the continental interior is the Altiplano, a high plateau (more than 3,500 meters, or almost 11,500 feet) running from the Peru-Bolivia border into northern Argentina. The Altiplano is the highest permanently inhabited subregion of any of the world's five continents. Perhaps its most famous geographical site is Lake Titicaca, South America's second-largest lake (8,300 square kilometers, or 3,200 square miles, compared to Venezuela's Lake Maracaibo, which covers about 12,950 square kilometers, or 5,000 square miles). Titicaca's elevation of 3,800 meters (12,500 feet) makes it the highest navigable lake in the world. The environmental harshness of the Altiplano is rivaled, if not surpassed, by the striking phenomenon of the Atacama Desert, which runs some 966 kilometers (600 miles) along the northern portion of Chile's Pacific Ocean coast. Its Peruvian counterpart, the Sechura Desert, is nearly as dry. A few desert oases are found in the Sechura, the most famous being the small village of Huacachina in southwestern Peru, with its unexpected clear-water lagoon. Rainfall levels in this region are very low (an average of 1 millimeter, or 0.04 inch, in the Antofagasta midcoastal region). This lack of appreciable rainfall is largely the result of an atmospheric inversion effect caused by the cold Humboldt Current that follows Chile's coast.

ECOSYSTEMS AND VEGETATION

Although the Altiplano forms a transitional ecosystem between the Andes and the Amazon basin, the vast majority of the plant and animal species of South America are found in the latter two zones. Plant species in the Andes generally correspond to levels of altitude. A broad layered zone referred to as the montane forest is found between about 600 meters (1,970 feet) and 3,500 meters (11,500 feet). Here the dominant (lower montane) tree species share ecological space with a wide variety of liverworts, mosses, ferns, and some flowering plants, including orchids. Many of these are epiphytes (plants that grow without roots in the soil). Other species found in the lower montane zone are Lauraceae (the laurel family), Melastomataceae (a family of perennial flowering herbs and shrubs), Rubiaceae (the family including

madder, bedstraw, and coffee varieties), and some Moraceae (the fig family).

Higher montane vegetation depends more on misty conditions than on direct rain. Typical plants up to elevations of 4,000 meters are ferns and lichens, with fewer species of, and more widely scattered, trees. Specialists have identified several other strata of Andean vegetation at elevations above the tree line, often called the tropical alpine zone. One important stratum, typical of the central Andes from northern Peru extending into northern Argentina, is the puna (which is itself divided into substrata ranging from 3,000 to 5,000 meters, or 9,800 to 16,400 feet, called humid, dry, and desert puna). Here the main forms of vegetation are grasses, with scattered flowering plants.

The contrast between the high mountains and Altiplano and the tropical areas of South America is almost total. The massive network of waterways that eventually form the main channel (the Solimões) of the Amazon River flow from two continental shields: the Guiana Shield to the north and the Brazilian (or Amazonian) Shield to the south. Because of major fluctuations in the volume of water carried by the Amazon (the result of seasonal rainfall in the Amazon basin—which averages upward of 230 centimeters, or about 90 inches, per year), the Solimões expands and recedes in floodplain known as the *varzea*. The average width of this floodplain is about 24 kilometers (15 miles). The total area of the Amazon basin (the largest river basin in the world) is almost 7.8 million square kilometers (3 million square miles), but vegetation within the basin, although considered tropical, varies considerably.

Despite a popular association of tropical rain-forest environments with dense jungle, problems linked with poor soil quality and irregular drainage patterns in the Amazon basin have produced different patterns of vegetation. Where there are very high annual rainfall levels and high average annual temperatures (around 26.7 degrees Celsius, or 80 degrees Fahrenheit), Amazon basin plant species are categorized as lowland tropical vegetation. In some areas, mostly in the *varzea* but also in poorly drained areas of Amazonia, vegetation has to survive long periods underwater. As one progresses toward and onto terra firma, actual bands of different plant species emerge, beginning with what are called tropical dry forests, which eventually blend into grassy savanna zones. Brazil's *cerrado* (Portuguese for "closed"—that is, interior), for example, covers about one-fifth of the coun-

(continued on page 1166)

Habitats and Selected Vertebrates of South America

North
Atlantic
Ocean

Coatimundi

Giant Anteater

Anaconda

Vampire Bat

Boa Constrictor

Sloth

Toucan

Giant Otter

Piranha

AMAZON RAIN FOREST

Peba

Marine Iguana

*Galápagos
Islands*

Giant Land
Tortoise

Darwin's
Finches

Puma

Vicuna

Capybara

Peccary

Jaguar

Tapir

Wooly Tree
Porcupine

Ocelot

Quetzal

Llama

Spectacled
Bear

Crab-Eating Fox

EASTERN HIGHLANDS

Gray Fox

Coypu Rat

**COASTAL
REGION**

Andean
Condor

Alpaca

Plains
Viscacha

Bushmaster

Condor

James
Flamingo

GRASSLANDS

Agouti

South
Pacific
Ocean

Chinchilla

Rhea

Guanaco

Armadillo

Maned Wolf

Great
Anteater

**COASTAL
REGION**

South
Atlantic
Ocean

Patagonian
Weasel

Cavy

Cormorant

Sea Lion

Penguin

A N D E S   M O U N T A I N S

try, with a characteristic vegetation of grass and short twisted trees. In Brazil, Colombia, and Venezuela alone, savannas occupy some 2.6 million square kilometers (about 1 million square miles), an area about one-quarter the size of Canada.

Beyond the world-renowned Amazon, a second major tropical forest zone is noteworthy, that associated with the Orinoco River, which is 2,100 kilometers (1,300 miles) long and receives waters from eleven tributaries in mountainous Colombia and Venezuela. Among these are the Apure, the Arauca and Guaviare (from Colombia), and the Caroní (from sources in the Guiana highlands) The Orinoco is the second-largest river in South America. Its enormous delta (covering almost 23,300 square kilometers, or 9,000 square miles) on the Atlantic coast of Venezuela is a maze of different waterways surrounded by marshy tropical lowlands. Given the variety of major tributaries coming from many different directions into the main channel of the Orinoco, the total waterway system served, until recent times, as the principal route of access to otherwise inaccessible inland areas, and even as far as the llanos (plains) of Colombia.

Two major environmental subregions of South America are Patagonia and the Pampas. The latter, flat and grassy plains mainly in Argentina and Uruguay, are well known as a natural environment for cattle raising. Patagonia, in contrast, is the spectacular mountainous region of southern Chile that ends at Cape Horn and the Tierra del Fuego (land of fire) island archipelago shared with southern Argentina. Patagonia's inland terrain is marked by very high craggy and glacier-clad mountains. The southern mountain zone (the Magellanes) contains Torres del Paine National Park, where peaks rise to more than 2,700 meters (9,000 feet). Another major mountainous preserve, Laguna San Rafael National Park, is located in the northern Patagonian region of Aisén.

ENVIRONMENTAL CONCERNS

The widespread building of dams to produce electricity has seriously affected the Amazon and other South American river environments. The natural habitats of flora and fauna have been altered, and isolated indigenous populations have been displaced, their traditional lifestyles threatened.

Moreover, the economies of the more developed areas in South America depend on intensive exploitation of either mineral resources or agriculture. In the latter case, overdependence on single crops (monoculture) has had negative effects. The continual expansion of the land area under cultivation—typical of large plantations, especially coffee plantations and corn/ethanol plantations in Brazil, for example—has contributed to deforestation, and repeated planting of the same crops has depleted soil fertility.

A major environmental issue facing tropical South America, one that is global in its implications, involves the repercussions that start with deforestation. As the surfaces of tropical forests recede, corresponding reductions occur in the

A child carries a baby turtle that is one of more than four thousand being released in the Orinoco River in Venezuela in 1999. The turtles were bred in captivity and then released in an attempt to save South America's largest freshwater turtle species and Venezuela's most endangered species. (AP/Wide World Photos)

level of oxygen (given off by all plants in the carbon dioxide-oxygen exchange process) released to the atmosphere. Of more immediate local environmental importance is the destruction of the natural ecosystem that sustains regionally typical vegetation. Concurrently, local animals, including rare species, are menaced with extinction, either because of the destruction of their natural habitat or because increased "commercial" access to their habitat encourages the illegal exportation of the animals to "exotic" pet markets abroad. A particularly alarming example of forest depletion involves South America's world-renowned big-leaf mahogany (*Swietenia macrophylla*), which was listed as endangered by the Convention on International Trade in Endangered Species in 2002. The total original virgin mahogany surface has been estimated to be nearly 300 million hectares (741 million acres); this amount has been reduced by more than 175 million hectares (432 million acres) through logging and intentional "forest conversions" (clearing for agriculture).

Programs aimed at promoting the conservation of Amazonian forests and animal species are quite numerous. Some private conservation groups in South America even seek individual contributions from abroad, presumably to purchase small plots of virgin forest to save them from irresponsible logging. In a different, clearly official arena, the Brazilian government has worked toward clarifying landownership titles (many falsified through the intimidation of small-scale landowners by large-scale investors), with the aim of implementing stricter regulations regarding the conditions that must be met before forest vegetation can be removed.

South American regions as far apart as the Amazon basin and the Andean highlands are experiencing equally alarming ecological threats, although of different types, as a result of large-scale economic development schemes. Environmentalists from around the globe have registered opposition to oil drilling in pre-Amazonian areas of Peru and in the Brazilian Amazon. Indigenous populations in these areas depend on the maintenance of ecosystems (water networks, vegetation, and fauna) that have long been the bases of their traditional local economies. A prototypical example of this problem appeared in 2004, when exploitation by foreign companies of the Camisea natural gas field in Amazonian Peru drew international attention from environmentalists who criticized what they considered to be irresponsible actions leading to

deforestation, declining medicinal plant resources, and degradation of local water supplies stemming from oil and gas pollution in surrounding soils.

Byron Cannon

FURTHER READING

Daly, Douglas C., and John D. Mitchell. "Lowland Vegetation of Tropical South America." In *Imperfect Balance: Landscape Transformations in the Pre-Columbian Americas*, edited by David L. Lentz. New York: Columbia University Press, 2000.

Luteyn, James L., and Steven P. Churchill. "Vegetation of the Tropical Andes." In *Imperfect Balance: Landscape Transformations in the Pre-Columbian Americas*, edited by David L. Lentz. New York: Columbia University Press, 2000.

Meggers, Betty J. *Amazonia: Man and Culture in a Counterfeit Paradise*. Rev. ed. Washington, D.C.: Smithsonian Institution Press, 1996.

Wirth, Christian, Gerd Gleixner, and Martin Heimann, eds. *Old-Growth Forests: Function, Fate, and Value*. New York: Springer, 2009.

SEE ALSO: Amazon River basin; Biopiracy and bioprospecting; Deforestation; Indigenous peoples and nature preservation; Monoculture; Rain forests; Slash-and-burn agriculture; Soil conservation.

Soviet nuclear submarine sinking

CATEGORIES: Disasters; nuclear power and radiation

THE EVENT: Explosion and fire in a missile tube of a Soviet nuclear-powered submarine that resulted in the sinking of the submarine, which carried two nuclear reactors and sixteen nuclear warheads

DATES: October 3-6, 1986

SIGNIFICANCE: The environmental effects of the radioactive materials that went down with the Soviet nuclear submarine K-219 remain unknown, but eventual leakage is inevitable, and depending on the leak rate and local water currents, high concentrations of radioactivity could persist around the wreckage for months or years.

On October 3, 1986, the Yankee-class Soviet nuclear submarine K-219 was cruising in the Atlantic Ocean north of Bermuda when an explosion of unexplained origin occurred in its fourth missile compartment. The explosion caused a leak in the compartment, and smoke from the missile fuel began to fill the submarine. The captain ordered the submarine to surface, but the explosion apparently breached the hull, and water began to leak into the vessel. Before abandoning the K-219, the captain ordered the crew to engage the shutdown mechanisms on the boat's two nuclear reactors; the control rods, which cause the nuclear reactions to cease, were manually inserted into the reactors. One crewman died in the effort to insert these control rods, and three others died during the firefighting and evacuation efforts.

Despite efforts to keep the submarine afloat by the crew and the crews of assisting ships over the next several days, the submarine sank into the Atlantic on October 6, taking its two nuclear reactors and the nuclear warheads on sixteen missiles to the seafloor. There was no evidence of immediate leakage of radioactive material from the wreckage, but the long-term environmental consequences of the sinking are unknown. Once a nuclear reactor has generated power, some of its fuel is converted into a wide range of fission products. Some of these fission products are ra-

dioactive and are readily concentrated by living organisms. At some point, the corrosive action of the seawater will eat through the confinement housings of the reactors and the warheads, releasing radioactive pollution into the Atlantic Ocean. Depending on the leak rate and the local water currents, high concentrations of radioactive material could persist around the wreckage for months or years after the leakage begins. During that time, fish and other organisms may ingest substantial quantities of radioactive material. Eventually, because the radioactive material will be diluted into the vast volume of the ocean, the concentration of radioactive material will be reduced to the point where it constitutes a negligible hazard.

The K-219 is one of a number of nuclear-powered submarines that have sunk with fully fueled nuclear reactors onboard. The Soviet K-8 sank in the Bay of Biscay on April 8, 1970, and the Soviet K-278 sank in the Norwegian Sea on April 7, 1989. The Soviet K-27 developed a nuclear reactor problem on May 24, 1968, and was intentionally sunk in the Kara Sea in 1981, when Soviet officials decided that salvaging the submarine would be too costly. Two U.S. Navy subma-

The Soviet nuclear submarine K-219 was photographed northeast of Bermuda by a U.S. Navy surveillance aircraft before the vessel sank. (AP/Wide World Photos)

rines, the USS *Thresher* (which sank on April 10, 1963, off New England) and the USS *Scorpion* (which sank on May 22, 1968, southwest of the Azores Islands) are both on the North Atlantic seafloor. Monitoring of the radiation around one or more of these sites could provide information on the potential long-term hazards posed by radioactive materials on the ocean floor.

George J. Flynn

FURTHER READING

Angelo, Joseph A., Jr. "Issues." In *Nuclear Technology.* Westport, Conn.: Greenwood Press, 2004.

Weir, Gary E., and Walter J. Boyne. *Rising Tide: The Untold Story of the Russian Submarines That Fought the Cold War.* New York: Basic Books, 2003.

SEE ALSO: North Sea radioactive cargo sinking; Nuclear accidents; Nuclear and radioactive waste; Nuclear weapons production; Ocean pollution; Radioactive pollution and fallout.

Soviet Plan for the Transformation of Nature

CATEGORY: Agriculture and food

IDENTIFICATION: An unsuccessful plan that was intended to increase Soviet agricultural production dramatically

DATES: 1948-1953

SIGNIFICANCE: Joseph Stalin's plan to increase his nation's agricultural production by applying the notions of Trofim Denisovich Lysenko, who rejected orthodox genetics, failed to produce any positive results for the Soviet Union. On the contrary, Lysenko's ideas set back the state of science in the Soviet Union for decades.

The pseudobiologist Trofim Denisovich Lysenko and his followers, known as Lysenkoists, came into prominence during the 1930's. In 1932, at the International Congress of Genetics held at Cornell University, prominent Soviet biologist Nikolai Vavilov praised Lysenko's experiments with adapting grain and other plants to unfavorable climates through a mysterious process known as "vernalization" (preheating the seeds). Vavilov's praise for Lysenko was met with skepticism by other scientists, and eventually

Trofim Denisovich Lysenko, originator of the ideas behind the Soviet Plan for the Transformation of Nature. (©Bettmann/CORBIS)

Vavilov also came to question Lysenko's results. After further inquiries, Vavilov became Lysenko's most outspoken critic in the Soviet scientific community.

Undaunted by rebukes from traditional scientists, the Lysenkoists made preposterous claims that wheat could be turned into barley, oats into rye, and oak trees into pine trees. The more bizarre the assertions, the more credibility they earned with Soviet Communist Party leaders. The Lysenkoists predicated all of their wild contentions on the conviction that orthodox genetics, and the whole notion of the primacy of genes, was a capitalist plot to hold back the advance of the Soviet Union. Environment, they insisted, could cause hereditary changes in plants. All that was necessary for the rapid improvement of Soviet agriculture was to assist crops in adjusting to different environments. There was no reason, therefore, that warm-weather crops could not be grown successfully in cold climates.

Those who spoke against Lysenko's ideas found themselves charged with being anti-Communist.

Many were removed from their posts, and some were purged from the party itself after 1935. Lysenko, meanwhile, rose quickly; in 1939 he replaced Vavilov as director of the Leningrad Plant Growing Institute. The following year, in a great irony, Lysenko was appointed head of the Genetics Institute. Shortly thereafter Vavilov was arrested and sent to Siberia, where he died in 1943.

During the period from 1941 to 1945, Soviet leader Joseph Stalin was preoccupied with the war effort, but when the fighting ceased, he turned his attention to rebuilding the Soviet Union. The country's greatest need was to increase agricultural production, and the Lysenkoists were promising great results with the abandonment of orthodox genetics. In July, 1948, Stalin and the Communist Party's powerful Central Committee gave the Lysenkoists an official endorsement. Party leaders liked the Lysenkoist notion of "proletarian" science, as opposed to the "degenerate, elitist" science practiced in the West.

On October 24, 1948, Stalin issued a decree for the Soviet Plan for the Transformation of Nature to begin. Peasant labor was to be used to establish huge agricultural strips (60 meters, or 200 feet, wide) in western Russia. Lysenko's methods would be used to improve grain yields. In addition, millions of trees would be planted in the tundra regions to help ease the harshness of the climate. The trees were to be planted in clusters in the belief that some would survive by adjusting to their new environment. Those that survived would then reproduce and thus alter the Soviet terrain. The plan also included the grafting and crossbreeding of plants to other plants; by these methods it was believed that ordinary weeds could be turned into wheat.

On January 1, 1949, Lysenko predicted that there would be a limitless growth in Soviet harvests. Soviet leaders proclaimed in 1950 that the plan was well on its way to success. In actuality, however, the opposite was true. In 1952 Stalin, embarrassed by reports of low crop production, attempted to deflect attention from the Plan for the Transformation of Nature by introducing a massive scheme for the building of dams and canals. Stalin also began to permit scientific criticism of Lysenko's methods. By the time of Stalin's death in March, 1953, Lysenkoism had fallen into disrepute.

With Stalin gone, the Soviet Plan for the Transformation of Nature was abandoned, and opposing scientific views were again recognized. Lysenko, however, continued to insist that he was right, and when Nikita Khrushchev emerged as Soviet leader in 1957, Lysenko made a brief return to prominence. Khrushchev, like Stalin, desperately sought a cure for his country's agricultural problems. This time, however, Lysenko's critics could not be silenced. Khrushchev's support for the discredited Lysenko contributed to the Soviet leader's removal from office in October, 1964.

The Soviet Plan for the Transformation of Nature produced no positive results for the Soviet Union. It did, however, have negative consequences. In particular, scientific advances in biology were disrupted by Lysenko's insistence that his critics be silenced and removed from their government posts. By the time Stalin realized that he had been misled by the Lysenkoists, this setback could not be repaired in a short time. Moreover, other areas of science, including medicine, were also affected in the Soviet Union. A large number of individuals, who later became known as "harebrained" scientists, attempted to gain favor from the Communist Party. The party, clearly more interested in proper ideology than in proper science, all too often endorsed dubious scientific methods during the late 1940's. It was not until the 1970's that the Soviet Union fully rejoined the world of legitimate scientific inquiry.

Ronald K. Huch

FURTHER READING

Joravsky, David. *The Lysenko Affair.* 1970. Reprint. Chicago: University of Chicago Press, 1986.

Schwartz, Katrina Z. S. *Nature and National Identity After Communism: Globalizing the Ethnoscape.* Pittsburgh: University of Pittsburgh Press, 2006.

Stevenson, Leslie, and Henry Byerly. "Scientists and the Totalitarian State." In *The Many Faces of Science: An Introduction to Scientists, Values, and Society.* 2d ed. Boulder, Colo.: Westview Press, 2000.

Ulam, Adam B. *Stalin: The Man and His Era.* 1973. Reprint. New York: Viking Press, 1987.

Weiner, Douglas R. "The Predatory Tribute-Taking State: A Framework for Understanding Russian Environmental History." In *The Environment and World History,* edited by Edmund Burke III and Kenneth Pomeranz. Berkeley: University of California Press, 2009.

SEE ALSO: Agricultural revolution; Biotechnology and genetic engineering; High-yield wheat; Sustainable agriculture.

Space debris

CATEGORIES: Atmosphere and air pollution; pollutants and toxins

DEFINITION: Human-made nonfunctional objects that are in orbit around the earth

SIGNIFICANCE: Space debris—which consists of non-functioning spacecraft, rocket bodies, refuse from missions, and fragments thereof—poses a hazard for space missions, satellite-based services, and people both in space and on earth. As space-faring nations have become more aware of the dangers of this debris, they have worked to minimize its generation during operations in space.

Since 1957 human beings have launched thousands of satellites and other spacecraft. Most of the spacecraft launched successfully achieve orbit. Those that explode after attaining orbit altitude and those that fail after achieving orbit become space debris (also known as orbital debris or space junk). Anything that reaches orbit altitude—about 300 kilometers (186 miles) above the earth's surface—becomes a satellite of the earth. Once in orbit, objects are constantly under the pull of the earth's gravity, and, in time, they slowly fall from orbit. The greater the distance from the earth, the longer an object will remain in orbit. Above 1,000 kilometers (621 miles) objects can remain in orbit for at least a century, objects orbiting at an altitude of 800 kilometers (497 miles) are likely to fall to earth within decades, and those at altitudes between 200 and 600 kilometers (124 and 373 miles) tend to remain in orbit for several years at best.

The U.S. Air Force Space Surveillance Network, which routinely tracks artificial objects orbiting the earth, has cataloged roughly 19,000 debris objects larger than 10 centimeters (4 inches) in diameter. An estimated 500,000 orbiting particles are between 1 and 10 centimeters (0.4 and 4 inches) in diameter. Particles measuring less than 1 centimeter in diameter probably number in the tens of millions. Most of the debris orbits within 2,000 kilometers (1,243 miles) of the earth's surface, with the greatest concentrations accumulating at altitudes between 800 and 850 kilometers (497 and 528 miles).

Each object in orbit runs the risk of running into another object. The volume of space surrounding the earth is immense, and the chances of a collision between two objects are relatively low; however, the likelihood of collision increases when the objects occupy the same orbit. Because certain orbits are particularly desirable for satellites used for communications and surveillance purposes, various nations and commercial interests place their satellites into these positions, thereby increasing the chances of collision.

Many different kinds of space debris orbit the earth. From the 1960's through the mid-1980's, nations deliberately destroyed orbiting satellites while testing weapons for antisatellite warfare. Other forms of space debris have less dramatic origins, such as astronaut Ed White's glove, which slowly drifted away from his Gemini spacecraft in 1965. Each item adds to the ever-increasing number of human-made objects orbiting the earth. Collisions between objects, and explosions of residual fuels in abandoned rocket engines, break existing debris into many smaller pieces.

Objects ranging in size from spent rocket boosters and nonfunctional satellites to small chips of paint, solid-fuel fragments, and coolant droplets have the potential to damage spacecraft. It is not merely the mass of an object that poses a danger but also its high velocity. At orbits below 2,000 kilometers, debris travels at speeds of 7 to 8 kilometers (4.3 to 5 miles) per second, so that even tiny particles can pit space shuttle cockpit windows and damage unshielded satellite components.

HAZARDS

The space debris population has grown great enough that it has become standard practice to shift unmanned satellites out of harm's way when large debris (objects larger than 10 centimeters) is detected. Space shuttle flights have to adjust course to avoid debris reported by the Space Surveillance Network. The International Space Station is heavily shielded against objects smaller than 1 centimeter, but it has the capability to maneuver away from larger tracked objects.

Only one collision between large, intact satellites has ever occurred. In February, 2009, an operational U.S. Iridium communications satellite accidentally struck a deactivated Russian Cosmos communications satellite. Both spacecraft were destroyed, and more than 1,500 large fragments were generated. The amount of large debris had already been dramatically increased two years earlier, when in January, 2007, China conducted an antiweapons test in which it used its aging Fengyun-1C weather satellite as a target. The resulting destruction created roughly 2,600 large debris fragments and hundreds of thousands of smaller particles.

Naturally Occurring Space Material

The term "space debris" is sometimes used to refer to naturally occurring material as well as that generated by human activity. However, as space technology consultant Mark Williamson notes in *Space: The Fragile Frontier* (2006), natural space objects such as meteoroids, meteorites, and cosmic dust are inherent to the space environment, unlike anthropogenic material. Humankind can guard against the flux of natural space material but cannot halt or lessen it; humans can increase it, however, by reducing a single larger body into countless smaller fragments.

An estimated 25 million bits of natural space material collide with the earth each day, the majority of it in the form of cosmic dust. These particles are so small that they do not even appear as meteors as they pass into the atmosphere. Most meteors that are seen are particles the size of a pea, and sometimes the larger ones reach the earth's surface as meteorites.

Occasionally an asteroid-sized object or comet collides with the earth. Such collisions have global implications. The impact destroys the comet and forms a huge crater. An enormous amount of gas and dust is carried into the atmosphere, creating a blanket of debris that blocks sunlight. This begins a "nuclear winter"-type effect, which can last anywhere from a few months to several years. During this time most life-forms will die as a result of the disruption of the food chain. Many scientists believe that such an event led to the mass extinctions of dinosaurs about 65 million years ago.

Although giant impacts are one cause of "nuclear winter," cosmic dust can produce the same effect. The solar system periodically runs into a cosmic dust cloud, thereby dramatically increasing the amount of dust that enters the atmosphere. A similar situation also results from periodic meteor storms.

As the earth runs into the debris of old comets, the number of meteors that enter the atmosphere increases. The earth occasionally encounters a particularly dense region of comet debris. During such meteor storms, thousands of meteors can be seen each hour. The most notable is the Leonid meteor storm, which occurs every thirty-three years. Many scientists fear that increases in comet debris could knock out hundreds of satellites as they are hit by microscopic particles and greatly affect global positioning and communication capabilities.

In 1978 a Soviet satellite with a nuclear power source survived reentry and strewed small amounts of radioactive material across Canada. The following year, large pieces of the Skylab space station withstood a fiery plunge through the atmosphere and scattered debris across western Australia. In 2001 a rocket upper stage that had been part of a 1993 global positioning satellite launch fell to earth in the Saudi Arabian desert. All of these incidents would have caused considerable damage if the debris had not landed in sparsely populated areas. Only one instance has been recorded of a person being struck by space debris: In 1997 a bit of woven metallic material from a Delta II rocket fuel tank hit an Oklahoma woman on the shoulder but did not injure her. On average, one piece of cataloged space debris falls out of orbit every day, usually burning up in the atmosphere.

Mitigation Measures

Careful design and operational measures can keep new space missions from contributing unnecessarily to the proliferation of space debris. For example, upper stages of launch vehicles can be placed at lower altitudes so that their orbits decay sooner. Since 1988 the United States has had an official policy of minimizing debris from governmental and nongovernmental operations in space, and the U.S. government approved a set of standard practices for the mitigation of space debris in 2001. The governments of France, the European Union, Japan, and Russia also have issued guidelines pertaining to space debris. Additional guidelines have been published by the United Nations Committee on the Peaceful Uses of Outer Space (COPUOS) and the Inter-Agency Space Debris Coordination Committee (IADC), a group established by the world's leading space agencies in 1993.

Cleaning up existing space debris remains an expensive and technologically challenging prospect. Proposed solutions have included hastening objects' fall to earth by using lasers to slow their orbits and conducting special robotic space missions to grab and haul debris. Solutions that are both technically feasi-

Efforts to minimize the problems associated with space debris include the boosting of geostationary satellites that have ended their missions out of their orbits (near 36,000 kilometers, or 22,369 miles, above the earth's surface) into a higher "disposal orbit." Similarly, deactivated satellites that operated at lower altitudes may be moved to even lower orbits that will decay more quickly, hastening the satellites' fall to earth. If a satellite fails to burn up in the atmosphere, however, it can present a threat to people and property on the earth's surface.

ble and economically viable have yet to be developed. In addition, the development of technologies for the cleanup of space debris is controversial because any methods capable of moving spacecraft have potential weapons applications.

A 2006 study sponsored by the U.S. National Aeronautics and Space Administration (NASA) Orbital Debris Program concluded that, if no new launches were conducted and no new objects introduced to earth's orbit, the number of objects falling out of orbit over the next half century would balance the number of new objects created through collisions. After 2055, however, the increasing number of collision-generated fragments—which would go on to create their own catastrophic collisions—would overtake the number lost through decaying orbits.

The international space community's concern in the wake of the 2006 NASA study, China's 2007 weapons test, and the 2009 satellite collision led to the first International Conference on Orbital Debris Removal, convened in December, 2009. Participants examined the many technical, economic, legal, and policy issues surrounding near-earth space cleanup, but they reached no conclusions regarding exactly how humankind might best address the worsening problem of space debris.

Paul P. Sipiera
Updated by Karen N. Kähler

FURTHER READING

Inter-Agency Space Debris Coordination Committee. *IADC Space Debris Mitigation Guidelines.* Vienna: United Nations, 2002.

Johnson, Nicholas L., and Darren S. McKnight. *Artificial Space Debris.* Updated ed. Malabar, Fla.: Krieger, 1991.

Klinkrad, Heiner. *Space Debris: Models and Risk Analysis.* New York: Springer, 2006.

National Research Council. *Orbital Debris: A Technical Assessment.* Washington, D.C.: National Academy Press, 1995.

Simpson, John A., ed. *Preservation of Near-Earth Space for Future Generations.* 1994. Reprint. New York: Cambridge University Press, 2006.

Smirnov, Nickolay N., ed. *Space Debris: Hazard Evaluation and Mitigation.* London: Taylor & Francis, 2002.

United Nations Committee on the Peaceful Uses of Outer Space. *Technical Report on Space Debris.* New York: United Nations, 1999.

Williamson, Mark. *Space: The Fragile Frontier.* Reston, Va.: American Institute of Aeronautics and Astronautics, 2006.

SEE ALSO: Nuclear winter; Solar radiation management; Weather modification.

Spaceship Earth metaphor

CATEGORIES: Ecology and ecosystems; resources and resource management

IDENTIFICATION: Figure of speech portraying the planet Earth as a fragile vessel in the vast expanse of space

SIGNIFICANCE: In using the Spaceship Earth metaphor, environmentalists and others emphasize the interrelationships of all living things on the planet and the need for worldwide cooperation in protecting Earth's resources.

Environmentalists, economists, scientists, and others have used the Spaceship Earth metaphor, which in essence compares the systems needed for the survival and continuation of the diverse life-forms on Earth to the systems needed on a spaceship to ensure the survival of the occupants. Major environmental systems needed for survival on Earth include air, water, food, shelter, economic goods and services, waste recycling and disposal, security from violence, and governmental justice, law, and order. Similar environmental systems are needed on a spaceship: sufficient air supply, clean water, adequate food, protection from outer space, adequate goods and services, effective waste disposal, recycling of used materials, and enforcement of appropriate safety rules and regulations by captain and crew.

In her book *Spaceship Earth* (1966), British economist Barbara Ward indicates that she borrowed the comparison of Earth to a spaceship from the visionary American inventor R. Buckminster Fuller. Ward argues that science and technology have created an intimate, spaceshiplike worldwide network of communication, transportation, and economic interdependence. This has resulted in a planetary fellowship, a close world community with the vulnerability of a spaceship. Inequities in power and wealth breed violence, which damages the general welfare in the crowded spaceship. Ward asserts that the divisions fostered by conflicting belief systems should be replaced with planetary unity, rational rules for survival, and common world institutions, policies, and beliefs, because such a worldwide system of order and welfare is needed if humankind is to avoid nuclear annihilation. According to Ward, the current system of national sovereignty is a divisive tribal system that stands as a major barrier to the system of planetary loyalty, patriotism, and citizenship needed for adequate living on Spaceship Earth. The instinct to kill strangers from other tribes or nations must be replaced by the vision of a single community on a spaceship carrying a single human species.

Kenneth E. Boulding, who was an economist and college professor, makes even more specific arguments in his essay "The Economics of the Coming Spaceship Earth," which was first published in 1966. Boulding labels the open-earth economy of exploitation, extraction, consumption, depletion, pollution, and violence the "cowboy economy" and calls it a "fouling of the nest" system. In order to sustain the human species, according to Boulding, the Earth needs a closed "spaceman" economy in which the planet is seen as a single spaceship; maintenance of resources and conditions for long-term quality living should be top priorities on Spaceship Earth.

Lynn L. Weldon

FURTHER READING

Boulding, Kenneth E. "The Economics of the Coming Spaceship Earth." 1966. In *Valuing the Earth: Economics, Ecology, Ethics*, edited by Herman E. Daly and Kenneth N. Townsend. Cambridge, Mass.: MIT Press, 1993.

Jasanoff, Sheila. "Heaven and Earth: The Politics of Environmental Images." In *Earthly Politics: Local and Global in Environmental Governance*, edited by Sheila Jasanoff and Marybeth Long Martello. Cambridge, Mass.: MIT Press, 2004.

Spowers, Rory. *Rising Tides: A History of the Environmental Revolution and Visions for an Ecological Age*. Edinburgh: Canongate, 2002.

SEE ALSO: Biosphere; Earth Day; Ecological economics; Ecosystems; Environmental economics; Sustainable development.

Speciesism

CATEGORY: Philosophy and ethics

DEFINITION: Prejudice in favor of one's own species, in particular the view among humans that the human species is more important than other animal species

SIGNIFICANCE: Advocates of animal rights often accuse their critics of speciesism, noting that such prejudice leads to the unthinking treatment of animals as inferior beings and to humans' use of animals for entertainment, scientific experimentation, clothing, and food.

Speciesism has deep roots in anthropocentric philosophies and theologies. The ancient Greek philosopher Plato was exclusively interested in human potential and regarded animals only as tools for human goals. The seventeenth century French philosopher René Descartes believed animals to be machines devoid of thought and feeling. Utilitarians have always prioritized human happiness over animal welfare. Even the eighteenth century German philosopher Immanuel Kant believed that the duties of human beings are restricted to their own species.

Throughout history, the subjugation of nonhumans has been accepted and even institutionalized. The motivations for speciesism may come from many sources, including fear of animals, religious dictates, the drive for survival, and simple arrogance. Superiority of the human species has been one of the most pervasive and persistent assumptions of human beings. Secular reasons for believing in human superiority to other animals include personal prestige and the desire to use animals for sport, as game, or in commerce. Although all religions place some emphasis on fair treatment of nonhuman life, most, including Christianity, give some speciesist privileges to humans. The Christian god created animals without souls for the purpose of service to humans, his best creation.

Critics view speciesism as unwarranted anthropocentric prejudice similar to sexism and racism. Generally, they contend that every species of sentient beings should have the same rights to life, liberty, and the pursuit of happiness. The antispeciesist debate about this contention began in the 1970's in England and spread around the world. In 1977 the key philosophers of the animal rights movement devoted an entire conference to speciesism at Trinity College in

Cambridge, England. The outcome was the drafting of a charter titled "A Declaration Against Speciesism," which became the basis for subsequent charters ratified by groups in Great Britain and many other nations. The original document confirms an evolutionary and moral kinship between humans and their "brother and sister animals" and expresses "total condemnation" for the inflicting of suffering on animals.

Antispeciesist philosophers find no moral, linguistic, cognitive, or divine basis for the claim of superiority of humans and point out that humans are similar to other sentient beings in many ways; even the making and use of tools, the use of language, and such moral sensibilities as altruism can be observed in animals. Antispeciesists such as Peter Singer assert that if people examine their beliefs, they are often left with the startlingly simple realization that whatever is morally wrong and abhorrent for humans is probably wrong and abhorrent for nonhumans. Pain and suffering are bad and should be prevented or minimized regardless of race, sex, or species. The logic is simple and similar to a stripped-down version of the "universalization principle" that Kant used to test human morality. Singer's interspecies version of the Kantian principle is this: If it is wrong to torture and kill humans, then it is wrong to do these things to nonhuman sentient beings. In the case of laboratory rats, for example, one should ask, Would it be right to inflict severe electrical shocks (or other experimental procedures) on unwilling men and women?

Almost all recent serious philosophical inquiries about speciesism have concluded that the candidates for moral consideration include other nonhuman sentient beings. The debates around speciesism are not easy to resolve and bring up many questions; for example, in what sense, if any, are people obligated to give equal moral standing to insects and rodents, populations of which, at times, wreak havoc on humankind? Such questions are difficult to answer.

Chogollah Maroufi

FURTHER READING

Cavalieri, Paola. "Speciesism." In *The Animal Question: Why Nonhuman Animals Deserve Human Rights*. New York: Oxford University Press, 2001.

Hursthouse, Rosalind. *Ethics, Humans, and Other Animals: An Introduction with Readings*. New York: Routledge, 2000.

Singer, Peter. *Animal Liberation*. 1975. Reprint. New York: HarperPerennial, 2009.

SEE ALSO: Animal rights; Animal rights movement; Animal testing; Anthropocentrism; People for the Ethical Treatment of Animals; Singer, Peter; Vegetarianism.

Stockholm Conference. *See* United Nations Conference on the Human Environment

Stockholm Convention on Persistent Organic Pollutants

CATEGORIES: Treaties, laws, and court cases; pollutants and toxins; human health and the environment

THE CONVENTION: International agreement banning or severely limiting the manufacture and use of certain substances linked to neurological, reproductive, and immune system damage in people and animals

DATE: Opened for signature on May 23, 2001

SIGNIFICANCE: The Stockholm Convention on Persistent Organic Pollutants represents an international effort to reduce the threat of many persistent organic pollutants, which have been linked to cancer, birth defects, and other neurological, reproductive, and immune system damage in people and animals. At high levels, these chemicals can damage the central nervous system, and many also act as endocrine disrupters, causing deformities in sex organs as well as long-term dysfunction of reproductive systems.

The United Nations Environment Programme's Governing Council in 1995 identified twelve persistent organic pollutants (POPs) as the subjects of an eventual ban on manufacture and use worldwide because these substances, known as the dirty dozen, damage the ecosphere and the diversity of life supported by it. While many of these pollutants had already been banned in the United States, other countries had continued to manufacture and use them. The international ban was negotiated in Stockholm late in 2000, and ratification of the protocol followed in 2001.

The substances that were targeted for elimination by the Stockholm Convention included the organochlorine pesticides (such as chlordane, mirex, hexachlorobenzene, endrin, aldrin, toxaphene, heptachlor, and dichloro-diphenyl-trichloroethane, or DDT) and industrial chemicals (including polychlorinated biphenyls, or PCBs, and the supertoxic dioxins and furans). DDT was allowed limited use because no other inexpensive alternatives were available to combat the mosquitoes that spread malaria. Because some of these substances (notably the PCBs and dioxins) are actually families comprising hundreds of chemicals, they could just as aptly be called the dirty hundreds as the dirty dozen.

Synthetic organochlorines such as dioxins and PCBs are perfect vehicles for worldwide pollution because they ignore boundaries, natural or artificial. These chemicals also bioaccumulate (biomagnify, or intensify in potency) along the food chain, sometimes to thousands of times their original toxicity, posing special perils to animals, including human beings, who eat meat and fish. Problems related to their toxicity are especially acute in places, such as the polar regions, where currents in the atmosphere and oceans cause organochlorines to accumulate. Organochlorines produced in the past for commerce and those created unintentionally can be found in the air and in lakes, oceans, soils, sediments, and animals, including humans, in every region of the planet.

POPs are not soluble in water, but they dissolve easily in fats and oils, accumulating in the bodies of living organisms and becoming more concentrated as they move along the food chain. Extremely small levels of such contaminants in water or soil can magnify into lethal hazards to predators who feed at the top of the food web, such as dolphins, polar bears, herring gulls, and human beings.

In some regions of the world, indigenous peoples whose diets consist largely of sea animals (whales, polar bears, fish, and seals) have long been consuming a concentrated toxic chemical cocktail. Abnormally high levels of dioxins and other industrial chemicals have been detected in Inuit mothers' breast milk. People in some villages in the Arctic experience higher concentrations of PCBs than anyone else on earth except victims of industrial accidents. They are at the top of a food chain composed mainly of PCB-laced polar bears, seals, and other animals.

Bruce E. Johansen

FURTHER READING

Colborn, Theo, Dianne Dumanoski, and John Peterson Myers. *Our Stolen Future: Are We Threatening Our Fertility, Intelligence, and Survival? A Scientific Detective Story.* New York: Penguin Group, 1996.

Cone, Marla. *Silent Snow: The Slow Poisoning of the Arctic.* New York: Grove Press, 2005.

Johansen, Bruce E. *The Dirty Dozen: Toxic Chemicals and the Earth's Future.* Westport, Conn.: Praeger, 2003.

SEE ALSO: Agent Orange; Biomagnification; Carcinogens; Chloracne; Dichloro-diphenyl-trichloroethane; Dioxin; Malathion; Pesticides and herbicides; Polar bears; Polychlorinated biphenyls.

Stormwater management

CATEGORY: Water and water pollution

DEFINITION: Methods of containing and channeling water from precipitation that flows over land or impenetrable surfaces and does not percolate into the ground

SIGNIFICANCE: When stormwater is uncontrolled, water resources may be wasted and destructive flooding may occur; in addition, pollution carried by stormwater can contaminate bodies of water and endanger human and animal health. Stormwater management systems are designed to minimize the environmentally destructive potential of precipitation runoff.

Stormwater management, which has its roots in ancient Greek and Roman cultures, has evolved into the vast systems now present in all developed and many developing nations. The management of stormwater has become increasingly important over time with human development, as the impervious surfaces of buildings, roads, parking lots, and so on do not allow rainwater and snowmelt to penetrate into the ground, resulting in much more runoff than occurs on undeveloped land. All developed and many developing nations have laws in place that set standards for stormwater management.

POLLUTION AND UNCONTROLLED RUNOFF

Stormwater management is necessary to address the problem of water pollution caused by runoff. Polluted stormwater runoff from impervious surfaces such as roads and roofs, as well as from agricultural

A drainage cascade in Seattle, Washington, is planted with sedges and rushes and uses a series of concrete dams with drains designed to slow the surge of stormwater carrying pollutants from street runoff. (AP/Wide World Photos)

fields, is a major source of water pollution around the world. Water flowing over human-built surfaces picks up pollutants such as gasoline, oil, trash, and heavy metals (mercury, lead, cobalt, copper, and zinc). Water flowing over lawns and farm fields collects nitrates and phosphates from the fertilizers used on these surfaces.

In the United States, the pollution of waterways became a nationally recognized issue in June, 1969, when the Cuyahoga River, flowing through Cleveland, Ohio, on its way to Lake Erie actually caught fire because it was so polluted. At that time, Lake Erie had become a victim of eutrophication, a process in which an increase in chemical nutrients alters plant growth in an ecosystem. Algae became the dominant plant species in the lake and absorbed all the oxygen in the water, killing most of the native fish and other aquatic species. For decades, far more nutrients entered the lake than it could handle; the main culprit was phosphorus, a component of fertilizers and detergents. Heavy metals and pesticides also contributed to the

pollution. Fish that managed to survive in the lake were unfit for human consumption.

Another problem addressed by stormwater management is uncontrolled runoff, which not only results in the waste of a valuable resource but also can cause extensive damage. When the amount of water generated by rain or snowmelt exceeds the capacity of waterways, flooding occurs. Floods are among the most frequent of all forms of natural disasters, and they are also among the most costly in terms of human and animal deaths, destruction of structures and habitats, and soil erosion.

Conditions that result in flooding include heavy or steady rains that last hours or days; such rains can exceed the capacity of the ground to absorb the water, and after the ground is saturated the excess flows to lower areas. In areas where forest or brush fires have recently taken place, denuding the landscape of plants that absorb groundwater and stabilize the soil, subsequent rainfall can result in flooding, often accompanied by mudslides.

METHODS OF MANAGEMENT

Methods of stormwater management address the problems of pollution and uncontrolled runoff. Bodies of water become contaminated by both point source pollution and nonpoint source pollution. Point source pollution occurs when pollutants enter a waterway through a specific entry point, such as a drainage pipe discharging into a river. This type of pollution is much more controllable than nonpoint source pollution, which originates from a variety of diffuse sources that are difficult to identify; pollution from stormwater is a form of nonpoint source pollution.

An example of pollution control involving stormwater management can be seen in the situation that was present in Lake Erie and the other Great Lakes during the 1960's. Recognition of the levels of pollution in the lakes led to the Great Lakes Water Quality Agreement (GLWQA), signed by the United States and Canada in 1972, and the U.S. federal Clean Water Act, also enacted in 1972. The GLWQA stressed the reduction of the amounts of phosphorus (then an ingredient in many laundry detergents) entering Lake Erie and Lake Ontario through stormwater runoff, and in 1977 maximum allowable levels of phosphorus were set by an amendment to the agreement. (Phosphorus was subsequently banned from detergents in both the United States and Canada.) By 2010 the phosphorus level in Lake Erie was found to be below the maximum allowed by the GLWQA, and the eutrophication of the lake had also abated. Despite these improvements, Lake Erie still has a number of problems, including contaminated sediments.

Ideally, stormwater runoff should be not only channeled to prevent damage but also collected for future use or decontamination. Stormwater collection methods range from rooftop cisterns used to collect and store rainwater for household uses to systems that channel stormwater to reservoirs that can supply water to large populations. Retention basins can both control runoff and allow treatment of the contained water. Stormwater is usually channeled to a retention basin from storm drains. Water entering the basin is restricted from flowing outward unless a major storm event occurs. These basins are often landscaped with varieties of plants that improve the water by removing excess nutrients and other pollutants. Another concept for runoff control is the swale or bioswale, which is a wide and shallow ditch designed to maximize the time water spends within it. Plants in the swale absorb and break down pollutants.

Even small stormwater runoff control systems can be effective. One example is the rain garden, which is a planted area designed to collect runoff from nearby impervious surfaces such as parking lots, driveways, roofs, and walkways. The hardy local plants used in the rain garden help the soil to absorb the runoff and any pollutants it contains.

REGULATIONS

All developed nations and some developing nations have well-established stormwater management policies. In the United States, the Environmental Protection Agency (EPA) oversees stormwater management under the provisions of the Clean Water Act. The EPA's National Pollutant Discharge Elimination System (NPDES) regulates stormwater discharges from municipal storm sewer systems, construction activities, and industrial activities. In addition to implementing NPDES requirements, many state and local governments in the United States have passed their own laws concerning stormwater management.

The treaty that created the European Union established environmental policy as one of the functions of that body. The aims of this policy are to protect and improve the quality of the environment, to protect the health of the population, to promote the careful and rational use of natural resources, and to promote measures at the international level to overcome regional and broader-scale environmental problems. By the early years of the twenty-first century all Western European nations had adopted legislation setting standards for the collection, treatment, and disposal of wastewater.

The stormwater policies of many Asian nations, including Japan and the People's Republic of China, are continually evolving. Australia, which has a well-developed stormwater policy, faces particular challenges because water resources are scarce on the Australian continent.

Stormwater management systems are generally much less advanced in developing nations than they are in the developed nations. In the cities of developing nations, the urban poor are most affected by inadequate stormwater management. Poor communities are often located on low-lying land, which is prone to flooding, or on steep hillsides, where heavy storms can wash away flimsy structures.

Robin L. Wulffson

FURTHER READING

Akan, A. Osman, and Robert J. Houghtalen. *Urban Hydrology, Hydraulics, and Stormwater Quality: Engineering Applications and Computer Modeling.* Hoboken, N.J.: John Wiley & Sons, 2003.

Committee on Reducing Stormwater Discharge Contributions to Water Pollution. *Urban Stormwater Management in the United States.* Washington, D.C.: National Academies Press, 2009.

Davis, Allen P. *Stormwater Management for Smart Growth.* New York: Springer, 2005.

Debo, Thomas N. *Municipal Stormwater Management.* 2d ed. Boca Raton, Fla.: CRC Press, 2002.

Gribbin, John E. *Introduction to Hydraulics and Hydrology with Applications for Stormwater Management.* 3d ed. Clifton Park, N.Y.: Thomson Delmar Learning, 2007.

Grigg, Neil S. *Water, Wastewater, and Stormwater Infrastructure Management.* Boca Raton, Fla.: CRC Press, 2002.

James, William. *Advances in Modeling the Management of Stormwater Impacts.* Boca Raton, Fla.: CRC Press, 1997.

Mays, Larry. *Stormwater Collection Systems Design Handbook.* New York: McGraw-Hill, 2001.

Sevbert, Thomas A. *Stormwater Management for Land Development: Methods and Calculations for Quantity Control.* Hoboken, N.J.: John Wiley & Sons, 2006.

SEE ALSO: Agricultural chemicals; Clean Water Act and amendments; Drinking water; Floods; Groundwater pollution; Pesticides and herbicides; Rain gardens; Rainwater harvesting; Runoff, agricultural; Runoff, urban; Urban planning; Wastewater management; Water pollution; Water quality.

Stringfellow Acid Pits

CATEGORIES: Human health and the environment; waste and waste management

IDENTIFICATION: Hazardous waste disposal facility in Southern California

DATES: Operated from August, 1956, to November, 1972

SIGNIFICANCE: The soil and groundwater pollution that resulted from years of dumping of industrial wastes at the Stringfellow Acid Pits was so severe that cleanup efforts continued for decades after the area was declared a top-priority federal Superfund site in 1983.

In the mid-1950's the state of California sought a location within a one-hour drive of Los Angeles that could serve as a dump for liquid industrial wastes. One promising site was Pyrite Canyon, located in a semirural area in the southern portion of the Jurupa Mountains near the community of Glen Avon in Riverside County. James Stringfellow, who operated a quarry in Pyrite Canyon, made an agreement with the state to allow waste disposal on a portion of his property. Disposal of liquid industrial wastes at the Stringfellow site commenced in August, 1956.

Over the next several years, roughly 128 million liters (34 million gallons) of liquid wastes were hauled to the 7-hectare (17-acre) state-licensed dump site. These wastes—including acidic pickling liquor from steel plants and other wastes from metal finishing, electroplating, and production of the insecticide dichloro-diphenyl-trichloroethane (DDT)—were discharged into unlined evaporation ponds. Hundreds of private and public entities in California and Nevada contributed wastes to the facility during its operation.

In 1969 heavy rains flooded the Stringfellow Acid Pits, causing contaminated runoff from the overflowing disposal ponds to flow into Pyrite Creek and Pyrite Creek Channel, and from there into the community of Glen Avon. Subsequently, the facility operator worked with engineers to improve site integrity so the facility could continue operations. However, when chromium was detected in the well of the quarry office roughly 1 kilometer (0.6 mile) from the site, Riverside County revoked Stringfellow's special land-use permit. The operator closed the facility in November, 1972, and ownership of the site later reverted to the state of California because of the owner's failure to pay back taxes. Between 1975 and 1980, the state conducted studies to determine the type and extent of contamination and initiated cleanup efforts.

Groundwater beneath and downgradient from the site was found to be contaminated with volatile organic compounds such as trichloroethane and heavy metals such as cadmium, chromium, manganese, and nickel. Elevated levels of groundwater radioactivity were later determined to be naturally occurring and unrelated to the acid pits. Heavy metals were also found in the soil, along with pesticides, polychlorinated biphenyls (PCBs), and sulfates.

Public objection to the site was galvanized in 1978, when another season of heavy rains led the California Regional Water Quality Control Board to authorize a controlled release of 3 million liters (800,000 gallons) of wastewater from the site to avert flooding and uncontrolled discharge. As the contaminated water flowed through the streets of Glen Avon, residents experienced dizzy spells, breathing problems, and other health effects. Chronic illnesses spurred the community to file a personal injury lawsuit in 1984 against the state of California and two hundred companies that had contributed wastes to the site. One decade later, a settlement of more than $114 million was reached.

In 1980 the federal government became involved in site cleanup. The Stringfellow Acid Pits were made a high-priority cleanup site under the federal Comprehensive Environmental Response, Compensation, and Liability Act, known as Superfund, in 1983 when the site was included on the U.S. Environmental Protection Agency's National Priorities List. Federal and state authorities continued to work with a group of private parties to remediate the contamination through the first decade of the twenty-first century. Total cleanup costs were projected to be between $250 million and $475 million.

Karen N. Kähler

FURTHER READING

Glazer, Myron Peretz, and Penina Migdal Glazer. "On the Trail of Courageous Behavior." In *Environmental Sociology: From Analysis to Action*, edited by Leslie King and Deborah McCarthy. 2d ed. Lanham, Md.: Rowman & Littlefield, 2009.

Gottlieb, Robert. "Grassroots and Direct Action: Alternative Movements." In *Forcing the Spring: The Transformation of the American Environmental Movement*. Rev. ed. Washington, D.C.: Island Press, 2005.

Hitt, Jack. "Toxic Dreams: A California Town Finds Meaning in an Acid Pit." *Harper's Magazine*, July, 1995, 57-66.

SEE ALSO: Groundwater pollution; Hazardous waste; Heavy metals; Soil contamination; Superfund.

Strip and surface mining

CATEGORY: Resources and resource management

DEFINITION: Mining methods in which rock and soil are removed to enable extraction of the substance being mined

SIGNIFICANCE: Surface-mining methods make possible the extraction of minerals and coal from narrow seams that could not feasibly be mined using underground techniques. These methods, however, cause many different forms of damage to the environment, radically changing landscapes, blocking streams, destroying habitats, and polluting waterways.

Humans have long scraped away the surface of the land to get at the minerals and other materials underneath. For centuries this type of mining created scars on the landscape but rarely did major harm to the environment. In the twentieth century, however, surface mines became much larger and began to cause lasting environmental degradation. In some cases surface mining is the only feasible way to extract very narrow seams of coal or minerals. In addition, surface mining tends to be less hazardous for miners than underground mining and requires fewer and often less skilled workers, reducing costs. Massive surface mining is a characteristic of modern industrial society.

OPEN-PIT MINING

Minerals have long been mined using underground tunnels, but it is often more feasible to remove the dirt and rocks covering mineral deposits (the overburden) to get at them. In some cases the ore grade is so low that extensive treatment with crushers and chemicals (such as the cyanide leach process used with gold) is required. Open-pit mines are gradually expanded and deepened over time. The sides of such a mine are cut in a spiral, with benches ringing the sides so that trucks can access the material at the bottom of the pit. As the mine is deepened, seepage often becomes a problem, requiring pumping or other forms of water control to keep the mine from filling with water.

Shallow open-pit mines (often called quarries) are used to extract granite, limestone, gypsum, and clay. Some gemstones, such as diamonds, are mined in open pits such as was done in Kimberly, South Africa. Copper, nickel, iron ore, gold, and uranium are some of the minerals mined in this fashion. One of the largest open-pit mines in the world, the Hull-Rust-Mahoning Mine in Hibbing, Minnesota, has been in operation since 1895. This iron mine is 4.8 kilometers (3 miles) long, 3.2 kilometers (2 miles) wide, and 163 meters (535 feet) deep.

Starting during the early twentieth century, large open-pit copper mines began to be opened in the American West. One of the earliest was the Bingham Canyon Mine in Utah, which has been in operation since 1906. The pit of this mine covers 770 hectares (1,900 acres) and is 1.2 kilometers (0.75 miles) deep and 4 kilometers (2.5 miles) wide. By the early twenty-first century the mine had produced more than 17 million tons of copper, 190 million ounces of silver, 23 million ounces of gold, and 850 million pounds of molybdenum.

Bingham Canyon Mine is the largest human-made excavation in the world, but other copper mines are also quite large, such as the Chuquicamata in Chile, the Nchanga mines in Zambia, and the Grasberg mine in Indonesia. Although copper is a major product of these mines, some also produce other minerals, such as gold and silver. Indeed, the Grasberg mine is the largest gold mine and the third-largest copper mine (behind Chuquicamata and Bingham Canyon) in the world. Other open-pit mines, such as Kumtor in Kyrgyzstan, produce gold and uranium. Massive open-pit mines are also found in Russia, Australia, Canada, Peru, Mongolia, and Namibia. Some of these mines exist in environmentally fragile surroundings; Kumtor, for instance, is more than 4,000 meters (14,000 feet) above sea level on a high plateau, and Grasberg is located on a mountaintop near rare equatorial glaciers

The Bingham Canyon Mine, located in Utah, is the largest human-made excavation in the world. (©Gary Whitton/iStockphoto)

and in a region subject to earthquakes that produce large landslides.

Open-pit mining produces several environmental hazards. The overburden removed at an open-pit mine, which often contains hazardous chemicals, must be placed nearby. As the mine expands, it consumes nearby land. The Berkeley Pit, for example, consumed several neighborhoods in Butte, Montana. In addition, the processing of the material taken from an open-pit mine often involves treating it with acid, which leads to extensive runoff. The acidic runoff, dissolved copper, and other materials from processing at the Grasberg mine, for example, wash into two rivers that are the water sources for several communities, and some of it ultimately reaches the ocean, a situation that will continue long after the mine is closed.

When mining ceases at an open-pit mine, the pit often becomes flooded, producing a large lake that is highly acidic. The Berkeley Pit in Montana provides an example of this process. The mine opened in 1955 and closed in 1982. Since 1982 water has filled the mine to within 46 meters (150 feet) of the natural groundwater level. The water contains several heavy metals, such as arsenic and cadmium, and is highly acidic (a pH of 2.5). A water diversion project has been established to keep the acidic water in the Berkeley Pit from entering the water supplies of Butte and surrounding communities. The U.S. Environmental Protection Agency has listed the Berkeley Pit as a site that requires massive cleanup efforts under the provisions of the Comprehensive Environmental Response, Compensation, and Liability Act of 1980, commonly known as Superfund.

STRIP MINING

The process of mining shallow deposits of coal and tar sands often involves the removal of the overburden. In the most common form of strip mining, the overburden is removed and placed in the excavation left by the previous strip. Contour mining involves removing land along the contour of the land and usually produces terraces on a mountainside. Massive pieces of equipment are used in strip mining, such as draglines, which remove the surface layer, and earthmovers, often two stories tall, which move the surface or coal.

Material Handled at U.S. Mines, by Type
(million metric tons)

	SURFACE MINING			UNDERGROUND MINING		
	CRUDE ORE	WASTE	TOTAL	CRUDE ORE	WASTE	TOTAL
2002	3,880	1,460	5,330	123	3	126
2003	3,930	1,430	5,360	121	1	122
2004	4,190	1,470	5,650	153	2	155
2005	4,300	1,420	5,720	156	2	158
2006	4,450	1,320	5,760	144	3	147

Source: Data from U.S. Geological Survey, *Minerals Yearbook, 2006.*
Note: Because data are rounded, they may not exactly total to figures shown.

Coal is strip-mined around the world. In the United States strip mining is used to remove surface coal seams in all parts of the country. The largest U.S. strip mines are found in the Powder River basin in northeast Wyoming and southeastern Montana. Unlike coal in other parts of the country, such as West Virginia, the Powder River coal lies close to the surface. The Powder River coal deposits are some of the largest in the world, and by 1988 Wyoming was producing more coal than any other U.S. state.

As energy companies seek out new sources of oil, oil-impregnated sands, known as oil sands or tar sands, are increasingly seen as sources of oil. Rising oil prices have made it cost-effective to, in essence, mine oil. One of the largest deposits of tar sands in the world is found in the Athabasca River basin in Alberta, Canada. In addition to endangering a fragile environment, the strip mining of these tar sands requires large amounts of energy to "crack" the oil and remove it from the sand.

MOUNTAINTOP REMOVAL

A variant on surface mining for coal that has come to be adopted in parts of West Virginia, Kentucky, and southern Ohio is mountaintop removal, a mining method in which explosives are used on mountainsides to remove the overburden, which is then pushed into adjoining valleys. This approach allows miners to reach deeper into the earth than does conventional strip mining. In some cases not only are whole mountains removed but also the remaining earth is excavated.

Mountaintop removal drastically changes landscapes as whole mountains are destroyed and valleys are filled with overburden. Streams are often buried in the process, and sometimes large lakes are created in what remains of the valleys; the water in such lakes is highly acidic and contains heavy metals. Critics of mountaintop removal point out that the plant and animal species that live in mountain valleys, some of which are quite rare, are endangered by this mining method. In parts of West Virginia and Kentucky the environment has been changed permanently as mountains have been leveled by mining. In the Appalachia region mountaintop removal has damaged the water sources of several communities. In addition, the explosions that are part of this kind of mining can be dangerous to local people and structures, and the sulfur compounds in the dust created by the explosions poses a health hazard.

The Surface Mining Control and Reclamation Act of 1977 requires that the owners of any surface-mined land in the United States restore the land after mining operations have ceased, but with mountaintop removal this is not completely possible, as large flat areas of land are created where mountains once existed. The reclamation of some of these areas has taken the form of the creation of golf courses or industrial sites.

Mining and Industrial Progress

Access to coal and to minerals such as copper and gold is essential for modern industrial society. Achieving this access cheaply is also an important ingredient for economic development. In some cases access helps to reduce environmental problems, such as the strip mining of the Powder River basin's low-sulfur coal to replace high-sulfur coal and reduce emissions of sulfur dioxide into the atmosphere. Some surface-mining approaches seem to be the only economically feasible way of acquiring certain minerals or energy sources that are essential to industrial society.

Nonetheless, an important drawback to surface mining is the extensive environmental degradation that often occurs during the mining process and after mines have ceased operations. In addition to the environmental damage posed by mining itself, many sites where deposits of desirable minerals or coal are located are heavily timbered and must be cleared before mining can begin; such deforestation destroys plant and animal habitats, contributes to global warming, and leads to soil erosion. Mining companies often opt for the cheapest approaches to resource extraction, even though these approaches may produce higher levels of environmental harm than other approaches.

Some countries have placed extensive regulations on surface mining of all kinds, with the aim of limiting the environmental damage such mining can do, whereas regulations in other nations are minimal. In the United States, the Surface Mining Control and Reclamation Act requires that owners of surface mine sites reclaim the land at least to some degree when mining has ended. Other laws limit the damage that mining operations are allowed to do to streams and endangered species. Debates are ongoing, however, regarding how effective these laws have been, especially after the presidential administration of George W. Bush weakened the application of some of the laws.

The societal dilemma posed by surface mining is that such mining is essential for economic progress yet often results in environmental harm. Further, the costs of this harm are not always shared equitably—industrialized nations frequently profit from surface mining that is conducted in developing countries, which often bear most of the costs of the environmental damage.

John M. Theilmann

Further Reading

Goodell, Jeff. *Big Coal: The Dirty Secret Behind America's Energy Future.* New York: Houghton Mifflin, 2006.

LeCain, Timothy J. *Mass Destruction: The Men and Giant Mines That Wired America and Scarred the Planet.* New Brunswick, N.J.: Rutgers University Press, 2009.

Lynch, Martin. *Mining in World History.* London: Reaktion Books, 2002.

McQuaid, John. "Mining the Mountains." *Smithsonian,* January, 2009, 74-85.

Montrie, Chad. *To Save the Land and People: A History of Opposition to Surface Coal Mining in Appalachia.* Chapel Hill: University of North Carolina Press, 2003.

Power, Thomas Michael. *Lost Landscapes and Failed Economies: The Search for a Value of Place.* Washington, D.C.: Island Press, 1996.

Shnayerson, Michael. *Coal River.* New York: Farrar, Straus and Giroux, 2008.

SEE ALSO: Acid mine drainage; Coal; Fossil fuels; Groundwater pollution; Heavy metals; Mine reclamation; Oil shale and tar sands; Restoration ecology; Superfund; Surface Mining Control and Reclamation Act.

Strip farming

CATEGORY: Agriculture and food

DEFINITION: The systematic planting of crops on narrow strips or bands of land

SIGNIFICANCE: Strip farming helps to alleviate several environmental problems associated with large-field agriculture at the same time it improves crop production: It reduces soil erosion from wind and rain; traps minerals, metals, fertilizers, pesticides, bacteria, and pathogens before they leave cultivated fields; and provides habitat for wildlife.

The origins of strip farming can be traced to the enclosure movement of postmedieval Great Britain. Landlords consolidated the small, fragmented strips of land farmed by tenant peasants into large block fields in an effort to increase agricultural production to meet the demands of growing human populations. A peasant plot was typically 0.4 hectare (1 acre) in size: 201 meters (220 yards), or 1 furlong (the distance a team of oxen could plow before resting), in length

and 20 meters (22 yards) in width (the total area being the amount of land one team of oxen could plow in one day). After enclosure, fields were 40.5 hectares (100 acres) or more in size. Larger fields were more productive, but the soil in such fields was also more exposed to wind and water erosion and nutritional exhaustion.

As agricultural production gradually shifted to new lands in the Americas and colonial Africa, farmers continued to use large-field farming techniques and developed large-field plantations. By the early twentieth century, all readily tilled lands had been opened by the plow and were suffering the effects of water and wind erosion. Strip farming, also known as strip cropping, was developed as a soil conservation measure during the 1930's. During the 1960's strip farming became an important tool to prevent water and air pollution and improve wildlife habitats.

Wind erosion begins when wind velocity at 0.3 meters (1 foot) above soil level increases beyond 21 kilometers (13 miles) per hour. Wind moves soil by saltation and surface creep. In saltation, small particles are

Rows of corn alternating with strips of grain follow the contours of a hillside in southern Wisconsin. (©Dreamstime.com)

lifted off the surface, travel ten to fifteen times the height to which they are lifted, then spin downward with sufficient force to dislodge other soil particles and break earth clods into smaller particles. Surface creep occurs when particles too small to be lifted move along the surface in a rolling motion. The wider the field, the greater the cumulative effects of saltation and surface creep, leading to an avalanche of soil particles across the widest fields even during moderate wind gusts.

Water erosion begins when rain or flowing water detaches and suspends soil particles above the surface and transports them downslope by splash or runoff. Water ice crystals expand, then contract when melted, dislodging soil particles and making them available for both water and wind erosion. Water also leaches nutrients and chemicals from the soil, causing the soil to experience both nutrient loss and an increase in salts and acids.

The U.S. Department of Agriculture computes annual soil loss from agricultural and developed land using the formula $A = RKLSCP$. In this formula, A equals annual soil loss, R equals the amount of rainfall on the plot, K equals the erosion factor for the type of soil on the plot, L equals the length of the slope on which the plot is located, S equals the angle of the slope, C equals the type of crop or soil cover on the plot, and P equals the presence of management conservation practices such as buffers, terraces, and strip farming. Soil loss tolerance—that is, the amount of soil that can be lost without any reduction in productivity—is computed for each plot. Farmers and developers reduce soil losses to tolerance levels by reducing soil exposure to wind and rain and by utilizing conservation practices such as strip farming.

Strip farming reduces field width, thus reducing erosion. Large fields are subdivided into narrow cultivated strips. The planting of crops along the contour lines around hills is called contour strip cropping. Another variation is field stripping, or the planting of crops in strips across the tops of predominant slopes. Crops are arranged so that a strip planted with hay plants (which include grasses, herbaceous plants, and legumes, such as clover or alfalfa) or a strip of close-growing small grain (such as wheat or oats) is alternated with a strip of cultivated row crop (such as tobacco, cotton, or corn). Rainwater runoff or blown dust from the row-crop strip is trapped as it passes

through the subsequent strip of hay plants or grain, thus reducing soil erosion and pollution of waterways. Contour or field strip cropping can reduce soil erosion by 65 to 75 percent on a 3 to 8 percent slope.

Cropping in the strips is usually rotated each year. In a typical four-strip field, each strip will be cultivated with a cover crop for one or two years, grain for one year, and row-crop planting for one year. Each strip benefits from one or two years of nitrogen replenishment from nitrogen-fixing cover crops such as alfalfa, and each strip benefits from one year of absorbing nutrient and fertilizer runoff from the adjacent row-crop strip.

Strip widths are determined by the slope of the land: The greater the slope, the narrower the strips. In areas of high wind, the greater the average wind velocity, the narrower the strips. The number of grass or small-grain strips must be equal to or greater than the number of row-cropped strips.

Terraces are often constructed to reduce the slope of agricultural land. To reduce erosion, at least one-half of the land between terrace walls is cultivated with grass or a close-growing crop. Diversion ditches are often used to redirect water from its downhill course across agricultural land. These ditches usually run through permanently grassed strips, through downhill grass waterways constructed across the width of the strips, and through grassed field borders surrounding each field.

Gordon Neal Diem

FURTHER READING

Gliessman, Stephen R., and Martha Rosemeyer, eds. *The Conversion to Sustainable Agriculture: Principles, Processes, and Practices.* Boca Raton, Fla.: CRC Press, 2010.

Hart, John Fraser. *The Rural Landscape.* Baltimore: The Johns Hopkins University Press, 1998.

Koepf, Herbert H. *The Biodynamic Farm: Agriculture in the Service of the Earth and Humanity.* Hudson, N.Y.: Anthroposophic Press, 2006.

Troeh, Frederick R., and Louis M. Thompson. *Soils and Soil Fertility.* 6th ed. Malden, Mass.: Blackwell, 2005.

SEE ALSO: Erosion and erosion control; Intensive farming; Organic gardening and farming; Runoff, agricultural; Soil conservation; Sustainable agriculture.

Subsistence use

CATEGORY: Land and land use

DEFINITION: Management of land in such a way that it yields enough essential foods and basic goods to cover the consumption needs of the people dwelling on the land

SIGNIFICANCE: From an aggregated global perspective, industrialization, urbanization, shifting values, and population growth have led to a constant decline in subsistence use, but a modern trend also exists among a minority of environmentally conscious individuals and groups who reject the overconsumption associated with modern mainstream lifestyles.

Subsistence use of land is frequently contrasted with commercial use or exploitation of land and with intensive (industrial) farming, which has historically succeeded subsistence land use, especially in developed counties. In many developing regions in Africa, Asia, Latin America, and Eastern Europe, subsistence farming remains common or even predominant, whereas in newly industrializing countries the practice of subsistence use is rapidly declining.

Subsistence use at the family level or in community contexts implies similar values, concepts of reality, and outlook perspectives as the pursuit of a self-sufficiency economy at a national level. Sustainability is often a part of subsistence use, but subsistence use does not necessarily always involve sustainable or organic methods—and commercial farming, on the other hand, can be sustainable and organic. Unsustainable subsistence use can occur, for example, if a growing number of people depend on and utilize a piece of land of fixed size.

The more intensive subsistence use is practiced in a country, the lower the national income. From a wealth-maximizing perspective, subsistence use is thus viewed as ineffective, as it secures only an existential minimum for direct consumption, with no or very limited barter. Usually subsistence farmers (or fishers, or hunter-gatherers) make no concrete attempts to produce surplus to gain greater incomes. Market mechanisms and competition also do not play a significant role in determining the value and distribution of what is produced, owing to the subsistence society's relative independence from the general market.

Individuals and communities may practice subsistence use for different reasons: They may lack alternatives, owing to unemployment, underemployment, poverty, or underdeveloped infrastructure (especially in developing countries), or they may make deliberate choices to engage in a self-sufficient, simple, and community-oriented lifestyle. Prominent supporters of subsistence use are, for example, the authors and environmentalists Ralph Waldo Emerson, Henry David Thoreau, and John Seymour; many Amish farms rely on subsistence use. In addition to food production, those who practice subsistence use often take care of the production of on-site energy and engage in water harvesting, waste management, building, and the production of textiles for clothing.

Roman Meinhold

FURTHER READING

Seymour, John, with Will Sutherland. *The Self-Sufficient Life and How to Live It: The Complete Back to the Basics Guide.* London: Dorling Kindersley, 2009.

Waters, Tony. *The Persistence of Subsistence Agriculture: Life Beneath the Level of the Marketplace.* Lanham, Md.: Lexington Books, 2007.

SEE ALSO: Back-to-the-land movement; Balance of nature; Bioregionalism; Carrying capacity; Composting; Eat local movement; Intensive farming; Organic gardening and farming; Overconsumption; Spaceship Earth metaphor; Sustainable agriculture.

Sudbury, Ontario, emissions

CATEGORY: Atmosphere and air pollution

IDENTIFICATION: Release of sulfur dioxide into the air by industry in Sudbury, Ontario, Canada

SIGNIFICANCE: After sulfur dioxide released during the processing of nickel and copper ore caused vast ecological devastation in the area in and around Sudbury, an environmental cleanup effort was undertaken that remains a model for the world at large.

The process of smelting involves heating sulfur-containing metal ores in air to convert the sulfide to the oxide form. In the process, the sulfur is removed in the form of sulfur dioxide. Without environmental controls, this gas goes into the atmosphere, where it reacts with water and oxygen to form sulfuric acid. This acid rain damages foliage, lakes, and structures, and causes difficulty for anyone with respiratory ailments. The unnatural acid levels (measured at a pH of less than 3 in some cases) may also lead to the leaching of aluminum from soil into groundwater, which has negative impacts on roots and aquatic life in streams and lakes.

In the late nineteenth and early twentieth centuries, nickel and copper sulfide ores were found in Sudbury, Ontario, Canada, just north of Lake Huron, and a metal mining and processing industry took root there. Extensive logging took place in the region as wood was needed for the growing settlements and for fuel for the smelter operations. Trees were unable to grow back in the deteriorating environment, and the denuded soil quickly eroded into waterways, adding to the devastation. At the same time, emissions from the smelting process included airborne metal particulates that deposited toxic levels of nickel and copper in the surrounding soil and water. By the early 1960's Sudbury was known around the world for its acidified, lifeless lakes and its blackened, treeless landscape.

Late in the 1960's the Canadian government began to respond to the growing worldwide environmental movement by ordering a reduction of sulfur dioxide and metal levels in the air around Sudbury. The ore-processing companies responded by constructing 380-meter (1,247-foot) "superstacks" that acted to reduce local air pollution by spreading it over a larger area. During this period, the world was beginning to recognize the transboundary nature of pollution, particularly air pollution. Dangerously acidic conditions were found in a 50-kilometer (31-mile) radius of Sudbury and even farther in the direction of the prevailing winds.

Citizen outcry made it clear that improvements also had to occur inside the industrial plants. Among the changes implemented were the use of higher-grade ore, recycling of sulfur gases to make and sell sulfuric acid, and treatment of stack gases to remove residual acid and particulates. Sudbury eventually became known for its remediation of local lakes and soil through the application of basic materials (such as lime) and for replanting and nurturing young trees and other plants. Over twenty years, soil and water pH levels rose measurably, and plant and aquatic life slowly began to return to the area. In 1992 Sudbury received commendation at the Earth Summit in Brazil for its unprecedented success in environmental cleanup; the city's efforts were noted as an example from which the rest of the world could learn.

Wendy Halpin Hallows

FURTHER READING

Jacobson, Mark Z. *Atmospheric Pollution: History, Science, and Regulation.* New York: Cambridge University Press, 2002.
Keller, W., et al. "Recovery of Acidified Lakes: Lessons from Sudbury, Ontario, Canada." In *Acid Rain: Deposition to Recovery,* edited by Peter Brimblecombe et al. New York: Springer, 2007.

SEE ALSO: Acid deposition and acid rain; Air pollution; Air-pollution policy; Soil contamination; Strip and surface mining; Sulfur oxides.

Sulfur oxides

CATEGORIES: Pollutants and toxins; atmosphere and air pollution
DEFINITION: Chemical compounds containing sulfur and oxygen
SIGNIFICANCE: Sulfur oxides emitted into the earth's atmosphere pollute the air and have negative effects on human and animal health. Sulfur dioxide is the major contributor to acid rain, which causes several types of environmental damage.

Sulfur oxides occur both naturally and as the result of human activities. On a global basis, natural sources, such as volcanoes, contribute about the same amount of sulfur oxides to the atmosphere as do human activities. In industrialized nations such as those of Europe and the Americas, however, human activities contribute 95 percent of the sulfur oxides emitted. The two most important sulfur oxides are sulfur dioxide (SO_2) and sulfur trioxide (SO_3). SO_2 is a colorless, dense, toxic, nonflammable gas with an intense odor; SO_3 is a liquid.

In 2005 the U.S. Environmental Protection Agency (EPA) estimated that anthropogenic (human-caused) emissions of SO_2 amounted to more than 14 million tons. Of these, 73 percent were the result of electricity generation, 15 percent came from fossil-fuel combustion (coal may contain from 1 percent to 4 percent sulfur, and when burned with oxygen in the air, it produces SO_2), almost 8 percent resulted from industrial processes, and smaller amounts came from nonroad equipment, on-road vehicles, fires, waste disposal, residential wood combustion, and solvent use.

SO_2 is an air pollutant and a lung irritant. Scientific

Global Sulfur Emissions by Source and Latitude

Atmospheric chemical inputs can vary greatly by region, as illustrated by the table below listing the vastly different sources of atmospheric sulfur in different parts of the globe.

LATITUDE	ANTHROPOGENIC %	MARINE %	TERRESTRIAL %	VOLCANIC %	BIOMASS BURNING %
90° south	0	0	0	0	0
75° south	0	80	0	19	1
58° south	2	97	0	0	1
45° south	22	72	0	9	1
28° south	67	28	0	1	4
15° south	21	47	1	22	10
0°	21	39	1	33	7
15° north	40	30	1	19	1
28° north	85	6	0	8	1
45° north	88	4	0	7	1
58° north	86	3	0	10	1
75° north	30	40	0	23	7
90° north	0	0	0	0	0

Source: Pacific Marine Environmental Laboratory, National Oceanic and Atmospheric Administration.

evidence links short-term exposures to SO_2 (5 minutes to 24 hours) with adverse respiratory effects, including bronchoconstriction (tightening of the bronchi in the lungs) and increased asthma symptoms. Studies also show a connection between short-term exposure to SO_2 and increased visits to hospital emergency rooms and hospital admissions for respiratory illnesses, particularly in at-risk populations: children (whose lungs are still developing and have elevated breathing rates while playing), asthmatics, and the elderly (who may also have compromised lung function). Sulfur oxides can react with other compounds in the atmosphere to form small particles that can penetrate deep into the lungs and cause or worsen respiratory diseases, such as emphysema and bronchitis, and aggravate existing heart disease.

Acid rain (or acid snow, sleet, or fog) is a direct result of the method that the atmosphere uses to clean itself. Tiny droplets of water in the atmosphere continuously capture suspended particles and gases; SO_2 and oxides of nitrogen are then chemically converted to sulfuric and nitric acids. SO_2 is the major gas producing acid rain. Both SO_2 and SO_3 contribute to the formation of acid rain, but SO_3 is found in the atmosphere at much lower concentrations than SO_2.

Industrial acid rain is a substantial problem in Europe, China, Russia, and areas downwind from them because sulfur-containing coal is burned to generate heat and electricity in these parts of the world. The use of tall smokestacks to reduce local pollution contributes to the formation of acid rain considerable distances downwind of the original emissions by releasing acid-forming gases high into the atmosphere. Acid rain has adverse impacts on forests, freshwater resources, soils, and the human-built environment, killing insects and aquatic life and causing damage to buildings. It also affects human health in ways similar to gaseous SO_2.

The EPA's air-quality standard for SO_2 is designed to protect against exposure to all sulfur oxides. SO_2 is of greatest concern and is used as the indicator for the larger group of sulfur oxides because emissions that lead to high concentrations of SO_2 generally lead to the formation of other sulfur oxides. Annual average ambient SO_2 concentrations across the United States have decreased by more than 70 percent since 1980, and by the early twenty-first century all areas of the United States had met the EPA's standard for SO_2.

Bernard Jacobson

FURTHER READING

Christiani, David C., and Mark A. Woodin. "Urban and Transboundary Air Pollution." In *Life Support: The Environment and Human Health*, edited by Michael McCally. Cambridge, Mass.: MIT Press, 2002.

McKinney, Michael L., Robert M. Schoch, and Logan Yonavjak. "Air Pollution: Local and Regional." In *Environmental Science: Systems and Solutions*. 4th ed. Sudbury, Mass.: Jones and Bartlett, 2007.

Vallero, Daniel. *Fundamentals of Air Pollution*. 4th ed. Boston: Elsevier, 2008.

SEE ALSO: Acid deposition and acid rain; Air pollution; Airborne particulates; Coal; Coal-fired power plants; Donora, Pennsylvania, temperature inversion; Odor pollution; Sudbury, Ontario, emissions; Water pollution; Yokkaichi, Japan, emissions.

Sun Day

CATEGORIES: Activism and advocacy; energy and energy use

THE EVENT: Day set aside by the U.S. government to increase awareness of solar energy and encourage the development of solar technologies

DATE: May 3, 1978

SIGNIFICANCE: Sun Day was successful in that it helped policy makers gather input for the development of national solar energy policies and raised awareness of the benefits of solar power, but progress in solar technologies continued at a relatively slow pace after the event, and the United States continued to be heavily dependent on fossil fuels.

During the 1960's and 1970's expanding demands for energy, increasing concerns regarding environmental quality, and limited domestic capacity to meet energy demands with traditional fossil fuels brought many Americans to the realization that the United States needed to place a higher priority on renewable sources of energy, particularly solar energy. The urgency of the problem was dramatically impressed upon the leadership of the nation with the Middle East oil embargo in 1973. It became clear to the American public that while the oil embargo would eventually pass, the nation, and even the world, could never again operate under the assumption that the traditional dependence on fossil fuels and other existing sources of energy could continue.

The need for a comprehensive program aimed at developing solar energy as a viable contributor to the future energy supply in the United States led to the creation of the Solar Energy Research Institute (SERI) in Denver, Colorado, in 1977 and the designation of May 3, 1978, as Sun Day. Solar energy awareness and development were emphasized throughout the week of May 1 through May 7, 1978. SERI provided technical support to the federal Sun Day Committee in its efforts to generate large volumes of information on solar energy for the public. A SERI-produced slide show on the technology and potential of solar energy was distributed throughout the nation to be shown at regular intervals in larger cities during the week. At the U.S. Customs House in Bowling Green, New York, solar energy displays were open to public view from May 3 to May 7.

On Sun Day, President Jimmy Carter visited SERI and gave an address on the future of solar energy in the United States in which he requested that every federal government agency consider more ways to help solar energy become a part of everyday American life. Carter pointed out the importance of developing renewable and essentially inexhaustible sources of energy in the future, particularly placing new emphasis on the importance of solar energy in the country's coming energy transition. He concluded that the costs associated with solar power technologies must be reduced so that solar power could be used more widely and would help establish a cap on rising fossil-fuel prices. In addition, Carter stated that he had just

California governor Jerry Brown, center, and an aide inspect a solar generator during 1978 Sun Day activities. (AP/Wide World Photos)

provided the U.S. Department of Energy with an additional $100 million for expanded efforts in solar research, development, and demonstration projects.

Following Carter's Sun Day address, a series of well-attended forums were conducted across the country. Participants included congressional representatives; state and local government officials; representatives of industries, labor organizations, public utilities, and special interest groups; and members of the general public. These public forums identified citizen groups interested in solar energy and provided input for the development of national solar energy policies.

Alvin K. Benson

FURTHER READING

Laird, Frank N. *Solar Energy, Technology Policy, and Institutional Values.* New York: Cambridge University Press, 2001.

Scheer, Hermann. *A Solar Manifesto.* 2d ed. London: James & James, 2001.

SEE ALSO: Alternative energy sources; Department of Energy, U.S.; Earth Day; Solar automobile racing; Solar energy; Solar One; Solar water heaters.

Superfund

CATEGORIES: Treaties, laws, and court cases; human health and the environment; pollutants and toxins

THE LAW: U.S. federal legislation designed to promote the cleanup of contaminated sites that threaten human health and the environment

DATE: Enacted on December 11, 1980

SIGNIFICANCE: The Superfund legislation established a regulatory structure for remediating contaminated sites as well as a massive fund to pay for cleanup. Slow-moving, controversial, and costly, Superfund has nevertheless restored hundreds of contaminated sites to productive use while encouraging industry to adopt more environmentally responsible business practices.

In the past, many inappropriate disposals of hazardous wastes occurred in the United States with at least tacit permission from regulatory authorities. Prior to enactment of the Resource Conservation and Recovery Act in 1976, federal regulations did not adequately address the proper management and disposal

of the millions of tons of hazardous wastes annually generated across the nation. Hazardous wastes were often poured into unlined ponds, poured into drums that were then buried in unlined pits or trenches, abandoned in containers aboveground, or incinerated without proper environmental controls. Contamination from such wastes affected local soil, groundwater, surface water, and air quality, as well as wildlife, ecosystems, and human health.

During the late 1970's a nightmarish environmental crisis erupted at Love Canal, a residential neighborhood built atop a former industrial waste dump in western New York State. Media coverage of the crisis showed Americans how the uncontrolled disposal of hazardous waste could threaten health, property, and quality of life. In response to public concern regarding abandoned waste sites, the U.S. Congress passed the Comprehensive Environmental Response, Compensation, and Liability Act (CERCLA), widely known as Superfund, in 1980.

The Superfund legislation includes a set of guidelines for assessing and remediating sites contaminated by past industrial, commercial, and municipal activities; a system for applying legal and financial liability to parties responsible for those sites; and a fund to finance all or part of specific site cleanup activities. Superfund sites include chemical manufacturing facilities, petrochemical plants, metal-related facilities, and old landfills or waste dumps in which industrial wastes were indiscriminately mixed with household refuse. Chemicals of concern at such sites include arsenic, cadmium, chromium, lead, mercury, benzene, dioxins, trichloroethylene, and polychlorinated biphenyls (PCBs).

RESPONSIBLE PARTIES

Moneys in the Superfund cleanup trust fund come primarily from an excise tax on the chemical and petroleum industries; a small portion of the funding is derived from individual income taxes. The fund is intended to cover the cleanup of abandoned or inactive sites for which the responsible parties cannot be determined. Whenever possible, however, the polluter pays. The U.S. Environmental Protection Agency (EPA) has formulated a system for the determination of potentially responsible parties (PRPs) and has established four classes of PRPs liable for the costs of investigation and cleanup at a Superfund site: the current owner or operator of the facility, all previous facility owners or operators (at the time of hazardous

Barrels sit at the site of the abandoned Anaconda Copper Company mine in Yerington, Nevada, in 2001. The huge open-pit copper mine is a Superfund site requiring cleanup of contaminated surface water, groundwater, and soil. (AP/Wide World Photos)

materials disposal), anyone who arranged for disposal or treatment of the hazardous substances, and transporters of the hazardous substances to disposal facilities. CERCLA allows a PRP that is liable to the government for cleanup costs to seek contribution from other PRPs. As a result, the payment process can drag on for years as PRPs seek out others who might offset some of the cleanup costs.

Under Superfund, PRPs are liable for all costs of removal or remedial action, costs incurred by other parties, damages for destruction of natural resources, and costs of any health assessment conducted. Liability under CERCLA is joint, which means that costs for environmental assessment and cleanup may be distributed among all PRPs at the discretion of the EPA. No excuses for liability are accepted. Finally, liability is retroactive: A party can be held liable for cleanup costs even if all suspect wastes were disposed of prior to the enactment of Superfund. In fiscal year 2008, financial commitments from PRPs accounted for almost $1.9 billion of the program's funding

SHORT- AND LONG-TERM RESPONSES

A CERCLA action at a contaminated site is designated as either a short-term removal action or a long-term remedial response. A removal action involves cleanup or other actions taken in response to emergency conditions or on a short-term basis. Actions may include installation of fences, evacuation of threatened populations, and construction of temporary containment systems.

Long-term remedial response, which addresses environmental problems that took years to develop, is time-consuming and complex. Some of its basic elements include preliminary investigation of the affected site, ranking of the hazard, selection of the cleanup remedy, preparation of the official record of decision, detailing of the remedial design, execution of the remedial action, and closing out of the project. Remedial designs range from the very simple to the highly intricate. Designs are prepared on a case-by-case basis because each site is unique in its characteristics, history, and complexity. Decisions on remedia-

Hazardous Waste Sites on the Superfund Priority List, 2007

	Total Sites	State Rank	Percent Distribution		Total Sites	State Rank	Percent Distribution
Total	1,311			Montana	15	26	1.2
				Nebraska	13	32	1.0
United States	**1,294**		**100.0**	Nevada	1	49	0.1
Alabama	15	24	1.2	New Hampshire	21	19	1.7
Alaska	5	45	0.4	New Jersey	116	1	9.3
Arizona	8	43	0.6	New Mexico	14	30	1.1
Arkansas	10	40	0.8	New York	87	4	7.0
California	96	2	7.7	North Carolina	31	14	2.5
Colorado	19	21	1.5	North Dakota	—	50	0.0
Connecticut	15	25	1.2	Ohio	38	10	3.0
Delaware	14	27	1.1	Oklahoma	11	39	0.9
District of Columbia	1	—	0.1	Oregon	12	36	1.0
Florida	49	6	3.9	Pennsylvania	96	3	7.7
Georgia	16	23	1.3	Rhode Island	12	37	1.0
Hawaii	3	46	0.2	South Carolina	25	17	2.0
Idaho	9	41	0.7	South Dakota	2	47	0.2
Illinois	49	7	3.9	Tennessee	14	31	1.1
Indiana	32	12	2.6	Texas	47	9	3.8
Iowa	12	33	1.0	Utah	19	20	1.5
Kansas	12	34	1.0	Vermont	11	38	0.9
Kentucky	14	28	1.1	Virginia	30	15	2.4
Louisiana	14	29	1.1	Washington	48	8	3.8
Maine	12	35	1.0	West Virginia	9	42	0.7
Maryland	18	22	1.4	Wisconsin	38	11	3.0
Massachusetts	32	13	2.6	Wyoming	2	48	0.2
Michigan	67	5	5.4				
Minnesota	25	18	2.0	Guam	2		
Mississippi	6	44	0.5	Puerto Rico	13		
Missouri	29	16	2.3	Virgin Islands	2		

Source: U.S. Department of Commerce, *Statistical Abstract of the United States, 2007.* Primary source, U.S. Environmental Protection Agency.

Note: Includes both proposed and final sites on the National Priority List for the Superfund program.

tion methods are influenced by technical issues, the extent to which the present hazard of the site affects local populations and ecosystems, and political considerations.

NATIONAL PRIORITIES LIST

In September, 1983, the number of sites on EPA's National Priorities List (NPL) of Superfund cleanups was 406. By 1990 this "final" count had increased to more than 1,200 sites and was still growing. By February, 2010, 1,269 sites (1,111 nonfederal and 158 federal) were included on the NPL, and an additional 63 (58 nonfederal and 5 federal) had been proposed. It

has been estimated that 36,000 severely contaminated sites exist in the United States; however, some federal agencies claim that the number is actually much higher. Cleanup costs for uncontrolled and abandoned U.S. hazardous waste sites have been estimated to exceed $350 billion.

Since the enactment of CERCLA, the EPA, Congress, and the American public have been concerned about the pace of cleanups. The average amount of time sites remained on the NPL increased during the 1990's, as did cleanup completion times. Nonfederal projects averaged more than ten years from the time of listing to the time of site completion. An EPA summary for fiscal year 2008 reported that in the decades since Superfund's inception only 343 sites had been made fit for return to productive use. Cleanup progress is often delayed by court battles, and determination of the optimum choice of remediation method can also slow the process. Over the program's history, the EPA has introduced various initiatives to speed the overall cleanup process.

REAUTHORIZATION

The Superfund program, originally budgeted for $1.6 billion and a period of five years, was reauthorized by the Superfund Amendments and Reauthorization Act (SARA) in 1986 with an additional $8.5 billion plus revenues from a corporate environmental tax. In 1990 Congress increased the fund to $11.5 billion. When Congress failed to reauthorize Superfund in 1995, the program was left to subsist off its $3.8 billion trust fund, which it exhausted in less than a decade. The program made do with general revenue from 2003 until 2009, when Superfund's long-term cleanup projects received $600 million in stimulus funds from the American Recovery and Reinvestment Act.

Controversial, costly, and seemingly never-ending, Superfund continues its slow progress. Despite the susceptibility of environmental cleanup projects to budget cuts and legal entanglements, the program has managed to make significant headway at more than 1,000 sites. It has also served to heighten public awareness of the need for proper hazardous waste management and has encouraged industries to minimize their generation of hazardous wastes and to dispose of wastes responsibly.

John Pichtel
Updated by Karen N. Kähler

FURTHER READING

Gerrard, Michael. *Whose Backyard, Whose Risk: Fear and Fairness in Toxic and Nuclear Waste Siting*. Cambridge, Mass.: MIT Press, 1994.

Gerrard, Michael, and Joel M. Gross. *Amending CERCLA: The Post-SARA Amendments to the Comprehensive Environmental Response, Compensation, and Liability Act*. Chicago: American Bar Association, 2006.

Macey, Gregg P., and Jonathan Z. Cannon, eds. *Reclaiming the Land: Rethinking Superfund Institutions, Methods, and Practices*. New York: Springer, 2007.

Probst, Katherine N. "Superfund at 25: What Remains to Be Done." *Resources* 158 (Fall, 2005): 20-23.

Sullivan, Thomas F. P., and Christopher L. Bell. *Environmental Law Handbook*. 19th ed. Lanham, Md.: Government Institutes, 2007.

SEE ALSO: Bioremediation; Environmental Protection Agency; Hazardous and toxic substance regulation; Hazardous waste; Landfills; Love Canal disaster; National Environmental Policy Act; Resource Conservation and Recovery Act; Right-to-know legislation; Rocky Mountain Arsenal; Solid waste management policy; Stringfellow Acid Pits; Times Beach, Missouri, evacuation; Waste management.

Superphénix

CATEGORIES: Nuclear power and radiation; energy and energy use

IDENTIFICATION: A fast-breeder nuclear reactor located near Lyon, France

DATES: Operated from December, 1985, to December, 1996

SIGNIFICANCE: The technological problems that arose during the years that the Superphénix nuclear reactor was in operation, as well as the financial consequences of those problems, contributed to a loss of confidence in the development of commercial fast-breeder reactors in other nations.

After twelve years of construction, the Superphénix nuclear reactor, a 1,240-megawatt fast breeder, went into operation in Creys-Malville in the Lyon area of France in December, 1985. Breeder reactors maximize the production of new fuel by using surplus neutrons not required to sustain the fission chain reaction to produce more fissionable fuels, such as plutonium. However, the Superphénix breeder was continually plagued by accidents and incidents during the years it was in use, and it operated at full power for a total of only 278 days. The reactor's cooling system, which used liquid sodium, repeatedly suffered costly shutdowns because of leaks. In addition, low uranium prices undercut the value of the plutonium fuel produced by Superphénix.

In 1994 it was decided that Superphénix would be converted from a breeder into a burner of plutonium and that the facility would be used only as a research tool. Consequently, on July 11, 1994, the license under which the facility operated was changed to reflect the fact that Superphénix was no longer a power reactor; it was licensed to operate as a research reactor for the demonstration of burning nuclear waste in breeder reactors. Superphénix was closed temporarily in December, 1996, for repair, maintenance, and reconstruction, with the plan to restart it in June, 1997. Based on procedural grounds, however, the reactor license was canceled in February, 1997. On June 6, Europeans Against Superphénix, a confederation of 250 environmental and antinuclear groups, demanded that Superphénix be shut down permanently. Subsequently, on June 19 newly elected French prime minister Lionel Jospin announced in his general policy statement to the parliament that operations at the Superphénix breeder reactor would be discontinued for economic reasons. Running Superphénix had cost France billions of dollars, and the nation had gained only approximately six months of electricity in return. The final announcement of the closure of Superphénix came on February 2, 1998. Many of the residents of Creys-Malville protested the shutdown; the facility had provided some thirteen hundred jobs in a town of twelve thousand people.

The dismantling of Superphénix was scheduled to begin in 2005 and was predicted to cost about $1.76 billion. However, since France's huge nuclear power industry generates 80 percent of the country's power, many groups lobbied for the Superphénix facility to remain standing. The closure of Superphénix made necessary a review and revision of the French breeder and plutonium recycling programs. The shutdown struck a serious blow to the French breeder program and raised questions about breeder programs in other countries, particularly Japan, India, and Russia.

Alvin K. Benson

FURTHER READING

Garwin, Richard L., and Georges Charpak. *Megawatts and Megatons: The Future of Nuclear Power and Nuclear Weapons.* Chicago: University of Chicago Press, 2001.

Murray, Raymond L. *Nuclear Energy: An Introduction to the Concepts, Systems, and Applications of Nuclear*

The Superphénix nuclear reactor in 1992. (AP/Wide World Photos)

Processes. 6th ed. Burlington, Vt.: Butterworth-Heinemann/Elsevier, 2009.

SEE ALSO: Alternative energy sources; Antinuclear movement; Nuclear accidents; Nuclear and radioactive waste; Nuclear fusion; Nuclear power; Nuclear regulatory policy; Radioactive pollution and fallout.

Surface Mining Control and Reclamation Act

CATEGORIES: Treaties, laws, and court cases; land and land use

THE LAW: U.S. federal legislation concerning regulation of the surface mining of coal and responsibility for the restoration of land after surface mining

DATE: Enacted on August 3, 1977

SIGNIFICANCE: The Surface Mining Control and Reclamation Act has been responsible for significant improvements in the environmental sustainability of coal mining in the United States and in the cleanup of pollution from abandoned mines. The Office of Surface Mining Reclamation and Enforcement, created by the act, conducts research and provides training in best practices for restoring land degraded by mining.

As the United States economy grew rapidly after World War II, the need for energy grew equally rapidly; one consequence of this rapid expansion was the mining of coal as quickly and as cheaply as possible, with little regard for the environment. Surface mining, or the mining of coal near the surface of the earth by removing the rock and soil on top of it, became increasingly common, and states desperate for mining jobs competed with one another to offer the least restrictive regulations on mining operations.

As concerns about the environment began to grow during the 1970's, the U.S. Congress attempted to set federal regulations on surface mining, passing legislation in 1974 and 1975; President Gerald Ford vetoed both bills, however. The country was still reeling from the oil embargo imposed by the Organization of Petroleum Exporting Countries (OPEC) in 1973-1974 and the ensuing energy crisis, and Ford did not want to make things worse by restricting the supply of coal. In 1977, Ford's successor, Jimmy Carter, signed the Surface Mining Control and Reclamation Act, over the strenuous objections of the coal industry but with the support of local advocacy groups including Save Our Cumberland Mountains.

The act created the new Office of Surface Mining Reclamation and Enforcement (OSM) in the Department of the Interior. The law began with the assumption that the technology already existed to make surface mining both economical and environmentally safe, and it established 114 performance standards for mining and reclamation described in eighty-eight pages. The new law forbade actions that were common in the past, such as the dumping of polluted soil and rock into streambeds.

Under the terms of the act, mining companies are required to submit plans in advance of beginning any new mining operations, explaining how they will prevent pollution to soil and water and how they will restore, or reclaim, the land after mining has ceased so that it can again be available for the same uses it could support before the mining. Certain environmentally sensitive areas, including wilderness areas, may not be mined. The law also requires coal mining companies to pay fees, based on the amount of coal they remove, into the Abandoned Mine Reclamation Fund. Each U.S. state collects up to half of these fees from its own mines, and the money is used to clean up abandoned mines that present environmental hazards.

The Surface Mining Control and Reclamation Act was amended by the Energy Policy Acts of 1992 and 2005, as new research and new economic pressures changed the perceived role of coal in U.S. energy production.

Cynthia A. Bily

FURTHER READING

Montrie, Chad. *To Save the Land and People: A History of Opposition to Surface Coal Mining in Appalachia*. Chapel Hill: University of North Carolina Press, 2003.

Otto, James M. "Global Trends in Mine Reclamation and Closure Regulation." In *Mining, Society, and a Sustainable World*, edited by J. P. Richards. New York: Springer, 2009.

Rowe, James F. *Coal Surface Mining: Impacts of Reclamation*. Boulder, Colo.: Westview Press, 1979.

SEE ALSO: Acid mine drainage; Coal; Department of the Interior, U.S.; Environmental economics; Environmental impact assessments and statements; Environmental law, U.S.; Land clearance; Mine reclamation; Strip and surface mining.

Sustainable agriculture

DEFINITION: The growing and harvesting of crops in a manner that has minimal impact on the environment

SIGNIFICANCE: The practices associated with sustainable agriculture help to protect the environment by preventing soil loss through erosion; minimizing the use of pesticides and chemical fertilizers, which can cause water pollution; conserving water; and enriching nutrient-depleted soils.

Most twentieth century agricultural practices were based on continued economic growth. This practice demonstrated dramatic increases in production but had negative impacts on the environment through the losses of plant and animal habitats, depletion of soil nutrients, and pollution of water supplies. The concept of sustainable development focuses on the use of renewable resources and working in harmony with existing ecological systems. The World Commission on Environment and Development described sustainable development as the ability "to meet the needs of the present without compromising the ability of future generations to meet their own needs." Practitioners of sustainable agriculture strive to manage their agricultural activities in such a way as to protect air, soil, and water quality, as well as conserve wildlife habitats and biodiversity.

Sustainable agriculture tries to match crops and livestock to the topography, soil characteristics, and climatic conditions that exist in a given region. The crops selected for cultivation must be well suited to the existing soil and site conditions and should also be resistant to known pests in the area.

PROBLEMS CAUSED BY AGRICULTURE

Water pollution is one of the most damaging and widespread effects of modern agriculture. The runoff from farms accounts for more than 50 percent of sediment damage to natural waterways, and the chemicals and nutrients associated with this runoff in the United States are estimated to cost between $2 billion and $16 billion per year to clean. Heavy application of nitrogen fertilizers, insecticides, and herbicides has raised the potential for groundwater contamination. Feedlots that concentrate manure production lead to further groundwater contamination. Several of the most commonly used pesticides have been detected

in the groundwater of at least one-half of U.S. states. In addition, monoculture, or the growing of a single crop on an area of land, requires heavy reliance on agricultural chemicals, because the practice of growing the same crop repeatedly depletes the natural organic nutrients that were formerly rich in North American topsoils.

Research has found that many of chemical agents, pesticides, fertilizers, plant-growth regulators, and antibiotics used in agriculture have ended up in the food supply. These chemicals can be harmful to humans at moderate doses, and chronic effects can develop with prolonged exposure at lower doses. Further, widespread pesticide use has been shown to result in severe stress in other animal populations, including bees. Pesticide use is often followed by occurrences of secondary pest outbreaks and resurgence of pests that have developed resistance to the pesticides previously used.

Because of these growing problems, many American farmers have turned to the practices associated with sustainable agriculture. The U.S. government offers guidance for this transition through the Food, Agriculture, Conservation, and Trade Act of 1990. This legislation describes sustainable agriculture as agriculture that, through an integrated system of plant and animal production practices, can, over the long run, meet human food and fiber needs, enhance environmental quality and natural resources, make the most efficient use of nonrenewable resources, maintain economic viability of farm operations, and enhance the quality of life for farmers and society.

WATER AND SOIL CONSERVATION

In the western United States, the ability to irrigate crops is an especially important factor in agriculture, as much of the region is naturally arid or semiarid. In California, the limited supply of surface water has caused overdraft of groundwater and the consequent intrusion of salt water, which causes the permanent collapse of aquifers. In order to counteract such negative effects, sustainable farmers in California have introduced improved water conservation and storage methods, selecting drought-resistant crop species, using reduced-volume irrigation systems, and managing crops to reduce water loss. Drip and trickle irrigation methods also dramatically reduce water usage and water loss while helping to avoid such problems as soil salinization.

Farmers can temporarily manage salinization and

contamination of groundwater with pesticides, nitrates, and selenium by using tile drainage to remove water and salt. This method often has adverse effects on the environment, however. Long-term solutions include conversion from the planting of row crops to production of drought-tolerant forages and the restoration of wildlife habitats.

One of the most important aspects of sustainable agriculture is soil conservation. In order to prevent excessive erosion, farmers might leave grass strips in the waterways of their fields to capture soil that begins to erode. A field with a 5 percent slope has three times the water runoff volume of a field with a 1 percent slope and eight times the soil erosion rate. Contour plowing, which involves plowing across a hill rather than up and down the hill, helps capture overland flow and reduce water runoff. Contour plowing is often combined with strip farming, where different kinds of crops are planted in alternating strips along the contours of the land. As one crop is harvested, another is still growing and helps recapture wind- and waterborne soil while preventing runoff from flowing quickly downhill. In areas of heavy rainfall, sustainable farmers might construct tiered ridges to trap water and prevent runoff. This method involves a series of ridges that are constructed at right angles to one another to direct runoff and slow it down so that the water has a chance to soak into the soil.

Another farming method that contributes to soil conservation is terracing, in which the land is shaped into level shelves of earth to hold in water and soil. Soil-anchoring plants are grown on the edges of the terraces to provide further stability for the soil. Terracing is costly, but it can make enable farmers to grow crops on steep hillsides that they otherwise could not use for production. To protect fragile or unstable soil on sloping sites or along waterways, farmers may need to plant perennial species of grasses every year.

Livestock and Animal Manure

Ruminant animals (sheep, cattle, and goats) can be raised on rangeland, pasture, cultivated forage, cover crops, shrubs, weeds, and crop residues. The breeds that have lower growth and milk production potential can adapt better to environments with sparse or seasonal forage. By growing row crops on level soil and growing

pasture on steeper slopes, farmers can help reduce soil erosion; growing pasture and forage crops in rotation can help improve soil quality. When ruminants and other farm animals are allowed to graze in pasturelands, those fields are fertilized naturally. Farmers can direct how their fields are fertilized with animal manure by using portable fencing to keep livestock grazing in one area or strip of pasture all the way down before moving the animals to another strip of the field while the first strip of pasture recovers.

Sustainable farmers also use so-called green manure—crops that are raised specifically to be plowed under—to introduce organic matter and nutrients into the soil. Green manure crops help protect against erosion, cycle nutrients from lower levels of the soil into the upper layers, suppress weeds, and keep nutrients in the soil rather than allowing them to leach out. Legumes such as sweet clover, ladino clover,

Terraced rice fields in Guilin, China. (©Shariff Che'Lah/Dreamstime.com)

Sunflowers and proso millet planted in alternating crop rotation plots. (USDA/David Nielsen)

and alfalfa are excellent green manure crops. They are able to extract nitrogen from the air into the soil and leave a supply of nitrogen for the next crop that is grown. Some crops, such as beans and corn, can cause high soil erosion rates because they leave the ground bare most of the year. One way sustainable farmers combat this is by leaving crop residues on the land after harvest. Residues help reduce soil erosion and even excessive soil temperatures in hot climates. Many farmers choose to use cover crops rather than residue crops. The decision about which cover crop to grow is based on the farm's geographical location and the purpose of the crop: to control erosion, to capture nitrogen for the soil, to release nitrogen to the crop, or to improve soil structure and suppress weeds.

COVER CROPS

Cover crops such as hairy vetch or clover are well suited to the needs of later crops with high nitrogen requirements, such as tomatoes or sweet corn. Both of these cover crops decompose and release nutrients into the soil within one month. To fight erosion, a farmer might choose a fast-growing cover crop, such as rye. Rye provides abundant ground cover and an extensive root system that can prevent soil erosion and capture nutrients. Alfalfa, rye, or clover can be planted after harvest to protect the soil and add nutrients and can then be plowed under at planting time to provide a green manure for the crop. Cover crops can also be flattened with rollers, and seeds can be planted in their residue. This gives the new young plants a protective cover and discourages weeds from overtaking them. Use of natural nitrogen also reduces the risk of water contamination by agricultural chemicals.

Sustainable agriculture emphasizes the use of reduced tillage systems, which are intended to disturb the soil as little as possible when preparing it for planting. Minimum-till farming involves using the disc of a chisel plow to make a trench in the soil where seeds are planted. Plant debris is left on the surface of the ground between the rows, which helps prevent erosion. Conservation tillage, or conser-till farming, uses a cutting tool (a coulter) attached to a plow to open slots in the ground just wide enough to insert seeds without disturbing the soil. No-till planting involves drilling seeds into the ground directly through any ground cover or mulch.

CROP ROTATION AND MONOCULTURE

Planting the same crop every year on a given field can result in depleted soil. In order to keep the soil fertile, practitioners of sustainable agriculture rotate nitrogen-depleting crops (such as sweet corn, tomatoes, and cotton) from year to year with legumes, which add nitrogen to the soil. By planting winter cover crops, such as rye grass, farmers can protect their land from erosion. Such cover crops will, when plowed under, provide nutrient-rich soil for the planting of cash crops. Crop rotation also improves the physical condition of the soil because the different crops vary in root depth and in the ways they are cultivated.

In nature, various kinds of plants grow in mixed meadows, and this helps them to avoid insect infestations. The agricultural practice of monoculture places great quantities of the food of choice in easy proximity to insect predators. Insect populations that feed on particular crops can multiply more than they otherwise would when those crops are grown in the same fields year after year. Since most insects are instinctively drawn to the same home area every year, they cannot proliferate and thrive if their crops of choice are not in the same fields every year.

In addition to helping farmers use fewer pesticides, crop rotation helps to control weeds naturally. Some crops and cultivation methods inadvertently allow certain weeds to thrive. Sustainable farmers often incorporate into their crop rotations successor crops that eradicate weeds. Some crops, such as potatoes and winter squash, work as cleaning crops because of the different style of cultivation that is used on them. Pumpkins planted between rows of corn will help keep weeds at bay.

INTEGRATED PEST MANAGEMENT

Most sustainable farmers control insect pests through the practice known as integrated pest management (IPM). In IPM, each crop and its pests are evaluated as an ecological system. A plan is developed to manage the damage that pests can do through the use of particular cultivation techniques, biological methods, and chemical methods at different timed intervals. Although effective, profitable, and safe, IPM techniques have been adopted widely only for a few crops, such as tomatoes, citrus, and apples.

The goal of IPM is to keep pest populations below the size where they can cause damage to crops. Fields are monitored to gauge the level of pest damage. If farmers begin to see crop damage, they put cultivation and biological methods into effect to control the pests; physical techniques such as vacuuming bugs off crops are also used. IPM encourages the growth and diversity of beneficial organisms that enhance the defenses and vigor of plants. Small amounts of pesticides are used only if all other methods fail to control pests. It has been found that IPM, when done properly, can reduce inputs of fertilizer, lower the use of irrigation water, and reduce preharvest crop losses by 50 percent. Reduced pesticide use can cut the costs of pest control by 50 to 90 percent and can increase crop yields without increasing production costs.

Toby Stewart and Dion Stewart

FURTHER READING

Chiras, Daniel D. "Creating a Sustainable System of Agriculture to Feed the World's People." In *Environmental Science*. 8th ed. Sudbury, Mass.: Jones and Bartlett, 2010.

Francis, Charles A., Raymond P. Poincelot, and George W. Bird, eds. *Developing and Extending Sustainable Agriculture: A New Social Contract*. Binghamton, N.Y.: Haworth Press, 2006.

Gliessman, Stephen R., and Martha Rosemeyer, eds. *The Conversion to Sustainable Agriculture: Principles, Processes, and Practices*. Boca Raton, Fla.: CRC Press, 2010.

Koepf, Herbert H. *The Biodynamic Farm: Agriculture in the Service of the Earth and Humanity*. Hudson, N.Y.: Anthroposophic Press, 2006.

Lyson, Thomas A. *Civic Agriculture: Reconnecting Farm, Food, and Community*. Lebanon, N.H.: Tufts University Press, 2004.

Troeh, Frederick R., and Louis M. Thompson. *Soils and Soil Fertility*. 6th ed. Malden, Mass.: Blackwell, 2005.

SEE ALSO: Agricultural revolution; Alternative grains; Green Revolution; Integrated pest management; Intensive farming; Monoculture; Organic gardening and farming; Runoff, agricultural; Strip farming.

Sustainable development

CATEGORY: Resources and resource management
DEFINITION: Development that meets the consumption needs of the current generation without compromising the ability of future generations to increase their economic production to meet future needs
SIGNIFICANCE: When the principles of sustainable development are followed, environmental benefits arise as a consequence of changes in human attitudes and behaviors, resource utilization, and applications of technology.

According to the 1987 report of the United Nations World Commission on Environment and Development, also known as the Brundtland Commission, humanity has the ability to make development sustainable—to ensure that the current generation meets the needs of the present without compromising the ability of future generations to meet their own needs. Sustainable development involves a process of change in which the exploitation of resources, the direction of investments, the orientation of technological development, and institutional change are all in harmony and enhance both current and future potential to meet human needs and aspirations. The Brundtland Commission envisioned the possibility of continued economic growth, population stabilization, improvements in global economic equity among all nations, and environmental improvement, all occurring simultaneously and in harmony. Since publication of the commission's report, titled *Our Common Future*, the goal of sustainable development—both environmental and economic development—has become the dominant global position.

Advocates of sustainable development hold a normative philosophy, or value system, concerned with equal distribution of the earth's natural capital among current and future generations of humans. They promote three core values. First, current and future generations should have equal access to the planet's life-support systems—including the earth's gaseous atmosphere, biodiversity, stocks of exhaustible resources, and stocks of renewable resources—and should maintain the earth's atmosphere, land, and biodiversity for future generations. Exhaustible resources, such as minerals and fossil fuels, should be used sparingly and conserved for use by future generations. Renewable resources, such as forests and fer-

tile soil, should be renewed as they are used to ensure that stocks are maintained at or above current levels and are never exhausted.

Second, all future generations should have the opportunity to enjoy a material standard of living equivalent to that of the current generation. In addition, the descendants of the current generation in underdeveloped regions should be permitted to increase their economic development to match that available to descendants of the current generation in the industrialized regions. Future development and growth in both developed and underdeveloped regions must be sustainable.

Finally, future development must no longer follow the growth path taken by the currently industrialized countries but should utilize appropriate technology. Development should also limit use of renewable resources to each resource's maximum sustained yield—that is, the rate of harvest of natural resources such as fisheries and timber that can be maintained indefinitely through active human management of those resources.

Weak sustainability requires that depletions in natural capital be compensated for by increases in human-made capital of equal value. For example, the requirements for weak sustainability are met when a tree (natural capital) is cut for the construction of a frame house (human-made capital). However, if the tree is cut and cast aside in a land-clearing project, the requirements for weak sustainability are not met. Strong sustainability requires that depletions of one sort of natural capital be compensated for by increases in the same or similar natural capital. For example, the requirements for strong sustainability are met when a tree is cut and a new tree is planted to replace it, or when loss of land in equatorial rain forests in Brazil is compensated for by an increase in the area of temperate rain forests on the Pacific coast of North America.

Sustainable development is promoted through a combination of public policies. First, to the extent possible, policy makers assign monetary values to elements in the earth's support system so that they can make the economic and financial calculations necessary to ensure that the requirements of weak sustainability are met. Second, economic development in the underdeveloped world is shifted away from high-resource-using, high-polluting patterns that have been seen in the developed nations and toward more sustainable or "appropriate" patterns. Suggested appropriate technologies and techniques for sustainability include solar energy, resource recycling, cottage industry, and microenterprises (factories built on a small scale). Third, objective and measurable quality standards for air, water, and other resources are established and enforced to ensure that a continuing minimum quality and quantity of natural capital is maintained and that certain stocks of natural capital are protected through the establishment of wilderness areas, oil and gas reserves, and other reserves. Finally, each individual human is encouraged to make a minimal personal impact on the earth's natural capital by adopting a commitment to a sustainable lifestyle.

Environmental improvement results from the changes in resource utilization that are part of sustainable development. For example, reductions in the use and waste of natural capital reduce the environmental impacts of resource extraction techniques such as strip mining and waste disposal methods such as incineration. The setting of environmental quality standards and policies requiring the maintenance of biodiversity lead to the implementation of antipollution efforts and ecosystem restoration projects.

Gordon Neal Diem

FURTHER READING

Bowers, John. *Sustainability and Environmental Economics: An Alternative Text.* Essex, England: Longman, 1997.
Dryzek, John S. "Environmentally Benign Growth: Sustainable Development." In *The Politics of the Earth: Environmental Discourses.* 2d ed. New York: Oxford University Press, 2005.
Landon, Megan. *Environment, Health, and Sustainable Development.* New York: Open University Press, 2006.
Lee, Kai N. *Compass and Gyroscope: Integrating Science and Politics for the Environment.* Washington, D.C.: Island Press, 1993.
Rogers, Peter P., Kazi F. Jalal, and John A. Boyd. *An Introduction to Sustainable Development.* Sterling, Va.: Earthscan, 2008.
Sitarz, Daniel, ed. *Agenda 21: The Earth Summit Strategy to Save Our Planet.* Boulder, Colo.: EarthPress, 1993.

SEE ALSO: Brundtland, Gro Harlem; Earth Summit; *Global 2000 Report, The*; Globalization; Intergenerational justice; *Our Common Future*; Sustainable agriculture; Sustainable forestry; United Nations Conference on the Human Environment.

Sustainable forestry

CATEGORIES: Forests and plants; resources and resource management
DEFINITION: A system of management that relies on natural processes to maintain forests' continuing capacity to produce stable and perpetual yields of harvested timber and other benefits
SIGNIFICANCE: Sustainable forestry offers an environmentally sensitive alternative to the logging practice of clear-cutting and to the technique of monoculture tree farming, but disagreements exist among advocates of sustainable forestry regarding issues of ecosystem maintenance versus high timber yields.

Forest management in the United States first became an issue in 1827 when the Department of the Navy and President John Quincy Adams saw the need for a continuous supply of mature timber for ship construction. In the 1860's the American Association for the Advancement of Science first discussed the need for sustained-yield forestry. In 1878 the members of the Cosmos Club, a group of Washington, D.C., intellectuals, proposed the wise use of natural resources for the greatest good, for the greatest number, and for the longest time, establishing the foundation of the conservation movement. The first national forest reserves were established by the U.S. government in 1891, and the first selective logging and marketing of U.S. government timber reserves occurred in 1897. Clear-cutting was the general method of timber harvesting. Continued clear-cutting during the twentieth century resulted in deforestation of both private lands and lands overseen by the U.S. Forest Service, leading to concerns about soil erosion, water pollution, loss of wildlife habitat, and the sustained availability of forest resources.

Forestry science developed the system of high-yield plantation tree farming in the 1930's. By the 1960's ecological concerns had led to restoration forestry, which emphasizes human intervention to reconstruct forest ecosystems and return forests to baseline conditions that existed before clear-cutting or plantation planting. By the 1980's new understandings concerning the complexity of forest ecosystems led to an emphasis on perpetually sustaining existing forest resources rather than relying on human efforts to reconstruct forests.

Sustainable forestry is an alternative to clear-cut-

A technician reads information transmitted from a microchip attached to the base of a tree with a GPS device during a presentation of the Monitoring System Electronic Tracking and Forestry project in midwestern Brazil in August, 2010. The microchip holds data about the tree's location, size, and who cut it down. This technology enables landowners using sustainable forestry practices to distinguish their wood from that acquired through illegal logging. (Ricardo Moraes/Reuters/LANDOV)

ting, the standard logging practice. Clear-cutting removes all timber in one harvest; a given area is harvested usually no more than once every sixty to one hundred years. During clear-cutting, both mature and immature trees are removed in one process. Logging roads are cut into the forest so heavy machinery can remove all trees from a large area, usually about 40.5 hectares (100 acres) at a time. Clear-cutting and the accompanying road construction lead to soil erosion and nutrient loss, topsoil loss, silting and pollution of waterways, loss of wildlife habitat, and loss of recreational benefits. Repeated cycles of growth and clear-cutting erode soil nutrition; destroy plants, animals, and microorganisms in the ecosystem necessary for healthy forest growth; and reduce the value of future harvests.

Sustainable forestry is also an alternative to monoculture plantation forestry. Plantation forestry requires active human intervention to plant tree seedlings, control diseases and pests, and nurture timber stands to maturity. Plantations usually feature grid plantings of single tree species, with all trees maturing simultaneously. The lack of species and age diversity makes tree plantations unsuitable for wildlife habitat or recreation and also makes the trees susceptible to diseases and pests. Monoculture plantations also deplete species-specific minerals and other nutrients in the soil, reducing its future productivity.

Sustainable forest management techniques seek a perpetual high yield of timber and pulpwood while maintaining biological diversity and natural forest ecosystems and permitting forests to restore their vitality through natural processes, such as foliage decomposition and fire. These techniques are designed to maintain a balance between natural environmental stresses and the human needs for timber, pulpwood, and a variety of harvested forest products, as well as recreation in natural settings. In spite of efforts to maintain this balance, however, various sustainable forestry methods often tend to favor either ecosystem maintenance or high timber yields.

Sustainable forestry with an ecosystem emphasis is the discipline of repeated thinning of natural tree stands to sustain a mixed-age, mixed-species forest that is naturally perpetuated by seeds from the mature trees. The forest is periodically thinned, usually every twenty years, to provide a steady income to the forest owners, permit the remaining trees to reach their full maturity, and provide space for new seedlings to grow. When the timber stand reaches full sustainable maturity, immature trees are continuously harvested for pulpwood, and mature trees more than one hundred years of age are continuously harvested for high-quality lumber. Natural processes promote the health of the forest and revitalize the forest soil. Diversity in both ages and species of trees makes the forest a suitable habitat for a variety of forest-dwelling animal species and for human recreation. The forest is able to recover quickly from natural disasters, fires, or drought.

Sustainable forestry with an emphasis on timber yield divides the forest into subplots, then manages each subplot to produce two sequential high-yield plantation crop cycles of eighty years each before permitting the plot to grow to maturity in a third four-hundred-year cycle. The third cycle permits the forest soil to restore its vitality and produces an old-growth forest suitable for wildlife and eventual timber harvesting. Once fully implemented, this system ensures that the forest has subplots at each stage of growth and harvesting, from newly planted plots to old-growth plots with trees at or near four hundred years of age.

Gordon Neal Diem

FURTHER READING

Berger, John J. *Forests Forever: Their Ecology, Restoration, and Protection.* Chicago: University of Chicago Press, 2008.

Bettinger, Pete, et al. *Forest Management and Planning.* Burlington, Mass.: Academic Press, 2009.

Colfer, Carol J. Pierce. *The Equitable Forest: Diversity, Community, and Resource Management.* Washington, D.C.: Resources for the Future, 2005.

Davis, Lawrence S., et al. *Forest Management: To Sustain Ecological, Economic, and Social Values.* 4th ed. Boston: McGraw-Hill, 2001.

List, Peter C., ed. *Environmental Ethics and Forestry: A Reader.* Philadelphia: Temple University Press, 2000.

Maser, Chris. *Our Forest Legacy: Today's Decisions, Tomorrow's Consequences.* Washington, D.C.: Maisonneuve Press, 2005.

SEE ALSO: Coniferous forests; Deciduous forests; Deforestation; Forest and range policy; Forest management; Logging and clear-cutting; National forests; Old-growth forests; Renewable resources; Restoration ecology; Sustainable agriculture; Wise-use movement.

Sydney Tar Ponds

CATEGORIES: Waste and waste management; pollutants and toxins

IDENTIFICATION: Hazardous waste site located at Cape Breton Island, Nova Scotia, Canada

SIGNIFICANCE: The site of a former steel factory, the Sydney Tar Ponds represents the largest toxic waste dump in North America, containing toxic by-products amassed from almost ninety years of coke burning. Attempts to clean it up have been fraught with false starts, missteps, and delays.

The Dominion Iron and Steel Company began construction on a steel plant in June, 1899, in Sydney Harbour, Nova Scotia, Canada. After its completion in 1901, this steel plant was the largest in North America. The Sydney plant used coke ovens to burn so-called slag coal mined from local coal mines to make coke. Coke fueled the oxygen blast furnace that smelted iron ore mined from Bell Island, Newfoundland. In 1903 a company called Domtar Incorporated began operating a refining plant and storage facility located directly adjacent to the coke ovens that processed the coal tar, the waste products left over from coal burning, and transported it through pipes to storage tanks.

By 1912 the Sydney plant was making half the steel used in Canada, and it continued to do so for almost a century, until its closure in 1988. Domtar abandoned its Sydney operation in 1962, leaving the storage tanks filled with coal tar, a waste disposal lagoon where overflow coal tar had been dumped, and the company's buildings, pipes, and equipment.

In 1982 Canada's Department of Fisheries and Oceans detected the presence of chemicals called polycyclic aromatic hydrocarbons (PAHs), which result from incomplete coal burning, in lobsters caught in Sydney Harbor. This finding indicated that the contaminants from the dump site were leaching into the surrounding ecosystem. Because exposure to particular PAHs can cause cancer and birth defects, the South Arm of Sydney Harbor was closed to lobster harvesting.

The Sydney Tar Ponds (STPs) contained more than 700,000 tons of toxic sludge. Furthermore, the network of approximately 161 kilometers (100 miles) of pipes beneath the coke oven, according to former coke oven workers, was never purged. The volatile organic chemicals in these pipes were potentially explosive. Soil sampling studies established that the STPs possess high concentrations of toxic chemicals—arsenic, lead, benzene, naphthalene, PAHs, polychlorinated biphenyls (PCBs), and many others. These toxic contaminants leaked into areas bordering the STPs, where homes, schools, playgrounds, and stores

are located. Sydney residents were found to have shorter life spans and higher rates of cancer than non-residents, and infants born to women who lived in Sydney showed higher rates of birth defects.

The Canadian government and the Nova Scotia provincial government signed the first agreement to clean up the Sydney Tar Ponds in 1986. Establishing a workable cleanup plan took more than twenty-two years, however, and the effort was marked by multiple failures, false starts, delays, and public protests. In 2001 Nova Scotia created the Sydney Tar Ponds Agency (STPA) to manage the cleanup. The STPA eventually settled on a ten-year plan that was put into effect in 2004; the plan's combination of treatments includes solidification, or stabilization, in which contaminated soil is mixed with Portland cement to prevent further leaching; bioremediation, which involves treating the soil with hydrocarbon-eating bacteria that degrade the toxins; and construction of an im-

Fencing topped with barbed wire surrounds the Sydney Tar Ponds. Tests of soils in a nearby neighborhood in 2005 found measurable levels of more than thirty toxins. (AP/Wide World Photos)

permeable encasement to contain the site. Landscaping of the site will further prevent dispersal of the contaminated soil.

Michael A. Buratovich

FURTHER READING

De Sousa, Christopher. "Brownfields Background." In *Brownfields Redevelopment and the Quest for Sustainability.* Boston: Elsevier, 2008.

Dodds, Linda, and Rosann Seviour. "Congenital Anomalies and Other Birth Outcomes Among Infants Born to Women Living Near a Hazardous Waste Site in Sydney, Nova Scotia." *Canadian Journal of Public Health* 92, no. 5 (2001): 331-334.

Guernsey, Judith Read, et al. "Incidence of Cancer in Sydney and Cape Breton County, Nova Scotia." *Canadian Journal of Public Health* 91, no. 4 (2000): 285-292.

SEE ALSO: Air pollution; Bioremediation; Coal; Coal ash storage; Environment Canada; Genetically altered bacteria; North America; Polychlorinated biphenyls; Water quality.

Synthetic fuels

CATEGORY: Energy and energy use

DEFINITION: Solid, liquid, or gaseous sources of energy that do not occur naturally

SIGNIFICANCE: Given the finite nature of the world's stores of natural petroleum, the development of economically viable, environmentally safe, and renewable synthetic fuels is important for human survival.

Synthetic fuels are normally produced from abundantly occurring natural resources such as coal, tar sands, oil shale, and biomass. One of the main objectives in the production of a synthetic fuel is to eliminate sulfur and nitrogen from the fuel compound, thereby creating an environmentally clean energy source. Oxides of nitrogen and sulfur dioxide are among the most undesirable of common air pollutants. Sulfur dioxide is one of the major causes of acid rain, which is created when sulfur dioxide combines with water vapor in the atmosphere to form sulfuric acid. Similarly, oxides of nitrogen produce nitric acid. These acids fall back to earth in rain and are detri-

mental to aquatic life as well as botanical life. Synthetic fuel manufacturers thus strive to eliminate these pollutants, as well as others such as carbon monoxide, hydrocarbons, particulates, and photochemical oxidants, from the fuel supply.

PRINCIPLES OF SYNTHETIC FUEL MANUFACTURE

The manufacture of liquid and gaseous synthetic fuels normally involves transforming naturally occurring carbonaceous raw material through a suitable conversion process. The techniques employed include hydrogenation, devolatilization, decomposition, and fermentation. The principal aim in the manufacture of synthetic fuel is to achieve a low carbon-to-hydrogen atomic mass ratio, or a high hydrogen-to-carbon atomic ratio, whenever possible. This results in a clean-burning fuel that releases by-products that are harmless to the environment. For example, pure methane (CH_4), with a molecular weight of 16, has a high hydrogen-to-carbon ratio of 4:1. Methane gas is a common component that is absorbed into coal. The method used to release the gas involves fracturing the coal and exposing it to low pressures. Coal-bed methane is one of the cleanest-burning fossil fuels; the by-products of burning it are simply carbon dioxide and water. Synthetically generated substitute natural gas is more than 90 percent methane. Natural gas (of which methane is the chief constituent) has a hydrogen-to-carbon ratio of approximately 3.4:1, which is also quite high. The ratios for liquefied petroleum gas and for naphtha lie between 2:1 and 3:1. (In comparison, the ratios for gasoline and fuel oil are less than 2:1. Bituminous coal has one of the lowest values, with ratio of much less than 1:1.)

COAL GASIFICATION AND LIQUEFACTION

Although coal is among the most abundant natural energy sources, it is also among the dirtiest. The composition of this solid fossil fuel is a major disadvantage; it consists of about 70 percent carbon and about 5 percent hydrogen, translating to a highly undesirable carbon-to-hydrogen mass ratio of 14:1. Coal-burning power-generating stations thus spew out large quantities of gases that are harmful to the environment. Despite the use of such emission-reducing devices as electrostatic precipitators, the levels of pollutants emitted by coal-burning plants remain high. Techniques such as coal gasification and coal liquefaction yield synthetic fuels that are safer for the environment.

A C-17 Globemaster III military transport jet flies over New York City after completing the first transcontinental flight on synthetic fuel in 2007. (USAF/CORBIS)

The process of coal gasification involves making coal react with steam at very high temperatures (in the range of 1,000 degrees Celsius, or 1,832 degrees Fahrenheit). This process produces synthetic gas. Three types of synthetic gas are in common use. Low-calorific-value gas (also called producer gas) is used in turbines. Medium-calorific-value gas (also called power gas) is used as a fuel gas by various industries. High-calorific-value gas (also called pipeline gas) is a very good substitute for natural gas and is well suited to economical pipeline transportation. Pipeline gas contains more than 90 percent methane; as a result, it has a high hydrogen-to-carbon ratio.

The process of coal liquefaction is employed to generate a liquid fuel with a high hydrogen-to-carbon ratio; it is also used to obtain low-sulfur fuel oil. Several methods are employed to accomplish coal liquefaction, including direct catalytic hydrogenation, indirect catalytic hydrogenation, pyrolysis, and solvent extraction. All of these methods produce fuels that are much safer for the environment than the original coal.

Tar Sands and Oil Shale

Naturally occurring tar sands contain grains of sand, water, and bitumen. Bitumen, a member of the petroleum family, is a high-viscosity crude hydrocarbon. A method known as hot water extraction is used to procure bitumen from tar sands. The bitumen is subsequently upgraded to synthetic crude oil in refineries. Synthetic crude oil (also called syncrude) is similar to petroleum and can be obtained through coal liquefaction as well as from tar sands and oil shale.

Large deposits of tar sands are found in Alberta, Canada; the United States has huge reserves of oil shale in Utah, Wyoming, and Colorado. Oil shale is probably the most abundant form of hydrocarbon on earth. Oil shale is a sedimentary rock that contains kerogen, which is not a member of the petroleum family. A popular method known as retorting is used to produce oil from shale. The process involves the method of pyrolysis, which reduces the carbon content in the raw hydrocarbon through distillation. Because the process is costly, however, the production of shale oil has not provided an economically feasible alternative to petroleum.

Biomass Fuels and Gasohol

Like oil and coal, biomass is derived from plant life. Oil and coal, however, are considered nonrenewable resources, as it takes vast periods of time for geologic processes to produce these materials naturally. Because biomass consists of any material that is derived from plant life, it is produced in far shorter spans—one hundred years or less—and is thus considered renewable. Wood is the most versatile biomass resource; farm and agricultural wastes, municipal wastes, and animal wastes are also considered to be biomass. Biomass can be processed into fuel using a variety of methods. Fermentation, for example, yields ethanol, or ethyl alcohol (sometimes called grain alcohol). Other methods include combustion, gasification, and pyrolysis.

Gasohol is a mixture of gasoline and small quantities of ethanol. The mixture burns cleaner than conventional gasoline; however, it can cause damage to plastic and rubber materials used in automobile engines. In the United States, therefore, the Environmental Protection Agency (EPA) permits the addition of only 10 percent ethanol by volume to gasoline to create gasohol. Methanol, or methyl alcohol (also called wood alcohol), can also be combined with conventional gasoline to produce cleaner fuel; however, the EPA limits the amount of methane in such mixtures to 3 percent.

Other Fuels

A nonpolluting rocket fuel based on alcohol and hydrogen peroxide has been developed by U.S. Navy research engineers at China Lake, California. This nontoxic homogeneous miscible fuel (NHMF) can be modified and used to drive turbines, which in turn drive alternators that produce electricity. Further developments based on what has been learned about this fuel may permit its use in automobiles. During World War II, moreover, Germany produced synthetic fuels in large quantities to meet its energy demands, employing coal gasification and also creating diesel oil and aviation kerosene using a reconstitution process; this process is still in use in many places.

Although the present abundance of natural petroleum limits the economic competitiveness of most synthetic fuels, the finite nature of the world's oil supply virtually ensures that synthetic fuels will become increasingly important energy sources. The U.S. Department of Energy and governmental agencies in many other countries thus provide funding to encourage research into the creation of economically viable, environmentally safe, and renewable synthetic fuels.

Mysore Narayanan

Further Reading

Deutch, John M., and Richard K. Lester. "Synthetic Fuels." In *Making Technology Work: Applications in Energy and the Environment.* New York: Cambridge University Press, 2004.

Lorenzetti, Maureen Shields. *Alternative Motor Fuels: A Nontechnical Guide.* Tulsa, Okla: PennWell, 1996.

Manahan, Stanley E. "Adequate, Sustainable Energy: Key to Sustainability." In *Environmental Science and Technology: A Sustainable Approach to Green Science and Technology.* 2d ed. Boca Raton, Fla.: CRC Press, 2007.

Miller, G. Tyler, Jr., and Scott Spoolman. "Nonrenewable Energy." In *Living in the Environment: Principles, Connections, and Solutions.* 16th ed. Belmont, Calif.: Brooks/Cole, 2009.

Speight, James G. *Synthetic Fuels Handbook: Properties, Process, and Performance.* New York: McGraw-Hill, 2008.

See also: Acid deposition and acid rain; Alternative energy sources; Alternative fuels; Alternatively fueled vehicles; Biomass conversion; Fossil fuels; Gasoline and gasoline additives; Refuse-derived fuel.

T

Tansley, Arthur G.

CATEGORY: Ecology and ecosystems
IDENTIFICATION: English botanist
BORN: August 15, 1871; London, England
DIED: November 25, 1955; Grantchester, England
SIGNIFICANCE: Tansley, who coined the term "ecosystem," published scholarly articles and books on natural processes that have become central to ecological theory.

Arthur G. Tansley was the only son of Amelia Lawrence Tansley and George Tansley, a businessman and teacher at the Working Men's College in London. Tansley was educated at University College London and Trinity College, Cambridge, where he studied natural science. He joined F. W. Oliver as a professor of botany at University College, where his work included the study of fernlike plants. He married one of his student collaborators, Edith Chick, in 1903; the couple had three daughters who went on to careers in physiology, architecture, and economics.

Tansley founded the journal *New Phytologist* in 1902 and served as its editor for thirty years. He directed about one dozen botanists known as the British Vegetation Committee, which published *Types of British Vegetation* in 1911. He formed the British Ecological Society in 1913, became its first president, and edited the society's *Journal of Ecology* from 1917 to 1938. In that journal's July, 1935, issue, Tansley published an essay titled "The Use and Abuse of Vegetational Concepts and Terms" in which he coined the word "ecosystem." This first use of the term was notable, although the general underlying concept of the ecosystem had existed prior to 1935. Tansley's essay stresses the integration and interdependence of succession (a continuous process of vegetation change), animal life, organic and inorganic matter, and climate and soil, with change usually occurring in a gradual manner. The climax is the highest stage of integration and the nearest approach to perfect dynamic equilibrium. Equilibrium, or stability, is never quite perfect, however.

Tansley is known for his emphasis on ecology as an "approach to botany through the direct study of plants in their natural conditions." Since plants exist in communities, he asserted, ecologists should concern themselves with the structure of communities. The study of a habitat should include a study of its parts, such as green plants, herbivores, carnivores, fungi, bacteria, dead and organic matter, solar energy, water, oxygen, carbon dioxide, nitrogen, heat, respiration, and nutrient losses.

The First Use of "Ecosystem"

Arthur G. Tansley introduced the term "ecosystem" in this passage in a 1935 journal article:

I have already given my reasons for rejecting the terms "complex organism" and "biotic community." Clements' earlier term "biome" for the whole complex of organisms inhabiting a given region is unobjectionable, and for some purposes convenient. But the more fundamental conception is, as it seems to me, the whole *system* (in the sense of physics), including not only the organism-complex, but also the whole complex of physical factors forming what we call the environment of the biome—the habitat factors in the widest sense. Though the organisms may claim our primary interest, when we are trying to think fundamentally we cannot separate them from their special environment, with which they form one physical system.

It is the systems so formed which, from the point of view of the ecologist, are the basic units of nature on the face of the earth. Our natural human prejudices force us to consider the organisms (in the sense of the biologist) as the most important parts of these systems, but certainly the inorganic "factors" are also parts—there could be no systems without them, and there is constant interchange of the most various kinds within each system, not only between the organisms but between the organic and inorganic. These *ecosystems*, as we may call them, are of the most various kinds and sizes. They form one category of the multitudinous physical systems of the universe, which range from the universe as a whole down to the atom.

Source: Arthur G. Tansley, "The Use and Abuse of Vegetational Concepts and Terms," *Ecology* 16 (July, 1935).

Tansley did not personally direct many research students, and some observers have criticized him for avoiding experimentation while concentrating on description, comparison, and synthesis in the sphere of ecological theory. Regardless of what some see as drawbacks in his work, Tansley's views have been central to most British and American ecological theory since the early twentieth century.

From 1907 to 1923 Tansley was a lecturer in botany at Cambridge. During that time he became interested in psychology and published *The New Psychology and Its Relation to Life* in 1920. He went to Vienna, Austria, and studied with Sigmund Freud in 1923-1924. Like the evolutionary naturalists Charles Darwin and Alfred Russel Wallace, Tansley had a side that also sought answers from a spiritist or unconscious realm.

He was appointed Sherardian Professor of Botany at Oxford in 1927, a post he held until 1937. Two years after his retirement from the university, he published his best-known book, *The British Islands and Their Vegetation* (1939). He entered the field of public policy when Great Britain was planning post-World War II conservation, and in 1949 he was instrumental in the founding of the British government agency the Nature Conservancy (which was later replaced by the Nature Conservancy Council); he served as the agency's first chairman from 1949 to 1953. Tansley was knighted in 1950.

Oliver B. Pollak and Aaron S. Pollak

FURTHER READING

Golley, Frank B. *A History of the Ecosystem Concept in Ecology: More Than the Sum of the Parts.* New Haven, Conn.: Yale University Press, 1993.

Pepper, David. *Modern Environmentalism: An Introduction.* 1996. Reprint. New York: Routledge, 2003.

Tansley, Arthur G. "The Use and Abuse of Vegetational Concepts and Terms." 1935. In *The Philosophy of Ecology: From Science to Synthesis*, edited by David R. Keller and Frank B. Golley. Athens: University of Georgia Press, 2000.

SEE ALSO: Ecology as a concept; Ecology in history; Ecosystems.

Taxol. *See* **Paclitaxel**

Taylor Grazing Act

CATEGORIES: Resources and resource management; land and land use

THE LAW: U.S. federal law imposing regulations on the use of the remaining public domain lands of the American West

DATE: Enacted on June 28, 1934

SIGNIFICANCE: The Taylor Grazing Act was passed in recognition of the need to protect public rangelands from destruction caused by overgrazing. Many cattle ranchers objected to the legislation, however, and debates continue regarding whether public lands should be controlled by the federal government or the states in which they are located.

When European immigrants first entered the New World, pigs, sheep, cattle, and horses accompanied them. These domesticated herbivores served multiple needs for dependable food supplies, clothing, and transportation. For the colonizers, the newfound world blessed with abundant natural resources meant limitless economic expansion in a variety of ways. A cattle-based economy easily thrived where patches or vast expanses of rich grasslands grew, and by the early nineteenth century, cattle grazing had become a growth industry that fed the rapidly expanding coastal population of a young country.

The first prime grazing areas in the United States where cattle fattened on public land for free before being herded to Baltimore for processing were in the Great Appalachian Valley, but incoming ranchers and farmers then began pushing their herds farther west. Ironically, as nutritious forage became less concentrated, herd sizes, and the demands for increased acreage to accommodate them, rose exponentially along with incomes. This westward-moving scenario repeated itself as herds crossed into arid country and cross-bred with hardy strays such as longhorns from the Spanish occupation period. Intermittent conflicts such as Native American uprisings, the Mexican War, and the U.S. Civil War depleted cattle populations in the battle zones. The arid West emerged as the nation's major cattle-producing area.

DEGRADATION OF RANGELANDS

After 1865 the demand for western beef accelerated. Large cattle companies prospered as their ever-larger herds grazed at will on public lands in the most arid region of the country. Competition for vanishing native grasses escalated, and violence erupted between sheepherders and cattle ranchers because grazing sheep ruined natural forage, leaving openings for invasive and often poisonous plant species. Both interests worked to eliminate browsing wildlife that shared the rangelands.

With the reduction of native grasses, soil erosion and compaction, as well as the pollution of rivers, accelerated. Long droughts and violent blizzards wiped out entire herds and the livelihoods of many owners. The sight of untold numbers of dead cattle and their rising stench following winter thaws horrified even tough ranch hands. Still the cattle business and its abuses continued into the twentieth century. The greed and lack of foresight that often accompany an extractive boom cycle blocked any concern for the degraded condition of the terrain and its declining carrying capacity for herbivores of all species. The catastrophic droughts of 1930 and 1934, combined with dwindling profits and the ranchers' inability to pay fees for grazing on private property, led to the grazing of more livestock on already stressed free public rangelands.

OUTCOMES OF THE ACT

The imminent collapse of the livestock industry forced bickering parties to negotiate. The Taylor Grazing Act of 1934 finally emerged under the sponsorship of a rancher and Colorado congressman named Edward T. Taylor. Two significant outcomes arose from the legislation, which was signed into law by President Franklin D. Roosevelt in June, 1934. First, it closed the remaining public lands of the United States to the free homesteading that had been offered since 1862. Second, it put the federal government in charge of the management of its own public lands.

The plan for managing grazing started with 32.4 million hectares (80 million acres) and later increased to 57.5 million hectares (142 million acres) of land unsuitable for farming. The Department of the Interior authorized the creation of grazing districts that would be overseen by local interests, usually ranchers; the Grazing Service, a federal agency that became the Bureau of Land Management (BLM) in 1946, was the entity ultimately in charge. Grazing permits were issued, for fees, to interested parties; the permits were good for ten years and were renewable. The fees were intended to be used to defray the costs of restoring rangelands to healthier conditions. Among these efforts were improvements to water systems carried out by the Civilian Conservation Corps and the reseeding of some of the most ravaged grazing lands.

The plan favored the big cattle-owning companies, which owned large tracts of land next to the publicly owned lands, and penalized small ranchers, but still the big corporate ranchers objected. Many pushed for an end to congressional funding of the Taylor Grazing Act as one tactic to stop the government oversight of grazing lands. A chorus of resentment toward governmental regulation and efforts to transfer control of public lands to the states ensued among many ranchers and has continued. Examples of such efforts include the Sagebrush Rebellion and the so-called county supremacy movement.

JoEllen Broome

FURTHER READING

Brick, Philip D., and R. McGreggor Cawley, eds. *A Wolf in the Garden: The Land Rights Movement and the New Environmental Debate.* Lanham, Md.: Rowman & Littlefield, 1996.

Fleishner, Thomas L. "Ecological Costs of Livestock Grazing in Western North America." *Conservation Biology* 8, no. 3 (September, 1994): 629-644.

Papanastasis, Vasilios P. "Restoration of Degraded Grazing Lands Through Grazing Management: Can It Work?" *Restoration Ecology* 17, no. 4 (July, 2009): 441-445.

Stout, Joe A., Jr. "Cattlemen, Conservationists, and the Taylor Grazing Act." *New Mexico Historical Review* 45 (October, 1970): 311-332.

White, Richard, "Animals as Enterprise." In *The Oxford History of the American West,* edited by Clyde A. Milner, Carol A. O'Connor, and Kartha A. Sandweiss. New York: Oxford University Press, 1994.

SEE ALSO: Bureau of Land Management, U.S.; Cattle; Conservation policy; Erosion and erosion control; Grazing and grasslands; Overgrazing of livestock; Sagebrush Rebellion.

Tellico Dam

CATEGORY: Preservation and wilderness issues
IDENTIFICATION: Hydroelectric dam on the Little Tennessee River near Knoxville, Tennessee
DATE: Completed on November 29, 1979
SIGNIFICANCE: Controversy over the transformation of a river valley into an artificial lake brought national attention to the conflict between wilderness preservation and human development.

As early as 1936, the Tennessee Valley Authority (TVA) had made plans to build a dam across the Little Tennessee River to facilitate navigation below Fort Loudoun, but the project was vetoed in 1942 because of a scarcity of steel. Building the dam remained a high priority for the TVA, and in 1959 the agency undertook a thorough study of how the region would be affected by the dam. The study also compared the projected cost of the dam to the benefits it would create, which included electricity as well as employment opportunities and recreational sites. The cost was estimated to be about equal to the possible benefits, and the TVA decided that building the dam would be economically feasible and beneficial to the area.

The appropriation of federal money for the dam's construction had to be approved by the president of the United States. On October 17, 1966, the TVA received $3.2 million to begin building the dam in 1967, with the completion date estimated to be 1970 or 1971. The final cost of the dam was actually $120 million, and it was not completed until November 29, 1979. The delay and extra costs were caused by many factors, including inflation and the diversion of federal money to support the Vietnam War.

Aside from economic factors, however, the construction of the dam was also delayed during the 1970's by legal action taken against the project by the Cherokee Nation, local residents, and environmentalists. The Little Tennessee Valley, which would be flooded upon completion of the project, contained many Cherokee historical sites, including sacred burial grounds and ruins of the Seven Towns, which were the center of the Cherokee Nation before the Cherokees were sent to reservations in Oklahoma and North Carolina. Residents of the area that would be flooded asserted their interest in maintaining homes that had been in their families for generations.

Prime farmland would also be lost once the valley flooded. Environmentalists pointed out that the dam was unnecessary given that, in comparison with the TVA's total electrical output, the dam would produce very little electricity. Within 96 kilometers (60 miles) of Tellico, twenty-four major dams already existed. The environmentalists also argued that the recreational opportunities offered by a new lake, such as boating, swimming, and fishing, were trivial compared to the greater wilderness activities associated with an untamed river near Great Smoky Mountains National Park. Building the dam would restrict the last free-flowing stretch of the Little Tennessee River.

After the discovery of an endangered species of fish, the snail darter, in the Little Tennessee River in 1973, environmentalists filed suit in 1977 to halt construction of the dam because operation of the dam would destroy the fish's habitat. After many legal battles, however, a special law was passed to exempt the TVA from complying with the Endangered Species Act of 1973, and Tellico Dam went into operation in January of 1980.

Rose Secrest

FURTHER READING
Murchison, Kenneth M. *The Snail Darter Case: TVA Versus the Endangered Species Act.* Lawrence: University Press of Kansas, 2007.
Palmer, Tim. "The Movement to Save Rivers." In *Endangered Rivers and the Conservation Movement.* 2d ed. Lanham, Md.: Rowman & Littlefield, 2004.
Wheeler, William Bruce, and Michael J. McDonald. *TVA and the Tellico Dam, 1936-1979: A Bureaucratic Crisis in Post-industrial America.* Knoxville: University of Tennessee Press, 1986.

SEE ALSO: Dams and reservoirs; Hydroelectricity; Snail darter; Tennessee Valley Authority; *Tennessee Valley Authority v. Hill.*

Tennessee Valley Authority

CATEGORIES: Organizations and agencies; energy and energy use

IDENTIFICATION: A federal corporation authorized to generate, transmit, and sell electric power through municipal distributors and rural electric cooperatives

DATE: Established in 1933

SIGNIFICANCE: The Tennessee Valley Authority's building of twenty-nine dams on the Tennessee River brought many improvements to the river basin while at the same time making unprecedented alterations to the natural environment.

Established by the Tennessee Valley Authority Act of 1933, the Tennessee Valley Authority (TVA) was, by 1945, the largest electrical utility in the United States. In addition to its original hydroelectric capability, the TVA later also invested in solar thermal-electric, nuclear, and other kinds of power plants, giving it the capacity by 2010 to produce 164 billion kilowatt-hours of electricity annually and provide electric power to some nine million customers.

The Tennessee River stretches for 1,049 kilometers (652 miles) and drains most of the state of Tennessee and parts of six other states. The TVA provided for the maximum development of the Tennessee River basin in what has been referred to as the most comprehensive environmental program in history. In addition to providing inexpensive and abundant electric power, the TVA undertook dam building to control the floods that periodically devastated the basin and to increase the navigation potential of the Tennessee River. With its system of locks and dams, the river is now navigable from Knoxville, Tennessee, to its junction with the Ohio River at Paducah, Kentucky. Other goals of the TVA were to encourage good conservation practices, to institute agricultural programs, to improve air and water quality, to attract industry and commerce, and to increase resource development, all of which would generally improve the quality of life for the residents of the region. During World War II and throughout the Cold War era, the TVA provided the energy for aluminum processing factories and uranium enrichment facilities.

To accomplish its purposes, the TVA built a total of twenty-nine dams on the Tennessee River and its tributaries; the largest is Kentucky Dam, and the highest, at 146 meters (480 feet), is Fontana Dam. The TVA dams are of two types: high dams with large reservoir capacities, which were constructed on the tributaries to provide flood protection and electric power generation, and low, broad dams on the Tennessee River designed to control navigation. To construct the first sixteen TVA dams between 1933 and 1944, the federal government purchased or condemned 445,000 hectares (1.1 million acres) of land and moved fourteen thousand families.

Donald J. Thompson

FURTHER READING

Andrews, Richard N. L. *Managing the Environment, Managing Ourselves: A History of American Environmental Policy.* 2d ed. New Haven, Conn.: Yale University Press, 2006.

Billington, David P., and Donald C. Jackson. *Big Dams of the New Deal Era: A Confluence of Engineering and Politics.* Norman: University of Oklahoma Press, 2006.

Black, Brian. "Referendum on Planning: Imaging River Conservation in the 1938 TVA Hearings." In *FDR and the Environment,* edited by David Woolner and Henry L. Henderson. New York: Palgrave Macmillan, 2005.

SEE ALSO: Dams and reservoirs; Hydroelectricity; Snail darter; Tellico Dam; *Tennessee Valley Authority v. Hill;* Watershed management.

Tennessee Valley Authority v. Hill

CATEGORIES: Treaties, laws, and court cases; animals and endangered species

THE CASE: U.S. Supreme Court decision regarding a possible violation of the Endangered Species Act

DATE: Decided on June 15, 1978

SIGNIFICANCE: In the case of *Tennessee Valley Authority v. Hill,* the U.S. Supreme Court ruled that in passing the Endangered Species Act, the U.S. Congress had made "a conscious decision to give endangered species priority over the primary missions of federal agencies."

Tellico Dam was part of a Tennessee Valley Authority (TVA) water resource and development project designed to control flooding, generate electric power, and promote industrial development in an economically depressed area. Opponents of the dam—a coalition of environmentalists and local farmers and landowners—initiated legal action to stop the dam's construction, which had begun in 1967, based on the presence of the snail darter, a small fish that had been placed on the federal endangered species list in October, 1975. The Little Tennessee River above the Tellico Dam project was considered the snail darter's critical habitat.

After the U.S. district court ruled that construction would be allowed to continue, opponents of the dam immediately appealed, and the U.S. Court of Appeals reversed the lower court's decision and enjoined construction. The TVA then appealed this decision to the U.S. Supreme Court, and in the case of *Tennessee Valley Authority v. Hill* the Court upheld the ruling of the U.S. Court of Appeals.

Chief Justice Warren Burger, writing for the Court, agreed with the Court of Appeals' opinion that the TVA would violate the Endangered Species Act (1973) if it completed the dam because the plain intent of Congress in enacting that law was to halt species extinction. The plain language of the statute, supported by its legislative history, revealed "a conscious decision to give endangered species priority over the primary missions of federal agencies." Since the statutory language and legislative history also revealed that Congress placed an incalculable value on endangered species, the Court stated that it would not engage in any "fine utilitarian calculations" and find that the loss of an almost completed dam at a cost of more than $100 million would outweigh the loss of the snail darter. The statute did provide hardship exemptions, but the Court noted that none applied to the Tellico Dam project, nor did continuing congressional appropriations for the dam constitute an implied repeal of the statute. Since the completion of the dam would destroy an endangered species, the Court held that the statute required an injunction forbidding its completion.

Justice Lewis Powell, joined by Justice Harry Blackmun, dissented. Condemning the Court's literalist interpretation of the Endangered Species Act, Powell argued that Congress could not have intended for the Court to give retroactive effect to the statute and disregard Congress's commitment for twelve years to complete the Tellico Dam project. He had no doubt that Congress would amend the statute so that the dam and its reservoir would serve their intended purposes instead of providing "a conversation piece for incredulous tourists."

Justice William Rehnquist, also dissenting, argued that the Endangered Species Act did not mandate the U.S. district court to use its equitable powers and enjoin the TVA from completing the dam, nor did the district court abuse its discretion in refusing to issue the injunction. In the face of conflicting evidence, the district court had quite properly decided that the public harm from the failure to complete the dam outweighed the need to preserve the habitat of the snail darter.

Congress responded swiftly to the Court's decision by attaching an amendment to a 1979 energy bill that exempted the Tellico Dam from all federal laws, including the Endangered Species Act. The dam was completed in November, 1979. The snail darter was not, as feared, extinguished by the impoundment of the Little Tennessee River; although the snail darter population there was destroyed, other colonies of the fish were discovered elsewhere.

William Crawford Green

FURTHER READING

Ferrey, Steven. "The Endangered Species Act." In *Environmental Law: Examples and Explanations.* 5th ed. New York: Aspen, 2010.

Murchison, Kenneth M. *The Snail Darter Case: TVA Versus the Endangered Species Act.* Lawrence: University Press of Kansas, 2007.

SEE ALSO: Dams and reservoirs; Endangered Species Act; Endangered species and species protection policy; Snail darter; Tellico Dam; Tennessee Valley Authority.

Teton Dam collapse

CATEGORIES: Disasters; water and water pollution

THE EVENT: The catastrophic failure of a large earth-fill dam in Idaho, leading to the flooding of a town and the deaths of eleven people

DATE: June 5, 1976

SIGNIFICANCE: The failure and collapse of the Teton Dam prompted government examination of the procedures involved in the selection of dam sites and recommendations by experts regarding ways to minimize the risk of dam failure, which poses potential hazards to the public and associated damage to the environment.

When the Teton Dam collapsed in 1976, the town of Rexburg and thousands of acres of farmland were flooded, and eleven people were killed. (AP/Wide World Photos)

The Teton Dam was located in a deep, narrow canyon on the Teton River, a tributary of the Snake River, in southeastern Idaho. It was 93 meters (305 feet) high and 975 meters (3,200 feet) long, forming a reservoir that extended 27 kilometers (17 miles) up the canyon. The U.S. Bureau of Reclamation used approximately 7.6 million cubic meters (10 million cubic yards) of clay, silt, sand, and gravel to build the multilayered structure, a construction technique that the bureau had previously used for approximately 250 other dams without a single failure. Extensive site investigations revealed that the fractured and porous bedrock of the area could be a problem, and in an attempt to produce a barrier impermeable to water seepage, grout (a cement-based filler) was pumped under high pressure into drill holes on both sides and across the floor of the canyon.

Construction was complete and the reservoir had been filled almost to capacity when two small springs were detected at 8:30 A.M. on June 5, 1976, in the lower wall of the canyon just below the dam. While attempts were being made to alleviate these flows, at 10:00 A.M. a large leak appeared in the dam itself, about one-quarter of the way up from the bottom and 4.6 meters (15 feet) from the canyon wall with the spring flow. This leak rapidly increased its discharge and began to erode material from the dam. Two 20-ton bulldozers were sent in to push boulders into the flow to stem it. By 11:00 A.M. the flow had become so rapid that a whirlpool developed on the upstream side of the dam, and the bulldozers had to be abandoned. At 11:57 the dam was breached, and a tremendous wall of water roared down the canyon. The flow was so powerful that one of the abandoned bulldozers was carried 11 kilometers (7 miles) downstream. The town of Rexburg and 777 square kilometers (300 square miles) of farmland were flooded. Eleven people were killed, and monetary estimates of the damage caused by the dam collapse were as high as $1 billion.

An independent panel of experts concluded that the failure of the dam was caused by water traveling through fissures in the canyon wall, penetrating the grout curtain, and then moving through and eroding the core of the dam, until it failed. The Bureau of Reclamation was criticized for poor design of the grout curtain and dam core and for overreliance on past design practice without giving sufficient consideration to the porous rock at the Teton Dam site. The bureau was also criticized for not including any way to collect and safely discharge leakage, which should have been anticipated because of the presence of porous bedrock.

Gene D. Robinson

FURTHER READING

Cech, Thomas V. "Dams." In *Principles of Water Resources: History, Development, Management, and Policy.* 3d ed. New York: John Wiley & Sons, 2010.

McDonald, Dylan J. *The Teton Dam Disaster.* Charleston, S.C.: Arcadia, 2006.

SEE ALSO: Dams and reservoirs; Hydroelectricity; Watershed management.

Thermal pollution

CATEGORY: Pollutants and toxins

DEFINITION: Adverse environmental effect caused by heat, particularly waste heat from steam-electric power plants

SIGNIFICANCE: When thermal effluent is discharged into waterways from steam-electric power plants, the resulting increases in water temperature can cause damage to the aquatic ecosystems of those waterways.

The overwhelming majority of waste heat in industrialized countries comes not from factories but from steam-electric power plants such as coal-burning and nuclear power plants. Steam-electric power plants convert thermal energy from the combustion of fossil fuels or nuclear reactions into mechanical work and then into electrical energy. While the generators that convert the mechanical work into electrical energy in such a plant are nearly 100 percent efficient, the rest of the plant is subject to maximum efficiencies imposed by the laws of thermodynamics and determined by the highest and lowest temperatures of the plant. Steam-electric power plants typically have efficiencies of about 40 percent or less, which means that 40 percent of the heat is converted into electrical energy, while the other 60 percent becomes unusable waste heat that must be removed.

As an example, consider a large electric power plant producing 1,000 megawatts (MW) of electricity. If its efficiency is 40 percent, the plant will have to produce 2,500 MW of heat (since 1,000 MW is 40 percent of 2,500 MW) to maintain this output. Waste heat will be produced at the rate of 2,500 MW − 1,000 MW = 1,500 MW. At a coal-burning plant, perhaps 200 MW of heat will be lost in and around the boilers and the rest of the plant, leaving 1,300 MW to be removed. Nuclear power plants usually run at lower maximum temperatures, so they have lower efficiencies and correspondingly produce more waste heat; in addition, less heat is lost around the plants themselves, so more of the waste heat needs to be removed. Common methods of disposing of this heat include dumping it into rivers or lakes and or using it to evaporate water in cooling towers. Such disposal methods can have adverse effects on aquatic ecosystems or generate fog and ice.

There are three major methods of removing waste heat. The least expensive is known as once-through cooling, in which water from a stream, lake, or other body of water is used to cool the steam, after which the water is returned to its source at a higher temperature. This method may result in temperature increases of several degrees in the body of water.

A second method is the use of artificial lakes or cooling ponds, which may be up to several square kilometers in size. The heated water from the power plant is discharged into one end of a pond, while water to be used for cooling is drawn from the bottom of the pond at the other end. The water in the pond cools naturally by evaporation; therefore, the pond's water source must be continuously replenished.

A third method is the use of cooling towers, either evaporative or nonevaporative. Evaporative towers, as their name suggests, cool water from the power plant by promoting evaporation. Some evaporative towers produce natural drafts, while others use fans to induce drafts mechanically. Natural draft towers may be more than 100 meters (328 feet) high, while mechanical draft towers are often much smaller. Nonevaporative cooling towers allow moving air to cool pipes containing the heated water from the power plants; these kinds of towers are less commonly used than evaporative towers because they are expensive to build and operate.

Since most cooling methods ultimately lead to the evaporation of substantial amounts of water, large power plants are usually located adjacent to rivers, which provide a source of water. The major ecological effects of thermal pollution occur in natural rivers and lakes and involve fish and other aquatic organisms. These organisms typically thrive when the temperature remains within a narrow range and may die if the water changes to lower or higher temperatures. For example, if a population of largemouth bass that are acclimated to a water temperature of 20 degrees Celsius (68 degrees Fahrenheit) are exposed to temperatures as low as 4.4 degrees Celsius (40 degrees Fahrenheit) or as high as 32 degrees Celsius (90 degrees Fahrenheit) for one or two days, about 50 percent will die. In addition, the sudden changes in temperature that are encountered by fish swimming into thermal effluent can produce thermal shock and almost instantaneous death if the changes are sufficiently large.

All chemical reactions are increased by heat, so thermal pollution can lead to more rapid physiological processes in fish and other aquatic organisms. In certain circumstances this can cause increased growth rates and shorter life spans, leading to decreased populations and less biomass in the ecosystem; in other circumstances it may lead to increased populations and biomass, or it may extend the growing system. Ecologists have established that a temperature change of a few degrees can have significant effects, both short-term and long-term, on aquatic ecosystems.

In addition, evaporative cooling methods inject large amounts of water vapor into the atmosphere. During humid weather this may lead to fog, which can cause dangerous visibility issues on nearby roads; during cold weather it may lead to damage by icing roads, trees, and buildings.

Laurent Hodges

FURTHER READING

Camp, William G., and Thomas B. Daugherty. "Water Pollution." In *Managing Our Natural Resources.* 4th ed. Albany, N.Y.: Delmar, 2004.

Goudie, Andrew. "The Human Impact on the Waters." In *The Human Impact on the Natural Environment: Past, Present, and Future.* 6th ed. Malden, Mass.: Blackwell, 2005.

Hill, Marquita K. "Water Pollution." In *Understanding Environmental Pollution.* 3d ed. New York: Cambridge University Press, 2010.

Laws, Edward A. "Thermal Pollution and Power Plants." In *Aquatic Pollution: An Introductory Text.* 3d ed. New York: John Wiley & Sons, 2000.

SEE ALSO: Coal-fired power plants; Hydroelectricity; Nuclear power; Power plants; Water pollution.

Thermohaline circulation

CATEGORY: Weather and climate

DEFINITION: Global system of oceanic currents driven by temperature and salinity gradients

SIGNIFICANCE: The global interchange of ocean water moderates climate at high latitudes, oxygenates the ocean depths, and aids fertility and productivity of the marine environment by recycling nutrients.

The existence of surface currents in the ocean has been recognized since antiquity. Some are wind-driven, but others are part of a vast interconnected system whereby warm water from tropical latitudes circulates northward close to the surface, while cold water from the poles sinks to the depths and flows toward the equator.

The driving forces behind this oceanic conveyor belt are differences in salinity and temperature. Surface water that is denser than the water below it, because of either increased salinity or lower temperature, will sink. In temperate-zone lakes, the water simply turns over in winter, but in the ocean, the displaced water flows "downhill" along the latitudinal density gradient. In the Tropics, surface evaporation increases the salinity, and therefore the density, of surface waters. In certain areas, particularly the northeastern Pacific Ocean and along the west coast of Africa, this leads to upwelling of nutrient-laden waters that support high oceanic productivity. The effect is too diffuse to enhance or retard the conveyor itself. Elevated temperatures may produce a "sluggish" conveyor, but the overall pattern has remained stable for at least the past fifty million years.

The cold, dense water driving thermohaline circulation forms principally in two regions: in the North Atlantic off the coast of Greenland and in a band encircling Antarctica. In the North Pacific, the narrow Bering Strait and shallow waters surrounding it prevent the southward flow of cold, dense water. In the Pacific, a deep current originating in Antarctica flows northward in the western Pacific to middle latitudes, while a southerly surface current loops around the west coast of North America, crosses the Pacific near the equator, passes north of Australia, merges with northerly surface currents of Antarctic origin as it rounds Africa, and continues northward along the western Atlantic margin and the east coast of Greenland.

Salinity and the temperature of surface currents both increase in the Tropics. In the present climate regime, these effects cancel each other out. To the extent that changes in global temperature affect the action of the conveyer, that effect is much more profound in polar regions.

Cold temperatures increase both the density and the salinity of seawater. Because sea ice is effectively fresh water, salinity beneath it increases. Any large-scale melting of polar sea ice can therefore be expected to reduce the rate of formation of cold water and thus slow the circulation of water in the oceanic conveyor belt. How much this would affect climate is uncertain. A widely held theory concerning a cold period, the Younger Dryas, and the end of the last ice

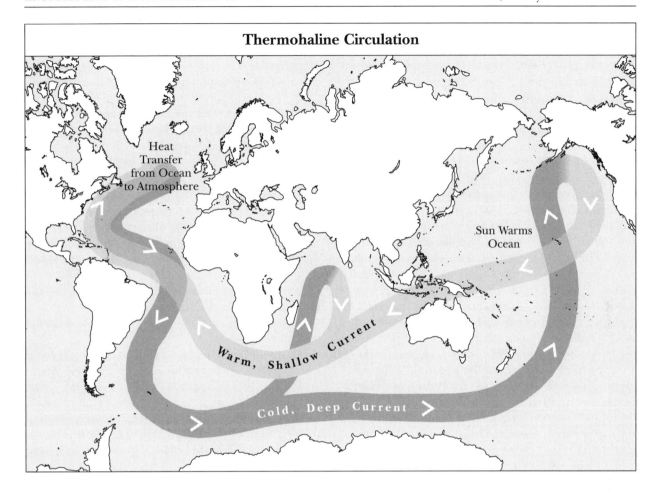

Thermohaline Circulation

Heat
Transfer
from Ocean
to Atmosphere

Sun Warms
Ocean

Warm, Shallow Current

Cold, Deep Current

age links it to rapid influx of fresh water into the North Atlantic and disruption of warm surface currents. It is uncertain whether a more gradual process would have a comparable effect.

In addition to affecting climate, the thermohaline circulation carries oxygenated water into the ocean depths, supporting life in the deep sea and preventing the buildup of organic matter under anoxic (oxygen-deprived) conditions on the ocean floor. Any significant disruption of these currents would, therefore, have undesirable implications for the diversity and productivity of marine life.

Martha A. Sherwood

FURTHER READING

Aken, Hendrik Mattheus van. *The Oceanic Thermohaline Circulation: An Introduction.* New York: Springer, 2007.

Marshall, John. *Atmosphere, Ocean, and Climate Dynamics: An Introductory Text.* Boston: Elsevier Academic Press, 2008.

Oliver, John E., ed. *Encyclopedia of World Climatology.* New York: Springer, 2005.

SEE ALSO: Climate change and oceans; Climatology; Dead zones; Global warming; Marine debris; Ocean currents.

Thoreau, Henry David

CATEGORY: Philosophy and ethics
IDENTIFICATION: American naturalist and philosopher
BORN: July 12, 1817; Concord, Massachusetts
DIED: May 6, 1862, Concord, Massachusetts
SIGNIFICANCE: Best remembered as a persuasive advocate of nonviolent civil disobedience to protest unjust laws, Thoreau was also an early advocate of environmentalism.

A lifelong resident of Concord, Massachusetts, Henry David Thoreau attended Harvard University. Soon after graduating and returning to his home in 1837, he met Ralph Waldo Emerson, a Transcendentalist philosopher who would play a significant role in his life. Emerson took a paternal interest in Thoreau, advising him, encouraging him to write and publish, and introducing him to like-minded writers and thinkers. Thoreau even moved into Emerson's home, where he lived for about four years, serving as an all-purpose helper and assistant. During that time, he adhered to Transcendentalism's loose and eclectic idealistic American philosophy, the most eloquent proponent of which was none other than Emerson.

As a philosopher himself, Thoreau always looked to nature for insight, inspiration, meaning, and sustenance. Given his acute environmental consciousness and his attunement to the human condition, he was in many ways ahead of his time. He created one of the most memorable meditations on simple living in an unspoiled environment in his most famous book. In 1845, he moved into a cabin he had built on land belonging to Emerson, along the shores of Walden Pond, not far from Concord. He lived there, alone, for more than two years, but in his account of his stay, the experience is translated into the natural cycle of a single year.

Thoreau did not rush into print with his book about his experience. Instead, between 1846, when he first began writing it, and 1854, when he finally published it, he reshaped his material through journal entries, essays, poems, lectures, and more than half a dozen successive manuscript drafts. The result of his efforts, *Walden: Or, Life in the Woods*, is his supreme achievement and one of the most accomplished works in American literature. Written in a baroque, epigrammatic style, *Walden* is not simply the record of one man's eccentric sojourn in a sylvan setting; it is an allegory of the deliberate life, a crafty provocation to its readers to awaken from the torpor and the quiet desperation of their lives. Thoreau describes his solitary existence beside Walden Pond as an experiment, and so, too, is his prose reenactment.

Walden reveals Thoreau's intention to clear his life of the unnecessary encumbrances that materialism and a lack of self-reliance encourage. The book, which concludes with the exuberance of spring revivifying the pond, would have readers undergo a similar process of purifying enlightenment. *Walden* demands a careful reader, one alert enough not to dismiss it as a naïve effusion over nature and one sensitive to its author's extravagant, incendiary wit compounded of puns, paradoxes, and hyperbole.

Thoreau also made substantial contributions to natural history and ecology and anticipated a number of methodological developments and findings in these areas. From a political point of view, Thoreau was a lifelong abolitionist and a proponent of the right of individuals to engage in civil disobedience or passive resistance in order to register their opposition on morally justified grounds against a government's unjust policies and measures. However, on pragmatic grounds, he called for improving government rather than abolishing it, though he favored its eventual abolishment.

Thoreau was a complex and multitalented person who worked strenuously to shape his philosophy and sculpt his life, as they amounted in his mind to the same thing. Although he was not always appreciated, his reputation has grown stronger since the late twentieth century. Thoreau wanted people to rethink their own lives creatively and boldly, to give up on waste and illusions in order to discover what they truly need.

Nader N. Chokr

FURTHER READING

Bloom, Harold, ed. *Henry David Thoreau*. Philadelphia: Chelsea House, 2003.
Hahn, Stephen. *On Thoreau*. Belmont, Calif.: Wadsworth/Thomson Learning, 2000.
Richardson, Robert D., Jr. *Henry Thoreau: A Life of the Mind*. Berkeley: University of California Press, 1986.
Shanley, J. Lyndon. *The Making of "Walden."* Chicago: University of Chicago Press, 1957.

SEE ALSO: Deep ecology; Ecocentrism; Environmental ethics; Environmentalism; Nature writing; Preservation; Sustainable forestry.

Three Gorges Dam

CATEGORIES: Preservation and wilderness issues; ecology and ecosystems

IDENTIFICATION: Tall dam on China's Yangtze River enabling operation of the world's largest hydroelectric power plant

SIGNIFICANCE: Construction of China's Three Gorges Dam and its hydroelectric power plant was opposed on environmental and social grounds. It was feared the dam would severely degrade the river's ecosystem and that the necessity of relocating more than one million people would have negative social impacts. Since 2007 the Chinese government has been especially active in attempts to mitigate any negative impacts of the dam's construction and operation.

The idea to dam the Yangtze (also known as the Chang), China's longest river, where it emerges through the three gorges of Qutang, Wu, and Xiling, to run east for about 1,900 kilometers (1,200 miles) to its estuary by Shanghai, originated in 1919. However, it was not until the early 1980's that the Ministry of Water Resources and Power of the People's Republic of China began to consider in earnest the construc-tion of a gigantic dam and a huge hydroelectric power plant at the Three Gorges site to control flooding and generate power.

CONTROVERSIES

From the beginning, the dam project was controversial in China and abroad for its possibly negative environmental and social impacts. Within the Ministry of Water Resources and Power, hydraulic engineers supported the idea for the benefits a dam would provide for flood control and improvement of river navigation. China's central Communist government and representatives of the provinces of Hubei, site of the dam, and Hunan, both affected by the Yangtze floods, agreed with this view. Electrical engineers were initially opposed, favoring a series of smaller, decentralized dams and hydroelectric power plants. More opposition arose in the province of Sichuan, where rising reservoir waters would force relocation of more than one million people and submerge unique ancient sites and artifacts. Concerns about negative environmental impacts on marine and riverbank life, soil erosion, and increased pollution created more opposition, which found a courageous spokeswoman in journalist Dai Qing.

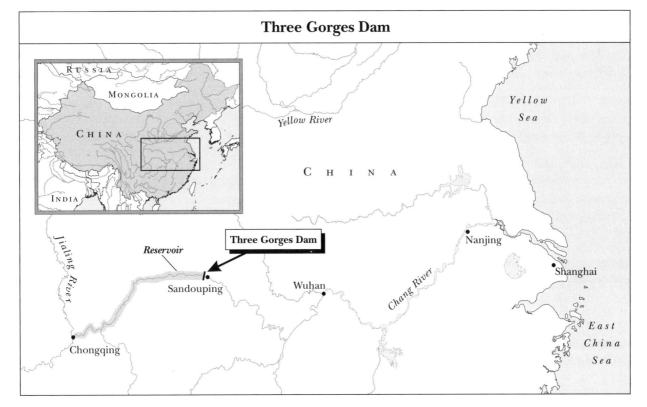

Aerial view of the Three Gorges Dam on the Yangtze River. (AP/Wide World Photos)

When Soviet-trained hydroelectric engineer Li Peng became prime minister of China in 1988, he enthusiastically supported the project. In April, 1992, China's National People's Congress approved the project by a surprisingly narrow margin: only 1,767 were in favor, with 177 nays, 644 abstentions, and 25 members not showing. Construction began on December 14, 1994.

DIMENSIONS AND OPERATION

Situated in the town of Sandouping in Hubei province, the Three Gorges Dam, with its hydroelectric power plant and two ship locks, represents a major engineering feat achieved at disputed environmental and social costs. Sitting on a massive concrete foundation, the dam wall is 101 meters (331 feet) tall. Its total length is 2,335 meters (7,661 feet). The top of the dam reaches a height of 185 meters (607 feet) above sea level. Below it, the Yangtze runs at about 65 meters (213 feet) above sea level. This means the dam and its above-surface foundation rise some 120 meters (394 feet) above the downstream river. Navigation is aided by two ship locks of five stages; the addition of a ship lift to this system is planned.

The dam's hydroelectric power plant is the world's largest. At peak operation, water can fall 113 meters (371 feet) to power gigantic generators. Since it became fully operational in October, 2008, the plant's twenty-six generators have been capable of producing 700 megawatts each, for a combined total of 18,200 megawatts; plans have been made to add six large generators and two smaller ones, bringing the plant's capacity up to 22,500 megawatts.

In 2008 and 2009, the hydroelectric plant provided 80 terawatt-hours (80 billion kilowatt-hours) annually, to increase to 100 terawatt-hours by 2012. With this, the Three Gorges Dam's power plant generated about 8 percent of China's power. Annually this saved the burning of about 50 million tons of coal to generate electricity, and thus the release of some 100 million tons of carbon dioxide into the atmosphere was avoided.

In the building of the dam, most of the natural beauty of the three gorges was preserved. As their cliffs rise between 800 and 1,000 meters (2,625 and 3,280 feet) above the Yangtze, the average rise of the water level by 75 meters (246 feet) there did not detract from the general visual experience. Kayaking the gorges was no longer an adventure, however.

ENVIRONMENTAL AND SOCIAL PROBLEMS

The gigantic reservoir behind the Three Gorges Dam is designed to run west of the dam for 663 kilometers (415 miles), creating a surface area of 1,045 square kilometers (403 square miles) with a capacity for 39.3 cubic kilometers (10.3 billion gallons) of water. By July, 2007, waterlogged soil had caused ninety landslides to crash into the reservoir as it filled up. Although the dam was designed to hold water up to 175 meters (574 feet) above sea level, or 113 meters (371 feet) above the Yangtze waters beneath it, filling of the reservoir stopped on November 2, 2009, at 171 meters (561 feet) above sea level. Before a decision was to be made regarding a higher water level, scientists needed to examine the risk of landslides and reservoir-induced seismic incidents, such as soil fissures, associated with the storage of more water in the reservoir.

The dam slowed the running speed of the Yangtze and threatened to withhold sediment, silting up areas upstream and depriving the estuary downstream. Increased pollution came from the flooding of 13 cities, 140 towns, and 1,350 villages, with their former factories, mines, and waste dumps. Pollutants from these as well as methane from rotting submerged vegetation were of concern.

Fish and other river species were negatively affected when their habitats were altered or destroyed. For the first time there was toxic algae bloom on the Yangtze. The government responded by providing funds for monitoring, particularly by the Institute of Hydrobiology at the Chinese Academy of Science in Wuhan. Beginning in 2007, the Chinese public began to show a marked increase in support for the assessment and mitigation of the environmental impacts of the dam.

The dam project necessitated the relocation of 1.24 million people, causing a number of social problems. Adding to the difficulties was the fact that some 12 percent of the money the government allocated to aid in the relocation efforts was embezzled. By 2009 more than three hundred officials had been convicted of corruption in relation to the relocation.

R. C. Lutz

FURTHER READING

Chetham, Deirdre. *Before the Deluge: The Vanishing World of the Yangtze's Three Gorges*. New York: Palgrave Macmillan, 2002.

Dai Qing, comp. *The River Dragon Has Come! The Three Gorges Dam and the Fate of China's Yangtze River and Its People*. Edited by John G. Thibodeau and Philip B. Williams. Armonk, N.Y.: M. E. Sharpe, 1998.

Heggelund, Gørild, ed. *Environment and Resettlement Politics in China: The Three Gorges Project*. Burlington, Vt.: Ashgate, 2004.

Hessler, Peter. "The Dam." In *River Town: Two Years on the Yangtze*. New York: Perennial/HarperCollins, 2002.

Lopez-Pujol, Jordi, and Ming-Xun Ren. "Biodiversity and the Three Gorges Reservoir: A Troubled Marriage." *Journal of Natural History* 43 (November, 2009): 2765-2786.

Stone, Richard. "Three Gorges Dam: Into the Unknown." *Science*, August 1, 2008, 628-632.

SEE ALSO: Asia; Dams and reservoirs; Ecosystems; Floods; Habitat destruction; Hydroelectricity; Public opinion and the environment.

Three Mile Island nuclear accident

CATEGORIES: Disasters; nuclear power and radiation

THE EVENT: Malfunction at the Metropolitan Edison nuclear power plant at Three Mile Island on the Susquehanna River near Harrisburg, Pennsylvania

DATE: March 28, 1979

SIGNIFICANCE: The accident at Three Mile Island exposed serious weaknesses in U.S. nuclear power plant operations, as well as in government oversight of the nuclear power industry, and prompted reforms intended to prevent such accidents in the future.

Metropolitan Edison's nuclear power plant at Three Mile Island (TMI) was designed with two pressurized water reactors, units 1 and 2, that generated electric power by boiling water into steam that spun the blades of a turbine generator. The heat to convert water to steam was produced by fission of uranium in the reactors' cores. These were submerged in water and encapsulated in a containment building 12 meters (40 feet high), with walls of steel 20 centime-

Three Mile Island, Pennsylvania

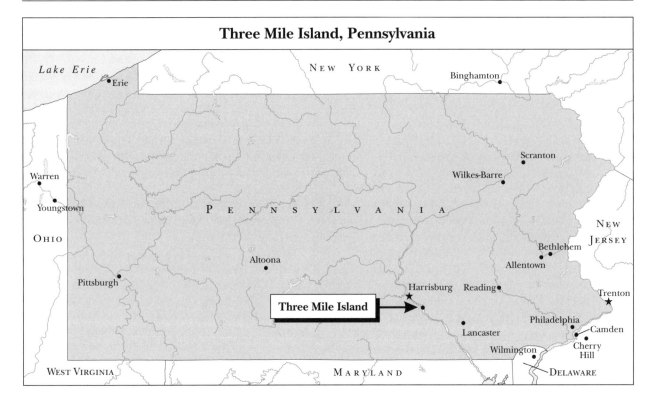

ters (8 inches) thick. Because the radioactive coolant water was under pressure, it could be superheated to 302 degrees Celsius (575 degrees Fahrenheit) without boiling. When it reached that temperature, it was pumped to a steam generator, where, in a secondary system and under less pressure, it heated cooler water to steam, which spun turbine blades and propelled a generator. The steam then passed through a condenser, where it changed back to water and began the circuit back to the steam generator.

The Accident

At 4:00 A.M. on Wednesday, March 28, a valve in the system inexplicably closed, interrupting the water supply to the steam system. The main water pumps automatically shut down, which decreased the steam pressure and shut down the steam turbine a few seconds later. This interrupted the transfer of heat from the reactor cooling system, where the pressure began to rise. A pressurizer relief valve opened, which reduced some of the pressure but also allowed radioactive water and steam to drain into a tank designed to hold excess water. The valve should have shut off after thirteen seconds; however, it remained open for more than two hours, during which time the primary coolant water continued to drain.

Less than one minute later, emergency backup pumps automatically engaged to maintain the water supply in the secondary system. No water was actually added, however, because two valves that controlled the flow had been closed for routine maintenance two days earlier. (Nuclear Regulatory Commission rules required that a plant be shut down if these valves remained closed for more than seventy-two hours.) Instead, two minutes into the crisis, as the temperature continued to rise and steam pressure declined, the emergency core coolant system began to operate, adding water to the reactor core.

There were no meters to measure the depth of water in the reactor core, but the technicians believed that sufficient water was present. To prevent the pressure in the primary cooling system from rising too high, they turned off one emergency pump and a few minutes later reduced the other one to half speed. This would have been proper procedure if the system had indeed been filled with water. In fact, the reactor core was not covered with water, and temperatures continued to rise.

At eight and one-half minutes into the crisis, technicians opened valves to fill the secondary system with water and draw heat away from the primary system. With the relief valve stuck open, however, the primary cooling water was still draining into the excess water tank, which overflowed and spilled its radioactive contents onto the containment building's floor. This activated suction pumps, which removed the water to a tank in the nearby auxiliary building. This tank too overflowed, and at 4:38 A.M., radioactive gases began to be released into the atmosphere.

By this time, the fuel rods in the reactor, which should have been covered with water, had become exposed. With no cooling system in operation, temperatures had continued to rise, leading to a partial melting of reactor fuel; the zirconium shields around the rods reacted with the steam and released radioactive debris and hydrogen, which collected in the containment building. At 6:50 A.M., a general emergency was declared.

Early Wednesday afternoon, some of the hydrogen in the containment building exploded. Hydrogen continued to be created by the exposed fuel rods, giving rise to fears that the hydrogen bubble at the top of the reactor building could self-ignite and result in a meltdown. Controlled and uncontrolled radiation

leaks from the plant continued through March 28 and 29. Lack of information, poor communication among the numerous agencies involved, and some degree of sensationalist news reporting fueled mounting public alarm. On Friday, March 30, Governor Richard Thornburgh ordered an evacuation of pregnant women and small children living within five miles of the facility. Administrators considered ordering a general evacuation but feared that doing so might set off a panic and result in more injuries than it would prevent. Finally, on Sunday, April 1, when President Jimmy Carter visited the facility, it was announced that the hydrogen bubble had shrunk and no longer posed a danger.

EFFECTS

The Three Mile Island accident exposed weaknesses in U.S. nuclear power plant design, management, and operation; in U.S. emergency preparedness; and in the workings of the Nuclear Regulatory Commission (NRC). The matter was investigated by a presidential commission and congressional commit-

Aerial view of the Three Mile Island nuclear power plant. The small dome in the center is where the 1979 accident took place. (AP/Wide World Photos)

tees and internally in the nuclear industry. An immediate response to the event was the closing of seven reactors similar to those at TMI and a delay in restarting others that had been shut down for maintenance. The NRC, which was eventually completely restructured, placed a temporary moratorium on the licensing of all new nuclear reactors; several reactor projects were canceled outright. Other countries reassessed their nuclear industries; Japan closed one reactor and postponed restarting nine others. In addition, the event mobilized the campaign of opponents to nuclear power worldwide. Longer-term results included changes to the design and operation of all nuclear power plants in the United States.

The TMI accident prompted widespread reconsideration of nuclear power and a reassessment of its relatively low economic cost in light of its risks, which were tragically demonstrated by the 1986 nuclear disaster at Chernobyl. Perhaps the most lasting effect of the TMI crisis was on the general public's faith in industry and government representatives. The actions of Metropolitan Edison and government officials during the crisis made it clear that their first priority had been not the health and safety of the public but rather the safeguarding of their own and the plant's reputations. Projections about the potential results of a nuclear disaster and uncertainty about the long-term health effects of radiation contributed to growing public mistrust and increasing militancy on the part of opponents of the nuclear industry.

John R. Tate

FURTHER READING

Cantelon, Philip L., and Robert C. Williams. *Crisis Contained: The Department of Energy at Three Mile Island.* Carbondale: Southern Illinois University Press, 1982.

Del Tredici, Robert. *The People of Three Mile Island.* San Francisco: Sierra Club Books, 1980.

Gray, Mike, and Ira Rosen. *The Warning: Accident at Three Mile Island.* New York: W. W. Norton, 2003.

Osif, Bonnie A., Anthony J. Baratta, and Thomas W. Conkling. *TMI Twenty-five Years Later: The Three Mile Island Nuclear Power Plant Accident and Its Impact.* University Park: Pennsylvania State University Press, 2004.

Walker, J. Samuel. *Three Mile Island: A Nuclear Crisis in Historical Perspective.* Berkeley: University of California Press, 2004.

SEE ALSO: Antinuclear movement; Chalk River nuclear reactor explosion; Chelyabinsk nuclear waste explosion; Chernobyl nuclear accident; Hanford Nuclear Reservation; Nuclear accidents; Nuclear power; Nuclear regulatory policy; Power plants; Windscale radiation release.

Tidal energy

CATEGORY: Energy and energy use

DEFINITION: Power generated during the rise and fall of the tides

SIGNIFICANCE: The energy generated during the rise and fall of the tides may be cleanly and safely converted into electrical power, but large-scale tidal power installations can have severe consequences for the environment, including decimation of fisheries; destruction of the feeding grounds of migrating birds; damage to shellfish populations; interference with ship travel, port facilities, and recreational boating; and disruption of the tidal cycle over a wide area.

Tidal power projects can be important sources of local electricity generation because they produce energy that is free, clean, and renewable; they produce neither air pollution nor thermal pollution, and they do not consume exhaustible natural resources. Only a limited number of places in the world offer the potential for such power installations, however, because a vertical tidal rise of 5 meters (16.4 feet) or more is required. Installations must also be near major population centers so that transmission requirements are minimized, and a natural bay or river estuary is required to store a large amount of water with a minimum of expense for dam construction. The seawater impounded behind the dam at high tide produces a hydrostatic head so that electricity is generated as the water passes through the dam's turbines when sea level falls. If the turbines in the dam are reversible, power can be generated on both incoming and outgoing tides.

Tidal power plants have been constructed on the Rance River near St. Malo, France (240 megawatts of power), on the Annapolis River in Nova Scotia, Canada (20 megawatts), on the Yalu River in the People's Republic of China (3.2 megawatts), in Kislaya Guba, Russia (1.7 megawatts), and on Strangford Lough in

Northern Ireland (1.2 megawatts). In the first decades of the twenty-first century, South Korea began construction of a plant on Sihwa Lake (254 megawatts) and made plans to build several more around the country. The Rance River plant, which has been in continuous operation since November, 1966, was for many years the world's largest tidal power installation. It bridges the estuary with a dam nearly 0.8 kilometer (0.5 mile) long and provides power for 300,000 people.

The environmental impacts of the Rance River dam have generally been limited to the modification of fish species distributions, the disappearance of some sandbanks, and the creation of high-speed currents near the sluices and the powerhouse. Tidal patterns have also changed, with the maximum average rise reduced from about 13.4 meters to 12.8 meters (44 feet to 42 feet) and a corresponding increase in the height of the mean low-tide level.

The environmental impacts of the smaller Annapolis River plant in Nova Scotia, which became operational in 1984, reportedly have included the generation of silt, which destroyed clam beds in the basin behind the dam, and increased erosion of the river's banks. The Nova Scotia Power Corporation reached a settlement with one nearby landowner whose house suffered a cracked foundation and shifted toward the river as a result of erosion.

Several tidal power projects were proposed for the United States during the early and mid-twentieth century but were never built because of environmental concerns. A proposed tidal power plant on the upper Saint John River in Maine was halted, for example, because damming the river would have destroyed a unique stand of a rare wildflower. The flower was later found growing elsewhere. Objections cited for other projects included possible effects on historic and archaeological sites, as well as presumed economic and

Government officials visit the SeaGen tidal energy converter in Strangford Lough, Northern Ireland. The SeaGen converter, which began operating in 2008, is the world's first operating commercial-scale tidal stream turbine. (AP/Wide World Photos)

social impacts on Native American communities such as the Passamaquoddy.

Shortly after the dramatic jump in world oil prices during the 1970's, the Tidal Power Corporation, a venture owned by the Nova Scotia government, proposed building a major tidal power project in the Bay of Fundy, which lies between Nova Scotia and New Brunswick in eastern Canada. This plant would have been the world's largest tidal power installation, producing 4,560 megawatts of power—nearly twenty times the output of the Rance River plant and more than three times the output of Hoover Dam on the Colorado River in the United States. A major feature of the project was to be a dam 8.5 kilometers (5.3 miles) long across the Bay of Fundy, which has the largest tidal range in the world, averaging more than 15 meters (50 feet). The enormous scope of the project forced scientists to pay close attention to its anticipated environmental consequences, and these appeared to be so severe that the project was never begun.

Disrupted bird migrations were predicted after the dam's completion because of the submersion of tidal mudflats where large numbers of semipalmated sandpipers and other shorebirds annually gorge on mud shrimp before beginning their fall migrations to wintering grounds in South America and the Caribbean. Damage to fish stocks was also predicted because of repeated passage of the fish through the dam's turbines as the tides rose and fell. Particularly affected would have been the American shad, a member of the herring family, which migrates to the Bay of Fundy each year from as far away as Florida in order to fatten itself on mysid shrimp living on the tidal mudflats. Oceanographers also used computer modeling to show that dam construction would alter tidal patterns over a broad area, resulting in tidal levels 10 percent higher and lower as far south as Cape Cod, Massachusetts, 400 kilometers (250 miles) away. They predicted that these tidal changes would flood coastal lands and threaten roads, bridges, waterfront homes, water wells, sewage systems, salt marsh areas, harbors, and docking areas along the entire coast.

Donald W. Lovejoy

FURTHER READING

Charlier, R. H., and C. W. Finkl. *Ocean Energy: Tide and Tidal Power.* London: Springer, 2009.

Cruz, João, ed. *Ocean Wave Energy: Current Status and Future Perspectives.* New York: Springer, 2008.

Hardisty, Jack. *The Analysis of Tidal Stream Power.* New York: John Wiley & Sons, 2009.

McKinney, Michael L., Robert M. Schoch, and Logan Yonavjak. *Environmental Science: Systems and Solutions.* 4th ed. Sudbury, Mass.: Jones and Bartlett, 2007.

Peppas, Lynne. *Ocean, Tidal, and Wave Energy: Power from the Sea.* New York: Crabtree, 2008.

SEE ALSO: Alternative energy sources; Dams and reservoirs; Hydroelectricity; Power plants.

Times Beach, Missouri, evacuation

CATEGORIES: Disasters; human health and the environment

THE EVENT: Relocation of the entire population of the town of Times Beach because of the threat posed by dioxin contamination

DATE: 1983

SIGNIFICANCE: The events leading up to the evacuation of Times Beach and subsequent evaluations of the evacuation reflect the American public's concern with environmental safety issues and serve to highlight the difficult decisions that government agencies must address in working to protect the public from environmental dangers.

During the 1960's a chemical plant in Verona, Missouri, produced Agent Orange, a chemical compound that was used during the Vietnam War. A by-product of this production was a highly toxic compound called dioxin, which was stored in large tanks. Syntex Agribusiness purchased the chemical plant in 1969 and then leased part of it to Northeastern Pharmaceutical and Chemical Corporation (NEPACCO), which manufactured hexachlorophene. The production of hexachlorophene, a popular skin cleanser at the time, also created the by-product dioxin, which was added to the same large tanks previously used.

When Independent Petrochemical Corporation (IPC), one of NEPACCO's suppliers, hired Russell Bliss to get rid of the compounds being stored in the tanks, Bliss, who also had contracted with local towns to spray unpaved roads to keep down the dust in summer, mixed the stored wastes with recycled oil and used the mixture in spraying the roads. Later, NEPACCO paid Bliss to dispose of additional wastes,

Environmental Protection Agency technicians take deep soil samples from Times Beach, Missouri, in January, 1983, to examine them for possible dioxin contamination. (©Bettmann/CORBIS)

which he did by spraying more roads. In 1971 thousands of gallons of recycled oil contaminated with dioxin were sprayed onto the unpaved streets of Times Beach, Missouri, a small town in the southwestern part of the state. At the time, the toxicity of dioxin had not been publicized and was not widely known.

Soon after the roads were sprayed, problems began. Horses, dogs, cats, chickens, rodents, and birds perished. Within a few months, children became ill. By 1981 government officials began recommending that people vacate the area.

Because dioxin binds tightly to soil and degrades very slowly, high levels of dioxin remained even ten years after the spraying. Soil tests conducted at Times Beach in November, 1982, verified the presence of high levels of dioxin in the town. Dioxin levels in some parts of the Times Beach area reached 100 to 300 parts per billion; at that time, the U.S. Centers for Disease Control (CDC) believed that dioxin levels above 1 part per billion posed a potential risk to human health. The CDC recommended the temporary evacuation of Times Beach until more tests could be conducted. In February, 1983, using $36 million allocated under the Comprehensive Environmental Response, Compensation, and Liability Act, known as Superfund, the federal government purchased Times Beach and relocated the entire population, about 2,300 to 3,000 people.

AFTER THE EVACUATION

Disagreements and debates continued to surround the events at Times Beach. In 1991 Vernon Houk, a top CDC official, stated that given the latest research findings on dioxin, he believed that the CDC overreacted by evacuating the population of Times Beach; he asserted that the danger was not as great as had been thought. Although dioxin had been shown to cause chloracne, a serious skin disease, and had been related to cancer and birth defects in humans, subsequent studies showed exposure to dioxin to be not a very great cancer threat unless the exposure was unusually high. In 1992 Edward Bresnick, chairman of the independent dioxin review panel of the U.S. Environmental Protection Agency (EPA), also stated that the government had overresponded at Times Beach.

Reports downplaying the toxicity of dioxin surfaced; some stated that significant increases in disease had not been documented at Love Canal (another Superfund site located in the state of New York) or at Seveso, Italy, the site of an accidental release of dioxin-contaminated vapor in 1976. To evaluate the toxicity of dioxin further, the EPA began a series of studies in 1991 that, the agency hoped, would prove once and for all that dioxin contamination need not be a source of concern. In fact, the opposite was proved. The first studies discovered more, rather than fewer, problems with dioxin than had been noted previously. Researchers found dioxin to be particularly damaging to animals exposed while in utero and that the chemical affects behavior and learning ability and acts like a steroid hormone. They also found extensive effects of dioxin on the immune system.

The citizens of Times Beach brought hundreds of lawsuits against Syntex, the company many thought was responsible for the dioxin contamination. None of these lawsuits succeeded, however. A jury in St. Louis in 1988 rejected the cases of eight plaintiffs, citing a lack of medical evidence to support the claims.

The Superfund cleanup of Times Beach continued to generate debate. In a controversial decision, federal and state environmental officials decided to incinerate the contaminated soil of twenty-seven sites in Missouri, including Times Beach. Environmentalists feared the dangers of smokestack emissions and accidents and vehemently opposed the incineration of contaminated soil. Their concern was justified: On March 20, 1990, a power outage led to the discharge of thousands of pounds of dioxin-contaminated pollutants into the air. Although the risks associated with such pollution were unknown, two environmental activist groups filed a federal lawsuit. A third organization, the Dioxin Incinerator Response Group, also opposed to incineration, argued that a feasible alternative to burning the dioxin-contaminated soil would be to store it in capped drums.

In July, 1997, the Environmental and Natural Resources Division of the U.S. Department of Justice announced that the cleanup of the Times Beach, Missouri, Superfund site was complete and that the land was once again fit for human use. Begun in 1984, the cleanup cost an estimated $200 million. A 165-hectare (409-acre) state park, named Route 66 State Park for the historical road that runs through it, was developed at the site.

Louise Magoon

FURTHER READING

Edelstein, Michael R. *Contaminated Communities: Coping with Residential Toxic Exposure.* Rev. ed. Boulder, Colo.: Westview Press, 2004.

Mason, Jan. "The Ordeal of a Poisoned Town." *Life,* May, 1983, 58-62.

Schecter, Arnold, and Thomas Gasiewicz, eds. *Dioxins and Health.* 2d ed. New York: Taylor & Francis, 2002.

SEE ALSO: Agent Orange; Centers for Disease Control and Prevention; Chloracne; Dioxin; Environmental Protection Agency; Italian dioxin release; Soil contamination; Superfund.

Tipping points. *See* Positive feedback and tipping points

Tobago oil spill

CATEGORIES: Disasters; water and water pollution

THE EVENT: Collision of two oil tankers off the coast of Tobago in the Caribbean Sea that caused significant loss of human life and a large oil spill

DATE: July 19, 1979

SIGNIFICANCE: Because the oil carried by the tankers *Aegean Captain* and *Atlantic Empress* contained a high percentage of gasoline, the environmental damage of the spill was minimized.

On the evening of July 19, 1979, the weather off the northern tip of the Caribbean island of Tobago was rainy with gusty winds. The Liberian-registered *Aegean Captain*, weighing 210,257 tons, was bound from the Netherlands Antilles to Singapore with a cargo of transshipped Arabian crude oil. The *Atlantic Empress*, also Liberian-registered, was en route from the Persian Gulf to Beaumont, Texas, with a cargo of Arabian crude; this vessel was operating under charter to Mobil Oil. At 325 meters (1,066 feet) long and 292,666 tons, the *Atlantic Empress* was a very large vessel.

The two tankers were both using radar for collision avoidance, but, as they later reported, the pictures were "fuzzy" because of the rain. Each vessel was unaware of the other's presence until it was too late. They did not sight each other until they were approximately 183 meters (600 feet) apart. Because of their size, such vessels require miles to stop. The *Aegean Captain* was on an easterly heading and, upon sighting the *Atlantic Empress*, began a sharp left turn, although navigation rules allow for turns to the right only. The *Atlantic Empress* was northbound toward Texas, yet at the time of the collision it was on a southerly heading for unknown reasons. At 7:15 P.M., the two vessels collided 29 kilometers (18 miles) north of the northern tip of the island of Tobago.

The bow of the *Aegean Captain* struck the starboard (right) side of the *Atlantic Empress* and drove deep into the center of the other ship. The two vessels were locked together as fires broke out on both ships and oil began to spill into the sea. On the *Atlantic Empress* twenty-six crew members were killed in the collision; three crewmen died on the *Aegean Captain*. The captain of the *Aegean Captain* then backed his vessel away from the *Atlantic Empress* even though the *Atlantic Empress*'s captain had asked him not to, fearing his ship would sink.

Salvage tugs stationed nearby responded almost immediately to the collision. The *Atlantic Empress*, on fire and leaking oil badly, was taken in tow. The tugs applied for permission to bring the tanker into several local ports, where fighting the fires would have been more efficient, but in all cases permission was denied. The vessel was towed out into the Atlantic Ocean, where it was racked by several explosions and sank. The *Aegean Captain* remained afloat, and after the fires were put out and the oil leaks were stopped, it was taken to a local shipyard.

The fact that both vessels were carrying high-quality Arabian crude oil was actually a benefit in this case, as such oil is almost 25 percent gasoline. This meant that in the collision a large proportion of the oil burned rather than fouling local beaches and fishing areas. A large percentage of the remaining oil evaporated, and dispersants were used to treat the rest before it could come ashore. In total, 270,000 tons (2.14 million barrels) of oil were lost, and damage to the two vessels and the environment came to $54 million in insurance claims.

Robert J. Stewart

FURTHER READING

Fingas, Merv. *The Basics of Oil Spill Cleanup.* 2d ed. Boca Raton, Fla.: CRC Press, 2001.

Laws, Edward A. "Oil Pollution." In *Aquatic Pollution: An Introductory Text.* 3d ed. New York: John Wiley & Sons, 2000.

SEE ALSO: *Amoco Cadiz* oil spill; *Argo Merchant* oil spill; *Braer* oil spill; *Exxon Valdez* oil spill; Oil spills; *Sea Empress* oil spill; *Torrey Canyon* oil spill.

Torrey Canyon oil spill

CATEGORIES: Disasters; water and water pollution

THE EVENT: Grounding of the oil tanker *Torrey Canyon* off the coast of England, resulting in the spilling of tons of crude oil into the sea

DATE: March 18, 1967

SIGNIFICANCE: The *Torrey Canyon* spill resulted in the deaths of some 75,000 seabirds and countless other marine animals. Attempts to remove the oil that was deposited on beaches in England and France also caused great environmental harm, as many of the detergents used were toxic.

The *Torrey Canyon* was built at Newport News shipbuilding yards in Virginia in 1959 and was modified at Sasebo, Japan, in 1964. The ship was 297 meters (974 feet) long, 38 meters (125 feet) wide, and had a draft of 16 meters (52 feet), with eighteen cargo tanks capable of carrying 120,000 tons of oil. The vessel was owned by Union Oil Company of Los Angeles, California. It was chartered to British Petroleum and flew the Liberian flag while carrying predominantly Italian officers and crew.

When the *Torrey Canyon* departed Mina ala Ahmadi in the Persian Gulf in early 1967, it was loaded with 119,193 tons of Kuwaiti crude oil. It was bound for Milford Haven in Wales. The last leg of the voyage was to go past the Scilly Isles and into the Bristol Channel. The Scilly Isles lie about 34 kilometers (21 miles) off Land's End along the Cornwall coast of England. The Seven Stones Shoal lies in between the Scilly Isles and Land's End. The *Torrey Canyon* normally passed outside the Scilly Isles. On this voyage, however, the captain chose to pass between the islands and the shoal. The vessel ran aground on the Seven Stones Shoal on March 18.

Once the *Torrey Canyon* was aground, it began leaking oil. A salvage effort was quickly undertaken to try to refloat the vessel. At the same time, detergent was sprayed over the spilled oil to help disperse it. After one week of work, the vessel was broken into pieces while salvagers were attempting to tow it off the shoal. By this time the oil had blanketed the sea within an 80-kilometer (50-mile) radius of the vessel; on some of the beaches where it had begun to wash ashore, the oil was 46 centimeters (18 inches) thick.

The salvagers were not equipped to deal with the volume of oil spilled. After twelve days of salvage efforts, the British Royal Air Force began to drop explosive bombs and incendiary devices around the vessel in an attempt to burn the oil. Some small fires were started, but this technique failed. The oil continued to spread over the coasts of England and France. Both governments called in troops to help remove oil from the beaches. Pump trucks and boats were used to pump oil off the beaches, and bulldozers and other heavy equipment were used to remove contaminated sand and rock. The largest single environmental problem that arose was caused by the use of detergents to remove oil from rocky and hard-to-reach places on the beaches in both England and France. Many of these detergents were toxic and killed everything in their path.

Many effects from the spill continued to be felt long after the oil was gone. An estimated 75,000 seabirds died as a result of the spill, as did unknown numbers of other animals of all types. Some of these deaths were directly caused by the effects of the oil, while others were caused by the detergents and heavy equipment used to remove the oil.

Robert J. Stewart

FURTHER READING

Clark, R. B. *Marine Pollution.* 5th ed. New York: Oxford University Press, 2001.
Fingas, Merv. *The Basics of Oil Spill Cleanup.* 2d ed. Boca Raton, Fla.: CRC Press, 2001.

SEE ALSO: *Amoco Cadiz* oil spill; *Argo Merchant* oil spill; *Braer* oil spill; *Exxon Valdez* oil spill; Oil spills; *Sea Empress* oil spill; Tobago oil spill.

The tanker Torrey Canyon *broke in two after running aground near Land's End, England.* (AP/Wide World Photos)

Tragedy of the commons

CATEGORY: Resources and resource management

DEFINITION: Situation in which the rational choices of individuals, acting independently and solely in their own self-interest, collide with the interests and needs of the larger community, resulting in the depletion of resources against the long-term interests of both individuals and the group

SIGNIFICANCE: Environmentalists, conservationists, and others concerned with the depletion of the world's shared natural resources have developed various approaches to averting the situation known as the tragedy of the commons.

Although the concept of the tragedy of the commons can be traced back to Aristotle, the modern application of the concept is closely associated with ecologist Garrett Hardin, who published an article about it in 1968. The commons can be defined as any resource to which all persons have open, free, and unrestrained access. Examples of such shared resources include the atmosphere, rain forests, outer space, oceans, fisheries, and public land. The tragedy of the commons is a situation in which each individual makes a rational decision concerning a shared resource based on his or her own best interest. For example, if farmers share a plot of land where they graze their cattle, individual farmers could increase their own individual profits by grazing additional animals. They would gain the sole benefit of each additional animal in their herds and only bear a fraction of the cost. When all the farmers sharing the land choose to increase their individual profits in this way, however, they harm the common resource by overgrazing the land and compacting the soil.

Although the tragedy of the commons is a rational choice for individuals, it can be overcome. Sometimes averting the tragedy can be as simple as educating individuals about their behavior and appealing to their consciences. Some commentators place faith in technology as a means to overcome the tragedy, asserting that scientific advancements can repair or counterbalance any potential harm to the commons. Another way to prevent people from overusing common resources is to create economic incentives: When individuals must pay taxes or fees to use a common resource, they may be inclined to reduce their use of that resource. For some commons, such as the seas, limits can be placed on the numbers or amounts of given resources, such as fish, that can be harvested within a given period. The use of taxes, fees, and quotas to avert the tragedy of the commons requires that someone or something, often a government, exists that will monitor the compliance of individuals.

Another way to prevent the tragedy of the commons from occurring is to privatize the resource, so that it is no longer a commons. If a farmer owns the land on which he grazes cattle, he still has to make choices about the number of cattle that should be grazed in particular areas. If the number of cattle is increased beyond a certain point, the carrying capacity of the land, then the land will be harmed. Each additional animal represents a benefit to the farmer, but because the farmer owns the land he bears the entire cost as well. Rational individuals will not compromise their own private resources.

Kathryn A. Cochran

FURTHER READING

Easton, Thomas A., ed. *Environmental Studies*. 3d ed. New York: McGraw-Hill, 2009.

Hanley, Nick, and Colin J. Roberts, eds. *Issues in Environmental Economics*. Malden, Mass.: Blackwell, 2002.

Hardin, Garrett. "The Tragedy of the Commons." *Science* 162 (December 13, 1968): 1243-1248.

Stavins, Robert N., ed. *Economics of the Environment: Selected Readings*. New York: W. W. Norton, 2000.

SEE ALSO: Environmental economics; Free market environmentalism; Hardin, Garrett; Land-use policy; Ostrom, Elinor; Privatization movements; Range management.

Trans-Alaska Pipeline

CATEGORY: Energy and energy use
IDENTIFICATION: Pipeline built to transport oil across Alaska from Prudhoe Bay to Valdez
DATE: Completed in July, 1977
SIGNIFICANCE: The plan to construct an oil pipeline across Alaska presented many technical challenges that exemplify the potential conflicts between supplying energy for human needs and protecting the environment.

In December, 1967, oil was first discovered during test drilling at Prudhoe Bay on the North Slope of Alaska. It soon became evident that this was the largest petroleum field in the United States. However, transporting the oil from Prudhoe Bay to a port at Valdez, on the southern coast of Alaska, would require the construction of a pipeline that would traverse 1,000 kilometers (620 miles) of federal land. The U.S. Geological Survey (USGS) was assigned the task of conducting an environmental impact assessment for the proposed project.

After an exhaustive investigation, the federal agency recommended against construction on the trans-Alaska route. Its objections were technical, geological, and ecological. The route crossed difficult terrain in an Arctic region where local environmental damage could be severe and long lasting. The geological hazards included active earthquake fault zones in southern Alaska, mountains (the Brooks Range in the north and the Alaska Range in the south), thirty rivers (many of which flood periodically), and unstable soil and permafrost (permanently frozen soil). Construction, subsequent pipeline operation, and potential accidental rupture or spillage would disturb the ground, water, fragile vegetation, and wildlife, including migratory routes of land species such as caribou.

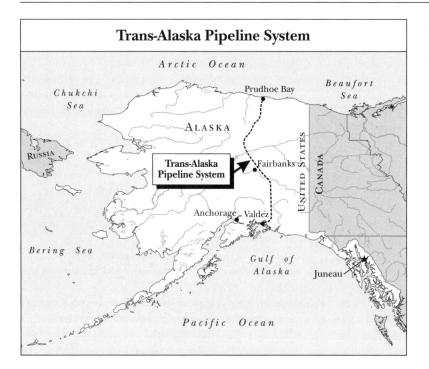

Trans-Alaska Pipeline System

Valdez. The first supertanker was filled with oil from the pipeline at Valdez in August, 1977—almost ten years after the North Slope discovery was made. The pipeline can deliver oil at a rate of 1.6 million barrels per day. The oil revenues for Alaska from state taxes have enabled the state to abolish its personal income tax and to distribute substantial annual cash dividends to all state residents.

Because the Arctic ecology, with its fragile plant and animal life and slow recovery, is particularly susceptible to damage from pipeline breaks and crude oil leaks, efforts were made during construction and have been made in subsequent pipeline monitoring to minimize any spillage. Despite these attempts, some incidents have occurred, a few caused by acts of vandalism. In June, 1981, a valve ruptured and spilled some 5,000 barrels of oil onto the soil. Corrosion in feeder pipelines caused a spill of some 6,000 barrels of oil in March, 2006. By far the greatest environmental and ecological damage related to the pipeline, however, although not caused by it directly, occurred in March, 1989, when the *Exxon Valdez* supertanker, loaded with more than 1.2 million barrels of crude oil from the pipeline, ran aground in Prince William Sound and spilled more than 240,000 barrels (10 million U.S. gallons) of oil into the water. This was one of the worst environmental disasters in the history of the United States, not only for the extent of the contamination and its impact on wildlife and fishing but also because of the remote location and Arctic climate, which made cleanup and reclamation particularly difficult.

Robert S. Carmichael

In addition, transporting the oil southward from Valdez would require oil tankers to face the hazards of docking in the Arctic and the possibility of spills and pollution. The USGS favored a longer inland pipeline route through Canada to refineries in Chicago, Illinois. Others, however, wanted to keep the construction and its economic benefits within Alaska and within U.S. territory. The USGS recommendation was overruled, and the U.S. Congress exempted the pipeline project from the law requiring a favorable environmental impact statement before work could begin.

Construction on the pipeline, which was projected to cost $900 million, was started in April, 1974, by a consortium of eight major oil companies named the Alyeska Pipeline Company. By the time the pipeline was completed in July, 1977, costs had reached nearly $8 billion, making it the most expensive privately financed construction project in history. Part of the overrun was caused by redesign and construction techniques adopted to minimize environmental impact.

The pipeline, which is 1.2 meters (4 feet) in diameter, extends 1,300 kilometers (800 miles) from Prudhoe Bay across Alaska to Valdez Arm, an inlet off Prince William Sound on the Pacific coast. The terminus of the pipeline is across the inlet from the town of

FURTHER READING

Coates, Peter A. *The Trans-Alaska Pipeline Controversy: Technology, Conservation, and the Frontier.* Bethlehem, Pa.: Lehigh University Press, 1991.

McBeath, Jerry. "Oil Transportation Infrastructure: The Trans-Alaska Pipeline System and the Challenge of Environmental Change." In *Smart Growth and Climate Change: Regional Development, Infrastructure, and Adaptation*, edited by Matthias Ruth. Northampton, Mass.: Edward Elgar, 2006.

Ross, Ken. *Environmental Conflict in Alaska*. Boulder: University Press of Colorado, 2000.

See also: Environmental impact assessments and statements; *Exxon Valdez* oil spill; Fossil fuels; Oil drilling; Oil spills; Trans-Siberian Pipeline.

Trans-Siberian Pipeline

Category: Energy and energy use

Identification: Pipeline built to carry natural gas from the Siberian Urengoy gas field to Western Europe

Date: Completed in 1984

Significance: The building of the Trans-Siberian Pipeline through formerly pristine and fragile ecosystems resulted in serious environmental damage to those systems, and environmentalists continue to be concerned regarding the quality of the ongoing maintenance of the pipeline system.

After World War II, Europe became increasingly dependent on the oil-producing countries of the Middle East for its fuel needs. However, the Arab-Israeli conflicts of 1973 alarmed European countries and prompted them to seek other sources of energy. One alternative was to utilize larger amounts of natural gas; the use of this form of fuel more than doubled during the late 1970's. Natural gas is cheaper than oil or coal, but transporting it requires enormous initial investments of capital and technology because pipelines are the only practical means of conveyance. In such pipelines, the natural gas is moved along the pipes in liquid form by the use of strategically located compressing stations.

In 1980 West German chancellor Helmut Schmidt, after a visit to Moscow, announced a plan for the Soviets to build a Euro-Siberian gas pipeline. It would start at Urengoy, east of the Ural Mountains and near the Arctic Circle, and extend westward to West Germany, France, and Italy. Objections to building the pipeline came largely from Europe's North Atlantic Treaty Organization (NATO) ally, the United States. It was feared that if an armed conflict were to break out, the Soviets would have an undue advantage over the Western Europeans.

The plan called for a 5,790-kilometer (3,600-mile) pipeline, four times the length of the controversial Trans-Alaska Pipeline, which had just been completed. Several small cities were built to accommodate the construction workers. Preparations for extraction and construction of the pipeline, which began in the early 1980's, required technical assistance, materials, and personnel provided by the potential European customer nations. In addition to West Germany, the largest purchaser, other countries were Austria, Belgium, France, Italy, the Netherlands, and Switzerland. Construction of the pipeline began in 1982 and was completed in 1984, with a finished length of 4,500 kilometers (2,800 miles). Despite the demise of the Soviet Union in 1991, the pipeline has operated for the most part without adverse economic or political consequences.

Although successful as a political and commercial venture, the Trans-Siberian Pipeline deserves attention because of its environmental impact. The resource-rich region of Siberia includes vast areas of tundra, taiga, and other fragile ecosystems, many of them in relatively pristine condition. As a result of the extraction process, the building of roads, and related activities, significant damage occurred to the environment that will require centuries for nature to correct. Many of these environmental concerns were compounded by the collapse of the Soviet Union and the resulting economic chaos. Given the new Russian republic's need of cash and weak environmental laws, some observers expressed concern that the pipeline would not be properly maintained, thus creating the potential for considerable environmental damage. Several sections of the Trans-Siberian Pipeline have burst over its years of operation, including a major oil seepage that occurred near the port of Archangel in 1994. In 2007 two explosions occurred at different places along the pipeline.

Thomas E. Hemmerly

Further Reading

Miller, G. Tyler, Jr., and Scott Spoolman. "Nonrenewable Energy." In *Living in the Environment: Principles, Connections, and Solutions*. 16th ed. Belmont, Calif.: Brooks/Cole, 2009.

Tusiani, Michael D., and Gordon Shearer. *LNG: A Nontechnical Guide*. Tulsa, Okla.: PennWell Books, 2007.

See also: Fossil fuels; Oil crises and oil embargoes; Trans-Alaska Pipeline.

Turtle excluder devices

CATEGORY: Animals and endangered species

DEFINITION: Grids inserted into shrimp trawl nets for the purpose of allowing sea turtles to escape

SIGNIFICANCE: Many shrimpers object to U.S. regulations requiring them to use devices in their nets for the protection of sea turtles, asserting that the practice hinders their ability to harvest shrimp. Some also argue that the regulations have resulted in the growth of shrimp farming, an industry responsible for a number of negative environmental impacts.

Each year some fifty thousand sea turtles are accidentally trapped in drift nets on shrimp boats, and an estimated eleven thousand of them die. The large majority of those turtles killed are loggerheads, with a small percentage being Kemp's ridleys. As sea turtles must go to the surface to breathe approximately every fifty minutes, shrimp boats trawling for more than one hour will drown sea turtles that are caught in their nets. When fitted properly with a turtle excluder device (TED), a shrimp boat net will catch only shrimp and small, nontarget fish of similar size. Larger marine life such as sea turtles, sharks, and large fish can escape through a hatch near the end of the net. A shrimp boat without a TED generally dumps 12 pounds of dead and useless bycatch overboard for each pound of harvested shrimp. The National Marine Fisheries Service (NMFS), a division of the U.S. Department of Commerce, requires TEDs on commercial shrimp boats operated out of the United States. TEDs are used during sea turtle nesting months (April to September) along the southeastern Atlantic coast and the Gulf of Mexico.

Many commercial fishers have protested the requirement that they use TEDs, claiming that they lose 20 percent of their catches because shrimp are dumped through the escape hatches. The situation is comparable to the earlier conflict between tuna fishers and advocates of dolphin protection. Citing 97 percent efficiency, the government has maintained strict regulations about the mandatory use of TEDs. The NMFS, noting that tuna fishers eventually accli-

A sea turtle escapes from a net equipped with a turtle excluder device. (NOAA)

mated to the use of dolphin-safe nets, requires shrimpers to modify their trawl nets to prevent a high mortality rate of sea turtles. Some environmentalists argue that shrimp boats using TEDs are more efficient and economical because they reduce fuel costs by excluding the drag and excess weight of large sea turtles and other marine animals and fish. TEDs additionally prevent shrimp from being crushed or damaged by the weight of sea turtles pushing against them in the nets.

The NMFS requirement that all shrimpers use TEDs went into effect in May, 1988, and in 1989 the U.S. government passed legislation banning the importation of shrimp from any country that does not use equipment on its shrimp boats to prevent turtle drownings. A major supporter of this legislation was the Center for Marine Conservation, one of the world's largest nonprofit agencies devoted to the protection of marine animals, plants, habitats, and resources. Earth Island Institute also led the way in convincing the federal government to enforce the ban by asking politicians to make saving sea turtles an international effort.

The countries most widely affected by U.S. TED laws are Belize, Colombia, Costa Rica, French Guiana, Guatemala, Honduras, Mexico, Nicaragua, Panama, Trinidad and Tobago, and Venezuela. All of these countries have agreed to comply with the TED laws and have also promoted shrimp farming as an alternative to ocean trawling. The growing shrimp agriculture industry, however, has sacrificed forest and freshwater supplies while depositing hazardous wastes and chemicals into ecosystems. Many shrimpers in the United States have protested the importation of the inexpensive shrimp produced by shrimp farming.

Dale F. Burnside with Welland D. Burnside

FURTHER READING

Kaczka, David. "Use of Turtle Excluder Devices to Save Sea Turtles Around the World." In *Foundations of Natural Resources Policy and Management,* edited by Tim W. Clark, Andrew R. Willard, and Christina M. Cromley. New Haven, Conn.: Yale University Press, 2000.

Kennelly, Steven J., ed. *By-catch Reduction in the World's Fisheries.* New York: Springer, 2007.

Safina, Carl. *Voyage of the Turtle: In Pursuit of the Earth's Last Dinosaur.* New York: Henry Holt, 2007.

SEE ALSO: Aquaculture; Commercial fishing; Dolphin-safe tuna; Gill nets and drift nets; Marine Mammal Protection Act.

U

UNESCO. *See* **United Nations Educational, Scientific, and Cultural Organization**

Union of Concerned Scientists

CATEGORIES: Organizations and agencies; activism and advocacy; human health and the environment

IDENTIFICATION: American nonprofit organization devoted to advocacy for science and opposition to the misuse of science and technology

DATE: Established in 1969

SIGNIFICANCE: The Union of Concerned Scientists has raised public awareness and influenced policy makers regarding many environmental issues with international impact, including global climate change, nuclear arms control, and agricultural biotechnology.

The Union of Concerned Scientists (UCS) was created during the Vietnam War era when faculty and students at the Massachusetts Institute of Technology (MIT) organized to call for the application of scientific research to pressing environmental and social problems rather than to military programs. Months of planning and preparation by eminent MIT scientists led to a voluntary research stoppage and day of education and discussion on March 4, 1969. News of the planned meetings at MIT spread to other universities, and meetings and protests took place on campuses elsewhere that same day.

UCS has since expanded its interests beyond protesting military research, and its membership has grown to become an alliance of scientists and citizens across the United States; by 2008 the organization had more than 78,000 members. It is headquartered in Cambridge, Massachusetts, and has offices in Washington, D.C.; Berkeley, California; and Chicago, Illinois.

UCS combines scientific research with legislative advocacy, grassroots organizing, and media coverage aimed at promoting a safer and healthier environment. A core group of researchers and policy experts collaborates with others to provide credible scientific information for use in citizen advocacy, consumer education, and expert testimony. The organization helps to coordinate and lobby for diverse projects on such issues and concerns as nuclear power safety, renewable energy sources, arms control, global climate change, international family-planning programs, agricultural biotechnology safety, clean vehicles, sustainable transportation policies, and scientific integrity.

UCS advocacy has included support for a plan that would advance the prospects for renewable energy under electric utility deregulation by increasing requirements for dollars spent on research for renewable energy sources by utilities and establishing a trust fund to support this research and development. UCS research and advocacy have also supported efforts to reduce reliance on single-passenger cars and provide incentives for the manufacture and purchase of cleaner vehicles.

U.S. Environmental Protection Agency policy regarding genetically engineered foods was influenced significantly by the book *The Ecological Risks of Engineered Crops* (1996), a revised and expanded version of a UCS report authored by Jane Rissler and Margaret Mellon. UCS has also worked on the issue of family planning on an international level and has helped convince the U.S. Congress to restore funding for family-planning programs. In addition, UCS has waged campaigns to pressure Congress and the president of the United States to take part in such international agreements as the Chemical Weapons Convention, the Comprehensive Nuclear-Test-Ban Treaty, and the Strategic Arms Reduction Treaty.

In 1992, a few months after the Earth Summit was held in Rio de Janeiro, Brazil, UCS issued a document titled "World Scientists' Warning to Humanity." More than 1,700 scientists signed the declaration, which calls for scientists, governments, business and industry, and citizens to work cooperatively to minimize carbon dioxide emissions and other stresses that human activity places on the ecosphere. In 1997 UCS organized the Science Summit on Climate Change, an event that clarified a growing consensus among scientists about global warming. This summit produced "World Scientists' Call for Action," a petition signed by 1,586 scientists (including 110 Nobel laureates) from sixty-three countries. The petition, which calls for an effective international climate treaty, helped shape the Kyoto Protocol. UCS has since issued a number of publications on greenhouse gas emissions, the possible impacts of projected climate change, and recommended emission reduction efforts.

During President George W. Bush's administration (2000-2008), UCS published reports claiming that scientific findings within several U.S. government agencies were being manipulated, suppressed, and distorted. UCS shared its recommendations for reversing the politicization of science with President Barack Obama's transition team, and in early 2009 the new president directed the Office of Science and Technology Policy to create a plan to ensure scientific integrity throughout the executive branch.

Anne Statham
Updated by Karen N. Kähler

FURTHER READING

Cleetus, Rachel, Steven Clemmer, and David Friedman. *Climate 2030: National Blueprint for a Clean Energy Economy.* Cambridge, Mass.: Union of Concerned Scientists, 2009.

Kendall, Henry. "Union of Concerned Scientists' Warning." In *Life Stories: World-Renowned Scientists Reflect on Their Lives and the Future of Life on Earth,* edited by Heather Newbold. Berkeley: University of California Press, 2000.

Moore, Kelly. *Disrupting Science: Social Movements, American Scientists, and the Politics of the Military, 1945-1975.* Princeton, N.J.: Princeton University Press, 2008.

Rissler, Jane, and Margaret Mellon. *The Ecological Risks of Engineered Crops.* Cambridge, Mass.: MIT Press, 1996.

SEE ALSO: Antinuclear movement; Earth Summit; Global warming; Kyoto Protocol; Population-control movement; SANE.

United Nations Commission on Sustainable Development

CATEGORIES: Organizations and agencies; resources and resource management
IDENTIFICATION: International organization created to monitor progress in the implementation of Agenda 21 as well as activities related to the integration of environmental and sustainable development goals
DATE: Established in December, 1992
SIGNIFICANCE: The United Nations Commission on Sustainable Development ensures that the issue of sustainable development has high visibility within the United Nations system and also helps to improve the coordination of United Nations activities related to the environment and development.

The United Nations Commission on Sustainable Development (CSD) was established by a United Nations General Assembly resolution in December, 1992, as a functional commission of the U.N. Economic and Social Council following the U.N. Conference on Environment and Development (UNCED), widely known as the Earth Summit, which was held in Rio de Janeiro, Brazil, in June, 1992. The Earth Summit resulted in the development of the plan of action called Agenda 21. As Agenda 21 was under preparation, participants in the summit discussed how the implementation of Agenda 21 programs should be monitored and agreed that a new commission was needed to oversee such monitoring. Chapter 38 identifies the creation of the Commission on Sustainable Development. The CSD has fifty-three member states, and about one-third of its members are elected on a yearly basis.

The CSD is charged with the implementation of the recommendations of Agenda 21, a landmark global agreement and plan of action that represents a comprehensive approach to addressing issues related to the environment and development. More specifically, the CSD is in charge of monitoring progress in the implementation of Agenda 21 as well as activities related to the integration of goals regarding environmental protection and sustainable development in the respective programs of governments, nongovernmental organizations, and United Nations agencies. The CSD is responsible for monitoring the implementation of Agenda 21, recommending strategies for overcoming barriers to achieving the goals of Agenda 21, addressing difficult and complex sustainable development issues, and charting new courses of action based on assessments of what works. The commission does not have authority to develop multilateral agreements.

The CSD meets annually in New York City. At each session, the CSD elects a bureau comprising a chair and four vice chairs. The CSD's multiyear activity program focuses on groups of specific thematic and sectoral issues related to sustainable development. The commission's major program areas include transport, chemicals, waste management and mining, and sustainable consumption and production patterns. Among the aims of CSD programs are poverty eradication, protection and management of the natural resources that serve as the base for social and economic development, and promotion of efforts to end unsustainable patterns of consumption and production.

As a result of the unprecedented involvement of a highly diverse group of stakeholders at the Earth Summit and the realization that numerous stakeholders play important roles in the implementation of Agenda 21, the CSD has opened its sessions to a broad range of stakeholders from governmental and nongovernmental agencies alike. The CSD is one of very few United Nations commissions that provides such opportunities for various stakeholders to be involved in its programs.

Lakhdar Boukerrou

FURTHER READING

Dodds, Felix, ed. *The Way Forward: Beyond Agenda 21.* London: Earthscan, 1997.

Landon, Megan. *Environment, Health, and Sustainable Development.* New York: Open University Press, 2006.

Robinson, Nicholas A. *Strategies Toward Sustainable Development: Implementing Agenda 21.* Dobbs Ferry, N.Y.: Oxford University Press, 2005.

Rogers, Peter P., Kazi F. Jalal, and John A. Boyd. *An Introduction to Sustainable Development.* Sterling, Va.: Earthscan, 2008.

Speth, James Gustave, and Peter M. Haas. *Global Environmental Governance.* Washington, D.C.: Island Press, 2006.

World Bank. *Advancing Sustainable Development: The World Bank and Agenda 21.* Washington, D.C.: Author, 1997.

SEE ALSO: Agenda 21; Earth Summit; Global Environment Facility; Sustainable development; United Nations Conference on the Human Environment; United Nations Environment Programme; World Summit on Sustainable Development.

United Nations Conference on Environment and Development (1992). *See* Earth Summit

United Nations Conference on the Human Environment

CATEGORIES: Ecology and ecosystems; resources and resource management

THE EVENT: Intergovernmental conference dedicated to the discussion of environmental issues

DATES: June 5-16, 1972

SIGNIFICANCE: The United Nations Conference on the Human Environment was the first international conference held to address global environmental issues. The proceedings helped to raise awareness around the world of the importance of international cooperation in the conservation of natural resources.

In 1967 Sweden began taking steps toward calling an international conference to discuss global environmental issues. The United Nations Conference on the Human Environment, also known as the Stockholm Conference, was convened by the General Assembly of the United Nations in Stockholm on June 5, 1972. In addition to being the first international conference dedicated to environmental issues, it was the first international conference ever held to focus on a single issue. The Stockholm Conference marked the beginning of coordinated global action to manage environmental issues. Delegates from 114 nations attended; however, the Soviet bloc nations did not participate. The United States played a lead role, as it was itself passing, for the first time, environmental laws with regulatory power to enforce them. In addition, representatives of 250 nongovernmental organizations

(NGOs) and 1,500 journalists attended the conference.

The conference itself was preceded by lengthy preparatory meetings at which delegates from the participating nations reached consensus on many issues, leaving a lesser number to be addressed at the conference. Most of the proposals written in advance of the conference emphasized the necessity for the universal involvement of every nation, and the principle of universality was implied in the conference proceedings. A series of regional meetings were held in Asia, Africa, and Latin America so that the conference participants would have a broad view of environmental problems and concerns. Among the considerations was the need to provide countries with scientific, technological, and planning information so that they could effectively address their national environmental problems.

The nations participating in the Stockholm Conference submitted papers describing how they were managing their own national environments. During the conference, differences of opinion between Northern Hemisphere countries and Southern Hemisphere countries surfaced. These differences continue to arise frequently when the global aspects of the environment are discussed in international forums. In Stockholm, representatives of developing countries in the Southern Hemisphere declared that environmental protection is an issue of interest to the wealthy, developed countries, not to the poor. They stated that their most important concern, economic development, outweighs their environmental concerns, and they asserted that the developed nations use issues related to the environment to retard poor countries' development. The developed Northern countries, in return, attempted to minimize the effects of many of the demands of the developing nations.

One economic issue that developing nations emphasized at Stockholm was the transfer of technology. They insisted that the world's advanced market economies should provide environmentally related technologies free of cost to nations with developing economies. A major part of this dispute was the demand by the developing states for access to biotechnology that uses species from the tropical forests, since such forests are found mainly in thirteen developing states. The final statement released by the conference minimized this demand, however. Another objection raised by the developing countries stemmed from

their fear that the proposed global environmental monitoring satellite system would be a means by which developed countries, especially the United States, would spy on them and their resources.

Two major documents resulted from the U.N. Conference on the Human Environment: the Stockholm Declaration and the Stockholm Action Plan for the Environment. The declaration was the first document in which the states agreed on the need for global action to protect the world's environmental assets. It lists twenty-six broad principles for managing the global environment, preceded by a preamble that expresses the urgency with which states must act, the magnitude of the task before them, and the complexity of the job of preserving the environment in which the world's population lives. According to the United Nations, more than 90 percent of the population of the world was represented at the Stockholm Conference.

The points on which the delegates had agreed were put into the Stockholm Action Plan for the Environment, which contains 109 recommendations for global cooperation on environmental issues and 150 proposals for action. The intent of the plan was to define and mobilize states in a "common effort for the preservation and improvement of the human environment." The major recommendations included a call for four U.N. conferences to be convened, one each on the topics of environmental education, human settlements, protection of the world's cultural and natural heritage, and international trade in endangered species. Also recommended was the establishment of an international fund to be used as seed capital for housing improvement, coordination of programs for integrated pest control, and the establishment of an international registry of data on chemicals in the environment, an international referral service, and a global atmospheric monitoring system.

The United Nations was given the mandate to provide leadership in carrying out the recommendations. In December, 1972, the United Nations Environment Programme (UNEP) was created to focus on and coordinate the environmentally related activities in which the U.N. system would be involved. One commitment made by the delegates in Stockholm was to regulate the trade in waste products. However, it was not until 1984-1985 that UNEP was able to get agreement for such regulation. Twenty years later, the 1992 United Nations Earth Summit was held as a follow-up to Stockholm to assess progress since 1972 and to identify and address environmental issues that had become important in the changed political, economic, and environmental situation of the 1990's.

Colleen M. Driscoll

FURTHER READING

Bodansky, Daniel. *The Art and Craft of International Environmental Law.* Cambridge, Mass.: Harvard University Press, 2010.

Clapp, Jennifer, and Peter Dauvergne. *Paths to a Green World: The Political Economy of the Global Environment.* Cambridge, Mass.: MIT Press, 2005.

Conca, Ken, and Geoffrey D. Dabelko, eds. *Green Planet Blues: Environmental Politics from Stockholm to Johannesburg.* 3d ed. Boulder, Colo.: Westview Press, 2004.

Speth, James Gustave, and Peter M. Haas. "From Stockholm to Johannesburg: First Attempt at Global Environmental Governance." In *Global Environmental Governance.* Washington, D.C.: Island Press, 2006.

SEE ALSO: Biodiversity; Earth Summit; United Nations Environment Programme; United Nations population conferences.

United Nations Convention on the Law of the Sea

CATEGORIES: Treaties, laws, and court cases; resources and resource management; water and water pollution

THE CONVENTION: International agreement regarding all matters pertaining to international waters

DATE: Opened for signature on December 10, 1982

SIGNIFICANCE: The United Nations Convention on the Law of the Sea provides a legal framework over the world's oceans designed to protect the seas and clarify the responsibilities of signatory nations; however, it has not been endorsed by several major nations, including the United States.

Law of the sea is a distinct area of international law that outlines rules governing the exploitation of the world's oceans. It was the subject of the first attempt by the International Law Commission to place a large segment of international law on a multilateral treaty basis. Four conventions resulting from the commission's work were produced by the first and second Geneva Conferences in 1958 and 1960. The pressure leading to the law of the sea conference, which lasted between 1974 and 1982 and involved a very wide range of states and international organizations, included a variety of economic, political, and strategic factors and resulted in the adoption of the United Nations Convention on the Law of the Sea (UNCLOS), also known as the Law of the Sea Treaty. This convention, with 320 articles and 9 annexes, was adopted in 1982 by 130 votes to 4, with 17 abstentions. It came into force on November 16, 1994, and has been ratified by 156 nations.

The convention, which provides a legal framework and specific regime over each area of the sea, is often referred to as the constitution of the sea. Among the areas covered by UNCLOS are the territorial sea, international straits, continental shelves, the high seas, exclusive economic zones, innocent passage, nationality of ships, collisions at sea, pollution, the deep seabed, and settlements of disputes. UNCLOS covers all of the ground of the four 1958 conventions and quite a lot more. The preamble to the convention states that one of its key purposes is "to contribute to the realisation of a just and equitable international economic order which takes into account the interests and needs of mankind as a whole and, in particular, the special interests and needs of developing countries, whether coastal or landlocked." This theme underpins the convention's approach to dealing with exclusive economic zones, continental shelves, the deep seabed, and other issues.

Article 136 of the convention provides that both the area that includes the seabed, ocean floor, and subsoil thereof beyond the limits of national jurisdiction or economic zone and its resources are "the common heritage of mankind" and that no sovereign or other rights to this area and its resources may be recognized. The regulation of the seabed is an area of controversy, as the developed countries have the technologies and financial resources to exploit the natural resources found on and beneath the seabed, but the developing countries claim a share in those resources. Dissatisfaction with the deep-seabed regime in UNCLOS led the United States to vote against the convention's adoption. The United States, Germany, and the United Kingdom are not signatories to this convention.

Because of the negative impacts of the disposal of toxic and other noxious materials at sea, UNCLOS devotes about fifty articles to the protection of the marine environment. Signatory nations are responsible for the fulfillment of their international obligations concerning the protection and preservation of the marine environment and are liable in accordance with international law. For example, in the *Torrey Canyon* incident of 1967, a Liberian tanker that ran aground along the British coast was bombed by the British military as a necessary measure to protect the coastline and its marine life. Parties to the convention are required to have adequate provisions to compensate when those acting under their jurisdictions cause environmental damage and pollution. The convention's articles concerning protection of the marine environment also provide for global and regional cooperation, technical assistance, monitory and environmental assessment, and the development of the enforcement of international and domestic law aimed at preventing pollution.

Josephus J. Brimah

FURTHER READING

Anderson, David. *Modern Law of the Sea: Selected Essays.* Boston: Martinus Nijhoff, 2008.

Freestone, David, Richard Barnes, and David Ong, eds. *The Law of the Sea: Progress and Prospects.* New York: Oxford University Press, 2007.

Nelson, Jason C. "The Contemporary Seabed Mining Regime: A Critical Analysis of the Mining Regulations Promulgated by the International Seabed Authority." *Colorado Journal of International Environmental Law and Policy* 16, no. 1 (2005): 27-76.

Sohn, Louis B., et al. *The Law of the Sea in a Nutshell.* 2d ed. St. Paul, Minn.: West, 2010.

SEE ALSO: Continental shelves; Ocean dumping; Ocean pollution; Seabed disposal; Seabed mining; *Torrey Canyon* oil spill.

United Nations Convention to Combat Desertification

CATEGORIES: Treaties, laws, and court cases; resources and resource management

THE CONVENTION: International agreement intended to address the problems of land degradation and desertification around the world

DATE: Adopted on June 17, 1994

SIGNIFICANCE: The United Nations Convention to Combat Desertification has brought to the attention of decision makers some of the most vulnerable ecosystems in the world: drylands and deserts. The degradation of these ecosystems has affected some of the most vulnerable people on earth, causing massive migrations and creating environmental refugees. The convention has helped bring increased attention also to particular issues related to desertification, such as climate change and trade liberalization.

It has been widely acknowledged that land degradation and desertification are serious problems in many regions of the world, with significant impacts on the economic, social, cultural, and environmental well-being of the populations they affect. Scientists and policy makers had been talking about how to address the problems of land degradation and desertification for many years before the United Nations Conference on Desertification in 1977 finally began strong international efforts to address these problems.

Participants at the 1992 United Nations Conference on Environment and Development, known as the Earth Summit, in Rio de Janeiro, Brazil, discussed the issue of desertification and decided to call on the U.N. General Assembly to establish a committee to prepare a convention addressing desertification by June, 1994. On June 17, 1994, the United Nations Convention to Combat Desertification (UNCCD) was adopted in Paris and subsequently opened for signature on October 14 of the same year. The convention entered into force on December 26, 1996, after it was ratified by 50 countries. The first Conference of the Parties (COP), the convention's governing body, was held in Rome, Italy, in October, 1997. By 2009 the number of countries that were parties to the convention had grown to 193.

A permanent secretariat is responsible for providing support to affected UNCCD member countries, assisting in the preparation for sessions of the COP, distributing information as it becomes available, and coordinating programs and activities with relevant international nonprofit environmental organizations. The United Nations Global Environment Facility is the official financing mechanism of the UNCCD. Since 2001 the Committee for the Review of the Implementation of the Convention has assisted the COP in conducting regular reviews of signatory nations' implementation of the convention.

At the country level, the desertification issues that a member nation faces must be detailed in a National Action Program, along with the measures being taken to address them. The UNCCD's approach is to support sustainable development at the community level, with the reasoning that in turn this will lead to overall reductions in land degradation and the protection of fragile ecosystems.

The UNCCD recognizes that human activities are the primary causes of land degradation and desertification. It further recognizes that land degradation and desertification result from a number of complex factors—political, physical, economic, social, cultural, and biological—and the interactions among them. Factors such as poverty, lack of food security, and lack of proper nutrition also play important roles, influencing human activities in ways that often contribute to land degradation and desertification.

Lakhdar Boukerrou

FURTHER READING

Delville, Philippe L. *Societies and Nature in the Sahel.* London: Taylor & Francis, 2007.

Geist, Helmut. *The Causes and Progression of Desertification.* Burlington, Vt.: Ashgate, 2005.

Goudie, Andrew. "The Human Impact on Vegetation." In *The Human Impact on the Natural Environment: Past, Present, and Future.* 6th ed. Malden, Mass.: Blackwell, 2005.

Middleton, N. *Global Desertification: Do Humans Cause Deserts?* Amsterdam: Elsevier, 2004.

United Nations. Convention to Combat Desertification Secretariat. *Desertification: Coping with Today's Global Challenges.* Eschborn, Germany: Author, 2008.

Williams, M. A. J., and Robert C. Balling, Jr. *Interactions of Desertification and Climate.* London: Arnold, 1996.

SEE ALSO: Africa; Deforestation; Desertification; Earth Summit; Environmental justice and environmental racism; Global Environment Facility; Grazing and grasslands; Kalahari Desert.

United Nations Educational, Scientific, and Cultural Organization

CATEGORY: Organizations and agencies

IDENTIFICATION: International body created to advance respect for justice, the rule of law, and human rights

DATE: Established in November, 1946

SIGNIFICANCE: The United Nations Educational, Scientific, and Cultural Organization has been a pioneering force for international cooperation in scientific research and the promotion of science education since its inception. Many of the organization's programs are directly involved in efforts to protect and preserve the world's natural resources.

The purposes of the United Nations Educational, Scientific, and Cultural Organization (UNESCO), as stated in the organization's constitution, are to contribute to worldwide peace and security by promoting international collaboration through education, science, and culture, and to advance universal respect for justice, the rule of law, human rights, and fundamental freedoms affirmed by world peoples. In addition to seeking to eliminate illiteracy and promote science education and cultural interdependence, UNESCO has made great efforts to articulate human rights and to promote conservation on a global scale.

UNESCO has especially backed science education as critical to various nations' cooperation with and ratification of key environmental treaties, including the Kyoto Protocol (1997) and the embattled Copenhagen Accord (2009) on climate change, which seeks to reduce emissions of greenhouse gases such as carbon dioxide. UNESCO has actively supported climate change activism through its advocacy of the United Nations Framework Convention on Climate Change. UNESCO has also promoted the idea that human rights include the right of access to clean water, as the lack of safe drinking water is a growing problem in some parts of the world, such as Africa.

UNESCO has spearheaded many international programs with direct environmental links. These include identifying biosphere preserves through the Man and the Biosphere Programme, established in 1977 to protect natural resources; areas designated as biosphere reserves are protected alongside sustainable development. UNESCO's Environmental Conservation Organization, established in 1986, focuses on raising environmental awareness among youth. UNESCO is active in hydrology and oceanographic research, and its International Hydrological Programme, established in 1965, addresses the global use and availability of water, including groundwater and riverine systems; the Intergovernmental Oceanographic Commission, established in 1960 to research and protect global oceans, coordinates the development of forecasts regarding ocean climate and short-term ocean-related climate change for affected populations. To highlight the alarming rate of species loss around the world, UNESCO declared 2010 the International Year of Biodiversity.

UNESCO also coordinates the monitoring of carbon in areas such as the Amazon rain forest and the alpine forest of the Great Himalayan National Park in India, because forests are vital to offsetting global carbon emissions and many of the world's forested areas are threatened by human activity. In addition, UNESCO promotes protection of the environment through the establishment of World Heritage Sites; the nations in which these sites of natural, cultural, or historical significance are located are obligated to maintain and preserve them, under the terms of the World Heritage Convention, as part of the universal heritage of humanity.

Patrick Norman Hunt

FURTHER READING

Di Giovine, Michael A. *The Heritage-scape: UNESCO, World Heritage, and Tourism.* Lanham, Md.: Lexington Books, 2009.

Hunt, Patrick Norman. "African Hydrology Crisis: When the Snows of Kilimanjaro Melt." *Diplomatic Journal,* April, 2010.

United Nations Educational, Scientific, and Cultural Organization. *Sixty Years of Science at UNESCO, 1945-2005.* Paris: Author, 2006.

SEE ALSO: Biosphere reserves; Deforestation; United Nations Environment Programme; United Nations Framework Convention on Climate Change; World Heritage Convention.

United Nations Environment Programme

CATEGORIES: Organizations and agencies; ecology and ecosystems; resources and resource management

IDENTIFICATION: International body created to coordinate worldwide environmental activities

DATE: Established in December, 1972

SIGNIFICANCE: The United Nations Environment Programme, the designated authority of the United Nations in regional and global environmental issues, monitors the status of the environment around the world and brings environmental concerns to the attention of the world's governments so that they may take action.

Creation of the United Nations Environment Programme (UNEP) resulted from the wave of concern that arose during the 1960's about all forms of pollution. UNEP was established by the General Assembly of the United Nations in 1972, several months after the U.N. Conference on the Human Environment was held in Stockholm, Sweden. Delegates attending that conference from 113 countries heard convincing evidence that expanding pollution was threatening the quality of human life. The delegates received reports from regional committees situated in every corner of the globe. In addition, an overall report, written by environmentalists Barbara Ward and René Dubos and based on the observations of prominent scientists and cultural experts in Europe, North America, South America, and Asia, was presented at the conference. The assessments, while not representing circumstances as threatening to humanity's very existence, emphasized the need for humankind to take immediate remedial action to preserve an acceptable quality of life. The Ward-Dubos report, later published as *Only One Earth: The Care and Maintenance of a Small Planet* (1972), raised issues relating to water and air pollution (especially acid rain), dwindling rain forests, and complex problems arising from urban drift.

It was clear that the world needed an agency that would coordinate and monitor the environmental information produced by existing organizations. Such an agency could identify trouble spots and suggest corrective measures before major environmental disasters could occur. Delegates to the conference from developing nations, fearing economic consequences, were not enthusiastic about tackling environmental issues. Nonetheless, the majority of delegates voted to recommend to the U.N. General Assembly that it create UNEP. In December, 1972, the General Assembly approved the recommendation and established UNEP's operations in Nairobi, Kenya. It was the first major undertaking of the United Nations to be based in an African country.

UNEP was set up with a fifty-eight-nation governing council responsible for determining the program's priorities and approving its budget. Maurice Strong, the North American businessman who had served as secretary general of the Stockholm conference, was appointed UNEP's first director. Succeeding executive directors have included Egyptian microbiologist and former government minister Mostafa Tolba, former Canadian assistant deputy minister of environment Elizabeth Dowdeswell, German politician and environmental politics expert Klaus Topfer, and German environmental politics expert Achim Steiner.

Nearly three years after its establishment UNEP began to implement Earthwatch, its mechanism for collecting and exchanging information on environmental issues. In 1975 UNEP established the Global Environment Monitoring System (GEMS), an Earthwatch system for collecting, collating, and dispensing environmental data received from hundreds of U.N.-associated agencies around the world. GEMS encourages governments to expand their monitoring activities and advises UNEP (through its Environmental Fund) as to which international conservation strategies are deserving of financial support. UNEP subsequently placed hundreds of monitoring stations around the world to record information on water quality in rivers, lakes, and oceans. Air-quality measuring systems were also constructed. By 1981 data on air and water quality were coming to GEMS from most large urban areas, including twenty cities in developing countries. GEMS also assists with the accumulation of information relating to food contaminants and toxic substances affecting marine life.

FURTHER EVOLUTION

In 1976 UNEP expanded the Earthwatch program with the addition of the International Register of Potentially Toxic Chemicals (IRPTC), which facilitates the distribution of information on chemicals, their effects, and the policies of the world's governments regarding their use and trade. In 1977 another

Earthwatch program was added, the International Referral System for Sources of Environmental Information, known as Infoterra. This decentralized information system links thousands of environmental information sources within the world's governments, research institutions, universities, U.N. agencies, and nongovernmental agencies.

UNEP teamed with the World Meteorological Organization in 1988 to establish the Intergovernmental Panel on Climate Change (IPCC). A cooperative effort that brings together the world's national governments and its scientific community, the IPCC assesses the mass of scientific data and policy alternatives pertaining to global climate change.

In 1990 GEMS, with the approval of UNEP, began a new Earthwatch program, the Global Resource Information Database (GRID). GRID received support from the U.S. National Aeronautics and Space Administration (NASA) for the development of necessary software. In addition, International Business Machines (IBM) contributed $6.5 million in computer equipment. As required by UNEP, control of the database was situated in Nairobi, but its most important computer centers were in Switzerland and the United States. The primary purpose of GRID was to provide technological support for the responsibilities imposed by UNEP on GEMS. GRID, through GEMS, greatly increased the efficiency of UNEP's operations.

In the spring of 2000, in anticipation of the Millennium Summit later that year, UNEP convened the first Global Ministerial Environment Forum in Malmö, Sweden. A primary purpose of the forum was to convey to the attendees of the upcoming summit that depletion of the environment and natural resources was advancing at an alarming rate, and that progress in sustainable development was lagging dangerously far behind the commitments that governments had made to sustainability. The Millennium Development Goals, which arose out of the subsequent summit, included environmental sustainability as an objective in itself and recognized such sustainability as fundamental for the realization of all the other goals.

A number of major global treaties regarding the environment have been negotiated under the auspices of UNEP. Among these are the 1987 Montreal Protocol on Substances That Deplete the Ozone Layer, the 1989 Basel Convention on the Control of Transboundary Movements of Hazardous Wastes, the 1998 Rotterdam Convention on the Prior Informed Consent Procedure for Certain Hazardous Chemicals and Pesticides in International Trade (a treaty regarding the import of banned and severely restricted hazardous chemicals, particularly into developing countries), and the 2001 Stockholm Convention on Persistent Organic Pollutants.

EFFECTIVENESS

Initially, UNEP's overall effectiveness did not prove to be all that its sponsors had hoped. Problems existed with its Nairobi location, where frequent power failures and lack of technical expertise caused serious disruptions. The program also found it difficult to convince developing countries to provide information and to take seriously the need for conservation of resources when those nations wanted to focus their limited financial resources on other, more immediate matters. The leaders of the developing countries showed little interest in environmental issues and frequently failed to cooperate with GEMS or any other UNEP agency. In general, Southern Hemisphere countries were slow to collect environmental information and slower still to file reports with UNEP. Most of the information UNEP collected and disseminated came from developed countries.

As governments, the public, and the United Nations itself became increasingly aware of environmental issues, and as nongovernmental organizations devoted to environmental issues mushroomed in developing nations, UNEP grew better able to gather and provide data of a global nature. UNEP has also facilitated partnerships between developing and developed nations to improve the participation of developing nations. During the 1990's, for example, the United States and Ireland partnered with countries in the Southern Africa subregion to establish and provide training in support of a regional Infoterra network. Interconnections among environmental issues and a greater understanding of the relationship between environment and development have made UNEP's ability to collate and share environmental data increasingly valuable. The advent of environmental issues with a global scope, notably climate change, has further reinforced UNEP's importance.

Ronald K. Huch
Updated by Karen N. Kähler

FURTHER READING

Birnie, Patricia, Alan Boyle, and Catherine Redgwell. *International Law and the Environment.* 3d ed. New York: Oxford University Press, 2009.
Downie, David L., and Marc A. Levy. "The UN Environment Programme at a Turning Point: Options for Change." In *The Global Environment in the Twenty-first Century: Prospects for International Cooperation,* edited by Pamela S. Chasek. New York: United Nations University Press, 2000.
Gosovic, Branislav. *The Quest for World Environmental Cooperation: The Case of the U.N. Global Environment Monitoring System.* New York: Routledge, 1992.
Speth, James Gustave, and Peter M. Haas. "From Stockholm to Johannesburg: First Attempt at Global Environmental Governance." In *Global Environmental Governance.* Washington, D.C.: Island Press, 2006.
Tolba, Mostafa Kamal, and Iwona Rummel-Bulska. *Global Environmental Diplomacy: Negotiating Environment Agreements for the World, 1973-1992.* Cambridge, Mass.: MIT Press, 1998.

SEE ALSO: Basel Convention on the Control of Transboundary Movements of Hazardous Wastes; Dubos, René; Earth Summit; Global Environment Facility; Intergovernmental Panel on Climate Change; Montreal Protocol; Stockholm Convention on Persistent Organic Pollutants; United Nations population conferences.

United Nations Framework Convention on Climate Change

CATEGORIES: Treaties, laws, and court cases; weather and climate

THE CONVENTION: Nonbinding international agreement concerning the reduction of greenhouse gas concentrations in the atmosphere to combat climate change

DATE: Opened for signature on May 9, 1992

SIGNIFICANCE: The signing of the United Nations Framework Convention on Climate Change demonstrated global concern about the problem of climate change, in particular human interference with the earth's climate system. The convention paved the way for the Kyoto Protocol in 1997.

In May, 1992, the United Nations Framework Convention on Climate Change (UNFCCC) was opened for signature after its text was finalized by an intergov-

Principles of the U.N. Framework Convention on Climate Change

Article 3 of the UNFCCC sets out the five principles that should guide all parties to the convention in all actions relating to the convention.

1. The Parties should protect the climate system for the benefit of present and future generations of humankind, on the basis of equity and in accordance with their common but differentiated responsibilities and respective capabilities. Accordingly, the developed country Parties should take the lead in combating climate change and the adverse effects thereof.

2. The specific needs and special circumstances of developing country Parties, especially those that are particularly vulnerable to the adverse effects of climate change, and of those Parties, especially developing country Parties, that would have to bear a disproportionate or abnormal burden under the Convention, should be given full consideration.

3. The Parties should take precautionary measures to anticipate, prevent or minimize the causes of climate change and mitigate its adverse effects. Where there are threats of serious or irreversible damage, lack of full scientific certainty should not be used as a reason for postponing such measures, taking into account that policies and measures to deal with climate change should be cost-effective so as to ensure global benefits at the lowest possible cost. To achieve this, such policies and measures

should take into account different socio-economic contexts, be comprehensive, cover all relevant sources, sinks and reservoirs of greenhouse gases and adaptation, and comprise all economic sectors. Efforts to address climate change may be carried out cooperatively by interested Parties.

4. The Parties have a right to, and should, promote sustainable development. Policies and measures to protect the climate system against human-induced change should be appropriate for the specific conditions of each Party and should be integrated with national development programmes, taking into account that economic development is essential for adopting measures to address climate change.

5. The Parties should cooperate to promote a supportive and open international economic system that would lead to sustainable economic growth and development in all Parties, particularly developing country Parties, thus enabling them better to address the problems of climate change. Measures taken to combat climate change, including unilateral ones, should not constitute a means of arbitrary or unjustifiable discrimination or a disguised restriction on international trade.

ernmental negotiating committee. The convention was produced at the United Nations Conference on Environment and Development, generally known as the Earth Summit, held in Rio de Janeiro, Brazil in June, 1992, and entered into force in March, 1994. Since then, the countries that are signatories to the UNFCCC have held annual conferences (Conferences of the Parties) to assess their progress in addressing climate change. During the mid-1990's they started negotiating the revision to the UNFCCC known as the Kyoto Protocol, which established legally binding obligations for the reduction of greenhouse gas emissions by developed countries.

The UNFCCC defines the climate system as the totality of the atmosphere, hydrosphere, biosphere, and geosphere and the interactions of these parts. It stresses the anthropogenic (human-caused) influences on climate, and it states the goal of stabilizing greenhouse gas concentrations at a level that does not pose a danger to the climate system. According to the UNFCCC, such a level needs to be achieved within a time frame sufficient to allow ecosystems to adapt nat-

urally to climate change, to ensure that food production is not threatened, and to enable economic development to proceed in a sustainable manner. The Conferences of the Parties have discussed plans of action to assist countries that are likely to be severely affected by global warming, such as small island nations and countries with low-lying coastal areas that could be inundated by rising sea levels and countries with arid and semiarid areas that could be devastated by drought and desertification.

The UNFCCC provides a framework for international cooperation in dealing with climate change. It stresses individual countries' sovereign rights to exploit their own natural resources but highlights the importance of nations' not causing any damage to the environments of states or other areas beyond their own borders. The UNFCCC requires signatory countries to develop national programs to combat climate change; these programs must take into consideration the ways in which the nations' activities and policies surrounding agriculture, industry, energy, and natural resources affect climate change.

COMMON BUT DIFFERENTIATED RESPONSIBILITIES

The UNFCCC places the heaviest burden for fighting climate change on developed or industrialized countries under the principle of "common but differentiated responsibilities," which recognizes the causes and the disproportionate impacts of climate change. The convention thus takes note of the role of the developed countries in causing the largest share of historical and current global emissions of greenhouse gases and holds them responsible for paying for the damage done to the natural environment.

The UNFCCC set up a national greenhouse gas inventory, or a count of each nation's greenhouse gas emissions and reductions of emissions, with 1990 as the base year. Under the convention developed countries were expected to reduce their emissions to 1990 levels by the year 2000. The developed nations are also required to provide financial assistance to developing countries to enable them to meet the costs of adaptation to changing climate; this is achieved through a system of grants and loans managed by the Global Environment Facility, a multinational organization. Developed countries are further expected to share environmentally friendly technologies with developing countries to aid them in reducing greenhouse gas emissions and implementing greater energy efficiency.

The UNFCCC recognizes that developing countries have a legitimate right to pursue sustained economic growth to eradicate poverty, and that this means that the share of greenhouse gas emissions produced by these countries is likely to grow for some time. The convention also notes, however, that with the financial and technical assistance of developed countries, developing nations should be able to limit emissions in ways that do not hinder the nations' economic growth.

RESEARCH AND PUBLIC AWARENESS

The UNFCCC sets no mandatory limits on greenhouse gas emissions for individual countries and contains no enforcement provisions, but it does include provisions for updates that set mandatory emission limits. As a result, representatives of many of the signatory countries met in the mid-1990's to discuss the possibility of establishing legally binding obligations for developed countries to reduce their greenhouse gas emissions. This led to the Kyoto Protocol—the principal update of the UNFCCC—in which many countries agreed to reduce their emissions of greenhouse gases an average of 6 to 8 percent below 1990 levels between the years 2008 and 2012.

The UNFCCC underlines the role of scientific research in deepening worldwide understanding of the "causes, effects, magnitude, and timing of climate change." The Intergovernmental Panel on Climate Change, under the UNFCCC Secretariat, gathers scientific evidence to facilitate global consensus on this issue. The UNFCCC also pays attention to research examining the possible social and economic consequences of various climate change response strategies and interventions. The convention stresses the importance of raising public awareness about climate change through educational and training programs, and it encourages the widest participation in the process, highlighting the important role that nongovernmental organizations can play in connecting governments, scientific communities, and the public.

WEAKNESSES AND CRITICISMS

Critics of the UNFCCC have objected that it arbitrarily divides the world into developed and developing countries without paying sufficient attention to the emergence of industrializing countries, such as China, that have become heavy greenhouse gas emitters and are not required to cap their emissions. Some developed countries, led by the United States, have argued that restricting the greenhouse gas emissions of developed countries is not an effective way to address climate change if the industrializing countries, especially China, are not required to share responsibility. This claim gave U.S. president George W. Bush a reason to reject the Kyoto Protocol in 2001, and the failure of the international community to set limits on the greenhouse gas emissions of China caused tensions between the United States and China during negotiations at the 2009 United Nations Climate Change Conference in Copenhagen, Denmark.

The UNFCCC has also been challenged for overemphasizing the role of emissions mitigation policies and sidestepping the significance of strategies focusing on adaptation to climate change. In addition, the convention's focus on reducing greenhouse gas emissions through the introduction of new technologies and through target setting has been criticized as demonstrating an inadequate understanding of the long-term support that developing countries need to modify their agricultural and land-use systems to cope with the changing climate.

Sam Wong

FURTHER READING

Bodansky, Daniel. "The United Nations Framework Convention on Climate Change: A Commentary." *Yale Journal of International Law* 18, no. 2 (1993): 451-558.

Sands, Philippe. "The United Nations Framework Convention on Climate Change." *Review of European Community and International Environmental Law* 1, no. 3 (1992): 270-275.

Schipper, Lisa. "Conceptual History of Adaptation in the UNFCCC Process." *Review of European Community and International Environmental Law* 15, no. 1 (2006): 82-92.

Speth, James Gustave, and Peter M. Haas. *Global Environmental Governance.* Washington, D.C.: Island Press, 2006.

SEE ALSO: Climate change and oceans; Development gap; Earth Summit; Greenhouse gases; Intergovernmental Panel on Climate Change; Kyoto Protocol; Sea-level changes; United Nations Commission on Sustainable Development.

United Nations population conferences

CATEGORY: Population issues

THE EVENTS: Series of periodic intergovernmental gatherings focusing on population issues

DATES: 1954, 1965, 1974, 1984, and 1994

SIGNIFICANCE: The United Nations population conferences have brought together representatives from around the world to consider the local and global consequences of population growth. The conferences have highlighted shared concerns about population issues and helped coordinate efforts to understand the relationships between population growth and quality of life.

The United Nations first held a World Population Conference in Rome, Italy, in 1954. Primarily an academic conference, it resulted in a resolution to generate more information on the demographic situations in developing countries. The Second World Population Conference, jointly organized by the United Nations and the International Union for the Scientific Study of Population, was held in Belgrade, Yugoslavia, in 1965. Experts in the field gathered to consider fertility as a factor in development planning policy.

The Third World Population Conference in Bucharest, Romania, in 1974 was the first intergovernmental conference of its kind. Official government delegates from 135 countries met to consider demographic issues under the auspices of the United Nations. Experts attended preliminary meetings devoted to the discussion of future trends, economic and social development, resources and the environment, and family units; they then formulated a draft for the World Population Plan of Action.

The United States advocated government sponsorship of family-planning services, which was a new concept. Not all delegates were convinced that population growth was a problem, however. The economic gap between rich and poor nations had widened during the 1960's, and developing nations wanted more financial assistance from developed nations. Oil prices were being set by the Organization of Petroleum Exporting Countries (OPEC), which, along with a bloc of African, Asian, and Latin American countries known as the Group of 77, wanted to reform international economic exchange. Conflict thus ensued at the Bucharest meetings over the relative importance of population versus development planning. The delegation from the People's Republic of China argued that population growth was not a problem under socialism, and the leader of the Indian delegation, Karim Singh, coined the slogan "Development is the best contraceptive." Despite these arguments, consensus was eventually achieved. The revised World Population Plan of Action stressed the importance of conserving and sharing the world's resources, reducing population growth, protecting women's rights, and supporting responsible parenthood.

INTERNATIONAL CONFERENCE ON POPULATION, 1984

Between 1974 and 1984 the world's population increased by 20 percent, and 90 percent of that growth occurred in developing areas. This population explosion resulted in a different climate at the International Conference on Population in Mexico City in 1984. Many governments of developing nations had reversed their positions on population growth, as exemplified by China's aggressive one-child policy. Furthermore, many of these nations were deeply in debt. The most successful East Asian and Latin American developing nations were pursing market-oriented investment strategies with the approval of wealthier countries such as the United States, now under more conservative political leadership. Reducing growth rates seemed useful with or without economic reform.

The biggest surprise in Mexico City was the change in position by the U.S. delegation led by James L. Buckley, a former U.S. senator and then president of Radio Free Europe, who described population growth as a "neutral factor." The U.S. delegates refused to support population programs that practiced coercion or provided abortions and also argued that fertility would decline in settings with free market economies. The impact of the change in the U.S. position was less than expected, however. Except for the Vatican, there was unanimous consent among the delegates that the primary elements of the World Population Plan of Action should be reaffirmed and extended. The new recommendations included strategies to integrate development and population planning and proposals for local and international efforts to eliminate hunger, illiteracy, unemployment, poor health and nutrition, and the low status of women. There were also recommendations about population distribution and movement, as well as condemnation of unlawful settlements, which reflected increased concerns about population growth's impact on global stability.

Soon after the 1984 conference, the United States ceased its financial support for the United Nations Fund for Population Activities (UNFPA) and the International Planned Parenthood Federation (this decision was partially reversed only after President Bill Clinton took office in 1993). By 1990, however, most developing nations had family-planning programs.

INTERNATIONAL CONFERENCE ON POPULATION AND DEVELOPMENT, 1994

The International Conference on Population and Development (ICPD) in Cairo, Egypt, in 1994 echoed the 1974 and 1984 gatherings, but this time the approximately twenty thousand attendees—who represented more than 180 governments, the United Nations, nongovernmental organizations (NGOs), and the news media—recognized the importance of reducing population growth. Emphasis was placed on the criticality of individual human rights in population and development concerns. The United States returned to its former position of supporting population growth control, but there were still contentious discussions before consensus was finally achieved on the last day of the conference.

The Vatican actively challenged the rest of the dele-

gates by opposing abortion, cohabitation without marriage, and sexual deviation, although many felt the Vatican's implicit, more significant goals were to limit female empowerment and access to birth control. Women's groups lobbied against the Vatican and representatives from some Muslim countries; in spite of the conflict, agreement was achieved on the worth of smaller families and slower growth rates.

The ICPD also produced many firsts. The Vatican approved the final Programme of Action, NGOs played a prominent role at the conference and helped draft the Programme of Action, environmental issues were placed into a population context, and the reproductive health and rights of individual women were accentuated at the expense of demographic goals or a serious consideration of the impact of consumption patterns on quality of life. An estimated $17 billion per year was suggested in order to implement the twenty-year Programme of Action.

In a follow-up debate in the United Nations General Assembly, developing nations in the Southern Hemisphere pointed out that they would need assistance from developed nations to implement the program. Leaders of Southern Hemisphere nations were concerned that increased economic demands might reduce the resources available for existing programs.

A 1999 special session of the U.N. General Assembly conducted a five-year review of the Programme of Action, examining the progress made and challenges encountered. The session established new benchmarks in the key areas of education and literacy, reproductive health care and contraception needs, maternal mortality reduction, and control of human immunodeficiency virus/acquired immunodeficiency syndrome (HIV/AIDS).

In 2004—ten years into the twenty-year Programme of Action—the UNFPA conducted a country-by-country analysis of implementation progress. It found that, in general, countries had progressed significantly in areas such as reproductive health, gender equality, and the integration of population considerations with development planning and policies. Remaining challenges included the treatment and prevention of HIV/AIDS and the collection and analysis of data. Another review conducted in 2009 found that, fifteen years into the program, poverty, maternal and infant mortality, and unmet needs for contraception were still urgent problems.

Joan C. Stevenson
Updated by Karen N. Kähler

FURTHER READING

Johnson, Stanley. *The Politics of Population: The International Conference on Population and Development, Cairo 1994.* Sterling, Va.: Earthscan, 1995.

_____. *World Population and the United Nations: Challenge and Response.* New York: Cambridge University Press, 1987.

Reichenbach, Laura, and Mindy Jane Roseman, eds. *Reproductive Health and Human Rights: The Way Forward.* Philadelphia: University of Pennsylvania Press, 2009.

Singh, Jyoti Shankar. *Creating a New Consensus on Population: The International Conference on Population and World Development.* Sterling, Va.: Earthscan, 1998.

United Nations Population Fund. *The World Reaffirms Cairo: Official Outcomes of the ICPD at Ten Review.* New York: Author, 2005.

SEE ALSO: Population Connection; Population-control and one-child policies; Population-control movement; Population growth; Sustainable development; World Fertility Survey.

Urban ecology

CATEGORIES: Urban environments; ecology and ecosystems

DEFINITION: Subfield within the science of ecology that is concerned with the interactions among human beings, plants, and animals in urban and metropolitan areas

SIGNIFICANCE: Since the beginnings of the modern environmental movement in the 1970's, the science of urban ecology has gradually become accepted as an indispensable part of the urban planning process.

In the 1920's sociologists Robert E. Park and Ernest Burgess developed a theory of urban ecology that postulates that cities are environments similar to those found in nature, regulated by principles analogous to those that govern nature and natural evolution. This theory views the overall structure of cities as the consequence of the struggle for limited urban land, which influences cities to evolve gradually into five concentric rings. The term "urban ecology" also refers generally to that part of the science of ecology that studies the interactions among human beings, plants, and animals within urban and metropolitan areas, as well as the effects that urban development has on natural ecosystems and on biodiversity in these areas.

IMPORTANCE OF BIODIVERSITY

High levels of biodiversity (genetic, species, and habitat diversity) are important not only for plant and animal species but also for the quality of life of human beings. The higher the level of biodiversity in an ecosystem, the better able the ecosystem is to face instability, whether of natural or anthropogenic (human-caused) origin.

In regard to biodiversity, human-nature interactions in urban and metropolitan areas face two challenges. First, the rapid and increasing pressure that will result from foreseeable demographic growth is likely to have substantial negative impacts on biodiversity. The United Nations estimates that in 2050 the world population will reach 9.1 billion, most living in urban areas; this trend will substantially increase the human pressure on natural ecosystems. Second, climate change, although still uncertain in its extent, requires adaptation and mitigation measures in urban and metropolitan areas that will increase the resilience of urban ecosystems. On the other side, changes in land cover affect the stocks of carbon, and through that effect these changes are also expected to have impacts on climate. For this reason, changes in land cover owing to anthropogenic factors constitute another facet of urban ecology that needs to become a central issue in urban planning.

CHANGES IN URBAN PLANNING

During most of the twentieth century the environment was not a central theme in urban planning, and most planners did not develop urban biodiversity strategies. Since the 1970's, however, protection of the environment has gradually become a key component of urban planning. Urban green spaces and green corridors are important elements in the urban planning paradigm, known as the new urbanism, that developed in part as a reaction against modern urbanism. Urban ecology also gradually became part of the core disciplines that constitute the multidisciplinary field of urban planning.

In addition to conducting biodiversity inventories (such as inventories of native plants in risk of extinction in specific urban areas), urban ecologists examine the impacts of urbanization on native species of plants and animals (both vertebrates and invertebrates) and on wildlife in general (including reptiles, aquatic vegetation, and native fish). They also evaluate the effects of streets, roads, and other urban infrastructures on plant and animal species.

The study of the urban ecologist ends, in most cases, in policy recommendations in the form of an urban biodiversity strategy, as part of a wider urban planning process. These policy recommendations address different types of actions, both direct and indirect. An urban biodiversity strategy may suggest, for example, that a city or metropolitan area create habitats or restore natural habitats within the area, clean waste from green spaces and green corridors, connect fragmented and isolated pieces of urban green spaces so that green corridors and networks are created, develop wildlife habitats in the backyard gardens of private homes, plant native species as part of the regeneration of brownfields sites, collect native species in places that are going to be developed and transplant them to other parts of the area, improve the water quality in rivers, or reduce soil erosion in particular areas. Indirect actions recommended might include the establishment of environmental education programs designed to change social values related to natural ecosystems and to increase public awareness about the environment.

Carlos Nunes Silva

FURTHER READING

Goudie, Andrew. *The Human Impact on the Natural Environment: Past, Present, and Future.* 6th ed. Malden, Mass.: Blackwell, 2006.

McDonnell, Mark J., Amy K. Hahs, and Jürgen Breuste, eds. *Ecology of Cities and Towns: A Comparative Approach.* New York: Cambridge University Press, 2009.

Marzluff, John M., et al., eds. *Urban Ecology: An International Perspective on the Interaction Between Humans and Nature.* New York: Springer, 2008.

Raven, Peter H., and Linda R. Berg. *Environment.* 5th ed. New York: John Wiley & Sons, 2006.

Selman, Paul. *Local Sustainability: Managing and Planning Ecologically Sound Places.* London: Paul Chapman, 1996.

SEE ALSO: Community gardens; Greenbelts; New urbanism; Open spaces; Urban parks; Urban planning.

Urban parks

CATEGORIES: Urban environments; land and land use

DEFINITION: Areas within cities that are set aside for recreation and green space

SIGNIFICANCE: Urban parks reflect efforts to reform cities and environmental consciousness by reintegrating nature into the urban landscape.

The appearance of the earliest urban parks in the mid- to late nineteenth century arose from a broader urban reform movement. Reformers responded to rapid industrialization, urbanization, and land development by calling on cities to create and develop parks within city limits. This reflected a new environmental ideology that sought to reconcile nature and the city and change the way urban residents related to the natural world. From 1860 to 1900, cities in the United States spent millions of dollars to build parks, including New York's Central Park, Boston's Emerald Necklace, San Francisco's Golden Gate Park, and Chicago's Lakefront.

Frederick Law Olmsted, one of the most influential landscape architects of the nineteenth century and leader of the urban park movement in the United States, participated in nearly every major urban park project in the nation. Olmsted believed that an urban park functions as the "lungs of the city." He argued that parks improve public health by increasing access to fresh air, trees, and sunshine—natural amenities that gritty, industrial cities lacked. According to Olmsted, parks promote spiritual restoration by providing an antithesis to the anxieties of the human-made, unnatural urban landscape. He also asserted that parks can prevent urban vice or riots by providing outlets for the urban underclass. Olmsted's park design legacy remains a cornerstone in urban planning.

During the twentieth century, the design and purpose of urban parks shifted from the Olmstedian emphasis on the park as pleasure ground to an emphasis on the park as multiuse recreation facility. Urban park designs began to include baseball fields, golf courses, tennis courts, ice-skating rinks, playgrounds, and even zoos.

Much of the impetus for the environmental movement of the late 1960's and early 1970's began in cities. Air pollution from automobiles and industrial manufacturing and water pollution from industrial dumping, sewage disposal, and toxic waste disposal were among the most pressing environmental issues. Urban residents began to mobilize around these issues, demanding change.

Although much of the resulting legislation addressed pollution problems in the cities, many citizens and environmental activists were also concerned that accelerating urban expansion and suburbanization were destroying the last remaining open spaces surrounding many cities. Nearly one hundred years after the first urban park movement, a new urban park movement emerged. It was tied to two developing political coalitions: the national environmental movement and the antigrowth movement. In some cities, local groups lamented the insufficiency of the existing parks and pressured civic leaders to protect open space, to increase the sizes of parks and open spaces, and to control urban development. Environmentalists argued that open spaces should be protected as unique and important ecosystems. In many cities, outlying open spaces became integrated into regional and city park systems.

The importance of urban parks became more than a local issue. The subject of parks and other areas for recreation became a national political issue. In the United States, for example, a congressional report released during the 1970's stated that U.S. cities lacked sufficient outdoor recreation opportunities. The National Park Service was charged with acquiring and maintaining national recreation areas (seashores, lakeshores, and other areas) located in or near urban areas. Among the new national urban parks that were designated were Cape Cod National Seashore (Massachusetts), Gateway National Recreation Area (New York and New Jersey), Golden Gate National Recreation Area (California), Cuyahoga Valley National Recreation Area (Ohio), and Santa Monica Mountains National Recreation Area (California).

Urban parks represent one way of reconnecting to nature and "greening" the city. Greening the city can take many forms, including the creation of urban farms, the planting of trees along street meridians, the cleanup of neighborhood parks, and the rehabili-

Central Park in Manhattan, New York City, provides open space and a relaxing environment for urban dwellers. (©William Perry/ Dreamstime.com)

tation of derelict or abandoned sites into small parks or gardens. Such projects are often carried out by neighborhood groups, nonprofit organizations, or local schools. These modest, small-scale efforts represent the fundamental cultural rediscovery of nature within the urban environment and can provide powerful learning experiences.

Urban parks are also integrated into local economic development projects. During the 1980's and 1990's many U.S. cities redeveloped their lakefronts or other waterfronts. Many redevelopment projects interspersed open spaces, small parks, picnic tables, and water fountains among office buildings, hotels, and residences. These projects served to reconnect urban residents to the natural environment while simultaneously revitalizing local economies. The integration of parks and nature in such developments is evidence of the degree to which environmental sensitivity has become important in urban design.

Urban parks continue to play an important role in city life. Many are mixed-use places that offer a range of activities from solitude and reflection in a quiet, shady grove to social interaction in outdoor cafés and aquariums. Such parks are symbolic of efforts to redefine broader relationships between humans and the environment through the reintegration of nature in the most human of all creations, the city. The environmental movement has become increasingly concerned with changing people's perceptions of and behaviors toward the natural environment. Urban parks, however large or small, whether self-contained or integrated into mixed-use sites, testify to the interdependence, not opposition, of nature and people.

Lisa M. Benton

FURTHER READING

Boone, Christopher G., and Ali Modarres. *City and Environment*. Philadelphia: Temple University Press, 2006.

Cranz, Galen. *The Politics of Park Design: A History of Urban Parks in America*. 1982. Reprint. Cambridge, Mass.: MIT Press, 1989.

Low, Setha, Dana Taplin, and Suzanne Scheld. *Rethinking Urban Parks: Public Space and Cultural Diversity*. Austin: University of Texas Press, 2005.

Platt, Rutherford H., ed. *The Humane Metropolis: People and Nature in the Twenty-first-Century City*. Amherst: University of Massachusetts Press, 2006.

Schmitt, Peter J. *Back to Nature: The Arcadian Myth in Urban America*. 1969. Reprint. Baltimore: The Johns Hopkins University Press, 1990.

SEE ALSO: Community gardens; Greenbelts; New urbanism; Olmsted, Frederick Law; Open spaces; Urban ecology; Urban planning; Urban sprawl.

Urban planning

CATEGORIES: Urban environments; land and land use

DEFINITION: The application of long-range, comprehensive decisions by governmental agencies to the growth and development of urban centers and surrounding suburbs

SIGNIFICANCE: Effective urban planning is increasingly important as a means to protect and improve the natural environment. It ensures that private development decisions result in public benefits as well as private profits and that the expansion of government services is efficient and effective.

Successful urban planning benefits a city in many ways. It ensures the timely availability of essential infrastructure and government services, including roads, schools, and utilities; a satisfactory quality of life for all residents; the compatibility of adjacent land uses, including buffers between incompatible uses; the rational allocation of land and the highest and best use for each parcel of land; the prevention of waste; the maintenance of public aesthetics, including open spaces; the promotion of social diversity and diversity in both private and public services and facilities, including housing and commerce; the orderly expansion of existing uses into new physical locations; the preservation of historically and culturally important buildings; and the protection of the natural environment.

HISTORY

Urban planning began in antiquity with the planned cities of Greece and pre-Columbian America. Hippodamus, a Greek lawyer who lived in the fifth century B.C.E., designed the first street layout on a gridiron pattern. The Greeks limited the size of their cities to the carrying capacity of the surrounding farmland. Once an existing site was maximized, the *paleopolis*, or old town, helped found a second *neopolis*, or new town, on a new site. Urban planning continued in Europe through Roman and medieval times into the Renaissance and Enlightenment periods. Many cities in the New World were planned cities, including Jamestown, Philadelphia, and Washington, D.C.

The technocrats of the nineteenth century developed the first planned urban renewal projects. For example, beginning in 1853, Napoleon III tore down nearly one-half of the buildings in Paris, France, in history's largest public works and urban renewal project in an effort to improve the street system. In the early twentieth century, the first zoning regulations to separate land uses were enacted in Rotterdam, the Netherlands; Frankfurt, Germany; and New York City. New York City enacted the first regional plan in 1929, regulating housing, highways, building height, and open spaces.

Before the twentieth century, urban planners' primary emphasis was on design aesthetics and the delivery of services to meet human needs, including sanitation, transportation, housing, and basic utilities. In the early twentieth century, social scientists such as Ernest Burgess and Homer Hoyt began to emphasize the expansion and succession of existing land uses as the basis for planning. They were concerned with the orderly growth and development of the urban area, the outward expansion of existing land uses into new plots of land, the orderly succession of one land use after another on a single plot of land, and the correction of urban social problems resulting from urban growth.

In the mid-twentieth century, Lewis Mumford and others in the "new communities" movement argued that piecemeal, incremental, successive patterns of growth should give way to totally planned, functionally integrated communities. Also in the mid-twentieth century, Ian L. McHarg, a European planner, began a new emphasis on the physical environment in urban planning. Using a series of geological, geographic, and current land-use overlays, he mapped those plots of land best suited for each of several types of land use.

In 1973 the Council on Environmental Quality recognized a growing consensus that control over land use is probably the most important single factor in improving the quality of the environment. Effective urban planning reduces environmental problems stemming from land-use patterns that impose conflicting or unsustainable demands on the environment. One suggested approach is to compare alternative uses of the land according to the relative demands each use places on the environment and to implement the one that minimizes human impact.

Public- and Private-Sector Planning

Many social scientists argue that effective urban planning requires a shift in power from the private to the public sector. Public support for urban planning in many European countries allows government agencies to counterbalance private-sector power in urban development. Nations with centralized authority, including many developing nations, and nations with socialist governments also tend to limit private power effectively. In the United States, private-sector freedom is equated with liberty, and government intrusions and regulations are kept to a minimum. Participants in the American private sector are allowed maximum leeway in a free market system to pursue their interests and to develop and dispose of their property as they see fit.

In addition, the pluralist nature of electoral politics and government decision making in the United States provides the private sector with multiple opportunities to influence public decisions through political appointments, campaign donations, and lobbying. Governmental jurisdictions tend to be fragmented into multiple units (cities, counties, and special districts), each with a unique set of functions, fiscal resources, expertise, and constituencies. These units of government often lack the time, resources, or interest to develop long-range, comprehensive, rational urban planning. Free market advocates argue against increased government involvement in urban planning and assert that the private self-interest of economically active individuals ensures that land uses are rationally allocated across the landscape and that urban development does not destroy the environmental life-support system on which the survival of the urban area depends.

Gordon Neal Diem

Further Reading

Boone, Christopher G., and Ali Modarres. *City and Environment.* Philadelphia: Temple University Press, 2006.

Freestone, Robert, ed. *Urban Planning in a Changing World: The Twentieth Century Experience.* New York: Routledge, 2000.

McHarg, Ian L. *Design with Nature.* 25th anniversary ed. New York: John Wiley & Sons, 1995.

Platt, Rutherford H., ed. *The Humane Metropolis: People and Nature in the Twenty-first-Century City.* Amherst: University of Massachusetts Press, 2006.

Riddell, Robert. *Sustainable Urban Planning: Tipping the Balance.* Malden, Mass.: Blackwell, 2004.

SEE ALSO: Council on Environmental Quality; Greenbelts; Land-use planning; Mumford, Lewis; Open spaces; Planned communities; Road systems and freeways; Sustainable development; Urban parks; Urban sprawl.

Urban sprawl

CATEGORY: Urban environments

DEFINITION: The uncontrolled and unregulated development that occurs outside the administrative boundaries of the zoning and land-use authority of municipalities and outside the conscious and deliberate direction of those authorities

SIGNIFICANCE: When urban settlements become overcrowded, some individuals, businesses, and industries migrate to the fringes of the urban area, where population concentrations are less dense but urban services remain available. The concentration and eventual diffusion of urbanization affect existing land uses, the physical environment, the survival of plant and animal species, and the aesthetics of the landscape.

Urban growth and development are usually controlled by two forces. First, municipal government authorities regulate growth through urban planning, zoning, and a variety of land-use ordinances. Second, social and economic forces combine to encourage outward growth of the urban area in one of three consistent patterns, described as concentric circle, sector, and multinuclear patterns. Urban sprawl, which is neither controlled nor regulated, often involves some break in this historic pattern. The unplanned nature of urban sprawl often results in a mix of incompatible land uses placed adjacent to one another on the same developed physical area.

The challenges presented by urbanization and urban sprawl are likely to intensify during the decades to come. In mid-2009 the world population officially became more urbanized than rural—of the earth's 6.83 billion people, 50.1 percent were found to be living in urban areas; 75 percent of the inhabitants of the more developed regions lived in urban areas, while such areas accounted for only 45 percent of the people living in less developed regions. According to United Nations projections for the year 2050, the world's population growth will continue to be concentrated within urban areas; of that increase, most is expected to be within the cities and towns of developing nations. By 2050 the number of people inhabiting urban areas is expected to be roughly 6.3 billion—the size of the total population of the earth in 2004.

PROBLEMS CAUSED BY SPRAWL

Sprawl results in discontinuous leapfrog or checkerboard patterns of development and strip development along transportation corridors, with skipped areas remaining undeveloped. This creates inefficiencies in providing urban services to the sprawl area. Sprawl also results in less-than-maximum utilization of existing developed land in the urban center. Instead of converting existing developed land to new uses, developers establish new developments on less expensive land in suburban areas adjacent to the existing urban area.

Urban sprawl generally increases the total amount of land affected by human activity, the amount of natural areas converted to recreational uses, the amount of wasteland, and the residential, industrial, institutional, and infrastructure uses of land. Wasteland is land disturbed by humans to such an extent that natural uses cannot be restored and future development on the land is restricted. Wastelands include soil borrow pits (areas where soil has been dug out to be taken to other locations), quarries, debris landfills, and construction material storage sites. Urban sprawl generally decreases agricultural uses and timber uses of land and decreases the amount of natural barren or rocky lands, river and stream floodplain areas, and native timber- and grasslands.

Sprawl is encouraged by a variety of social and economic forces. First, it is often a consequence of increasing heterogeneity of the urban population and is a means to limit or escape from interactions between dissimilar social, economic, and ethnic groups. Second, sprawl allows individuals and businesses to escape from the negative effects caused by concentrations of undesirable land uses—such as industrial, commercial, and low-income residential zones—and relocate to suburban areas with more pristine and af-

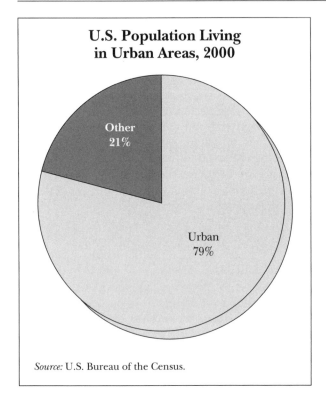

U.S. Population Living in Urban Areas, 2000

Other 21%

Urban 79%

Source: U.S. Bureau of the Census.

fordable real estate. Third, sprawl is encouraged by landowners in suburban areas seeking to maximize profits from investments in land. Finally, sprawl permits individuals and businesses to escape taxes and regulations imposed in core urban areas.

Some researchers contend that the sprawl problem is overstated. In the United States, for example, urban land in 2002 represented less than 3 percent of the nation's total land area and housed some 79 percent of the population. Urban land together with rural residential land occupied 7 percent of the total land area. By contrast, about 5 percent of the nation's land is protected from development under the 1964 Wilderness Act.

ENVIRONMENTAL IMPACTS

The first significant impact of urban sprawl on the environment is the abnormal greening of the physical area. In most cases, total greening is reduced through destruction of forests, grasslands, and floodplains to make way for development. This decreases the number and variety of species able to occupy the space and creates microclimate effects, such as regional warming. However, in some cases total greening is increased through irrigation and the introduction of cultivated lawns, orchards, and other plantings in ar-

eas that are naturally arid or barren. This results in the introduction of new species into the environment, increased pollen counts, and a variety of microclimate effects, such as increases in regional humidity. In either case, the presprawl natural ecology is dramatically changed, with resulting negative impacts on plant and animal species displaced by or unable to adapt to the new environment.

In the course of sprawl development, existing natural ecosystems are destroyed while presprawl plants and animals are killed, displaced, or replaced. Most presprawl wildlife retreats in the face of development. However, some species are able to adapt to the sprawl environment and find the mix of land uses, the dispersion of human activities, and the residue of human activity conducive to their survival. Among the wildlife that benefits are scavengers, such as pigeons, rats, raccoons, and opossums; vermin hunters, such as falcons, foxes, and coyotes; and foragers, such as deer, squirrels, and rabbits. Benefiting plant species include all those that homestead on disturbed soil or wasteland or that thrive in the open sunlight of most new suburban developments.

Sprawl reduces the amount of productive agricultural land and economically important forests. Productive agricultural land and forests are lost either by development of the land for residential, industrial, or institutional uses or by the transformation of productive farms and forests into hobby farms and parklands in an effort to preserve green space without preserving the productive purpose of the agricultural lands or forestlands.

Finally, sprawl increases the total amount of land surface that is impervious to rainfall penetration. Roadways, parking lots, slab building foundations, and other paved areas increase rainwater runoff and the possibility of flooding, soil erosion, and ecosystem destruction.

IMPACTS ON AIR AND WATER RESOURCES

Other consequences of urban sprawl include increases in noise, light pollution, land area devoted to highways and roads, and public-utility impacts as land is cleared for underground water, sewer, and utility pipes and for aboveground utility cables.

The impact of sprawl on air and water resources is both negative and positive. The increase in human population that accompanies sprawl increases the concentration of significant amounts of unnatural substances in the soil, water, and air and also produces abnormally high concentrations of natural substances at levels that may cause undesirable health effects, corrosion, and ecological change. However, studies also indicate that the population dispersal associated with sprawl actually reduces air pollution by dispersing both the mobile and stationary sources of pollutants. Increases in air pollution from automobiles associated with sprawl may be less than the air pollution produced by traffic gridlock, mass-transit buses, and trains in denser urban areas.

Subsurface water supplies and surface watercourses are less affected by sprawl than by denser patterns of development. Denser urban development increases the demands on water resources, runoff and the possibility of flooding, and the likelihood that watercourses will be channeled and hardened by concrete and other construction materials. Denser urban development, and the increase in paved surfaces that comes with it, also make it more difficult for subsurface water to replenish itself.

Limiting Sprawl

Many municipalities attempt to limit sprawl by refusing to extend essential services such as water lines, sewer lines, and road systems outside their municipal boundaries. Rural areas may attempt to limit sprawl through zoning restrictions on development, farmland protection ordinances, environmental impact regulations, and special development-impact fees levied on new development to recover the public costs associated with constructing roads, schools, and other facilities necessary to provide services to the newly developed areas.

In North America, a movement toward "smart

Urban sprawl generally increases the total amount of land affected by human activity, the amount of natural areas converted to recreational uses, the amount of wasteland, and the residential, industrial, institutional, and infrastructure uses of land. (©Ron Chapple Studios/ Dreamstime.com)

growth" has emerged as a sustainable middle ground between sprawl and zero growth. Smart growth emphasizes high-density neighborhoods and reduced car use. It occurs not on a city's periphery, but at its heart, where urban land is redeveloped to provide a mix of residential, retail, and office uses. The intent of this clustered development is to enable residents to work and shop near their homes. Reduced travel distances between people's residences and their workplaces and the sites of their leisure activities enable city dwellers to walk, bike, or use public transportation to reach their destinations.

Gordon Neal Diem
Updated by Karen N. Kähler

FURTHER READING

Frumkin, Howard, Lawrence D. Frank, and Richard Jackson. *Urban Sprawl and Public Health: Designing, Planning, and Building for Healthy Communities.* Washington, D.C.: Island Press, 2004.

Gillham, Oliver. *The Limitless City: A Primer on the Urban Sprawl Debate.* Washington, D.C.: Island Press, 2002.

Miller, Debra A., ed. *Urban Sprawl.* Detroit: Greenhaven Press, 2008.

Pugh, Cedric, ed. *Sustainability, the Environment, and Urbanization.* 1996. Reprint. Sterling, Va.: Earthscan, 2002.

Soule, David C., ed. *Urban Sprawl: A Comprehensive Reference Guide.* Westport, Conn: Greenwood Press, 2006.

Squires, Gregory D., ed. *Urban Sprawl: Causes, Consequences, and Policy Responses.* Washington, D.C.: Urban Institute Press, 2002.

SEE ALSO: Air pollution; Automobile emissions; Greenbelts; Mumford, Lewis; New urbanism; Open spaces; Planned communities; Road systems and freeways; Urban ecology; Urban planning; Zoning.

U.S. Climate Action Partnership

CATEGORIES: Organizations and agencies; activism and advocacy; weather and climate
IDENTIFICATION: Alliance of business corporations and leading environmental organizations that promotes the passage of federal legislation to reduce greenhouse gases
DATE: Established on January 22, 2007
SIGNIFICANCE: The U.S. Climate Action Partnership represents a shift in direction among many corporations toward working with environmentalists and others to address the problem of global warming.

Members of the U.S. Climate Action Partnership (USCAP) pledge to work with the Congress and the president of the United States, as well as other stakeholders, in enacting climate protection programs that are fair, economically sustainable, and environmentally effective, with the goal of significantly reducing greenhouse gas emissions. Fourteen businesses and four environmental organizations were the initial founding members of USCAP. Since its founding, the organization has both gained and lost members; environmental organizations that were members in 2010 included the Environmental Defense Fund, the Natural Resources Defense Council, and the Nature Conservancy, and business members included Alcoa, Chrysler, the Ford Motor Company, PepsiCo, and the Dow Chemical Company.

USCAP outlines a series of design principles and recommendations for legislation to address the problem of global warming in a document titled *A Call for Action.* In 2009 USCAP published *A Blueprint for Legislative Action: Consensus Recommendations for U.S. Climate Protection Legislation*, which addresses how the United States can reduce greenhouse gas emissions without hindering the strength and productivity of the American economy. Among the recommendations offered in this document are that businesses be allowed to invest in forest conservation programs to meet their required emission reductions and that businesses be offered incentives to engage in the development and deployment of emissions-reducing technologies.

USCAP asserts that policy changes based on its recommendations would result in innovation, energy security, and economic growth. Furthermore, such changes would improve the U.S. balance of trade and demonstrate much-needed U.S. leadership on the issue of climate change. Some critics have argued, however, that USCAP supports self-serving regulations and policy changes that would be beneficial to its members, such as a cap-and-trade system to reduce carbon dioxide emissions; many environmentalists assert that cap-and-trade systems in effect reward polluters by protecting them from paying the cost of compliance with regulations.

Prior to joining USCAP, some of the group's members were strongly opposed to any regulation aimed at reducing greenhouse gases and lobbied against U.S. ratification of the Kyoto Protocol. Some scientists and environmentalists have expressed concerns that the same companies that are members of USCAP are also engaged in lobbying against USCAP's stated goals.

Lakhdar Boukerrou

SEE ALSO: Climate change and oceans; Climate models; Coal; Global warming; Greenhouse effect; Greenhouse gases; *Inconvenient Truth, An*; Intergovernmental Panel on Climate Change; Kyoto Protocol; United Nations Framework Convention on Climate Change; U.S. Conference of Mayors Climate Protection Agreement.

U.S. Conference of Mayors Climate Protection Agreement

CATEGORIES: Urban environments; weather and climate

IDENTIFICATION: Agreement among mayors of U.S. cities to take steps in their local areas to combat global warming

DATE: November, 2007

SIGNIFICANCE: The Climate Protection Agreement entered into by the members of the U.S. Conference of Mayors has led to the initiation of many local projects that have increased energy efficiency and reduced the harmful emissions linked to global warming.

In November, 2007, the U.S. Conference of Mayors held a meeting described as a climate protection summit in Seattle, Washington. At the summit, more than one hundred U.S. mayors joined together to call for a partnership between cities and the federal government to work against energy dependence and global warming. The meeting had been initiated by

Miami, Florida, mayor Manny Diaz plants one of one thousand Florida native trees during a mass planting at Virginia Key Beach Park in 2008, part of his city's commitment to climate protection. Diaz was chairman of the U.S. Conference of Mayors in 2008-2009. (AP/Wide World Photos)

Greg Nickels, mayor of Seattle, who also proposed the U.S. Mayors Climate Protection Agreement. Ironically, Nickels was voted out of his post in 2009 by voters angry at the city's inability to deal with unusually heavy snow and ice the previous winter.

By the time the summit had ended, more than seven hundred mayors had signed the agreement, pledging to take action to combat global warming locally; by 2010 more than one thousand mayors had signed. In Los Angeles, for example, Mayor Antonio Villaraigosa, in partnership with the Los Angeles City Council and environmental leaders, unveiled a plan called Green L.A., described as "an action plan to lead the nation in fighting global warming." Villaraigosa pledged to reduce his city's carbon footprint to 35 percent below 1990 levels by 2035, the most ambitious goal set by a major American city. Los Angeles also planned to increase its use of renewable energy to 35 percent by 2020, much of it through changes at its municipal electrical utility, the largest in the country.

In Austin, Texas, energy-efficiency standards were raised for homes, requiring a 60 percent reduction in energy use by 2015. Chicago undertook a program to replace traditional urinals with waterless models and also planted several thousand trees. Philadelphia began to replace the black tar paper on the roofs of old row houses with snow-white, high-reflection composites. Keene, New Hampshire, required parents waiting for their children in their cars outside of schools to turn off their engines. Portland, Oregon, had already reduced its carbon emissions to 1990 levels by 2007; in addition, water flowing through Portland's drinking-water system is used also to generate hydroelectricity. Mayors of at least 134 U.S. cities have instituted the use of energy-efficient lighting in public buildings, in parks and on streets, in traffic signals, and other places in their cities. Many city governments' auto fleets have converted to alternative fuels or hybrid-electric technology.

The Chicago Climate Action Plan, announced during September, 2008, aims to cut greenhouse gas emissions in that city by 25 percent, to 1990 levels, by 2020. The plan requires retrofitting of commercial and industrial buildings, increased energy efficiency in residences, and more use of electricity from renewable sources. Buildings, which emit 70 percent of Chicago's carbon dioxide, are the major target of the Climate Action Plan. Chicago's city hall already has a green roof, designed as a model for as many as six thousand buildings citywide. The city's Smart Bulb Program has distributed more than half a million free energy-saving compact fluorescent lightbulbs to residents.

Several cities have targeted poor neighborhoods with subsidies and grants for the insulation of older homes that often leak heat in winter. Such programs also allow some people to acquire replacements for older, energy-inefficient basic electrical appliances, such as refrigerators, washers, and dryers. Nonprofit groups such as NeighborWorks America have helped to coordinate community efforts along this line. Workers with Greenprint Denver, for example, have gone door-to-door in low-income neighborhoods offering energy audits and help with goods and services. Rays of Hope in Austin, Texas, offers basic services as well as solar panels.

Bruce E. Johansen

FURTHER READING

El Nasser, Haya. "'Green' Efforts Embrace Poor." *USA Today*, November 24, 2008, 3A.

Faiola, Anthony, and Robin Shulman. "Cities Take Lead on Environment as Debate Drags at Federal Level." *The Washington Post*, June 9, 2007, A1.

Gore, Christopher, and Pamela Robinson. "Local Government Response to Climate Change: Our Last, Best Hope?" In *Changing Climates in North American Politics: Institutions, Policymaking, and Multilevel Governance*, edited by Henrik Selin and Stacy D. VanDeveer. Cambridge, Mass.: MIT Press, 2009.

SEE ALSO: Carpooling; Compact fluorescent lightbulbs; Global warming; Green buildings; Landscape architecture; U.S. Climate Action Partnership; Waste management.

U.S. Geological Survey

CATEGORIES: Organizations and agencies; land and land use; water and water pollution; resources and resource management

IDENTIFICATION: U.S. federal agency that provides scientific information on the conditions of the nation's natural resources

DATE: Established on March 3, 1879

SIGNIFICANCE: By providing scientific information on water, biological, energy, and mineral resources to the public, legislators, and policy makers, the U.S. Geological Survey carries out its mission of enhancing and protecting Americans' quality of life.

In 1879 President Rutherford B. Hayes signed the congressional bill providing funding for the establishment of the U.S. Geological Survey (USGS) within the Department of the Interior. Industrial growth in the years immediately following the Civil War had produced a significant strain on the nation's natural resources. In an 1866 report Joseph Wilson, commissioner of the General Land Office, indicated that proper management of mineral resources in the West was vital to further development of the United States. Following up on Wilson's recommendations, Congress authorized a geological survey of the West, largely following the path of the newly finished transcontinental railroad. Clarence King and Ferdinand Hayden were placed in charge of the project and by

1870 had presented a plan to Congress for the survey. Additional surveys were privately sponsored as well.

Downturns in the American economy resulted in Congress looking for more efficient alternatives for mapping the West. In 1878 Congress requested that the National Academy of Sciences, which had been established in 1863 by President Abraham Lincoln, develop a plan for surveying and mapping western territories. The academy's recommendations included the establishment of the USGS, the purpose of which would be to oversee the study of geological and mineral resources in the public domain.

King was appointed first director of the USGS, and in 1879 a comprehensive study of mining districts in Nevada and Colorado was begun, as well as similar studies of iron and copper resources in other parts of the country. The study was completed prior to King's resignation from the position of director in 1881.

The duties assigned to the USGS underwent significant expansion during subsequent decades. In 1882 Congress authorized the creation of a comprehensive geological map of the United States, with the result that topographic mapping became the largest program within the USGS. The agency's geological studies also benefited with the inclusion of scientific research into the origins of ore deposits as well as newly introduced fields such as glacial ecology and studies of rock classes. Western droughts during the 1880's resulted in the addition of research concerning irrigation and water utilization within the USGS before the beginning of the new century. By 1904 the USGS had completed topographic maps covering more than 25 percent of the United States and Alaska. That year Congress also authorized mapping of areas of potential fossil fuels, including both coal and oil deposits.

Although the divisions within the USGS have undergone changes over the years, the agency has remained largely unchanged in its focus. Its areas of responsibility have grown over time to include the monitoring of seismic and magnetic activity throughout the world and the examination of the geological features of worldwide earthquake zones and volcanoes.

Richard Adler

FURTHER READING

Cech, Thomas V. "Federal Water Agencies." In *Principles of Water Resources: History, Development, Management, and Policy.* 3d ed. New York: John Wiley & Sons, 2010.

National Research Council. *Toward a Sustainable and Secure Water Future: A Leadership Role for the U.S. Geological Survey.* Washington, D.C.: National Academies Press, 2009.

_____. *Weaving a National Map: Review of the U.S. Geological Survey Concept of the National Map.* Washington, D.C.: National Academies Press, 2003.

SEE ALSO: Department of the Interior, U.S.; Earth resources satellites; Everglades; Geographic information systems; Geothermal energy; Kesterson Reservoir; North America; Oil drilling; Powell, John Wesley; Runoff, agricultural; Trans-Alaska Pipeline.

V

Vegetarianism

CATEGORIES: Philosophy and ethics; agriculture and food

DEFINITION: The practice of refraining from eating meat

SIGNIFICANCE: Many vegetarians believe that the processes necessary to the raising of animals to be eaten by humans—which include feeding, slaughtering, and disposing of the livestock and their waste products—constitute an inefficient use of the planet's limited resources. Others believe that eating lower on the food chain reduces their chances of ingesting environmental pollutants and other potentially harmful substances.

Human beings practice many different forms of vegetarianism; not all people who refer to themselves as vegetarians eat the same diet. It has been estimated that more than twelve million Americans consider themselves vegetarians, although many of them sometimes eat meat. Traditional vegetarians, also known as lacto-ovo vegetarians, eat no meat but do consume dairy products and eggs. Vegans, sometimes known as pure vegetarians, consume no animal products at all and also do not wear or otherwise use products of animal origin, such as leather and wool. The terms "pescovegetarian" and "pollovegetarian" refer to people who eat no mammalian meat but do eat fish and chicken, respectively.

The Indian nationalist and spiritual leader Mohandas Gandhi once said, "Live simply, so that others may simply live." For many, adopting a vegetarian lifestyle is the embodiment of his message. According to Frances Moore Lappé, author of *Diet for a Small Planet* (1971), it takes 1.3 hectares (3.25 acres) to support the average American meat-rich diet. Feeding a lacto-ovo vegetarian requires 0.2 hectare (0.5 acre), while a pure vegan diet requires only 0.07 hectare (0.17 acre). Put another way, if the amount of grain fed to cattle were converted into bread, it would be enough to provide every human in the world with two loaves each day.

The issue of water resources is another important consideration for some vegetarians. Producing one pound of beef requires up to one hundred times more water than producing one pound of wheat. In addition, groundwater contamination from animal waste is a growing problem. Livestock on U.S. feedlots produce twenty times as much excrement as the country's human population—more than one billion tons per year. Animal waste contains high levels of nitrogen, which can be beneficial to the soil, but it converts to ammonia and nitrates if not processed right away. If not treated properly, the runoff from agricultural feedlots can leach into the groundwater supply, cause algae overgrowth and oxygen depletion in the nation's lakes and rivers, and eventually pollute reservoirs and wells used for drinking water.

Animals not confined to feedlots pose different problems for the environment. Overgrazing and disagreements about land use in the western United States have become major battleground issues between ranchers and environmentalists. The U.S. Bureau of Land Management spends many millions of dollars maintaining federal grazing lands, yet collects only a fraction of that cost back in fees. Cattle and sheep grazing on public lands compete with wildlife for grass and water, and more than one million predators (including wolves, bears, and coyotes) are killed each year to protect free-ranging livestock.

The impacts of raising animals for food are even more dramatic in South and Central America. Cattle ranching is one of the leading causes of the destruction of the rain forests, as ranchers employ slash-and-burn agriculture to clear grazing areas for their livestock. Unfortunately, this newly created range lacks the topsoil necessary for sustained grazing and must be abandoned after only a few years.

"You are what you eat" is a well-worn axiom, but it contains a simple truth about the place of humans on the food chain. A diner sitting down to a plate of filet mignon is eating not only part of a steer but also everything that steer ingested during its life cycle. In the case of animals raised on feedlots, this goes well beyond grain and hay. Grain and corn destined to become cattle feed are sprayed with pesticides and larvacides, and the animals are injected with any combination of hormones, growth stimulants, tranquilizers, antibiotics, and appetite stimulants. Only minute

The menu at Otarian vegetarian restaurant in New York City provides information on the carbon footprint of every dish served, along with a comparison to the likely carbon footprint of a similar dish containing meat. (AP/Wide World Photos)

traces of these products can be found in meat sold for public consumption, but the long-term effects on humans are not known.

The scenario is no different for marine life. Fish and mollusks literally "breathe" their environment, and contaminated water leads to contaminated seafood. This is especially true of shellfish, such as oysters and scallops, which feed by filtering water. Predators that eat fish—for example, bears, eagles, and humans—consume the toxins accumulated by the fish they eat as well as the toxins of all the smaller fish those fish have eaten. In addition, ground fish meal is used as both fertilizer and a component of some livestock feeds.

For many people, choosing vegetarianism and eating lower on the food chain is their way of reducing the amount of toxins in their diets, but there is no perfect solution to the problem. Most fruit, vegetables, and grain products have also been chemically treated, especially those grown in countries where farming regulations are less strict than in the United States. Even buying "organic" products does not guarantee

safety, because no industry standards have been established for products carrying such a label.

P. S. Ramsey

FURTHER READING

Lappé, Frances Moore. *Diet for a Small Planet.* 20th anniversary ed. New York: Ballantine Books, 1991.

Maurer, Donna. *Vegetarianism: Movement or Moment?* Philadelphia: Temple University Press, 2002.

Preece, Rod. *Sins of the Flesh: A History of Ethical Vegetarian Thought.* Vancouver: University of British Columbia Press, 2008.

Sapontzis, Steve F., ed. *Food for Thought: The Debate over Eating Meat.* Amherst, N.Y.: Prometheus Books, 2004.

Tracy, Lisa. *The Gradual Vegetarian.* 1985. Reprint. New York: Dell, 1993.

SEE ALSO: Animal rights; Animal rights movement; Biomagnification; Cattle; Food chains; Organic gardening and farming; Overgrazing of livestock; Rain forests; Slash-and-burn agriculture.

Vernadsky, Vladimir

CATEGORY: Philosophy and ethics
IDENTIFICATION: Russian geochemist and mineralogist
BORN: March 12, 1863; St. Petersburg, Russia
DIED: January 6, 1945; Moscow, Russia
SIGNIFICANCE: Vernadsky developed the concepts of the biosphere and the noosphere, and his 1926 book *The Biosphere* inspired a new vision of humankind's role in shaping the earth's environment.

Vladimir Vernadsky, who became a pioneer in the field of biogeochemistry, was given professional direction early in his life. An older cousin who was a retired army officer and an independent man of extensive reading remarked to Vernadsky that "the world is a living organism." Profoundly impressed, Vernadsky within a few years began his scholarly studies of the earth's physiology—the ways in which its matter and biota, including humankind, interact and affect one another and their common planetary environment.

Vernadsky graduated from St. Petersburg University in 1885 and earned his Ph.D. from the University of Moscow in 1897. He was professor of crystallography and mineralogy at Moscow University from 1890 until 1911. Following the 1917 Russian Revolution, he spent three years at the Sorbonne in Paris, where he wrote extensively on the subjects of geochemistry and biochemistry, crystallography and mineralogy, geochemical activity, marine chemistry, the evolution of life, and futurology, displaying all the signs of a polymath. From 1926 to 1938, Vernadsky directed the State Radium Institute in Leningrad; he was among the earliest scientists to recognize the tremendous importance of radioactivity as a source of thermal energy. He established the first Soviet national scientific academy, the Ukrainian Academy of Science, in 1928 and served simultaneously as its president and as the director of the Academy of Science's Leningrad biogeochemistry laboratory. Vernadsky founded the field of biogeochemistry, and it was the principal field in which he gradually gained international distinction.

A man of broad scholarly talents based on his mastery of several scientific specialties, Vernadsky became best known outside the Soviet Union for his 1926 publication *La Biosphère* (*The Biosphere*, 1929), a study in which he elaborated on his theory of the biosphere.

"Biosphere" was a term Vernadsky borrowed from Eduard Suess (1831-1914), a Viennese professor of structural geology and eminent scholar who suggested the existence of an ancient supercontinent. Suess—who had first used the word at the end of a monograph about the Alps—and Vernadsky used the term "biosphere" to refer to the total mass of living organisms that process and recycle the energy and nutrients available in the environment. This activity occurs inside a thin layer of life that circles the globe.

Vernadsky was concerned that the importance of life in the entire structure of the earth's crust had been underestimated—when it was not ignored altogether—by his scientific colleagues. He elaborated an imaginative theory that, like the lithosphere, the atmosphere, the hydrosphere, and the sphere of fire (the earth's reliance on the sun), the biosphere is one of the concentric spheres enveloping the earth.

In the early 1940's, having fixed the word "biosphere" in the scientific lexicon, Vernadsky added another concept: that of the noosphere. The term comes from *noös*, the Greek word for "mind"; Vernadsky believed that the "sphere of the mind" represents a new power altering the face of the earth. Defined precisely, in the manner of science, the noosphere is neither a sphere, like the atmosphere or lithosphere, nor a physical phenomenon, yet the noosphere has physical consequences, for the human mind, in Vernadsky's words, had become, for the first time, "a large-scale geological force" reshaping the planet.

As scholars familiar with Vernadsky's scientific achievements have noted, his imaginative conceptions of the biosphere and noosphere predate James Lovelock's inspired Gaia hypothesis and parallel some fundamental ideas integral to it. Lovelock, who first expounded the Gaia hypothesis in 1972 and published further specifics during the 1980's, fully acknowledged Vernadsky's importance to his work.

Alexander Scott

FURTHER READING

Krebs, Robert E. "Biosphere: Envelope of Life." In *The Basics of Earth Science*. Westport, Conn.: Greenwood Press, 2003.
Smil, Vaclav. *The Earth's Biosphere: Evolution, Dynamics, and Change*. Cambridge, Mass.: MIT Press, 2002.

SEE ALSO: Biogeochemical cycles; Biosphere; Gaia hypothesis; Lovelock, James.

Volatile organic compounds

CATEGORIES: Pollutants and toxins; atmosphere and air pollution

DEFINITION: Broad class of natural and human-made organic compounds that evaporate readily and have low solubility in water

SIGNIFICANCE: In indoor environments high concentrations of volatile organic compounds emitted by household products and construction materials can lead to reduced air quality and potential health problems. In urban and industrial environments these compounds play an important role in the development of tropospheric ozone and photochemical smog, and they have also been shown to play a role in groundwater contamination, the depletion of stratospheric ozone, and global climate change.

Volatile organic compounds (VOCs) are highly reactive pollutants that are emitted by both human-made and natural sources. The definitions as to what constitutes a VOC vary among scientific organizations and regulatory agencies in different countries. In general, however, VOCs are characterized as organic, or carbon-based, compounds that have a high vapor pressure, meaning that they evaporate easily under normal atmospheric conditions. Abundant organic compounds such as carbon monoxide (CO), carbon dioxide (CO_2), and methane (CH_4), along with other less reactive compounds, are normally not considered to be VOCs.

VOCs are typically present in the atmosphere in trace quantities. Plants and trees are the largest natural sources of VOCs; however, they are also emitted by soil microbes, forest fires, animals, and the oceans. Anthropogenic sources—that is, those related to human activities—include the use of liquid and gaseous fossil fuels for transportation and power production, biofuel burning for heating and cooking, and agricultural pesticide use. VOCs are also emitted by ingredients that are commonly used in building materials and such household products as cleaning supplies, paints, cosmetics, and disinfectants.

Natural sources account for much more of total VOC emissions than do anthropogenic sources. The impacts of the natural emissions on air quality, however, are minimized by the fact that natural sources are widely distributed across space. Human-caused emissions, in contrast, often have far greater impacts on air quality because they are typically concentrated in urban and industrial regions or, at a smaller scale, inside buildings. In urban regions VOCs, particularly unburned gasoline or hydrocarbons, react with nitrogen oxide (NO) and nitrogen dioxide (NO_2) in the presence of sunlight to form tropospheric ozone (O_3); VOCs and ozone are primary ingredients in the formation of photochemical smog. By modifying the concentration of ozone (O_3), a greenhouse gas, VOCs not only pose a health concern in urban regions but also have an indirect effect on global warming. In addition, the improper disposal of VOCs, including gasoline and industrial solvents, can contaminate groundwater and adversely affect drinking-water sources.

In indoor environments VOCs pose a particular health concern because their concentrations may build to higher levels than are typically found outdoors. The widespread presence of VOCs in household products used indoors has the potential to expose people to a variety of VOC compounds at relatively high concentrations for long exposure periods.

Scientific and regulatory efforts to identify and describe the health impacts of VOCs are complicated by the fact that many different types of VOCs are released into indoor and outdoor environments and their impacts are quite variable; some VOCs have no health effects, whereas others are known cancer-causing agents. Because of this variability, efforts to regulate VOC emissions are diverse. Notable efforts to reduce VOCs have included the development of the catalytic converter, the regulation of toxic emissions under the U.S. Clean Air Act, and the regulation of municipal and industrial waste handling, drinking-water quality, and indoor air quality by the Environmental Protection Agency.

Jeffrey C. Brunskill

FURTHER READING

Ahrens, C. Donald. *Meteorology Today: An Introduction to Weather, Climate, and the Environment.* Belmont, Calif.: Brooks/Cole Cengage Learning, 2009.

Girard, James E. *Principles of Environmental Chemistry.* Sudbury, Mass.: Jones and Bartlett, 2005

Koppmann, Ralf, ed. *Volatile Organic Compounds in the Atmosphere.* Hoboken, N.J.: Wiley-Blackwell, 2007.

SEE ALSO: Air pollution; Automobile emissions; Catalytic converters; Clean Air Act and amendments; Fossil fuels; Gasoline and gasoline additives; Groundwater pollution; Hazardous and toxic substance regulation; Indoor air pollution; Sick building syndrome; Smog.

Volcanoes and weather

CATEGORY: Weather and climate

SIGNIFICANCE: Volcanoes have the ability to send vast amounts of chemicals, in the form of gases and particles, into the earth's atmosphere, sometimes with resulting impacts on weather patterns.

A general pattern of the distribution of volcanoes can be observed on the earth around the Pacific Rim region known as the Pacific Ring of Fire and in other "hot spots" associated with active and past tectonic margin zones. The significance of volcanic eruptions for weather is both direct and indirect. Water vapor plays an important role in the formation of humidity in the air and the ability of the air to absorb and release heat, at least at local levels. Dust particulates, extremely prevalent in these eruptions, can form condensation nuclei, the bases for the formation of precipitation. Additionally, as these aerosols spread into the upper atmosphere and around the world, they tend to diffuse incoming solar radiation, producing a general, well-documented cooling effect on weather.

Some scientists think that volcanoes contributed to the early formation of the earth's atmosphere. Effluents from volcanoes and steam vents have probably always been similar in nature to those seen today: about 80 percent water vapor and 10 percent carbon dioxide, along with components of sulfur dioxide and various dust particulates.

In a paper he read before the Philosophical Society of Manchester England, on December 22, 1784, Benjamin Franklin put forth the idea that volcanoes play a role in modifying climate. Franklin reported that in the wake of the volcanic eruption of Laki crater in Iceland in 1783, he had observed a reduction in the intensity of sunlight. He suggested that a "dry fog" produced by the volcano was preventing sunlight from entering the lower atmosphere. This in turn, he suggested, was the contributing factor in producing a very severe winter in western Europe and the eastern United States during the 1783-1784 season. Franklin's proposal was based on the idea that volcanic dust was the major factor contributing to the reflection of sunlight. Additionally, it was believed that this effect could be observed only during violent eruptions. With later technological advancements, scientists

Mount Pinatubo spews volcanic ash and steam during its 1991 eruption. (USGS/USAF/R. Batalon)

were able to sample the upper stratosphere and found that volcanic dust plays less of a role than do volcanic emissions of sulfate particles, which have been shown to affect surface temperatures.

VOLCANIC EMISSIONS

The aerosols emitted from volcanoes that could have influences on weather are very complex. Aerosols are not simply the dust of crushed rock; they are composed of various chemicals and also maintain various shapes and sizes. It is known that volcanoes are a source of sulfate aerosols. These are sulfur-based particles that have highly reflective surfaces that have the capability of scattering sunlight back into space. Additionally, they can act as condensation nuclei for the formation of clouds and affect the structure of clouds, adding to their reflectivity and their ability to block sunlight.

Sulfur, in particular, after about two months in the atmosphere, combines with water vapor to form a haze of sulfuric acid. Such a volcanic haze may blanket the stratosphere for many years. The effect of this haze on weather and climate is found in its ability to reflect and absorb energy from the sun as well as to absorb infrared energy emitted from the earth. The results of these processes are a general cooling of the lower atmosphere and a warming of the stratosphere.

It has been suggested that the second-largest contribution to chlorine in the atmosphere, next to its mixing into the air from the oceans, is volcanoes. During noneruptive stages as well as during violent eruptions, volcanoes emit chlorine gas. Volcanoes also contribute to injecting this chlorine high into the stratosphere, where it has the potential to interact with natural ozone as well as to add more aerosols into the atmosphere. It has been found, however, that volcanic gases can be infused high into the stratosphere without violent eruptions. Initially, scientists believed that only the largest of the particles erupted from a volcano made it into the stratosphere.

MAJOR ERUPTIONS

During the twentieth century several major eruptions contributed to the injection of massive amounts of volcanic gases and particulates into the atmosphere. It has been estimated that in April, 1982, the eruption of the volcano El Chichón in Mexico injected about 8 tons of sulfur dioxide gas into the atmosphere to an altitude of 32 kilometers (20 miles). In contrast, in May, 1980, Mount St. Helens in Washing-

Composition of Air and of Volcanic Gases

| | PERCENT BY VOLUME | |
GAS	AIR	VOLCANIC* GAS
N_2 (nitrogen)	77	5.45
O_2 (oxygen)	21	
H_2O (water vapor)	0.1 to 2.8	70.8
Ar (argon)	0.93	0.18
CO_2 (carbon dioxide)	0.033	14.07
Ne (neon)	0.0018	
CH_4 (methane)	0.00015	
NH_3 (ammonia)	0.000001	
SO_2 (sulfur dioxide)		6.4
SO_3 (sulfur trioxide)		1.92
CO (carbon monoxide)		0.4
H (hydrogen)		0.33

*Kilauea volcano, Hawaii.

ton State produced almost a ton of sulfur dioxide, but, unlike in the case of El Chichón, its eruption did not cause any noticeable cooling.

In June, 1991, Mount Pinatubo in the Philippines erupted, the second-largest eruption in the century since the Mount Katmai Novarupta eruption in Alaska in 1912. Pinatubo sent nearly 20 million tons of sulfur dioxide into the atmosphere, producing a noticeable worldwide cooling of about 0.5 degree Celsius (0.9 degree Fahrenheit) by 1992. Most models suggest that temperatures can drop from 0.2 to 0.5 degree Celsius (0.36 to 0.9 degree Fahrenheit) after a major eruption.

M. Marian Mustoe

FURTHER READING

Blanchard, Duncan C. *From Raindrops to Volcanoes, Adventures with Sea Surface Meteorology.* Mineola, N.Y.: Dover, 1995.

Bourseiller, Philippe, and Jacques Duriex. *Volcanoes.* New York: Harry Abrams, 2008.

Lopes, Rosaly. *The Volcano Adventure Guide.* New York: Cambridge University Press, 2005.

Simkin, Tom, and Richard S. Fiske. *Krakatau, 1883: The Volcanic Eruption and Its Effects.* Washington, D.C.: Smithsonian Institution Press, 1983.

Stommel, Henry, and Elizabeth Stommel. *Volcano Weather: The Story of 1816, the Year Without a Summer.* Newport, R.I.: Seven Seas Press, 1983.

SEE ALSO: Aerosols; Krakatoa eruption; Mount Tambora eruption; Sulfur oxides.

W

Waste management

CATEGORY: Waste and waste management

DEFINITION: Policies and procedures for handling the physical by-products of human activity that cannot be reintegrated into the ecological biomass cycle

SIGNIFICANCE: Waste includes solid, liquid, and airborne substances that are potentially harmful to living organisms. As the human population has grown and the use of manufactured materials has expanded, the disposal of waste has become more challenging, and strategies of waste management have come under increasing scrutiny by environmentalists.

According to the Worldwatch Institute, world production of manufactured materials (not counting recycled materials) increased nearly two and one-half times between the early 1960's and the end of the twentieth century. In industrialized countries the increase was far greater: The United States, for example, saw an eighteenfold increase in materials production from 1900 to 2000. It has been estimated that the average U.S. citizen throws away 2 to 8 pounds of garbage daily, and although studies demonstrate that the per-person production of waste has remained approximately the same since the beginning of the twentieth century, the sharp rise in population and expanding industrial base have meant greater total accumulations of waste. Furthermore, the types of waste have changed.

Waste is commonly categorized as domestic, or solid, waste and industrial, or liquid, waste, although the distinction is not absolute. Both may contain toxic substances, but the percentage of toxins in industrial waste is likely to be higher, and the two types of waste are disposed of in different ways. The smoke emitted from industrial processing of materials and vehicle exhaust are additional types of waste, although these are commonly thought of as pollution rather than waste.

SOLID WASTE

Solid waste is the familiar garbage that households and businesses in the United States have sent to dumps since garbage collection began late in the nineteenth century. The largest portion, more than 40 percent, consists of paper products, especially newspaper and containers. Yard waste, food debris, plastic containers and wrappings, bottles, metals, and appliances are also regularly thrown away. About 1 percent of this waste involves hazardous materials, typically insecticides, beauty aids, and cleaning products. Construction waste accounts for a large share—about 12 percent—of solid waste and may contribute a higher proportion of hazardous materials, such as solvents and paint.

Although most of these materials are solid, when dumped together they can soak up rainwater and then ooze chemical-laden liquids. This leachate may filter down into the groundwater and pollute nearby streams and wells. If it contains toxic elements, such as lead or mercury from batteries, the leachate can be dangerous to health. The odor from rotting garbage may also foul the air, seldom enough to be harmful but still repellent to people living nearby. It can attract animal scavengers, which may become infected with diseases from the garbage and spread them to other animals or even humans, especially if feces are part of the waste.

In order to combat these effects, sanitary landfills dump waste on top of plastic linings designed to contain leachate; they then cover each day's load of garbage under a thin layer of soil. Pipe systems in such landfills disperse methane gas produced by rotting organic materials. Sanitary landfills are therefore less dangerous to human health or the environment than most dump sites established in earlier times, but many old, abandoned sites and illegally dumped materials can still pose problems. They may continue to dribble harmful chemicals into groundwater for decades and emit methane, which is flammable.

Measures instituted in attempts to reduce the amount of waste deposited in landfills have had partial success. Recycling has drastically cut the total paper, metal, and glass waste in many U.S. states and in other industrialized countries. The use of garbage disposals and the practice of composting have caused the proportion of organic materials in landfills to decline. However, such reductions have not eliminated solid waste. By the early years of the twenty-first century, cities were finding it increasingly difficult to find room for new landfill sites, even when the space was urgently needed. Stringent regulations concerning the geological composition of landfills reduced the number of usable sites, and increased objections from citizen action committees, known as NIMBY (from "not in my backyard") groups, also eliminated sites near populated areas.

Facilities that incinerated waste, some of which used the resulting heat energy to power electrical generators, also faced objections from environmentalists and others because burning can release health-threatening materials, such as dioxins, into the air. Moreover, a significant proportion of waste, such as appliances and concrete, cannot be eliminated by burning. Tires, too hazardous to burn, float to the surface in landfills, causing continuous problems for waste managers; tires often end up stacked in immense piles that, if accidentally ignited, can burn out of control and create large clouds of black fumes.

INDUSTRIAL WASTE

The effluent streams of by-products from factories, as well as chemical and petroleum refineries, are made up of water, solid filings and cuttings, liquid solvents and oil derivatives, and semisolid sludge. The solid components are usually no more hazardous than household wastes, although medical wastes—particularly tainted blood and used "sharps," such as needles and scalpels—may pose the additional danger of spreading disease. However, liquids and semiliquids sometimes contain a high proportion of hazardous chemicals. Rain also leaches chemicals, such as cyanide and mercury, out of the smelted tailings from mines. Agricultural fertilizers and pesticides can enter groundwater and streams as well. Because these liquid wastes spread rapidly through waterways and groundwater, they are often collectively known as toxic waste.

Industries in developed countries use all ninety-two naturally occurring elements on the periodic table, and the isotopes of some of these are radioactive. Nuclear weapons manufacturing in particular leaves radioactive debris, but medical procedures that use radioactive tracers and scientific instruments may also create radioactive wastes. This nuclear waste continues to

emit radiation for thousands or hundred of thousands of years, and improperly stored radioactive materials have been associated with increased risk of disease for people, animals, and plants.

In the United States federal and state regulations brought industrial waste management under rigorous control during the 1980's and 1990's. Facilities known as secure landfills are designed to contain nonradioactive industrial wastes in tightly lined, self-contained areas. Incinerators reduce the waste to harmless ash while releasing few or no harmful particulates into the atmosphere. Separate repositories store nuclear wastes deep underground in leak-proof containers.

The public is seldom reassured by such measures, however. Leakage occasionally occurs from secure landfills. Near-zero toxic emissions from incineration means that some toxins do, in fact, escape into the atmosphere. In addition, nuclear repositories may not be catastrophe-proof; for example, an earthquake

Materials Generated* in the U.S. Municipal Waste Stream
(thousands of tons)

MATERIALS	1960	1970	2005
Paper and paperboard	29,990	44,310	83,950
Glass	6,720	12,740	12,750
Metals	10,820	13,830	18,720
Plastics	390	2,900	28,910
Rubber and leather	1,840	2,970	6,700
Textiles	1,760	2,040	11,140
Wood	3,030	3,720	13,930
Other**	70	770	4,570
Total materials in products	54,620	83,280	180,670

OTHER WASTES	1960	1970	2005
Food scraps	12,200	12,800	29,230
Yard trimmings	20,000	23,200	32,070
Misc. organic wastes	1,300	1,780	3,690
Total other wastes	33,500	37,780	64,990
Total waste by weight	88,120	121,060	245,660

Source: U.S. Environmental Protection Agency, *Municipal Solid Waste in the United States: 2005 Facts and Figures*, October, 2006.
Notes:
*Generation before materials recovery or combustion. Does not include construction and demolition debris, industrial-process wastes, or certain other wastes.
** Includes electrolytes in batteries and fluff pulp, feces, and urine in disposable diapers.

could crack open containers, releasing radioactive material into groundwater supplies. Although waste managers insist that these dangers are minimal, the news media bring potential problems to public attention, and NIMBYism regularly leads to resistance to the opening of new secure landfills and radioactive waste repositories. State governments often object as well, as was the case when the Nevada legislature stalled the construction of a nuclear waste repository at Yucca Mountain. Many old facilities, built before strict government oversight was instituted, remain in use and could leak toxic materials into the environment undetected. Memories of deadly chemical leaks, such as that discovered at Love Canal in New York in 1976, and of released radioactive material, such as the plutonium that escaped the Hanford Nuclear Reservation in Washington State, make the public wary of hazardous wastes.

As a result of citizen concern, most new hazardous waste disposal sites are now located far from population centers. This has created a new peril, however: The waste must be transported, primarily by trucks and trains, to disposal sites. Traffic accidents and train derailings en route can dump extremely dangerous chemicals straight into bodies of water or the atmosphere. Evacuations of residents near such accidents, while not common, increased during the 1990's. Even if people are rescued, however, plant and animal life is not safeguarded.

ENVIRONMENTAL CONSEQUENCES

Many critics of existing waste management policies insist that only source reduction—a drastic decrease in the use of raw materials—will make waste disposal safe. Accordingly, during the 1990's some countries, notably Denmark and Germany, sought to reduce virgin material use as much as 90 percent by intensifying recycling programs. In the United States, the Comprehensive Environmental Response, Compensation, and Liability Act of 1980, widely known as Superfund, sets aside federal funds to pay for cleanups of the most dangerous hazardous waste sites. Other industrialized nations have similar projects. Still, only a fraction of sites receive attention, and until source reduction goals are met, household and industrial wastes will continue to swell landfills with environmentally haz-

ardous substances. Illegal dumping of hazardous waste exacerbates the danger.

Scientists disagree about how severely wastes damage the environment, but there is agreement that repercussions are evident and likely to increase. Methane from dumps, smokestack emissions, and vehicle exhaust contain greenhouse gases, which are implicated in global warming. Nutrients released from sewers, as well as runoff from agriculture and mining, degrade the environment of rivers and streams, harming aquatic life and leaving the water unusable without special treatment. Waterborne wastes that reach the ocean, supplemented by ocean dumping of toxic materials, alter and sometimes destroy offshore ecosystems such as coral reefs.

Roger Smith

FURTHER READING

Almorza, D., et al., eds. *Waste Management and the Environment.* Boston: WIT Press, 2002.

Cozic, Charles P., ed. *Garbage and Waste.* San Diego: Greenhaven Press, 1997.

Hickman, H. Lanier, Jr. *American Alchemy: The History of Solid Waste Management in the United States.* Santa Barbara, Calif.: Forester Communications, 2003.

Melosi, Martin V. *Garbage in the Cities: Refuse, Reform, and the Environment.* Rev. ed. Pittsburgh: University of Pittsburgh Press, 2005.

Pichtel, John. *Waste Management Practices: Municipal, Hazardous, and Industrial.* Boca Raton, Fla.: CRC Press, 2005.

Rathje, William L., and Cullen Murphy. *Rubbish! The Archaeology of Garbage.* 1992. Reprint. Tucson: University of Arizona Press, 2001.

Rogers, Heather. *Gone Tomorrow: The Hidden Life of Garbage.* New York: New Press, 2005.

Whitaker, Jennifer Seymour. *Salvaging the Land of Plenty: Garbage and the American Dream.* New York: William Morrow, 1994.

SEE ALSO: Electronic waste; Hazardous waste; Landfills; Leachates; NIMBYism; Nuclear and radioactive waste; Sewage treatment and disposal; Solid waste management policy; Superfund; Waste treatment; Wastewater management.

Waste treatment

CATEGORY: Waste and waste management

DEFINITION: Processing that alters the physical, chemical, or biological composition of waste in order to concentrate or neutralize it

SIGNIFICANCE: Given the potential environmental hazards posed by improper disposal of wastes and the ever-increasing problem of lack of sufficient landfill space, the need for cost-efficient and effective methods of waste treatment is widely recognized.

Rapid advances in technology and industrialization have resulted in the discharge of increased quantities of wastes into the environment. Adverse environmental effects develop if the concentrations of wastes exceed the natural capacity of air, water, and land systems to assimilate them. In the United States, where pollution control is exercised by the federal government, virtually all of the environmental control legislation has been written and passed since the end of World War II, when dramatic increases in urban density and industrialization occurred.

The principal wastes associated with industrial and municipal facilities can be categorized as organic and inorganic, solid (suspended and dissolved), acid and base, and hazardous. Organic wastes are oxygen-demanding and lower the amount of dissolved oxygen in the receiving waters. Suspended solids settle and cause benthic deposits. Acid and base wastes destroy natural buffers, and acid rain has the potential to have devastating impacts on aquatic life and forests. Hazardous wastes—which include explosive, flammable, volatile, radioactive, toxic, and pathological wastes—can cause serious damage to people or property. The storage, collection, transportation, treatment, and disposal of such wastes require special caution.

Waste treatment processes can generally be divided into three categories: physical, chemical, and biological. Physical treatment is used to concentrate wastes, reduce volume, and separate different components for further treatment or disposal. Chemical treatment is used to precipitate, detoxify, or destroy the hazardous properties of wastes. Biological treatment utilizes microorganisms to stabilize organic wastes.

The degree of treatment required for any waste stream depends on the characteristics of the waste, the maximum discharge limit, and the final disposal requirements imposed by regulatory agencies. Waste treatment facilities utilize a number of processes to achieve the desired degree of treatment. More than twenty types of physical treatment processes—also called unit operations—are commonly used for handling wastes.

Removal of wastes is achieved by physical forces. Common examples of physical treatment processes are sedimentation, centrifugation, flotation, evaporation, drying, distillation, stripping, carbon adsorption, ion exchange, membrane processing, freeze crystallization, and solidification. Sedimentation, centrifugation, and flotation processes are typically used to remove suspended solids. Evaporation, drying, and distillation are utilized to concentrate wastes. Stripping removes ammonia and volatile organic compounds. Carbon and resin adsorption is used for removal of organic solute from aqueous waste streams. Their applications include separation or removal of phenols, fats, colors, pesticides, carcinogens, and chlorinated hydrocarbons. Membrane processes such as reverse osmosis, ultrafiltration, and electrodialysis utilize synthetic membranes to concentrate industrial and hazardous wastes; such processes are also used in the desalination of brackish water. In freeze crystallization, the waste stream is cooled so that pure water (in the form of ice crystals or solid ice) can be separated from the contaminants, which concentrate in liquid. Solidification is the transformation of hazardous waste into a nonhazardous solid product through fixation or encapsulation.

Chemical processes are used for the treatment of industrial wastes, usually in conjunction with other methods to achieve desired end results. Common chemical treatment processes used for waste treatment include neutralization, oxidation, reduction, precipitation, hydrolysis, catalysis, chlorinolysis, electrolysis, photolysis, and incineration. Neutralization is adjustment of pH level through the introduction of either acids or bases. This process has wide application in the treatment of wastes from many industries. Oxidation is used mainly for detoxification of hazardous wastes. Chlorine, ozone, hydrogen peroxide, and potassium permanganate are excellent oxidizing agents often used in the presence of ultraviolet light. Reduction is achieved through the use of a reducing agent such as sulfur dioxide. As an example, hexavalent chromium, a very toxic substance, is reduced to trivalent chromium, which is much less toxic, and then precipitated. Hydrolysis, catalysis, chlorinolysis, electrolysis, and photolysis are destructive processes used to break chemical bonds. Incineration is the thermal destruction of hazardous organic wastes.

Biological waste treatment processes involve biochemical reactions that take place in or around microorganisms. Generally, organic compounds are decomposed in suspended or attached growth reactors. The most common biological waste treatment processes use biological reactors to stabilize organic matter, followed by the separation of solids and liquid. The waste solids, referred to as sludge, are combined with other solids and treated separately. Anaerobic sludge digestion is used for stabilization of sludge and high-strength wastes. With proper controls, biological waste treatment processes are reliable and environmentally sound; no chemicals are added to the wastes, and operational costs are relatively low.

Syed R. Qasim

FURTHER READING

Chiras, Daniel D. "Hazardous and Solid Wastes: Sustainable Solutions." In *Environmental Science*. 8th ed. Sudbury, Mass.: Jones and Bartlett, 2010.

LaGrega, Michael D., Philip L. Buckingham, and Jeffrey C. Evans. *Hazardous Waste Management*. 2d ed. New York: McGraw-Hill, 2001.

Manahan, Stanley E. *Fundamentals of Environmental Chemistry*. 3d ed. Boca Raton, Fla.: CRC Press, 2009.

SEE ALSO: Hazardous waste; Landfills; Refuse-derived fuel; Sewage treatment and disposal; Sludge treatment and disposal; Solid waste management policy; Waste management.

Wastewater management

CATEGORIES: Water and water pollution; waste and waste management

DEFINITION: Treatment and discharge of water supplies that have been already used for consumption, bathing, washing, industrial processes, and irrigation

SIGNIFICANCE: Not enough fresh water is available on the planet to meet human water needs without the reuse of wastewater. Because human uses of water almost always result in its contamination, managing the treatment and discharge of wastewater is crucial for protecting this vital resource.

Water is arguably the single most important resource on the planet, and one for which there is no alternative. All life depends on water. Drinking water is a basic human necessity, yet more than one billion people do not have access to safe drinking water. Humans use about half the freshwater that is available for agricultural (about 70 percent), industrial (about 20 percent), and municipal (about 10 percent) purposes. Because these uses add contaminants such as nutrients, chemicals, and pathogens to water, managing the collection, use, treatment, and distribution of wastewater is of utmost importance. Additionally, healthy ecosystems depend on adequate supplies of clean water.

Wastewater management involves the rules, best practices, and technology of collecting, treating, and distributing wastewater. Wastewater is a major source of nutrient pollution, which leads to the eutrophication of aquatic ecosystems, harmful algal blooms, fish kills, and dead zones. Contaminants such as pesticides and hormones can interfere with the endocrine systems of animals, resulting in reproductive, developmental, and immunological problems observed in fish and amphibians. Some health experts are concerned that these contaminants are affecting humans, too.

MUNICIPAL WASTEWATER

Municipal wastewater contains a mixture of contaminants added to the water by every residence and business in the city. Common contaminants in municipal wastewater include detergents, urine, feces, pharmaceuticals, pesticides, food waste, fats, and chemicals from several kinds of personal care products. Gray water—wastewater from showers or laundry that does not contain human waste—is often combined with black water (water that comes from toilets containing urine and feces) and sent to a common sewage system or septic tank. It is becoming increasingly popular for gray water to be reused for applications that do not require potable water, such as irrigation and toilet flushing.

In settlements of low population density, domestic wastewater can be treated with septic systems. Household wastewater is collected in a tank, where anaerobic bacteria digest solid waste and the liquid component is released into a leach field, where water percolates into the soil or is transpired by plants. Water released from septic systems contains high concentrations of nitrogen and phosphorus, which con-

Aeration basins at a wastewater treatment plant in Wilmington, Delaware. (AP/Wide World Photos)

tribute to eutrophication of aquatic ecosystems. Septic tanks must be periodically pumped to remove solids that are not decomposed by bacterial digestion.

In higher-density communities, domestic wastewater is collected in networks of pipes and transported to central wastewater treatment plants. Some collection systems combine stormwater with sewage for treatment. (Stormwater by itself is not considered wastewater, because it has not been used by people.) Municipal sewage comprises a complex mixture of contaminants, but the main contaminants that wastewater treatment plants remove include solids, wastes that impose a high biochemical oxygen demand (BOD), and nitrogenous waste. Municipal wastewater treatment includes physical separation (primary treatment) of large solids, such as rags, condoms, and tampon applicators, using a bar screen; the removal of grit; and gravity separation of smaller solids that sink and fats and oils that float. Biological removal (secondary treatment) of BOD wastes, ammonia, nitrates, and other nutrients takes place in aeration basins, trickle filters, and other kinds of bioreactors.

Although different secondary treatment processes are used, they all engage both aerobic bacteria, to metabolize organic material and carry out nitrification to reduce BOD and ammonia, and anaerobic denitrifying bacteria, to convert nitrates to harmless nitrogen gas and reduce BOD. Protozoans, fungi, and small animals such as rotifers and insects are found in these biological systems, further reducing the amount of waste solids in this bacteria-based food web. Solids and liquids are again separated by gravity in a clarifier. The solids are treated with aerobic or anaerobic digestion, and the resulting biosolid is incinerated, stored in a landfill, or used as a soil amendment. Anaerobic digestion produces methane gas, which can be used for power generation. The effluent is "polished" (a tertiary treatment) through processing that removes additional contaminants. Finally, disinfection is carried out before the treated water is discharged back into the environment. Many communities are increasingly using such "reclaimed water" for irrigation.

Municipal wastewater treatment plants do not remove all contaminants, and many of the chemical contaminants that remain in water discharged from wastewater treatment plants have negative environmental consequences. There is mounting evidence, for example, that treated effluent has endocrine-disrupting compounds that create reproductive problems in fishes. Caffeine is another compound that is not removed by standard wastewater treatment, making it a commonly used tracer compound for waters affected by human excrement.

AGRICULTURAL WASTEWATER

In the United States, raising animals such as cattle, pigs, chickens, and fish for human consumption creates more than three times as much urine and feces as are produced by humans. This type of agricultural wastewater is often retained in ponds, which allows for some settling of solids and microbial degradation of waste before it is released into the environment. During rainstorms, however, large amounts of this minimally treated wastewater can escape into the larger environment. The solids that are retained are typically composted and used as soil amendments. Although it is possible to treat this wastewater using the processes described above for municipal wastewater, such treatment rarely occurs.

Waste irrigation water is the largest source of wastewater. This wastewater is contaminated by whatever materials are applied to the land, such as fertilizers, insecticides, and herbicides. Irrigation wastewater gets contaminated by the 2.5 million tons of pesticides that are applied annually across the globe. One of the most effective "best practices" to protect aquatic ecosystems is to maintain a riparian buffer between crop fields and streams. Riparian vegetation slows the flow of runoff, trapping some of the sediment and contaminants in the irrigation wastewater. Nutrient farming is a growing practice of treating irrigation wastewater using the ecosystem service of nutrient removal. With this practice, coupled nitrification/denitrification is enhanced, removing nitrogen pollution from the water and lessening the problem of eutrophication.

INDUSTRIAL WASTEWATER

Industrial processes contaminate water with acids, bases, metals, solvents, oil, grease, organics, and BOD. Urban industries work closely with municipal wastewater treatment plants to devise plans to pretreat wastewater before it is discharged into sewage treatment systems designed to remove wastes, such as metals, that municipal treatment plants do not treat well.

Larger-scale industries, notably the mining industry, produce large amounts of metal-contaminated wastewater that cannot be combined with municipal wastewater. This wastewater is generally held in settling ponds before it is discharged into surface waters. Inadequate treatment can often lead to problems with acid mine drainage and therefore significant environmental impairment. Many Superfund sites (sites of abandoned or uncontrolled hazardous wastes that have been identified by the Environmental Protection Agency and placed on its National Priorities List for eventual cleanup) have had to be established because of inadequate treatment of mining wastes.

Mining operations constitute the largest single source of waste in the United States, estimated to total 50 billion tons located at more than 500,000 sites. Active treatment plants can remove metals from mining wastewater by adjusting the pH level of the waste to around 10, precipitating the metal hydroxides with polymers, separating the precipitates in a clarifier, filtering the water, and adjusting the pH back to neutral (about 7) for discharge into the environment. The metal hydroxides are dewatered and discarded. Passive treatment of mining and other wastewaters can be achieved with constructed wetlands, which carry out the ecosystem services that improve water quality. Passive treatment is cheaper than active treatment, making it more applicable for most contaminated sites in remote locations.

EMERGING TECHNOLOGIES

The diversity of materials that contaminate water presents a huge challenge for effective treatment. Advances are being made in wastewater treatment to help correct some of the shortcomings in current practices mentioned above. Removal of endocrine disrupters from municipal wastewater is being achieved using activated carbon, wetlands, reverse osmosis, and even aquaponic systems. Microbial fuel cells are being tested to convert organic waste in water directly to electricity. Some dairy farmers are generating methane from manure and using it to heat buildings or produce electricity. Wastewater effluent flows downgradient to surface waters in nearly all wastewater treatment plants. This hydraulic head can be used to drive turbines for electrical generation. When wastewater is viewed as a resource instead of an environmental hazard, wastewater treatment plants can become power plants.

Greg Cronin

FURTHER READING

Grigg, Neil S. *Water, Wastewater, and Stormwater Infrastructure Management.* Boca Raton, Fla.: CRC Press, 2002.

Hammer, Mark J. *Water and Wastewater Technology.* Upper Saddle River, N.J.: Pearson, 2004.

Miller, G. Tyler, Jr., and Scott Spoolman. "Water Pollution." In *Living in the Environment: Principles, Connections, and Solutions.* 16th ed. Belmont, Calif.: Brooks/Cole, 2009.

Qasim, Syed A. *Wastewater Treatment Plants: Planning, Design, and Operation.* 2d ed. Lancaster, Pa.: Technomic, 1999.

Roseland, Mark. "Water and Sewage." In *Toward Sustainable Communities: Resources for Citizens and Their Governments.* Rev. ed. Gabriola Island, B.C.: New Society, 2005.

U.S. Environmental Protection Agency. Office of Wastewater Management. *Primer for Municipal Wastewater Treatment Systems.* Washington, D.C.: Author, 2004.

SEE ALSO: Chlorination; Clean Water Act and amendments; Ecosystem services; Eutrophication; Leachates; Odor pollution; Septic systems; Sewage treatment and disposal; Sludge treatment and disposal; Stormwater management; Waste treatment; Water conservation; Water pollution; Water treatment; Water use.

Water conservation

CATEGORIES: Water and water pollution; resources and resource management

DEFINITION: Management of water consumption in ways that minimize waste, maximize efficiency, and help to maintain adequate supplies of high-quality water

SIGNIFICANCE: Water conservation is a comparatively simple and inexpensive means of addressing water shortages and making the most of existing water supplies. Effective conservation programs can postpone or prevent the need to construct additional water-supply infrastructure and can also reduce wastewater discharges. As the earth's population increases, conservation measures become an increasingly important tool for meeting society's water needs.

In 2010 the world population reached 6.9 billion and was projected to rise to 9.5 billion by the year 2050. Because water is a basic human necessity, it follows that the amount of water needed for human existence will increase by a factor of roughly 1.4 over this forty-year period. While such an expansion in water supply is theoretically possible to achieve, the likelihood that it can or will happen is low for several reasons. First, although there is enough water available globally, it is not uniformly distributed around the world when and where it is needed. Second, the costs involved in such an expansion would be astronomical. Third, even if all needed moneys were readily available, it would be difficult to build the necessary facilities and have them operational by the time they would be needed.

DOMESTIC, INDUSTRIAL, AND AGRICULTURAL MEASURES

The greatest potential for water conservation in residential settings is in bathrooms. Traditional toilets use 19 to 26 liters (5 to 7 gallons) per flush. However, low-flow toilets using 13 or fewer liters (3.5 gallons) per flush—and ultra-low-flush models using as little as 6.1 liters (1.6 gallons)—are increasingly being installed in new homes and as replacements for traditional toilets in older homes. Baths and showers also use large amounts of water, and low-flow showerheads can contribute to water conservation efforts.

A great deal of water is also wasted in kitchens. Water can be conserved at kitchen sinks through simple means such as collecting food scraps for compost piles rather than putting them through garbage disposals. Low-flow sink faucets are also available. Running automatic dishwashers only for full loads is another water conservation technique; the same is true for clothes washers. Older models of both types of washers can be replaced with newer, more water-efficient ones.

Another potential source of water conservation in residential areas is found outside the homes, in plant and lawn watering. Methods of saving water include decreasing lawn sizes, landscaping with plants that require low amounts of water, mulching to reduce moisture evaporation from the soil, minimizing the frequency of watering, and using drip systems instead of sprinklers.

Business and industrial settings use copious amounts of water. In some cases, water conservation measures may involve simple changes in the way

U.S. Daily Water Use, 1940-2000
(billions of gallons)

Year	Total	Per Capita (Gallons)	Irrigation	Public Supply	Rural	Industrial & Misc.	Steam Electric Utilities
1940	140	1,027	71	10	3.1	29	23
1950	180	1,185	89	14	3.6	37	40
1960	270	1,500	110	21	3.6	38	100
1970	370	1,815	130	27	4.5	47	170
1980	440	1,953	150	34	5.6	45	210
1990	408	1,620	137	41	7.9	30	195
1995	402	1,500	134	40	8.9	29	190
2000	408	1,430	137	43	9.2	23	196

Source: U.S. Department of Commerce, *Statistical Abstract of the United States, 2004*, 2004.
Note: Per capita figures are gallons; all other values are in billions of gallons.

things are done, such as substituting sweeping for the washing down of floors. In most cases, however, in these settings water conservation on a noticeable scale results from process changes of some type. For example, by converting from a water-cooled ice machine to an air-cooled one, a restaurant may improve water efficiency in its ice making by 25 to 50 percent.

The largest amounts of water are used in agriculture for irrigation. In the United States, a history of relatively cheap and readily available water led farmers to waste great amounts through improper or excessive application. Water can be and has been saved through computerized timing of water application. Modifications of irrigation procedures have also helped. For example, drip irrigation, which applies water slowly and uniformly at or below soil level adjacent to plants through mechanical water outlets, has produced significant water savings over traditional methods of simply spraying water onto the soil.

OTHER WAYS OF CONSERVING

As is true of the consumption of other commodities, water usage can be affected by pricing. Water-conserving rates exhibit increasing unit costs as volume used increases. One simple model determines a certain rate for the average amount of water a household might be expected to use and a much higher rate for all water used over that average amount.

Yet another way in which municipalities as well as home and business owners can conserve water is by finding and repairing leaks that occur anywhere in water distribution systems; water audits—careful reviews of water usage over time—can help to identify leaks. In some water distribution systems, documented leakages have been found to amount to losses of more than 50 percent of the water sent through the systems.

Water reuse may be thought of as another form of water conservation. Ample opportunities for water reuse exist during industrial processes. For example, wastewaters that were being treated on-site and discharged into receiving streams might be reused as cooling water. Households might recycle water used for dishwashing or bathing as irrigation water for outdoor plants. While it is possible to treat sewage for reuse by households, most people are not yet ready to accept such a conservation measure.

The negative aspects of water conservation are generally related to monetary issues. For example, successful water conservation efforts can lead to reduced revenues for water utility companies, which may then be forced to increase their rates. Conserving water over the long term may also reduce the "slack" in the system, making short-term drought savings difficult to achieve and reducing the amount of water available for water rationing. Some water conservation efforts can be expensive to put in motion, while the actual savings from reduced water use are achieved more slowly, over a period of time. Such water conservation is not always cost-effective.

Jack B. Evett
Updated by Karen N. Kähler

Further Reading

American Water Works Association. *Water Conservation Programs: A Planning Manual.* Denver, Colo.: Author, 2006.

Asano, Takashi, et al. *Water Reuse: Issues, Technologies, and Applications.* New York: McGraw-Hill, 2007.

Chiras, Daniel D. "Water Resources: Preserving Our Liquid Assets and Protecting Aquatic Ecosystems." In *Environmental Science.* 8th ed. Sudbury, Mass.: Jones and Bartlett, 2010.

Glennon, Robert Jerome. *Unquenchable: America's Water Crisis and What to Do About It.* Washington, D.C.: Island Press, 2009.

Postel, Sandra. *Last Oasis: Facing Water Scarcity.* New York: W. W. Norton, 1992.

Unger, Paul W. *Soil and Water Conservation Handbook: Policies, Practices, Conditions, and Terms.* Binghamton, N.Y.: Haworth Press, 2006.

U.S. Environmental Protection Agency. *Cases in Water Conservation: How Efficiency Programs Help Water Utilities Save Water and Avoid Costs.* Washington, D.C.: Author, 2002.

Vickers, Amy. *Handbook of Water Use and Conservation: Homes, Landscapes, Businesses, Industries, Farms.* Amherst, Mass.: Waterplow Press, 2001.

SEE ALSO: Drinking water; Irrigation; Rainwater harvesting; Resource depletion; Sewage treatment and disposal; Wastewater management; Water quality; Water-saving toilets; Water use.

Water pollution

CATEGORY: Water and water pollution

DEFINITION: Degradation of natural water as the result of human activities

SIGNIFICANCE: The pollution of water supplies, which can be caused by activities such as mining, agriculture, and improper waste disposal, can have negative health impacts on plant and animal life.

Before the Industrial Revolution in the nineteenth century, humans produced only minimal amounts of refined metals and organic materials. The production of various alloys of copper, tin, lead, and zinc by heating the mineral ores or by using natural copper metal was also minor. During the Industrial Revolution, however, people began to produce cast iron by burning charcoal to heat iron ores at high temperatures. They also used other methods to produce additional metals; for example, nickel, aluminum, titanium, cobalt, platinum, chromium, niobium, and molybdenum were discovered during this time. The extensive use of such metals resulted in massive increases in exploration, mining, and production, as well as increased use of energy, resulting in waste disposal problems and contamination of water supplies. In addition, growing urban populations produced concentrations of untreated human and animal wastes and associated disease-producing organisms in natural waters.

Inorganic Constituents

There are both natural and human sources of water contamination. Humans may increase natural contamination by, for example, mining natural resources and disposing of the waste, which may leach out dangerous constituents. Animal and plant health may be affected by enrichment or deficiency of certain dissolved constituents. Rocks and soils may have low concentrations of substances such as selenium, potassium, phosphorus, copper, cobalt, molybdenum, zinc, or iodine, which cause health problems in animals. Although humans may need to supplement their diets with these constituents for optimal health, substances such as selenium, radioactive elements, copper, and zinc can be concentrated enough in some drinking waters to be harmful.

Humans mine many elements and use them for manufacturing and other purposes faster than these elements weather out of natural rocks, producing a variety of atmospheric and water pollutants. For example, fertilizers have high concentrations of soluble nitrogen and phosphorus compounds, and animal wastes contain high nitrate. The use of fertilizers thus can produce high nitrate (a nitrogen-oxygen compound) and phosphate (a phosphorus-oxygen compound) in natural waters. Nitrate can combine with hemoglobin so that oxygen transport in the body is inhibited. This is a potentially serious threat to infants; they can literally turn blue, become sick, and die if they consume water high in nitrate over a long period of time. Water high in phosphorus can stimulate the growth of organisms such as algae. As the abundant algae die and drop to the bottom of a body of water, they may use up the dissolved oxygen in the water, which may, in turn, cause fish to die.

The use of table salt (sodium chloride) to melt the ice on roads in winter can result in high dissolved sodium and chloride concentrations in natural waters. In the past, deep groundwater with high salt concentrations that was brought up during petroleum extraction was placed in "evaporation pits" on the surface until the water evaporated. This would allow the salt to leak slowly into the water supply in the ground. Now petroleum companies are required to inject these salty waters back into the ground at the level from which they came.

Another major pollutant in water results from the acidity produced by acid mine drainage and acid rain. Acid mine drainage results from the chemical reaction of sulfide minerals such as pyrite (iron sulfide) with water and oxygen from the atmosphere to produce sulfuric acid and dissolved metals in water. Most

Workers attempt to plug a copper mine tunnel that spilled millions of tons of mine tailings into the Boac River in the Philippines. Mining operations of all kinds are major contributors to water pollution around the world. (AFP/ Getty Images)

acid mine drainage comes from small amounts of pyrite in coal mines or waste piles from coal miners. Some acid mine drainage results from metal mines and wastes such as those found in lead and zinc mines in southeastern Kansas. These acid waters can readily dissolve other metals, so acid mine waters may contain high concentrations of many poisonous metals.

Acid rain results from high concentrations of sulfur dioxide, carbon dioxide, and nitrogen oxide gases spewed out into the atmosphere by industry. These gases dissolve in the water in the atmosphere to produce acidity. The worst acid rain thus occurs in industrial areas. Acid rain produces acid lakes and streams in areas that have little natural capacity to neutralize the acid. This results in the destruction of organisms that cannot live in such acidic waters. In some areas that have abundant limestone (a rock composed of calcium carbonate), acid rain reacts with the limestone to neutralize the acidity, so there is little problem with acidity of the natural waters. Areas without rocks that can neutralize the acid continue to have problems with acidity.

ORGANIC COMPOUNDS

Organic compounds consist of carbon in chemical combination with hydrogen, oxygen, sulfur, chlorine, or nitrogen. Many thousands of organic compounds are currently manufactured, and their classification is complex, but they may be grouped simplistically as alkanes, benzene derivatives, chlorinated hydrocarbons, and pesticides. Alkanes are straight chains of carbon atoms combined with hydrogen. Benzene derivatives consist of six-membered rings of carbon combined with other constituents chemically attached to the carbon atoms. Alkanes and benzene derivatives with six to ten carbon atoms are the organic compounds found in gasoline, diesel fuel, and other fuels. Alkanes in gasoline are not very soluble in water, but benzene derivatives are. Alkanes are also more easily degraded by bacteria than are benzene derivatives. The leakage of gasoline into the groundwater system can be a major pollution problem in areas around gasoline stations.

The term "groundwater" refers to any water found under the earth's surface. The upper portions of groundwater contain both air and water in the pore space between mineral grains; the lower portions of groundwater contain only water in the pore space. The surface between the upper and lower zone is called the water table. Wells are usually drilled into the water table so some of the more soluble hydrocarbons of gasoline that has leaked into the groundwater system can dissolve in water and move in the saturated water zone. The insoluble alkanes and benzene derivatives of gasoline can also move as a separate fluid plume above the water table since they are lighter than water. The maximum allowable concentrations of these benzene derivatives in water are much lower than the allowable concentrations of the individual elements discussed previously.

Benzene derivatives are also fairly volatile. If they move in a liquid plume under buildings, they can move as vapors into basements or sewers, where they then may produce explosions or may cause illness in occupants of the affected buildings.

Chlorinated hydrocarbons contain the element chlorine in one or more parts of the compound. Many of these—such as dichloroethane, tetrachloroethane, and chloroform—are among the most common organic pollutants found in waste disposal sites in the United States. Many are carcinogens (cancer-producing substances), and they become increasingly toxic at higher concentrations.

PESTICIDES

Pesticides are complex organic compounds that kill unwanted organisms such as insects. Examples of pesticides include dichloro-diphenyl-trichloroethane (DDT), malathion, alachlor, atrazine, and chlordane. Pesticides may be carcinogenic, and some may not decompose readily in the food chain. DDT, for example, has long been banned in the United States because of its harmful effects and its slow decomposition in nature.

Pesticides vary greatly in the time it takes them to decompose naturally and move to the water table. At one extreme, pesticides such as prometon can last a long time and quickly move to the water table, thus rapidly contaminating the groundwater. At the other extreme, methyl parathion decomposes more readily and moves more slowly to the water table, and is thus less likely to contaminate the groundwater. Among the pesticides most often found in groundwater are dibromochloropropane (DBCP), aldicarb, carbofuran, chlordane, alachlor, and atrazine. In addition, a pesticide applied during times of high rainfall can rapidly move to streams, where the stream water can soak into the ground and contaminate the groundwater supply.

Another problem is that even if the original pesticide has been naturally destroyed by bacteria, the degradation products from the decomposition of the pesticide can be even more harmful to humans than the original pesticide. Few studies of these kinds of problems have been undertaken, and decay products from pesticides are often only poorly understood. The degradation products of a few pesticides, such as aldicarb, however, have been studied for a number of years. Such products may have entirely different movement and stability than the original pesticides.

RADIOACTIVITY AND HEAT POLLUTION

Humankind's use of radioactive materials has created special problems in the areas of waste disposal and water pollution, in particular because radiation cannot be detected by the senses and can be very damaging if ingested. At one extreme, the radioactive element plutonium has a half-life (the time it takes for one-half of the radioactivity to decay) of about twenty-four thousand years, and it concentrates in the bones of vertebrates. This means that plutonium that has leaked into groundwater must be removed for hundreds of thousands of years. At the other extreme, radioactive materials with short half-lives that do not concentrate in organisms may not be of much concern, since the radioactivity will decay before it can be ingested.

Radioactive wastes are divided into low-level, intermediate-level, and high-level wastes. High-level wastes may be more than one million times more radioactive than what is considered acceptable for human exposure. Low-level wastes may contain radioactivity up to one thousand times more radioactive than what is considered acceptable. Intermediate-level wastes have radioactivity between these ranges. A wide variety of low-level radioactive wastes are produced by hospitals, the nuclear industry, and research laboratories. These wastes are often sealed in drums and buried under a thin layer of soil or diluted in water to acceptable levels of radioactivity and flushed into sewer systems.

High-level radioactive wastes are produced by nuclear fuel generation in fairly small volumes and account for about 95 percent of radioactive waste materials. Many high-level wastes have been stored for decades in double-walled stainless-steel tanks that are air-conditioned because of the intense heat given off by the radioactivity. Some of these tanks have leaked, and radioactive fluids have moved into nearby groundwater systems.

Thermal pollution is the heating of natural waters caused by the activities of industry, the burning of fossil fuels, or nuclear power production. Dumping heated waters directly into a body of water can kill many temperature-sensitive organisms living there. One method of reducing thermal pollution is to allow heated water to cool in ponds before it is returned to the source body of water.

PREVENTION AND REMEDIATION

In the long run, it is easier and less expensive to prevent the pollution of natural waters than to try to clean a polluted water supply or remediate sources of pollution. Preventive approaches include keeping contaminants contained so they cannot escape into water systems and banning the production and use of dangerous substances such as certain pesticides. One way of disposing of potential polluting substances would be to transport them to an area with a dry climate and low population density and place them in geologic materials that are impermeable to water flow. This would minimize the chances that moving water would carry hazardous constituents to groundwater. The costs of transporting waste materials are high, however, and this means that municipal wastes are usually disposed of locally. In areas that have high rainfall, waste management facilities must be especially careful to keep the waste contained in geologic materials, such as unfractured mudstone, that are impermeable to movement of water.

The expense of both prevention and remediation of water pollution is a source of ongoing conflict between environmentalists and the industries that produce polluting materials. In general, industries are interested in avoiding the costs associated with proper waste disposal, whereas environmentalists' priority is to keep pollution to surface water and groundwater to a minimum.

Robert L. Cullers

FURTHER READING

Goudie, Andrew. "The Human Impact on the Waters." In *The Human Impact on the Natural Environment: Past, Present, and Future.* 6th ed. Malden, Mass.: Blackwell, 2005.

Hill, Marquita K. "Water Pollution." In *Understanding Environmental Pollution.* 3d ed. New York: Cambridge University Press, 2010.

Manahan, Stanley E. "Water Pollution." In *Fundamentals of Environmental Chemistry.* 2d ed. Boca Raton, Fla.: CRC Press, 2001.

Montgomery, Carla W. "Water Pollution." In *Environmental Geology.* 9th ed. New York: McGraw-Hill, 2010.

Smol, John P. *Pollution of Lakes and Rivers: A Paleoenvironmental Perspective.* 2d ed. Hoboken, N.J.: Wiley-Blackwell, 2008.

Spellman, Frank R. "Water Pollution." In *The Science of Water: Concepts and Applications.* 2d ed. Boca Raton, Fla.: CRC Press, 2008.

Thomas, Sarah V., ed. *Water Pollution Issues and Developments.* New York: Nova Science, 2008.

Viessman, Warren, et al. *Water Supply and Pollution Control.* 8th ed. Upper Saddle River, N.J.: Pearson/Prentice Hall, 2009.

Vigil, Kenneth M. *Clean Water: An Introduction to Water Quality and Water Pollution Control.* 2d ed. Corvallis: Oregon State University Press, 2003.

SEE ALSO: Acid deposition and acid rain; Acid mine drainage; Hazardous waste; Heavy metals; Landfills; Lead; Mercury; Nuclear and radioactive waste; Runoff, agricultural; Runoff, urban; Thermal pollution.

Water-pollution policy

CATEGORY: Water and water pollution

DEFINITION: High-level governmental plan of action for protecting waters from environmental degradation that would render them unfit for desired uses

SIGNIFICANCE: Water-pollution policy is determined by laws and regulatory agencies that deal with a society's interactions with waterborne contaminants, including infectious disease organisms. Included are policies that prevent the entry of these agents into or lower their levels in aquatic ecosystems. Two primary goals pervade such policies: protection of human health and protection of natural aquatic resources.

Water pollution can be defined as any physical, chemical, or biological change in water quality that adversely affects living organisms or makes water unsuitable for desired uses. There are natural sources of water contamination, such as oil seeps and toxic algal blooms, but water pollution is generally caused by human activities, and natural processes that produce contamination are often triggered or exacerbated by human actions. Categories of water pollutants include infectious agents such as bacteria and viruses; organic chemicals such as pesticides, plastics, and oil; inorganic chemicals such as acids, salts, and metals; radioactive materials; sediments; plant nutrients such as nitrates and phosphates; oxygen-demanding wastes such as manure and plant residues; and heat. Each of these categories presents unique scientific and technical problems that must be addressed through specific policies.

While the term "water pollution" evokes images of industrial or sewage discharge flowing into pristine waters, most pollution sources are not so easy to identify or control. Nonpoint sources—diffuse sources of a variety of contaminants—contribute substantially to pollution problems and pose a major challenge to pollution-control efforts. Wherever rainfall, snowmelt, and irrigation flow across the land or through the ground, they pick up contaminants. Ultimately, those contaminants migrate into rivers, lakes, coastal waters, and groundwater.

WATER QUALITY IN THE UNITED STATES

Since the 1970's water-pollution policy has been highly effective at increasing the quality of water in the United States. However, surveys indicate that a significant percentage of the surface waters and estuaries in the United States still do not meet water-quality goals. The National Water Quality Inventory Report is the primary method for informing Congress and the public about the conditions of water quality. The report, required under section 305(b) of the Clean Water Act, periodically characterizes the nation's water quality, identifies problem with water quality of national significance, and describes various programs implemented to restore and protect water. During 2004 a total of 16 percent—about 907,600 kilometers (564,000 miles)—of the river miles in the United States were surveyed for water quality. Of the surveyed miles, 56 percent had good water quality; 3 percent of these, however, were considered threatened. Some form of pollution affected the remaining miles. Pathogens, habitat alterations, and organic enrichment were cited as the leading pollutants entering rivers, with agricultural runoff, hydromodification (alterations such as dam construction and channeli-

U.S. Drinking Water: Maximum Allowed Concentrations of Key Toxic Compounds	
CONSTITUENT	MILLIGRAMS PER LITER
Arsenic	0.010
Atrazine (pesticide)	0.003
Benzene (volatile organic)	0.005
Cadmium	0.005
Chromium	0.1
Cyanide	0.2
Lead	0.015
Mercury	0.002
Pentachlorophenol	0.001
Selenium	0.05

zation), and unspecified nonpoint sources being the leading sources of the impairments.

The 2004 inventory assessed some 29 percent (approximately 65,780 square kilometers, or 25,400 square miles) of the nation's estuarine waters; of these, 70 percent were found to have good water quality. The 30 percent regarded as impaired were most affected by pathogens, organic enrichment, and mercury. Atmospheric deposition and unspecified nonpoint sources were major sources of pollutants. Despite the construction of new sewage treatment plants, improvements in older plants, and an increasing focus on improving the quality of stormwater runoff, municipal discharge was found to be another leading pollutant of estuaries.

In the same inventory, 39 percent of lake acreage in the United States was surveyed. Of this, 35 percent was reported to have good water quality, and an additional 1 percent was deemed good but threatened. The remaining 64 percent was impaired to the greatest extent by mercury, polychlorinated biphenyls, and nutrients (such as nitrogen and phosphorus). Atmospheric deposition, unspecified nonpoint sources, and agriculture were reported to be the leading sources of these pollutants.

U.S. WATER-POLLUTION LAWS

Numerous laws have been passed in the United States that have some influence on water quality in the country. In response to increasing public concern about water pollution, Congress passed the Federal Water Pollution Control Act (FWPCA) of 1948, the first federal legislation to deal directly with the issue. Its primary goal was to provide funding for the research and implementation of state programs to control water pollution. Additional legislation and funding occurred through a 1956 amendment to the FWPCA, which was drafted to combat water-quality problems associated with increasing industrialization. The early emphasis of state control of water quality was extended in a 1965 amendment to the FWPCA. Called the Federal Water Quality Act, it required states to adopt water-quality standards and implementation plans. However, this act did not provide for sufficient enforcement mechanisms, and by 1972 only about one-half of the states had set water-quality standards. Furthermore, many of the states did a poor job of enforcing the standards, particularly in the case of individual dischargers fouling state waters.

Congress passed the Federal Water Pollution Control Act Amendments of 1972 with the objective of restoring and maintaining the chemical, physical, and biological integrity of the nation's waters. Some of the specific goals of the act were to eliminate all discharges of pollutants into the navigable waters by 1985; protect fish, shellfish, and wildlife, and provide for recreation by 1983; prohibit the discharge of toxic pollutants in toxic amounts; and provide financial assistance for the construction of publicly owned waste treatment works. The amended act sought to accomplish these goals by combining state water-quality standards with the technology-based approach of setting effluent limitations. The Clean Water Act amendment of 1977, which addressed various technical issues that had become apparent since passage of the 1972 amendments, gave the legislation its current name.

In 1987 the Clean Water Act underwent further amendment. Congress recognized that, although significant progress had been made, substantial water-quality problems persisted. New provisions established a comprehensive program for controlling toxic pollutant discharges beyond that already provided in the act, added a program requiring states to develop and implement programs to control nonpoint sources of pollution, and authorized a total of $18 billion in aid for wastewater treatment assistance.

While the ambitious goals of the Clean Water Act have not been entirely met, there are many success stories. Notable among them is the improved management of the nation's municipal wastewater. The number of people in the United States served by sew-

age treatment plants increased from 85 million in 1972 to some 208 million by 2000. Discharges of untreated sewage were eliminated by 1996.

OTHER IMPORTANT POLICIES

In order to protect the environment and human health in the United States, Congress has passed a number of other water-quality bills. For example, in response to suggestions that drinking-water policies and enforcement of those policies were too lax, especially in rural water districts and small towns, the Safe Drinking Water Act of 1974 was passed. This law regulates water quality in commercial and municipal drinking-water systems by establishing minimum standards for drinking-water quality for every community. Among the contaminants regulated are bacteria; nitrates; metals such as arsenic, cadmium, chromium, lead, mercury, and silver; pesticides; radioactivity; and turbidity. The act also contains limited provisions for the protection of groundwater and aquifers.

In 1976 Congress passed the Toxic Substances Control Act, which categorizes toxic and hazardous substances, funds a research program, and regulates the use and disposal of poisonous chemicals. Before a new chemical can be manufactured in bulk, the manufacturer must submit a premanufacturing report to the Environmental Protection Agency (EPA) in which the environmental impacts are assessed, including those associated with disposal of the chemical.

Another important environmental protection law passed in 1976 was the Resource Conservation and Recovery Act (RCRA). The law, which regulates hazardous waste and nonhazardous solid waste, takes a "cradle-to-grave" approach to hazardous waste management. Under RCRA, once hazardous waste is generated, it must be tracked, handled, stored, transported, and disposed of responsibly. The law also encourages recycling and reuse as a means for reducing the volume of waste requiring disposal.

In response to water-quality problems associated with toxic waste dumps, Congress passed the Comprehensive Environmental Response, Compensation, and Liability Act (CERCLA) in 1980. This law is also known as the Superfund act because it established a fund that has provided billions of dollars toward cleaning abandoned toxic waste dump sites. Amendments to CERCLA include a 1986 act establishing a community's right to know about the presence of toxic materials in their area. Additional legislation, the Oil Pollution Act of 1990, was passed to toughen cleanup requirements and penalties for oil discharges after the highly publicized *Exxon Valdez* oil spill of 1989.

During the early twenty-first century, U.S. water-pollution policy under President George W. Bush's administration (2001-2009) relaxed to become more industry-friendly. Supreme Court decisions in 2001 and 2006 rolled back wetlands protections, funding lagged for wastewater treatment infrastructure, enforcement cases were stalled or dropped, and in a 2008 rule the transfer of contaminated water from one water body into a cleaner receiving water body was deregulated. Nine states sued the EPA over this water-transfer rule. During the presidency of Barack Obama in 2010 Congress began considering an amendment to the Clean Water Act that would restore protections to all natural water bodies put at risk by the Supreme Court decisions of 2001 and 2006.

INTERNATIONAL AGREEMENTS

The Convention on the Prevention of Marine Pollution by Dumping of Wastes and Other Matter, better known as the London Convention of 1972, is an international treaty ratified by the United States that calls for the cessation of the dumping of industrial wastes, effluents from cargo tank washing, and plastic trash into the world's oceans. The United States passed the Marine Protection, Research, and Sanctuaries Act in 1972 to support the provisions of the London Convention. Pollution from oceangoing vessels was addressed in another international agreement, the 1973 International Convention for the Prevention of Pollution from Ships, and a 1978 protocol that amended it. MARPOL 73/78, as the amended treaty is known, was further amended in 1988 to prohibit the dumping of plastics anywhere in the ocean.

Another key international agreement is the 1972 Great Lakes Water Quality Agreement between Canada and the United States. This agreement and subsequent amendments in 1978 and 1987 affirmed the two countries' determination to restore and enhance water quality in the Great Lakes system, which includes the entire lake basin and the St. Lawrence River. The initial agreement focused on the eutrophication of the Great Lakes and the need to reduce loadings of phosphorus. Since the signing of the agreement in 1972, additional objectives have focused on the virtual elimination of persistent and toxic substances.

Roy Darville
Updated by Karen N. Kähler

FURTHER READING

Copeland, Claudia. *Clean Water Act: A Summary of the Law.* Washington, D.C.: Congressional Research Service, 2008.

Finkmoore, Richard J. *Environmental Law and the Values of Nature.* Durham, N.C.: Carolina Academic Press, 2010.

Gross, Joel M., and Lynn Dodge. *Clean Water Act.* Chicago: American Bar Association, 2005.

Lazarus, Richard J. *The Making of Environmental Law.* Chicago: University of Chicago Press, 2004.

Milazzo, Paul Charles. *Unlikely Environmentalists: Congress and Clean Water, 1945-1972.* Lawrence: University Press of Kansas, 2006.

United States Environmental Protection Agency. *National Water Quality Inventory: Report to Congress, 2004 Reporting Cycle.* Washington, D.C.: Author, 2009.

SEE ALSO: Clean Water Act and amendments; Groundwater pollution; London Convention on the Prevention of Marine Pollution; Ocean dumping; Ocean pollution; Safe Drinking Water Act; Superfund; Water pollution; Water quality; Watershed management.

Water quality

CATEGORY: Water and water pollution

DEFINITION: Characteristics of water as defined by the solutes and gases dissolved in it, as well as the matter suspended in it

SIGNIFICANCE: Because safe drinking water is crucial to human and other animal life, scientists have developed methods for determining the quality of water supplies, and governments have established minimum standards for water quality.

Only a tiny fraction of the earth's abundant water supply is available as fresh water for consumption. Once a water source becomes contaminated or polluted, it must be restored before it can be returned to its original desired use. The U.S. government has passed laws to help ensure that natural water resources are protected from contamination and that water meets certain quality standards before it is consumed. The 1977 amendments to the Clean Water Act directed each state to establish water-quality standards for bodies of surface water. The Safe Drinking

Concentrations of Main Constituents in Some Natural Waters
(milligrams per liter)

CONSTITUENT	ESTIMATED AVERAGE WORLD RIVER WATER	AVERAGE SEAWATER	SHALLOW GROUNDWATER	DEEP GROUNDWATER, MIDLAND, MICHIGAN
Calcium	13.0	410	48.0	93,500
Magnesium	3.4	1,350	3.6	12,100
Sodium	5.2	10,500	1.0	28,100
Potassium	1.3	390	1.2	11,700
Bicarbonate	52.0	142	152.0	low
Sulfate	8.3	2,700	3.0	17
Chloride	5.8	19,000	8.0	255,000

Water Act of 1974 mandated the U.S. Environmental Protection Agency (EPA) to establish drinking-water standards for all public water systems serving twenty-five or more people or having fifteen or more connections.

When contaminants enter a water supply, the quality of the water is often compromised. The contaminants affect the water in such a way as to alter one or more quality parameters. Several parameters are used to characterize a given body of water; these can be broadly classified into physical or chemical parameters. Physical parameters include turbidity, color, temperature, taste, odor, and amount of suspended solids. Chemical parameters include pH and hardness, as well as amount of dissolved solids, fluoride, metals, organics, nutrients, pathogens, and dissolved oxygen.

Water quality is perceived differently by different people. For example, public health officials are concerned with the viral and bacterial safety of water used for bathing and drinking, fishers are concerned that the quality of a body of water provides the best habitat for fish, and aquatic scientists are concerned about the habitats of all aquatic organisms. The state of the water and the nature of the concerned party will often determine which water-quality parameters must be measured. For instance, raw wastewater entering a wastewater treatment plant does not need to be tested for dissolved oxygen, but parameters such as amount of organics and metals are often important. In contrast, dissolved oxygen is extremely important for the health of rivers and lakes.

In an attempt to devise a standard system for comparing river water quality in various parts of the United States, the National Sanitation Foundation de-signed a water-quality index (WQI) in 1970 that became one of the most widely used tools of its kind. The WQI can be used to compare the water quality of different rivers, to monitor water-quality changes in a particular river section, or to compare the water quality of different sections within a river. To determine a WQI score, nine tests are performed. These include measures of dissolved oxygen, biochemical oxygen demand, pH, temperature, total solids, turbidity, nitrates, total phosphorus, and fecal coliform. The results from the tests are given numerical values, and the sum of the nine values yields the overall WQI score. Values range from 0 to 100: A score of 0-25 is very bad, 26-50 is bad, 51-70 is medium, 71-90 is good, and 91-100 is excellent.

POLLUTANTS

A host of different pollutants can enter a water body and affect its quality. The principal water pollutants include disease-causing agents (pathogens), oxygen-demanding wastes, inorganic chemicals, organic chemicals, and sediment or suspended matter. The disease-causing agents include bacteria, viruses, protozoa, and parasitic worms. These pathogens enter the water from domestic sewage and animal wastes. They can cause a variety of diseases, including cholera, dysentery, giardiasis, hepatitis, and typhoid fever.

Oxygen-demanding wastes are organic wastes capable of being decomposed by aerobic (oxygen-requiring) bacteria. In the process of decomposing this waste, the bacteria consume oxygen. If aquatic plants and contact with air do not replenish the oxygen at a rate that is equal to or greater than the rate at which it is depleted, then the oxygen level will drop.

This can be measured by a decrease in the dissolved oxygen content. If the level drops low enough, it will affect all aquatic organisms that depend on oxygen. The quantity of oxygen-demanding wastes can be determined through measurement of the biochemical oxygen demand, or the amount of oxygen needed by aerobic bacteria to decompose the organic materials over a five-day period at 20 degrees Celsius (68 degrees Fahrenheit).

Inorganic chemicals that may pollute water include toxic metals, such as mercury and lead, and plant nutrients, such as nitrates and phosphates. When nutrients enter the water, they can cause extensive algal growth. When the algae die and decay, oxygen is depleted, and aquatic organisms are killed. This process is known as eutrophication. The source of these nutrients is often agricultural runoff containing fertilizers. Water-polluting organic chemicals include pesticides and petroleum products. These threaten human health as well as aquatic life. They are often resistant to microbial decomposition and can persist within the environment for long periods of time. Sediment and suspended matter are insoluble particles of soil and other solids that become suspended in water. By weight, such particles—which enter bodies of water most often as the result of soil erosion of land—are by far the largest water pollutant; they cloud the water, disrupt food chains, and often contain substances harmful to plant and animal life.

John P. DiVincenzo

FURTHER READING

Cech, Thomas V. "Water Quality." In *Principles of Water Resources: History, Development, Management, and Policy.* 3d ed. New York: John Wiley & Sons, 2010.

Gray, N. F. *Drinking Water Quality: Problems and Solutions.* 2d ed. New York: Cambridge University Press, 2008.

Udeh, Patrick J. *A Guide to Healthy Drinking Water: All You Need to Know About the Water You Drink.* Lincoln, Nebr.: iUniverse, 2004.

Vigil, Kenneth M. *Clean Water: An Introduction to Water Quality and Water Pollution Control.* 2d ed. Corvallis: Oregon State University Press, 2003.

SEE ALSO: Clean Water Act and amendments; Cultural eutrophication; Drinking water; Eutrophication; Groundwater pollution; Water pollution; Water treatment.

Water rights

CATEGORY: Water and water pollution
DEFINITION: Legal interests regarding the use of water resources
SIGNIFICANCE: Laws concerning water rights are generally intended to safeguard the use, quality, availability, and enjoyment of water found underground and in streams, rivers, lakes, and ponds.

In the early years after the establishment of the United States, most Americans considered water to be an abundant resource that needed scant government regulation. As the population of the nation grew, however, this view changed. Growing numbers of disputes over access to fresh water prompted lawmakers and courts to establish various doctrines of water rights to protect water as a precious and limited resource.

English common law provided the legal foundation for decisions about water rights in the eastern states. One of its chief provisions recognized riparian, or riverside, rights. These rights entitled landowners whose lands abutted natural, free-flowing streams and rivers to use the water from those sources for ordinary purposes such as bathing and drinking. They did not have the right to pollute, stem, or divert waters from their natural paths in ways that would interfere with the water rights of other landowners downstream. With the arrival of the Industrial Revolution, riparian rights came under attack as factory owners tried to divert water from rivers for industrial purposes. Though many courts struck down such attempts, others rendered legal decisions that allowed riparian owners "reasonable use" of water for industrial purposes, providing it did not excessively harm the rights of other riparians.

Even more radical changes took place in the American West during the late nineteenth century, when growing numbers of miners, farmers, and ranchers competed for scarce water sources with little regard for riparian rights as recognized in the East. As a result, disputes over water rights multiplied. One of the more famous feuds took place along the Poudre River in Colorado when one group of settlers diverted water for irrigation purposes and deprived others downstream of sufficient water. The Colorado state legislature took up the matter in 1876 and established the Colorado doctrine, also known as the prior appropriation doctrine, which held that all water rights belonged to the first user to claim them.

Other western states quickly adopted versions of Colorado's "first in time, first in right" approach. Eventually, however, state courts modified the doctrine by applying a "reasonable use" standard test to water disputes. Under this provision, senior water users could be forced to reduce their water consumption if a challenger proved the senior user took more than a reasonable share of water. Unlike in the East, western laws concerning water rights hold that water is a thing separate from the soil rather than an ingredient of it. This principle permits western landowners to remove water from their land and sell it as a commodity. In 1982 the U.S. Supreme Court ruled in *Sporhase v. Nebraska* that an individual even has the right to export water from one state to another.

Though water rights exist in many—and often conflicting—forms across the United States, they are all subject to state and federal authority. Laws concerning water rights have increasingly needed to address water disputes arising from water shortages. While most of the feuds over water rights in the United States still take place in the semiarid West, conflicts have also occurred in water-rich states with fast-growing populations, such as Florida and Georgia.

Water shortages are not unique to the United States. In 1997 the United Nations reported the results of a study that found that one-fifth of the world's population lacked clean drinking water. In many developing nations, people face daily water rations and have no running potable water. The growing water scarcity is a result of human population growth, increasing agricultural demands, industrialization, urbanization, and the continuous degradation of a finite reserve of fresh water. The United Nations has predicted that if present trends continue, two-thirds of the world's population will face critical water shortages by the year 2025.

Sovereign nations have the legal means and the authority to resolve disputes over water rights within their own borders. Such conflicts are harder to settle when several countries share a given body of water, however. Serious international quarrels arose among riparian nations along the Ganges, Niger, Mekong, and several other river basins during the late twentieth century. Squabbles over access to the Jordan River in the parched Middle East have, on some occasions, provoked talk of war. International law offers only limited help in solving these problems. For example, nations have the right to exercise authority over any resources under their immediate control, but a United Nations convention of 1972 and the Helsinki Rules of 1966 oblige countries to share water rights with other riparians as long as they show regard for the needs of local populations and traditional water consumption practices. Generally, riparian nations rely on international treaties and agreements to establish mutual water rights. Without effective enforcement powers, however, they have few options other than persuasion, sanctions, and even military force to ensure treaty compliance.

Proposed solutions to the world's growing water shortage include improved water conservation and pollution reduction efforts, a community approach among riparian nations to establish equitable water rights, and the privatization of water delivery systems that are currently under government control. Water experts at the World Bank and the United Nations have called upon nations to abandon the concept of water as an abundant and cheap resource to be subsidized by governments. Instead, they suggest, water should be viewed as an "economic good," much like oil or gold, that is subject to free market mechanisms. They assert that under such an approach water's proper price would match its value and assure its own protection. Critics, however, contend that water is too precious to be turned over to commerce. All humans, they argue, have a right to a fair share of this scarce resource that bestows life.

John M. Dunn

FURTHER READING

Andrews, Richard N. L. *Managing the Environment, Managing Ourselves: A History of American Environmental Policy.* 2d ed. New Haven, Conn.: Yale University Press, 2006.

Cech, Thomas V. "Water Allocation Law." In *Principles of Water Resources: History, Development, Management, and Policy.* 3d ed. New York: John Wiley & Sons, 2010.

Ferrey, Steven. "Rights to Use Water." In *Environmental Law: Examples and Explanations.* 5th ed. New York: Aspen, 2010.

Hodgson, Stephen. *Modern Water Rights: Theory and Practice.* Rome: Food and Agriculture Organization of the United Nations, 2006.

SEE ALSO: Drinking water; Riparian rights; Water conservation; Water quality; Water use.

Water-saving toilets

CATEGORY: Water and water pollution

DEFINITION: Toilets designed to use minimal amounts of water to move human waste products into sewer lines or septic tanks

SIGNIFICANCE: Because flush toilets account for the greatest amount of water used by residences in industrialized nations, toilets designed to minimize the water needed for each flush offer excellent opportunities for water conservation.

It has been estimated that an average family of four in the United States uses 340,000 liters (90,000 gallons) of water per year. Modest but conscious changes in water use and modifications of plumbing fixtures in and around homes can save thousands of gallons each year. Such savings reduce water and energy costs in individual homes and also help protect water resources for future generations.

Toilets are made to suit a wide range of budgets, decors, and functions, and choosing the right toilet involves informed decision making. Among the characteristics of toilets that consumers should consider are trap size, bowl-water surface, tank lining, and toilet footprint. Sufficient trap size ensures the ability of a toilet to flush well without clogging; larger traps are less likely to clog. The amount of bowl-water surface influences how well a toilet remains clean between regular scrubbings; toilets with larger bowl-water surface areas generally require cleaning less often. Whether or not a toilet's tank is lined can be particularly important in a bathroom that is not adequately air-conditioned; lined tanks do not "sweat" during hot weather. The toilet footprint is the space that the toilet pedestal takes up on the floor.

A further consideration in selecting a toilet is how much water it uses. The toilet in a residence housing four people is flushed, on average, thirty times per day. Toilet flushing is generally considered to be the single largest source of water loss in the home because it accounts for about 38 percent of the water used each day.

The amount of water used by a toilet is called the flush rate, which is measured in gallons per flush (gpf) in the United States and liters per flush (lpf) in most other parts of the world. Typical toilets rate at 3.5 gpf (13 lpf) or higher. However, designs have been introduced that rate at 1.6 gpf (6 lpf). Some of the best available rate at 1.5 gpf (5.7 lpf). Toilets that perform at these levels are called low-consumption, low-flush, ultra-low-flush, or water-saving toilets. They come in a variety of engineering designs. The four designs that are the most common are gravity-tank, flushometer, pressurized-tank, and vacuum-assist toilets.

It has been estimated that water consumption in the home can be cut by 25 percent or more through the replacement of an old-style toilet with a low-consumption model. This reduction is both automatic and permanent. In order to achieve these savings on a larger scale, some communities in Canada and the United States now require that all new or replacement installations of two-piece tank-type and floor-mounted flushing toilets rate at no more than 1.6 gpf. Water management officials in these communities also provide incentives to owners of existing homes to install low-flush toilets.

Josué Njock Libii

FURTHER READING

Cech, Thomas V. *Principles of Water Resources: History, Development, Management, and Policy.* 3d ed. New York: John Wiley & Sons, 2010.

George, Rose. *The Big Necessity: The Unmentionable World of Human Waste and Why It Matters.* New York: Henry Holt, 2008.

SEE ALSO: Sewage treatment and disposal; Water conservation; Water use.

Water treatment

CATEGORY: Water and water pollution

DEFINITION: Processing of raw water to make it safe for drinking

SIGNIFICANCE: To be safe for drinking, water must be free of disease-producing bacteria, undesirable tastes and odors, color, turbidity, and harmful chemicals. The proper treatment of water supplies addresses all such possible problems and is thus essential to providing the safe drinking water that is necessary to human health and welfare.

Many substances may occur naturally in raw water that are either harmful or unpalatable to people. Human discharge of many substances into the environment also contaminates water supplies. Contaminants, either natural or anthropogenic (human-caused) in origin, can be divided into three groups:

organoleptic substances that pertain to the senses of vision, taste, and odor; inorganic and organic chemical substances, which could be toxic or aesthetically undesirable, or could interfere with water treatment processes; and harmful microorganisms, which usually result from human and animal wastes.

The organoleptic parameters must be reduced to very low levels for drinking water to be acceptable for public use. Color, turbidity, and particulate matter represent visual problems. Color results from organic matter that leaches from soil or decaying vegetation. Turbidity results from suspended clay or organic matter that imparts a muddy and therefore undesirable appearance to the water. Particulate matter floating in water is not only aesthetically undesirable but also may provide food for certain organisms. Decomposed organic material and volatile chemicals result in unpleasant tastes and odors in water.

Iron, manganese, and aluminum are metals that are commonly found in water. Other metals such as lead, copper, cadmium, and silver are occasionally present, as are the nonmetals nitrate, fluoride, and phenols. These chemicals have both natural and anthropogenic origins. Chemically synthesized compounds such as pesticides, herbicides, and polychlorinated biphenyls (PCBs) are particularly dangerous as they can enter the food chain and accumulate in animal tissue.

Although most bacteria are harmless and indeed essential to life, some varieties (pathogens) can cause illness and death. These waterborne diseases include cholera, typhoid, and bacillary dysentery, all of which are common in areas without properly treated water. Viruses are pathogenic organisms that are much smaller and much harder to control than bacteria. Common viral diseases include poliomyelitis and infectious hepatitis. *Cryptosporidium* and *Giardia* are protozoan waterborne parasites that are found in surface waters. They cause severe forms of gastroenteritis that can be deadly in people who have immune-suppressed systems, such as those living with acquired immunodeficiency syndrome (AIDS).

A water treatment plant operated by the Albuquerque Bernalillo County Water Utility Authority in New Mexico. (AP/Wide World Photos)

The origin and characteristics of the raw water source govern the type of treatment necessary to provide safe drinking water. For example, groundwater may require only pH adjustment and minor disinfection if the source is relatively pristine. In heavily fertilized agricultural areas and locations where soluble iron and manganese are naturally present, however, ion exchange for nitrate removal and chemical treatment for iron and manganese removal may be needed.

Surface water generally requires many more types of treatment, such as screening, sedimentation, chemical treatment, clarification, filtration, and disinfection. Installation of bar screens to block fish and debris is a standard first step in treating raw surface water. The screens must be strong enough to prevent wood, game fish, and even shopping carts from getting into the treatment plant and damaging the machinery. The next step is usually a sedimentation basin, where the larger suspended particles can settle out by gravity. This process may be accelerated through the mixing of chemicals with the water to form a flocculate precipitate, which helps settle the suspended particles. The chemical coagulation process removes natural color originating from peat, animal and vegetable debris, plankton, and other organic substances.

Even after sedimentation, some of the finer particles in the water may still be in suspension and have to be removed by filtration. Sand filters provide an inexpensive and effective medium for the removal of fine solids in either raw water or partially treated water. Many facilities in the United States have replaced sand filters with granular activated carbon (GAC) filters because these can remove a wide variety of undesirable organic compounds such as herbicides, pesticides, and chemical compounds that form naturally. They are also useful in the treatment of taste and odor. Indeed, many beverage manufacturers that make products (such as soda or beer) in which water is a major component use GAC filters. Residential point-of-use kitchen filters for drinking water incorporate GAC filters as the major treatment technique.

The final treatment process is disinfection, since pathogenic bacteria can pass through both the sedimentation basin and filtration. Government-set standards for drinking-water quality in the United States require the absence of the indicator organisms fecal streptococci and the coliform group of bacteria, specifically fecal *Escherichia coli*, in the distributed water. Disinfection, which is the killing of harmful bacteria, is usually accomplished through chlorination. Chlorine is a very effective biocide, but one major disadvantage of its use is that it is very reactive and can produce compounds, such as trihalomethanes, that are potentially carcinogenic. Other compounds produced by chlorine have taste and odor problems. Ozone and ultraviolet light are also powerful disinfectants, but they do not have the residual properties of chlorine, which protects water from contamination as it travels through the distribution system.

Some water treatment plants built since the late twentieth century use a combination of ozonation for its effectiveness against *Cryptosporidium*, *Giardia*, and viruses; GAC filters for taste and odor control; and small amounts of chlorine as a residual biocide for the treated water in the distribution system.

Robert M. Hordon

Further Reading

Ford, Tim. "Water and Health." In *Environmental Health: From Global to Local*, edited by Howard Frumkin. 2d ed. Hoboken, N.J.: John Wiley & Sons, 2010.

McKinney, Michael L., Robert M. Schoch, and Logan Yonavjak. "Water Pollution." In *Environmental Science: Systems and Solutions*. 4th ed. Sudbury, Mass.: Jones and Bartlett, 2007.

Manahan, Stanley E. "Water Treatment." In *Fundamentals of Environmental Chemistry*. 2d ed. Boca Raton, Fla.: CRC Press, 2001.

Sullivan, Patrick J., Franklin J. Agardy, and James J. J. Clark. "Water Protection." In *The Environmental Science of Drinking Water*. Burlington, Mass.: Elsevier Butterworth-Heinemann, 2005.

SEE ALSO: Chlorination; Clean Water Act and amendments; Desalination; Drinking water; Fluoridation; Water quality.

Water use

CATEGORIES: Water and water pollution; resources and resource management

DEFINITION: Consumption of fresh water by residences, businesses, industries, governments, and food-production interests in support of human populations

SIGNIFICANCE: Human beings' use of fresh water has numerous impacts on the environment, in part because available freshwater resources, like mineral resources, are unevenly distributed, necessitating the transport of water over long distances.

Although water is the earth's most abundant liquid and can be found almost everywhere, 97 percent of it is too salty for human use. Of the remainder, about 68.7 percent is frozen in icecaps and glaciers. Raw fresh water can come from either surface or ground sources. Surface sources include river systems, lakes, and reservoirs. Groundwater sources vary from unconsolidated materials, such as the sandy deposits along the Atlantic and Gulf Coastal Plain and the stratified sands and gravels of glaciated areas, to consolidated rocks, such as sandstone and shale, where water is obtained from the fractures within the formations. Water-supply systems that rely on groundwater can vary in size from a few wells serving a small community to a network of many wells serving a larger area, such as the Suffolk County Water Authority in Long Island, New York.

TYPES OF WATER USES

Water for public supply needs is defined as water that is delivered to multiple users for domestic, commercial, and industrial purposes, as well as for firefighting, street washing, municipal parks, and public swimming pools. The purveyor may be either public, such as a city-run utility, or private (an example is American Water, an investor-owned company that serves communities in several states). In the United States public systems that deliver potable

Yemeni women collect water from a reservoir and load it onto the backs of donkeys to carry it to their village. Yemen's groundwater supply is insufficient to support the nation's growing population. (Khaled Abdullah/Reuters/LANDOV)

water to a variety of users must comply with federal and state standards for safe drinking water. In public systems, the water's source and quality are subject to routine tests to ensure regulatory compliance.

Domestic water use is defined as the use of water for normal household purposes, such as drinking, food preparation, bathing, washing dishes and clothes, toilet flushing, lawn and garden watering, and home car washing. Households that obtain their water from on-site wells are not part of the public potable water system infrastructure. The number of households in the United States that fall into this self-supplied category is substantial. The federal Environmental Protection Agency (EPA) reports that approximately 15 percent of the U.S. population—more than 15.8 million housing units, according to 2009 census data—are served by their own private drinking-water supplies. Although these supplies are not regulated by the EPA, some state and local governments set standards for private wells. Self-supplied domestic water systems are rarely metered, and minimal data exist as to the amounts withdrawn.

Industrial water use includes the use of water necessary for processing, washing, diluting, cooling, and sanitation in factories that make a variety of products. Industries that use large amounts of water include the steel, chemical, paper, and petroleum-refining industries. Of the industrial uses of water, thermoelectric power constitutes a substantial water-use category in itself. It includes water used for electric power generation from fossil-fuel, nuclear, or renewable energy sources. Most of the water used by thermoelectric plants goes for condenser and reactor cooling. Only a small fraction of the water used in this category comes from public water systems. Another industry, mining, is also in a category by itself. Mining water use includes quarrying, crushing, washing, and other activities associated with mineral extraction operations.

Irrigation water use includes water employed to sustain plant growth in agriculture and horticulture. It encompasses the use of water not only for watering plants but also for applying agricultural chemicals, controlling weeds, preparing fields, cooling crops, suppressing dust, protecting against frost, and harvesting. Larger-scale nonagricultural irrigation—of golf courses, parks, nurseries, cemeteries, and the like—are included in this category.

Livestock water use is that associated with the farming of dairy and beef cattle, sheep and lambs, goats, hogs and pigs, horses, and poultry. It includes water devoted to livestock watering, feedlots, dairy operations, cooling of animal facilities, facility sanitation and washdown, and animal waste disposal systems. Similarly, aquaculture water use is that associated with sustaining organisms that live in water, notably finfish and shellfish. Aquaculture, which involves controlled feeding, sanitation, and harvesting of aquatic organisms, may be conducted for food, restoration, conservation, or sport purposes.

According to the U.S. Geological Survey, the estimated percentage distribution for total water withdrawals in the United States in 2005 was as follows: 49 percent thermoelectric power, 31 percent irrigation, 11 percent public supply, 4 percent self-supplied (not public supply) industrial, 2 percent aquaculture, 1 percent self-supplied domestic, 1 percent mining, and less than 1 percent livestock. (These percentages include water lost through system leakage.) Some 1.552 billion cubic meters (410 billion gallons) per day were withdrawn in 2005 for all these categories.

Water Consumption Trends and Conflicts

In the United States, more than one-fourth of the total water used in 2005 was withdrawn in California, Texas, Idaho, and Florida. California alone was responsible for 11 percent of all the nation's withdrawals that year. Thermoelectric power and irrigation were the largest water-use categories; along with public supply, they accounted for more than 90 percent of the nation's withdrawals. Generally, per-capita water consumption in the United States tends to be highest in the West, presumably reflecting the region's lower precipitation and higher evapotranspiration rates in comparison with other regions.

Water use in most countries is a function of population served. Consequently, as the population increases, water consumption also increases, which means that water purveyors continually need to expand their water-supply sources. The need for additional supplies of water has resulted in innumerable sociopolitical disputes over the years. In arid areas such as the Middle East, water is crucial for general use and irrigation; thus the decision by Turkey in the late twentieth century to build large reservoirs in the headwaters of the Tigris and Euphrates rivers caused friction between that nation and the downstream states of Syria and Iraq, which objected to the prospect of being deprived of a portion of the flow on which they had historically depended. The allocation of the Jordan River among Israel, Jordan, Syria, and

Lebanon in another politically sensitive arid area is intimately related to the possibility of sustained peace in the region. Egypt is totally dependent on the Nile River, which originates in Ethiopia and Lakes Albert and Victoria in East-Central Africa. Thus, when upstream African countries began negotiating a cooperative framework to revise water-sharing arrangements in 2010, concerns over the impact of major irrigation and hydropower projects upstream caused Egypt to oppose the agreement.

Balancing water use and the availability of water requires serious effort at local, state, national, and international levels. Even with prudent planning and cooperation among parties, drought, desertification, water pollution, and population expansion increase the likelihood that water-use conflicts will arise.

Robert M. Hordon
Updated by Karen N. Kähler

FURTHER READING

Chiras, Daniel D. "Water Resources: Preserving Our Liquid Assets and Protecting Aquatic Ecosystems." In *Environmental Science.* 8th ed. Sudbury, Mass.: Jones and Bartlett, 2010.

Correljé, Aad, and Thorsten Schuetze. *Every Drop Counts: Environmentally Sound Technologies for Urban and Domestic Water Use Efficiency.* Nairobi: United Nations Environment Programme, 2008.

Dzurik, Andrew Albert. *Water Resources Planning.* 3d ed. Lanham, Md.: Rowman & Littlefield, 2003.

Gleick, Peter H., et al. *The World's Water, 2008-2009: The Biennial Report on Freshwater Resources.* Washington, D.C.: Island Press, 2009.

Glennon, Robert Jerome. *Water Follies: Groundwater Pumping and the Fate of America's Fresh Waters.* Washington, D.C.: Island Press, 2002.

Kenny, Joan F., et al. *Estimated Use of Water in the United States in 2005.* Reston, Va.: U.S. Geological Survey, 2009.

Shiklomanov, I. A., ed. *World Water Resources at the Beginning of the Twenty-first Century.* New York: Cambridge University Press, 2003.

Vickers, Amy. *Handbook of Water Use and Conservation: Homes, Landscapes, Businesses, Industries, Farms.* Amherst, Mass.: Waterplow Press, 2001.

SEE ALSO: Aquifers; Clean Water Act and amendments; Desalination; Drinking water; Groundwater pollution; Safe Drinking Water Act; Water conservation; Water quality; Water rights; Wells.

Watershed management

CATEGORIES: Water and water pollution; land and land use

DEFINITION: Policies governing the use of land areas bounded by drainage divides within which precipitation drains to particular watercourses or bodies of water

SIGNIFICANCE: Human activity can cause unanticipated changes in watersheds, affecting the hydrologic balance. Careful management of watersheds is important because land use alters the balance between storage and dispersal of precipitation, in many cases increasing erosion, stream sedimentation, and flooding hazards.

Watersheds are defined at many scales: The Mississippi River watershed contains the Ohio River watershed, which in turn contains smaller watersheds. A fundamental part of the hydrologic cycle, the watershed collects and stores precipitation in soils, lakes, wetlands, or aquifers and disperses water by evaporation, plant transpiration, surface runoff, springs, and base flow to streams. Watersheds of different geographic regions have distinctive characteristics based on climate, topography, and soil type; therefore, the natural variability among watersheds is predictably large. In arid regions, precipitation occurs as intense, infrequent storms, with most of the water rapidly running off and eroding soil with little protective vegetation. Watersheds in humid areas are characterized by frequent, usually gentle rain that replenishes aquifers and sustains streams, springs, and wetlands.

Ecologically, the watershed provides habitat and nutrients for plants and animals, including humans. Land use can disrupt a watershed's ecology by disturbing habitat and nutrient cycling through soil loss and removal of native vegetation. The role of the watershed in environmental problems such as flooding, erosion, sedimentation, and ecological disruption has led to increased emphasis on the watershed as the basic unit for environmental management, rather than political units such as states or counties.

The 1954 Watershed Protection and Flood Prevention Act authorized the secretary of the U.S. Department of Agriculture (USDA) to manage watersheds in cooperation with states and local organizations, such as soil and water conservation districts. The driving idea behind the act is that floods are better controlled

Characteristics of Selected Major Drainage Basins

RIVER	CONTINENT	OUTFLOW	LENGTH	AREA	AVERAGE ANNUAL SUSPENDED LOAD
Amazon	South America	180.0	6,300	5,800	360
Congo	Africa	39.0	4,700	3,700	—
Yangtze	Asia	22.0	5,800	1,900	500
Mississippi	North America	18.0	6,000	3,300	296
Irawaddy	Asia	14.0	2,300	430	300
Brahmaputra	Asia	12.0	2,900	670	730
Ganges	Asia	12.0	2,500	960	1,450
Mekong	Asia	11.0	4,200	800	170
Nile	Africa	2.8	6,700	3,000	110
Colorado	North America	0.2	2,300	640	140
Ching	Asia	0.06	320	57	410

Note: Rivers are ordered by outflow; outflow is multiplied by 1,000 cumecs (cubic meters of water per second); length is measured in kilometers; area is measured in square kilometers multiplied by 1,000; average annual suspended load is measured in millions of metric tons.

through management of runoff upstream in the watershed than through downstream engineering projects. The act requires local interests to contribute up to 50 percent of the costs to ensure local support for watershed projects. In contrast, Army Corps of Engineers flood-control projects originally were funded entirely by the federal government. The Watershed Protection and Flood Prevention Act is generally administered through the USDA's Natural Resources Conservation Service, formerly the Soil Conservation Service.

Recognizing the need for basinwide planning, the federal government created the Water Resources Council through the 1965 Water Resources Planning Act. This council created river basin planning commissions but fell into disfavor and lost funding because the river basins were too large for effective planning.

A major step in watershed management was taken with the 1972 Clean Water Act. With this act, land management began to include water-quality control. Nonpoint sources of pollution were targeted, among them agriculture, forestry, mining, and waste disposal. Most states passed laws directing the use of certain widely accepted methods of preventing soil and water problems (known as best management practices, or BMPs) to protect or rehabilitate watershed functions. The Clean Water Act provided for regulation of land use, initially through incentives. The 1985

Food Security Act provided incentives for landowners to control erosion on highly erodible croplands. The act's "swampbuster" provisions directed protection of existing wetlands and also provided incentives for wetlands restoration. The 1986 amendments to the 1974 Safe Drinking Water Act encouraged public suppliers of drinking water to protect wellheads. The 1987 amendments to the Clean Water Act encouraged states to address nonpoint source pollution.

State regulations on nonpoint source pollution range from voluntary compliance with BMPs to strict enforcement of BMPs with fines for noncompliance. In general, however, such regulations have become increasingly detailed and comprehensive. The concept of the total maximum daily load (TMDL) permissible for nonpoint source pollutants has been introduced, but determining TMDL is costly and difficult because appropriate loads vary with land use and with watershed.

Over time, Americans' perspectives on land and water management have broadened. Whereas management initially focused on single farms or individual fields, the watershed view has come to be widely accepted. This change has been influenced in part by concerns about the greenhouse effect, in which atmosphere-biosphere-hydrosphere-terrasphere interactions are critical. The term "ecosystem management" may better reflect watershed management focus in the future.

Watersheds are managed for a spectrum of land uses, including water supply, settlement, grazing, crop production, forestry, and recreation. Management focuses on water, sediment, and wastes. Water management generally seeks to reduce runoff; exceptions are landfills and mine spoils in which infiltration is minimized. Sediment management seeks to prevent soil erosion or to trap eroded sediment. Waste management seeks to distribute the waste load properly and to prevent it from reaching water. The appropriate strategies for achieving these management goals vary from problem to problem. For example, in forestry one strategy might include revegetating logged areas, diverting water from logging roads, and closing logging roads after use. The overall approach to watershed management is to identify the problem and its source and then select and implement BMPs. While the law requires that BMPs be considered, no definitive catalog of such practices exists. Many state agencies have written and assembled their own collections of BMPs for various land uses, which are available to the public. Public education and public participation in decision making have played increasingly important roles in sustainable watershed management in the United States.

Mary W. Stoertz

FURTHER READING

Black, Peter E. *Watershed Hydrology.* 2d ed. Chelsea, Mich.: Ann Arbor Press, 1996.

France, Robert L., ed. *Facilitating Watershed Management: Fostering Awareness and Stewardship.* Lanham, Md.: Rowman & Littlefield, 2005.

Heathcote, Isobel W. *Integrated Watershed Management: Principles and Practice.* 2d ed. Hoboken, N.J.: John Wiley & Sons, 2009.

Newson, Malcolm. *Land, Water, and Development: Sustainable and Adaptive Management of Rivers.* 3d ed. New York: Routledge, 2009.

Satterlund, Donald R., and Paul W. Adams. *Wildland Watershed Management.* 2d ed. New York: John Wiley & Sons, 1992.

SEE ALSO: Ecosystems; Erosion and erosion control; Floods; Forest and range policy; Forest management; Runoff, agricultural; Runoff, urban; Sedimentation; Wetlands.

Watson, Paul

CATEGORIES: Activism and advocacy; animals and endangered species
IDENTIFICATION: Canadian animal rights and environmental activist
BORN: December 2, 1950; Toronto, Ontario, Canada
SIGNIFICANCE: A dissident Greenpeace member and experienced sailor, Watson founded the Sea Shepherd Conservation Society, one of the world's most aggressive environmental organizations. He and his organization have mounted vigilante (but deliberately nonlethal) attacks against the efforts of seal hunters, whalers, and drift-net fishers.

A onetime officer in the Canadian Coast Guard, Paul Watson helped to found Greenpeace in 1971. By the mid-1970's, however, he found himself increasingly at odds with the group's philosophy. Initially a small band of dedicated activists who placed themselves at personal risk to interfere with such environmentally negative practices as nuclear weapons tests, Greenpeace had grown into an international concern that directed much of its effort toward fundraising. Watson believed that the group's evolution had dissipated its effectiveness, and he also chafed at the organization's commitment to passive protest in pursuit of its goals.

Watson believed that doing harm to living things is wrong, but the use of nonharmful force against property could be justified to protect life. Moreover, he became increasingly convinced that passive protest was not working. In March, 1977, while participating in a Greenpeace protest of the slaughter of baby harp seals by hunters using clubs in Canada's Gulf of St. Lawrence, Watson seized a club from a seal hunter and threw it into the water. Such direct action conflicted with Greenpeace's expressed policy, and he was expelled from the group. When Greenpeace's board of directors charged Watson with vigilantism, he replied that, in the absence of an environmental police force to oppose environmental crimes, environmental vigilantes were bound to appear. Greenpeace soon distanced itself from Watson and his unapologetically extreme methods.

In the summer of 1977 Watson and several friends established an organization they at first called Earthforce; it later became the Sea Shepherd Conservation Society. Headquartered in Vancouver, Canada, the group was dedicated to the use of direct action to protect the world's animals. Although Watson and his band of activists would later achieve notoriety for defending marine life, on their first mission they traveled to East Africa to document the killing of elephants for ivory. Earthforce presented its findings, including film of the illegal slaughter of elephants, to the U.S. government in support of a legislative ban on the importation of African ivory.

For Watson, however, the preservation of marine mammals was a recurring goal. With the aid of a grant from the writer and animal rights activist Cleveland Amory, Watson purchased a retired fishing boat, which he christened the *Sea Shepherd*, and hired a crew. On the *Sea Shepherd*'s first voyage, Watson and the crew took the ship through some 644 kilometers (400 miles) of ice to the Gulf of St. Lawrence. There they sprayed hundreds of baby seals with a harmless red dye that made the animals' white pelts valueless to hunters. Soon, however, they were stopped and arrested by Canadian police. Undeterred, Watson and

his allies would return to the gulf in 1982 and 1983, after which the gulf hunt was discontinued.

AGGRESSIVE TACTICS

Publicity stemming from the group's sometimes violent encounters with sealers and from its brushes with the law had led to steady membership increases, and the scope and ambition of the group's activities expanded as well. In the spring and summer of early 1979, Watson and other members of the Sea Shepherd Conservation Society organized what was essentially an espionage ring that tracked the movements of the *Sierra*, a notorious "pirate" whaling ship whose crew had killed an estimated four hundred whales per year since the 1960's. In July, 1979, Watson and two assistants aboard the *Sea Shepherd* located the *Sierra* off the coast of North Africa and followed the whaling ship to the harbor in Leixões, Portugal. Outside the port, the larger and faster *Sea Shepherd* twice rammed the *Sierra*, rupturing the whaler's side but causing no injuries to the crew. The Portuguese navy took the *Sea Shepherd*

Animal rights and environmental activist Paul Watson at a 2010 demonstration against whaling held in Paris, France. (AP/Wide World Photos)

The Sea Shepherd Conservation Society's Mission

The Sea Shepherd Conservation Society's mission statement reads as follows:

Established in 1977, Sea Shepherd Conservation Society (SSCS) is an international non-profit, marine wildlife conservation organization. Our mission is to end the destruction of habitat and slaughter of wildlife in the world's oceans in order to conserve and protect ecosystems and species.

Sea Shepherd uses innovative direct-action tactics to investigate, document, and take action when necessary to expose and confront illegal activities on the high seas. By safeguarding the biodiversity of our delicately-balanced ocean ecosystems, Sea Shepherd works to ensure their survival for future generations.

into custody, but the *Sierra* had incurred enormous damage and had to be towed to Lisbon for repairs. After Portuguese authorities threatened to turn the *Sea Shepherd* over to the *Sierra*'s owners as compensation, Watson and two friends sneaked into the Lisbon harbor and scuttled the ship by opening its pipes.

Watson and the Sea Shepherd Conservation Society undertook similar actions many times in ensuing years, actively confronting whalers, sealers, and driftnet fishers, attacking their vessels and equipment, and often ending up in court. By the 1990's, the organization had grown to include thousands of members and was operating a fleet of antiwhaling vessels. It continues its radical practices in the early twenty-first century and remains a source of controversy.

Watson is a professional lecturer and the author of several books on protecting wildlife and marine life. In 2000 *Time* magazine named him one of the environmental heroes of the twentieth century. He is the subject of the 2008 biographical film *Pirate for the Sea* and has appeared in a number of other documentaries. He is also a central figure in *Whale Wars*, a television reality series that debuted on the Animal Planet channel in 2008. The program documents the exploits of Watson and the Sea Shepherd fleet as they fight to deter Japanese whaling vessels hunting off the Antarctic coast.

Opponents of Watson's extreme methods have portrayed him and his followers as fanatics and have accused them of using terrorist tactics. Watson has countered that he and his organization have never in-

jured a human being, committed an act of violence against a living creature, been convicted of a felony, or been sued. He has maintained that he does not break the law, but rather upholds it, and he has condemned society's failure to save imperiled creatures from slaughter as being in itself an act of violence.

Robert McClenaghan
Updated by Karen N. Kähler

FURTHER READING

Heller, Peter. *The Whale Warriors: The Battle at the Bottom of the World to Save the Planet's Largest Mammals.* New York: Free Press, 2007.

Morris, David B. *Earth Warrior: Overboard with Paul Watson and the Sea Shepherd Conservation Society.* Golden, Colo.: Fulcrum, 1995.

Watson, Paul. *Ocean Warrior: My Battle to End the Illegal Slaughter of Marine Life on the High Seas.* London: Vision, 2003.

_____. *Seal Wars: Twenty-five Years on the Front Lines with the Harp Seals.* Buffalo, N.Y.: Firefly Books, 2003.

Watson, Paul, Warren Rogers, and Joseph Newman. *Sea Shepherd: My Fight for Whales and Seals.* New York: W. W. Norton, 1982.

Weyler, Rex. *Greenpeace: How a Group of Ecologists, Journalists, and Visionaries Changed the World.* Emmaus, Pa.: Rodale Press, 2004.

SEE ALSO: Amory, Cleveland; Animal rights movement; Greenpeace; International whaling ban; Save the Whales Campaign; Sea Shepherd Conservation Society; Seal hunting; Whaling.

Watt, James

CATEGORY: Preservation and wilderness issues
IDENTIFICATION: American attorney who served as U.S. secretary of the interior
BORN: January 31, 1938; Lusk, Wyoming
SIGNIFICANCE: Watt was labeled a major antienvironmentalist during his tenure as secretary of the interior; he was frequently accused of using his office to weaken environmental policies that fell under his domain of authority.

James Watt's ancestors were nineteenth century homesteaders in Wyoming, laying claim to a large tract of land for ranching. Watt was both a rancher and a successful lawyer. As a child, he became familiar with the harsh, barren land of the family ranch, pumping water for the cattle, repairing fences, and performing other difficult chores. He recalled later in life that his early experiences trained him to challenge a hostile environment.

In 1962 Watt obtained a degree in law from the University of Wyoming; shortly thereafter, he moved to Washington, D.C., to assume a position as legislative assistant and counsel to Senator Milward Simpson of Wyoming. From 1966 to 1969 Watt worked with the U.S. Chamber of Commerce's Washington, D.C., office in natural resource and environmental pollution policy. He lobbied for pro-development business interests in such areas as the use of public lands for mining, energy, and water resource development. During the presidential administrations of Richard Nixon and Gerald Ford, Watt served in the Department of the Interior as an assistant secretary responsible for water and energy resources and as director of the Bureau of Outdoor Recreation.

James Watt, secretary of the U.S. Department of the Interior under President Ronald Reagan. (AP/Wide World Photos)

During the 1970's Watt became closely associated with the Sagebrush Rebellion, a movement of ranchers and entrepreneurs in the American West who opposed numerous federal regulations that they asserted were inhibiting the profitable exploitation of natural resources. In 1977 Watt assumed the presidency of the Mountain States Legal Foundation, an organization founded by brewing magnate Joseph Coors to provide assistance for individuals who challenge government restrictions on strip mining, oil and gas exploration, mineral extraction, and grazing lands.

Watt was a logical choice to become secretary of the interior in President Ronald Reagan's administration, which was committed to economic expansion and resource development with minimal government intrusion. From the beginning of his tenure in January, 1981, Watt worked to cut the department's budget and eliminate agencies and programs. He promoted measures to ease restrictions on oil and gas exploration on federal lands and in offshore waters, open more federal land for grazing and timber cutting, facilitate the construction of dams and reservoirs to improve irrigation of farmland, and restrict expansion of the national park system. Watt's actions were consistent with his ideology of economic growth with minimal government interference, and he felt compelled by his religious convictions to "follow the Scriptures which call upon us to occupy the land until Jesus returns."

Environmental groups strongly opposed Watt's appointment as secretary of the interior, arguing that his ideology was contrary to the mission of the department—to manage federal lands in the public interest. During his tenure as secretary, Watt refused to meet with environmental group leaders and made statements suggesting that they were subversive and were weakening the United States. He resigned from his position in November, 1983, after he was criticized for remarks he made about Senate members who had been appointed to a coal advisory panel. At the time of Watt's resignation, the Senate was working on a resolution calling for his dismissal.

Ruth Bamberger

FURTHER READING
Andrews, Richard N. L. *Managing the Environment, Managing Ourselves: A History of American Environmental Policy.* 2d ed. New Haven, Conn.: Yale University Press, 2006.

Davis, Charles, ed. *Western Public Lands and Environmental Politics.* 2d ed. Boulder, Colo.: Westview Press, 2001.

Short, C. Brant. "Conservation Reconsidered: Environmental Politics, Rhetoric, and the Reagan Revolution." In *Green Talk in the White House: The Rhetorical Presidency Encounters Ecology,* edited by Tarla Rai Peterson. College Station: Texas A&M University Press, 2004.

SEE ALSO: Antienvironmentalism; Department of the Interior, U.S.; Grazing and grasslands; Sagebrush Rebellion; Wise-use movement.

Weather modification

CATEGORY: Weather and climate

DEFINITION: Changes to weather, intentional or unintentional, caused by human activities

SIGNIFICANCE: Unintentional weather modifications such as changes in rainfall patterns and the creation of fogs have had negative impacts in many parts of the world. Experiments in intentional weather modification, most of which have focused on creating conditions favorable for agriculture, have often produced questionable results.

Inadvertent weather modification, including fog formation and increases or decreases in precipitation downwind from large industrial sites, creates problems in some locales. Scientific attempts to modify weather deliberately have been pursued since World War II. The most popular techniques involve cloud seeding, the injection of cloud-nucleating particles into likely clouds to alter the physics and chemistry of condensation. Proponents of this technique claim that it may enhance precipitation amounts by 5 to 20 percent. Some scientists believe, however, that deliberate efforts to enhance precipitation often yield questionable results, even in favorable situations. In 1977 the United Nations passed a resolution prohibiting the use of weather modification for hostile purposes because of the threat to civilians. The United States signed the resolution but has continued defense research on operational weather modification in battlefield situations, as summarized in the 1996 U.S. Air Force position paper *Weather as a Force Multiplier: Owning the Weather in 2025.*

Various methods of weather modification have been field-tested, and the results have varied widely. Weather modification has been attempted in many countries around the world, by government agencies, agricultural cooperatives, private companies, and research consortiums. In some agricultural areas farmers are convinced that hail suppression and precipitation augmentation have been achieved through weather modification. In some of these same locales, meteorologists have been unable to determine whether activities intended to modify the weather have produced any changes from what would have occurred without intervention. Attempts to duplicate weather modifications that have apparently been successful in other locales have often met with questionable results. Meteorologists occasionally disagree among themselves as to whether specific attempts at weather modification have succeeded. Reexamination of data from American studies undertaken in the past has led many scientists to conclude that the efficacy of cloud seeding has been overstated.

Scientists agree that it is impossible to change the climate of an entire region for a desired outcome through weather modification. It is also impossible to end a drought by seeding clouds. Cloud seeding for agricultural purposes assumes that some enhancement of regional rainfall amounts over the course of the growing season will increase crop yields. Weather modification for hail suppression assumes that reduction in regional crop losses over the growing season is an attainable goal.

INADVERTENT WEATHER MODIFICATION

Pulp and paper mills produce huge quantities of large- and giant-diameter cloud condensation nuclei; downwind of these mills, precipitation appears to be enhanced about 30 percent above what was observed prior to construction of the mills. It is also thought that the heat and moisture emitted by these mills may play an active role in precipitation enhancement. One specific study of a kraft paper mill near Nelspruit in the eastern Transvaal region of South Africa found that storms modified by the mill emissions lasted longer, grew taller, and rained harder than other nearby storms occurring on the same day. Radar measurements supported the theory that hygroscopic (moisture-retaining) particulates released by this mill accelerated or amplified growth of unusually large-diameter raindrops.

An egregious example of inadvertent weather

modification is the formation of ice fog over Arctic cities in Siberia, Alaska, and Canada. During winter, cities such as Irkutsk, Russia, and Fairbanks, Alaska, experience drastic reductions in visibility as particles released by combustion act as nuclei for the formation of minute ice crystals. No techniques have been developed to modify ice fogs.

During an investigation of the meteorological effects of urban St. Louis, Missouri, conducted during the 1970's, it was found that urban summer precipitation was enhanced by 25 percent relative to the surrounding area. Most of the increased precipitation occurred in the late afternoons and evenings as a result of convective activity. The frequency of summer thunderstorms was enhanced by 45 percent, and the frequency of summer hailstorms was higher by 31 percent over the city and adjacent eastern and northeastern suburbs. During the late 1960's, studies demonstrated that widespread burning of sugarcane fields in tropical areas released large numbers of cloud condensation nuclei. Downwind, rainfall decreases of about 25 percent were noted.

Cloud Seeding

For millennia, human beings attempted to influence the weather by using prayers and incantations. Sometimes rain followed, and sometimes no rain fell for extended periods. Scientists began attempting various techniques to modify weather during World War II. In 1946 Vincent Schaefer of the General Electric Research Laboratory observed that ice crystals were formed when dry ice was put into a freezer with supercooled water droplets. On November 13, 1946, Schaefer demonstrated that pellets of dry ice dropped from an aircraft into stratus clouds caused liquid water droplets to change to ice crystals and fall as snow. Bernard Vonnegut, a coworker of Schaefer, determined that silver iodide (AgI) particles also caused ice crystals to form. Project Cirrus involved apparently successful scientific attempts to seed clouds with ground-based AgI generators in New Mexico. These researchers then tried seeding a hurricane on October 10, 1947. The hurricane changed direction and made landfall in Georgia, resulting in a number of lawsuits against General Electric.

Early experiments in cloud seeding were empirical. AgI was dropped from aircraft, shot into clouds by rockets, or dispersed from ground-based generators. Researchers would selectively seed a pattern such as an "L" into a supercooled stratus cloud and see a visible "L" appear, thus "proving" that they could achieve results. When any rain occurred near a seeded area, it was attributed to the intervention. The apparent success of cloud seeding using AgI led

Selected Scientific Attempts to Modify Weather

LOCATION	WEATHER SITUATION	MODE OF DEPLOYMENT	RESULTS
New Mexico	no special selection of conditions	ground-based silver iodide generator	possible rain enhancement
Atlantic Ocean near Georgia	hurricane	airborne seeding	inconclusive
Colorado Rockies	no special selection of conditions	ground-based silver iodide generators	doubtful enhancement
Atlantic Ocean (Project Stormfury)	hurricanes	airborne seeding	inconclusive
Vietnam	military	airborne seeding	rain enhancement
Switzerland	thunderstorm (hail suppression)	rocket seeding	inconclusive
Bulgaria	thunderstorm (hail suppression)	rocket seeding using lead iodide	claimed 50 to 60 percent reduction in crop losses
Soviet Union	thunderstorm (hail suppression)	rocket seeding	claimed 50 to 95 percent reduction in crop losses

to the modification and adoption of the technique in France, Canada, Argentina, Israel, and the Soviet Union. Wine-growing regions such as the south of France and Mendoza, Argentina, installed ground-based AgI generators. The Soviet Union opted for rocket-borne AgI, which was launched in agricultural areas during thunderstorms in an effort to suppress hail.

In 1962 the U.S. Navy and Weather Bureau began an ambitious cooperative plan, named Project Stormfury, to modify hurricanes. Only a few hurricanes were seeded in attempts to reduce the intensity of the storms. Proponents of Stormfury suggested that the seeding of Hurricane Debbie in 1969 caused a reduction of 30 percent in wind speed on one day. The following day, no seeding was done, followed by another seeding attempt. The second seeding was thought to have caused a 15 percent reduction in wind speeds. Proponents believed that 10 to 15 percent reductions in wind speeds might result in a 20 to 60 percent reduction in storm damage if similar results could be achieved with the seeding of other hurricanes. Project Stormfury was terminated in the late 1970's, however, with no definitive results.

During winters from 1960 through 1970, the Climax I and Climax II randomized cloud seeding studies were conducted in the Colorado Rockies. Although it was initially thought that precipitation enhancements on the order of 10 percent may have resulted, more recent examination of the results appears to indicate that cloud seeding had no statistically discernible effect on precipitation. During the Vietnam War, the U.S. military attempted to increase precipitation along the Ho Chi Minh Trail in an effort to impede enemy forces. In the United States during the 1970's, some entrepreneurs deployed ground-based AgI generators in selected agricultural regions, billing farmers for their services. Aircraft delivery of AgI became increasingly popular. By the late 1990's, a number of private companies were delivering airborne cloud-seeding services in various areas worldwide.

Cloud physicists have explored why cloud seeding might be effective. The evidence suggests that seeding increases the size of droplets or ice crystals, allowing them to fall as precipitation. Two concepts have emerged: a static mode theory, which assumes that the natural clouds seeded were deficient in ice nuclei, and a dynamic mode theory, which assumes that enhancement of vertical movement in clouds increases precipitation. The static mode theory assumes that a "window of opportunity" exists for seeding cold continental clouds during which clouds must be within a particular temperature range and contain a certain amount of supercooled water.

Fog Dissipation and Hail Suppression

During World War II, when improvements in visibility were crucial for military operations, weather scientists became involved in efforts to dissipate fog. Fog can be dissipated through a reduction of the number of droplets, a decrease in the radius of droplets, or both. Decreasing droplet radius by a factor of three through evaporation can provide a ninefold increase in visibility. Possible methods of fog removal include using dry ice pellets or hygroscopic materials, heating the air, and mixing the foggy air with drier air. Airports that are plagued by supercooled fog in winter, such as those in Denver, Colorado, and Salt Lake City, Utah, can dissipate the fog by dropping dry ice pellets. Dry ice causes some liquid water droplets to freeze and grow, evaporating the remaining liquid droplets and allowing the larger frozen ice crystals to fall. Military airports sometimes use helicopters to provide mixing when they need to clear shallow radiation fog that is close to the ground. One expensive technique for dissipating fog at airports that has been employed in France is to use jet engines as heaters.

Farmers worldwide fear damaging hailstorms that can devastate crops. Three approaches to suppressing hail have been developed: converting all liquid water droplets to snow to prevent hail formation, seeding to promote growth of many small hailstones instead of larger damaging hail, and introducing large condensation nuclei to reduce the average hailstone size. Most proponents of weather modification believe that hailstone size can be substantially reduced through cloud seeding with lead iodide or AgI, which they assert causes many small hailstones to form, so that fewer large and potentially damaging ones are present. This technique may not eliminate crop losses, but it could reduce them significantly. It has been claimed that rocket-borne lead iodide seeding in Bulgaria reduced crop losses from hail by 50 to 60 percent. Similar seeding operations in the former Soviet Union were said to have reduced crop damage by 50 to 95 percent. The findings of a randomized study in North Dakota conducted over four summers indicated that seeding helped to reduce hail severity.

Anita Baker-Blocker

FURTHER READING

Ahrens, C. Donald. "Cloud Development and Precipitation." In *Essentials of Meteorology: An Invitation to the Atmosphere.* 5th ed. Belmont, Calif.: Thomson Learning, 2008.

Cotton, William R., and Roger A. Pielke. *Human Impacts on Weather and Climate.* 2d ed. New York: Cambridge University Press, 2007.

Goudie, Andrew. "The Human Impact on Climate and the Atmosphere." *The Human Impact on the Natural Environment: Past, Present, and Future.* 6th ed. Malden, Mass.: Blackwell, 2006.

House, Tamzy J., et al. *Weather as a Force Multiplier: Owning the Weather in 2025.* Maxwell Air Force Base, Ala.: Air University, Air Command and Staff College, 1996.

Kwa, Chunglin. "The Rise and Fall of Weather Modification: Changes in American Attitudes Toward Technology, Nature, and Society." In *Changing the Atmosphere: Expert Knowledge and Environmental Governance*, edited by Clark A. Miller and Paul N. Edwards. Cambridge, Mass.: MIT Press, 2001.

National Research Council. *Critical Issues in Weather Modification Research.* Washington, D.C.: National Academies Press, 2003.

SEE ALSO: Aerosols; Cloud seeding; Geoengineering; Volcanoes and weather.

Wells

CATEGORY: Water and water pollution

DEFINITION: Holes bored into the ground to extract or inject fluid

SIGNIFICANCE: Wells serve an important purpose in enabling the extraction of fluids from beneath the earth's surface, but sometimes the processes involved in boring wells can contribute to environmental degradation, such as through habitat destruction and water pollution.

The earliest wells were excavated by hand, and over time various mechanical methods have been developed for boring into the ground. The machinery used to create wells in the modern world depends on the depth desired and the types of rock through which the drill must bore. The borehole is cased with plastic pipe in shallow wells and steel pipe in deep wells to prevent caving of the walls. The casing is perforated at the depth from which production occurs. This section, called a screen, allows an exchange of fluid between the casing and surrounding rock. The space between the screen and the surrounding rock is filled with gravel to allow fluid to flow freely between the well and the aquifer. The space between the casing and the exposed rock in the upper part of the well is tightly sealed with impermeable grout to prevent contaminants from entering the bore from the surface.

Wells were originally used only to extract underground water. Today they serve many purposes, including extraction, injection, and monitoring of fluid below the surface. Water wells and oil wells are examples of producing wells that extract fluids from the subsurface. A typical water well draws water either from the surface aquifer or from deeper, confined aquifers. Oil and gas wells produce from deep rock strata. Wells are also used to remove contaminated groundwater and dewater saturated zones in which construction or other activity extends below the water table.

Injection wells are used to introduce fluids into the subsurface. They are used as a way to store water that could otherwise be lost to high evaporation rates or runoff. Oil is occasionally pumped into subsurface, impermeable salt caverns for storage. To increase production in an oil well, water or gas may be pumped into oil-bearing strata to displace the oil. One means of disposal of hazardous wastes is to inject them into deep levels of the earth's crust. Monitoring wells are used to determine variations in depth to the water table and to provide early warning of the migration of hazardous fluids. Frequent tests of monitoring wells are required around sites of potential groundwater contamination.

Groundwater is a major source of water for domestic, agricultural, and industrial use. It exists in the subsurface, filling pores and cracks in consolidated rocks and loose, unconsolidated sand, gravel, clays, and mixtures of these materials. The surface aquifer is the saturated zone that receives water by percolation down from the surface. This is the zone most susceptible to contamination by toxic substances from industrial and municipal wastes, feedlots, septic tanks, crop fertilizers, pesticides, and herbicides. Confined aquifers are less susceptible to contamination because they are sealed from surface percolation by overlying impermeable beds. Confined aquifers can be contaminated when they are exposed to direct recharge or by boreholes that reach them from the surface. Disposal wells with corroded casings may serve as conduits for hazardous waste into subsurface water supplies. Improperly grouted wells may allow surface contamination to infiltrate water supplies through seepage along the outside of the casing.

René A. De Hon

FURTHER READING

Misstear, Bruce, David Banks, and Lewis Clark. *Water Wells and Boreholes.* Hoboken, N.J.: John Wiley & Sons, 2006.

Sipes, James L. *Sustainable Solutions for Water Resources: Policies, Planning, Design, and Implementation.* Hoboken, N.J.: John Wiley & Sons, 2010.

SEE ALSO: Aquifers; Groundwater pollution; Oil drilling; Water pollution.

Rescue workers dig in the pile of rubble at the Pantglas Junior School after the collapse of the coal tip at Aberfan on October 21, 1966. (AP/Wide World Photos)

Welsh mining disaster

CATEGORIES: Disasters; waste and waste management

THE EVENT: Avalanche of coal sludge from the side of Merthyr Mountain in Wales onto the village of Aberfan below, causing the deaths of 144 people

DATE: October 21, 1966

SIGNIFICANCE: The disaster that took place at Aberfan led to research in how colliery tips are influenced by their environmental surroundings and to the establishment of programs aimed at increasing the understanding of coal tips and improving their regulation.

In 1870 workers at Merthyr Vale, a coal mine near the village of Aberfan, Wales, began dumping the remains of processed coal onto nearby Merthyr Mountain, forming a coal tip. When the area was struck with heavy rainfall in October of 1966, the black sludge loosened from its base; in addition to the rain, a hidden underground spring contributed to the saturation of the inert materials. On the morning of October 21, a massive amount of the sludge slid down the side of the mountain and crushed eighteen homes, a farm, and the Pantglas Junior School. Of the 144 people who died in the disaster, 116 were children—about half the school's students.

On October 22, the day after the disaster, the National Coal Board (NCB) acknowledged its accountability for the collapse of the coal tip and instantly sought procedures to eliminate similar tragedies. To aid the NCB, the Welsh secretary of state appointed a tribunal of inquiry, to be headed by Lord Justice Edmund Davies, to investigate the causes of the slide. The first measure taken to ensure Aberfan's safety was securing the coal tip on Merthyr Mountain. Inspection of 477 of the 1,753 coal tips used by the colliery was the next step in searching for evidence of instability that required prompt action. At the same time, information concerning the state of each tip was gathered to assist in the creation of regulations and correct supervision of the tips. The research findings led to the creation of a more explicit set of individual obligations and responsibilities, including detailed methods of management. One of the outcomes of the investigation was information on the wide diversity of characteristics distinguishing each coal tip. Consequently, the NCB decided to turn its attention to keeping a vigilant eye on the tips for any changes occurring in them.

Despite allegations made by Aberfan's local newspaper that the slide was similar to previous slides, the NCB determined that the avalanche was the result of unusual environmental factors. The newspaper, however, did point out that the years leading up to the Aberfan disaster displayed neglect of coal tip safety. The NCB, the Mines Inspectorate, the legislature, and other countries had disregarded the importance of maintaining definitive laws for colliery tips. To rectify the problem, specifically at Aberfan, drains and boreholes were constructed for improved drainage at the tips. In addition, the spring that caused most of the saturation was enclosed to prevent further water emissions. The people of Aberfan, however, opted for complete removal of the coal tip; the removal was funded by the NCB, local agencies, and the Welsh government.

Several new practices arose from the disaster at Aberfan that involved Wales as a whole. The NCB, for instance, started a program to research how colliery tips are influenced by their environmental surroundings. Similar policies and programs were created with the aim of increasing the understanding of coal tips and improving their regulation.

Carolynn A. Kimberly

FURTHER READING

McLean, Iain, and Martin Johnes. *Aberfan: Government and Disasters.* Cardiff: Welsh Academic Press, 2000.

Smith, Keith. *Environmental Hazards: Assessing Risk and Reducing Disaster.* 3d ed. New York: Routledge, 2001.

SEE ALSO: Coal; Coal ash storage; Solid waste management policy; Strip and surface mining.

Wetlands

CATEGORY: Preservation and wilderness issues

DEFINITION: Transitional areas between terrestrial and aquatic ecosystems that exhibit characteristics of both

SIGNIFICANCE: Wetlands are widely considered to be among the world's most important ecosystems because of their high biodiversity and productivity. They also perform important functions related to the maintenance of surface and groundwater quality and quantity, prevention of saltwater intrusion, control of coastal erosion, and regulation of climate.

Wetlands are often distinguished by having three major components: water, hydrophytic vegetation, and hydric soils. All wetlands have water present for at least part of the year, though the depth and duration of flooding vary considerably. Some wetlands have water-saturated soil, whereas others are characterized by permanent flooding. At least periodically, wetlands support a predominance of hydrophytic vegetation—that is, plant life adapted to thrive in saturated soil conditions. Wetlands are also characterized by having undrained, or hydric, soils. These are soils in which an anaerobic condition (an absence of free oxygen) has developed because of long periods of saturation, flooding, or ponding during the growing season.

DEFINING WETLANDS

No single, formal definition of wetlands has been established, because no single description is appropriate for all of the diverse wetland types that exist over a large geographic scale with diverse climatic conditions. Dozens of definitions have been written, however, for specific reasons by specific interest groups and various regulatory agencies. The problem of defining wetlands is one of critical consequence to those persons who are subject to restrictions and limitations placed on them by various national, regional, state, and local laws concerning wetlands. Inconsistent definitions place a severe burden on private landowners who may be subject to such laws and do not have adequate technical or legal knowledge about wetlands.

In the United States, a regulatory definition of the term "wetland" has been developed so that the Army Corps of Engineers and the Environmental Protection Agency (EPA) can administer the permitting of dredging and filling of wetlands as prescribed in section 404 of the Clean Water Act. For this purpose, wetlands are considered to be

> those areas that are inundated or saturated by surface or ground water at a frequency and duration sufficient to support, and that under normal circumstances do support, a prevalence of vegetation typically adapted for life in saturated soil conditions. Wetlands generally include swamps, marshes, bogs, and similar areas.

Thus jurisdictional wetlands in the United States—those that are subject to section 404 permitting—must possess all three key characteristics: hydrology,

The wetlands of Nisqually National Wildlife Refuge in Washington State. (AP/Wide World Photos)

hydrophytic vegetation, and hydric soils. However, because many wetlands are not permanently wet and because water may not be seen during a single site visit, positive hydrology indicators must be found, which must be supported by wetland vegetation and soils. These strict requirements for wetland identification have caused many wetlands to fall into uncertain categories. For example, some wetlands can have the appropriate hydrology but fail to develop the appropriate wetland soils and vegetation; often two of the three characteristics can be confirmed but not the third. These situations continue to cause confusion among governmental agencies and private landowners.

Another significant issue concerning wetlands is that they form ecotones (ecological transition zones) between upland and aquatic ecosystems. Thus even if an area is identified as a wetland, determination of its exact boundaries may be extremely difficult because of the gradual, perhaps imperceptible, changes in soil and vegetation characteristics. The problem of identifying areas as wetlands and defining the boundaries of those wetlands is known as wetland delineation. The ability to perform delineations is acquired only through extensive training, especially in the areas of soils and botany.

FUNCTIONS AND VALUE

Not all wetlands perform the same set of functions or perform their functions at the same rate or efficiency. Often the size of a wetland and its location in the watershed determine its functions. Wetlands generally have extremely high biodiversity and rates of productivity. They provide food, shelter, and water for various invertebrates and vertebrates, many of which may be endangered or threatened. According to the U.S. Department of Agriculture's Natural Resources Conservation Service, nearly 5,000 plant species, 190 amphibian species, one-third of all bird species, and all wild ducks and geese depend on the nation's wetlands. Endangered species in the United States that have wetland habitats include the bald eagle, red wolf, whooping crane, fatmucket mussel, and swamp rose. It has been estimated that wetlands provide essential habitat for 40 percent of the nation's endangered and 60 percent of its threatened species.

Another function associated with wetlands is the maintenance of the quantity and quality of both surface water and groundwater. Many wetlands serve to recharge aquifers. Wetlands also accumulate sediments, nutrients, and many forms of water pollutants from their watersheds. By removing these materials, wetlands serve to clean the water.

Wetlands are sometimes referred to as nature's sponges because of their ability to ameliorate the effects of stormwater runoff and reduce floodwater damage. Stormwater enters wetlands and spreads out over large areas and then is slowly released. Increased property damage from flooding has been shown to occur following the destruction of wetlands. The Army Corps of Engineers noted in a 1976 report that if the Charles River wetlands near Boston, Massachusetts, were destroyed, flood damage in the river basin would increase by as much as $17 million annually. Other wetland functions are prevention of saltwater intrusion into groundwater and surface-water supplies, protection against coastal erosion from storms, and regional and global climate stabilization. Concerns regarding global climate change have sparked interest in the ability of wetlands to function as carbon reservoirs.

A wetland value is any product, characteristic, or function of a wetland that has worth or is beneficial to the environment or to people. Wetland products such as timber, fiber, food, and fish have commercial value and are easily measured. Some other values of wetlands, however, are not as easily quantified. For example, wetlands may have sociocultural significance and provide sites for recreation, research, and education. Further, it is impossible to assign a dollar value to the fact that wetlands provide habitat for a high percentage of endangered and threatened species around the world.

The value placed on a wetland's functions is, in many cases, the most important factor that determines whether the wetland is preserved or converted to some other use. As society's needs and perceptions change over time, the value assigned to wetland functions also changes. The values associated with wetlands are often in conflict because of the large number of functions these ecosystems can perform. For example, if the water level in a wetland is raised, waterfowl production may increase while timber production decreases. Managers of wetlands may thus make decisions that are popular with one user group but unpopular with other user groups.

LOSS AND DEGRADATION

"Wetland loss" refers to a decrease in wetland area caused by the conversion of wetland to nonwetland. "Wetland degradation" refers to the impairment of one or more wetland functions because of human activity. In most cases, wetland loss is difficult or impossible to reverse because of the complexity of wetland structure and function. Wetland degradation, in contrast, is more easily reversed through a variety of applied science and conservation tools. Wetland creation—the formation of wetlands in formerly nonwetland areas—has become an increasingly common strategy for combating wetland loss.

Wetlands are found on every continent. Even beneath Antarctica's mantle of ice, there are wetlands that support life. The exact size of global wetland areas is difficult to assess because of differences in wetland definitions and lack of documentation in many countries, but wetlands have been estimated to cover about 6 percent of the land area of the earth. The largest wetland areas are found in tropical, subtropical, and boreal regions. Since the beginning of the twentieth century, the world has lost more than 50 percent of its wetland area.

A 1990 report issued by the U.S. Fish and Wildlife Service estimated that the area of wetlands in the United States decreased from about 158 million hectares (391 million acres) during the 1780's to about 111 million hectares (274 million acres) in the mid-1980's—a 30 percent overall loss of wetland area. Within the lower forty-eight states, the estimated loss was 53 percent. The Fish and Wildlife Service has been monitoring the nation's wetlands trends since the 1950's. During the period from the mid-1950's to the mid-1970's an estimated 185,400 hectares (458,000 acres) of wetlands in the coterminous United States were lost every year. From the mid-1970's to the mid-1980's, by which time the value of wetlands had begun to be recognized, the estimated rate of loss decreased to 117,400 hectares (290,000 acres) per year. From 1986 to 1997 the loss rate dropped by 80 percent, to 23,700 hectares (58,500 acres) annually. Between 1998 and 2004 net wetlands gains in the coterminous United States surpassed net losses for the first time since the survey began, thanks to wetlands creation and restoration efforts. The net gain for the period was 77,598 wetland hectares (191,750 acres), or an average net gain of 12,950 hectares (32,000 acres) per year. As of 2004 there were an estimated 43.6 million hectares (107.7 million acres) of wetlands in the coterminous United States.

The underlying causes of wetland loss and degradation are numerous. These include poverty and economic inequality; population pressures from growth, immigration, and mass tourism; social and political conflicts; high demand for wetland resources such as timber; drainage; diking and damming; air and water pollution; introduction of exotic species; natural events such as hurricanes; and economic policies. In the United States, approximately 80 percent of wetland losses from the mid-1950's through the mid-1970's were the result of agricultural practices. Since the 1970's growing awareness of the importance of wetland functions has slowed the destruction of wetlands, but wetland loss and, more frequently, degradation continue, caused by development and other stressors.

President George W. Bush, speaking on Earth Day, 2004, at the Wells National Estuarine Research Reserve in Maine, promised that the federal government would restore or protect as much as three million acres of wetlands in the next five years. (AP/Wide World Photos)

CONSERVATION AND PROTECTION

Wetlands in the United States are protected through regulation, economic programs, and acquisitions. At the federal level, a confusing mix of programs and legislation simultaneously encourages and discourages wetland conservation. As early as 1903, President Theodore Roosevelt recognized that wetland loss had become significant. By executive order, he established Pelican Island in Florida as the nation's first wildlife refuge. The federal government also protects wetlands through several laws, including the Clean Water Act of 1972, which created a plan to control the discharge of dredged or fill materials into wetlands and other waters of the United States. The Army Corps of Engineers and EPA share responsibility for implementing the program. The "swampbuster" program is part of the Food Security Act of 1985 and 1990. It seeks to remove federal incentives for the agricultural conversion of wetlands to nonwetlands. In conjunction with this act, the 1990 Farm Bill created a voluntary Wetland Reserve Program, which provides financial incentives to farmers to restore and protect wetlands through the use of long-term easements.

The North American Waterfowl Management Program represents another milestone in the conservation of important wetland habitat. This plan was signed between Canada and the United States in 1986 to restore declining waterfowl populations through habitat acquisition, development of economic incentives to change land-use practices, and improvement of water management. Mexico became a signatory to the agreement in 1994. At the global level, the most significant wetland conservation work has resulted from the Convention on Wetlands of International Importance, or Ramsar Convention, in 1971. This global treaty provides a framework for the international protection and wise use of wetlands.

U.S. presidents became active in wetland protection during the 1970's. President Jimmy Carter signed two executive orders that provide guidance for wetland and floodplain management and protection of these areas by federal agencies. President George H. W. Bush extended these efforts to recommend that the United States establish a national goal of "no net loss" of wetlands. This policy became a major force for wetland conservation in the United States.

Despite all of this activity, several difficulties remain. Wetlands, because of the complexity of their values and functions, continue to be managed for a variety of purposes, many of them conflicting. The policy of no net loss of wetlands applies to the loss of wetland acreage only, not to wetland functions, values, or quality. President Bill Clinton took a compromise position in wetlands protection by reaffirming the "no net loss" policy and supporting the Wetland Reserve Program; however, he created section 404 exemptions for 21.4 million hectares (53 million acres) of previously converted wetlands and for small plots of land owned by families who wanted to build single-family houses.

On Earth Day 2004, President George W. Bush announced an initiative to achieve an overall increase in the quantity and quality of U.S. wetlands. While there was a net gain in wetland area between 1998 and 2004, there were still wetland losses; urban and rural development was responsible for about 61 percent of the net freshwater wetlands losses during this period. Two notable U.S. Supreme Court cases during the Bush administration, *Solid Waste Agency of Northern Cook County v. Army Corps of Engineers* (2001) and *Rapanos v. United States* (2006, consolidated with *Carabell v. Corps of Engineers*), found that the Clean Water Act did not protect the isolated wetlands addressed in each case from being developed.

Roy Darville
Updated by Karen N. Kähler

FURTHER READING

Batzer, Darold P., and Rebecca R. Sharitz, eds. *Ecology of Freshwater and Estuarine Wetlands.* Berkeley: University of California Press, 2006.

Fowler, Theda Braddock, and Lisa Berntsen. *Wetlands: An Introduction to Ecology, the Law, and Permitting.* 2d ed. Lanham, Md.: Government Institutes, 2007.

Keddy, Paul A. *Wetland Ecology: Principles and Conservation.* 2d ed. New York: Cambridge University Press, 2010.

Mitsch, William J., and James G. Gosselink. *Wetlands.* 4th ed. Hoboken, N.J.: John Wiley & Sons, 2007.

Spray, Sharon L., and Karen L. McGlothlin, eds. *Wetlands.* Lanham, Md.: Rowman & Littlefield, 2004.

Tiner, Ralph W. *Wetland Indicators: A Guide to Wetland Identification, Delineation, Classification, and Mapping.* Boca Raton, Fla.: CRC Press, 1999.

SEE ALSO: Clean Water Act and amendments; Coastal Zone Management Act; Dredging; Everglades; Great Swamp National Wildlife Refuge; Ramsar Convention on Wetlands of International Importance; Watershed management.

Whaling

CATEGORY: Animals and endangered species
DEFINITION: Commercial harvesting of whales
SIGNIFICANCE: Until the mid-1980's, when the International Whaling Commission enacted a moratorium on whaling, commercial exploitation resulted in a serious decline in the total whale population. This led to concerns over biodiversity and the possibility of the extinction of some species of whales. Some countries have continued whaling despite the moratorium.

Humans have hunted whales for thousands of years. Whales have provided meat, oil (used for lighting and as an industrial lubricant), and ambergris (used in the manufacture of perfume). Blubber was converted into soap, baleen was fashioned into objects such as stays for women's corsets, whale-tooth ivory was carved and scrimshawed, and whale bones were ground into fertilizer. The abundant uses that humans made of whales quickly led to overhunting of certain species. Bowheads were increasingly scarce by the late seventeenth century, and the Atlantic gray whale was hunted to extinction sometime during the early eighteenth century. One hundred years later, the U.S. whaling industry went into decline as the sperm whale became increasingly difficult to find.

The advent of new technologies during the late nineteenth and twentieth centuries hastened the decline of whale populations on a global scale. As steam and diesel engines replaced sails, ships gained the speed necessary to pursue faster species of whale. Harpoons fired from cannons were tipped with grenades that exploded on contact, immediately killing the whale and ending the era of epic struggles often depicted in art and literature. With the exception of right whales, all whales sink when dead, so engineers crafted inflation devices to float the carcasses. Factory ships introduced during the 1920's allowed fleets to process the whales at sea. Each of these developments increased the efficiency of whalers, leading to larger catches and placing more stress on whale populations.

Whaling decreased during World War II, but at war's end several nations aggressively competed for the remaining whales. In 1946 fifteen nations signed the International Convention for the Regulation of Whaling in an effort to conserve whale stocks for commercial harvesting and develop the industry in a controlled fashion. To this end, the convention established the International Whaling Commission (IWC). Between 1940 and 1980, an estimated 1.5 million whales were harvested, nearly double the approximately 794,000 killed between 1900 and 1940. The peak year was 1961, when whalers captured more than 66,000 whales. Several species, including the world's largest mammal, the blue whale, neared extinction.

While early expressions of concern over the fate of whales focused on the need to conserve remaining stocks for future commercial exploitation, new attitudes toward the environment prompted some people to call for an end to commercial whaling. As the environmental movement gained momentum during the 1960's and 1970's, whales became a powerful symbol of humankind's obligation to preserve wildlife. The United States passed the Marine Mammal Protection Act in 1972 with the aim of keeping whales and other marine mammals from being killed, injured, captured, or harassed.

THE MORATORIUM

In 1982 the IWC voted to impose a commercial whaling moratorium, which went into force in 1986.

The first year or so after the IWC imposed the moratorium, the Soviet Union, Norway, Iceland, and Japan lodged objections and persisted in their commercial hunting. Norway continued commercial hunting through the twentieth century and into the twenty-first, when it was joined by Iceland. Additional, permitted whale harvesting has occurred since 1986 in the form of aboriginal subsistence whaling, which native populations in Greenland, Russia, the United States, Canada, and St. Vincent and the Grenadines have carried out. Every year, aboriginal subsistence whaling kills a few hundred whales worldwide.

During the moratorium Norway, Iceland, and especially Japan (where whale meat is part of the nation's traditional diet) have harvested still more whales under special permits that allow whaling for scientific purposes. Through scientific whaling, researchers claim to gain an understanding of the roles that whales play in ecosystems, the factors that affect natural whale mortality, how environmental changes affect whales, and how humans can improve whale management efforts. Whether scientific whaling programs are actually scientific means for assessing whale

Demonstrators take part in a rally against commercial whaling in Washington, D.C., on Earth Day, April 22, 2010. (AP/Wide World Photos)

stocks or veiled commercial whaling efforts is a matter of intense, ongoing controversy.

By the mid-1990's several nations, including Japan and Norway, were arguing that populations of some whale species had increased enough to permit a renewal of commercial whaling. Problems in assessing the number of whales, many of which travel over wide areas, made conflicts over the resumption of whaling difficult to resolve. Nonetheless, rough estimates indicated that most species remained in grave danger of extinction as the twentieth century drew to a close. Right whales, long the favorite of whalers because they float after they are killed, had been reduced from a prehunting population of approximately 100,000 to between 2,000 and 3,000 survivors in 1990. Nearly 250,000 blue whales had once graced the world's seas, but in 1990 fewer than 12,000 remained. Other species had experienced similar declines.

Ironically, the devastation benefited at least one species of whale. Because the minke whale faced less competition for food from other whales whose populations were shrinking, the number of minkes actually doubled, reaching an estimated 880,000 by 1990. It was this growth that prompted Japan and Norway to request a return to minke whaling, in part because they claimed that the minke was responsible for the decline of the mackerel and other fish on which their fishing industries depended. In addition to the minke whale, wildlife that benefited from the disappearance of whales included seals and birds that eat krill. The reduction in the number of baleen whales consuming krill led to an increase in food supply for many animals, which resulted in population growth.

Obstacles to Recovery

Many scientists fear that some whale species will never recover and will eventually become extinct. The pressing issue regarding the survival of most species of whale is that of biodiversity. With the small populations existing, species such as the blue whale might lack the genetic diversity to respond to environmental changes. Present populations might persist and even grow but may be unable to survive any significant alterations in the environment. Concerns about biodiversity are linked to the fact that the whales' habitat, the world's oceans, is an increasingly inhospitable and even dangerous place. Like land animals, whales suffer the consequences of habitat loss and human encroachment. They are injured when struck by ships or entangled in abandoned fishing equipment. Global climate change, sound and chemical pollution, and declines in fish populations also affect whales. Minute tissue samples collected through nonlethal means from almost one thousand free-ranging sperm whales between 2000 and 2005 were found to contain astonishingly high levels of heavy metals. Whales that have washed ashore on Canada's Gulf of St. Lawrence have been found to be so loaded with chemicals that their carcasses have been designated as toxic waste.

In June, 2010, during the sixty-second IWC meeting, member nations considered suspending the commercial hunting ban and implementing limited whaling in its place. The rationale for such a suspension was twofold: Some whale species had recovered sufficiently, it was argued, that they might be hunted sustainably; and limited, permitted whaling would have less of an impact on whale populations than unlimited hunts in the name of scientific research. The meeting was tainted with allegations that Japan had bribed other IWC member nations to vote against continuing the moratorium. Ultimately, the IWC agreed to table two contentious proposals—one lifting the ban to allow Japan to hunt whales in its coastal waters, the other creating a whale sanctuary in the South Atlantic where no hunting would be allowed—and refer the matters to a multinational working group to sort out.

Given the pressures that pollution places on whales and the resulting toxicity of whale products, many observers have argued against any renewal of whaling. Calls for an end to the moratorium indicate that many people still regard whales as a commercial resource to be exploited. The ongoing debate between these two groups has revealed that international cooperation on environmental issues remains problematic, especially when it concerns the care of the oceans and their inhabitants. Although whales live far from most humans, human management has become necessary to ensure their continued survival.

Thomas Clarkin
Updated by Karen N. Kähler

Further Reading

Dolin, Eric Jay. *Leviathan: The History of Whaling in America.* New York: W. W. Norton, 2007.

Ellis, Richard. *Men and Whales.* 1991. Reprint. New York: Lyons Press, 1999.

Estes, James A., et al., eds. *Whales, Whaling, and Ocean Ecosystems.* Berkeley: University of California Press, 2006.

Friedheim, Robert L., ed. *Toward a Sustainable Whaling Regime*. Seattle: University of Washington Press, 2001.

Heazle, Michael. *Scientific Uncertainty and the Politics of Whaling*. Seattle: University of Washington Press, 2006.

Hoare, Philip. *The Whale: In Search of the Giants of the Sea*. New York: Ecco, 2010.

Kalland, Arne. *Unveiling the Whale: Discourses on Whales and Whaling*. New York: Berghahn Books, 2009.

Mulvaney, Kieran. *The Whaling Season: An Inside Account of the Struggle to Stop Commercial Whaling*. Washington, D.C.: Island Press, 2003.

Stoett, Peter J. *The International Politics of Whaling*. Vancouver: University of British Columbia Press, 1997.

Tønnessen, J. N., and A. O. Johnsen. *The History of Modern Whaling*. Berkeley: University of California Press, 1982.

SEE ALSO: Extinctions and species loss; International Convention for the Regulation of Whaling; International whaling ban; International Whaling Commission; Marine Mammal Protection Act; Save the Whales Campaign; Sea Shepherd Conservation Society; Watson, Paul.

White, Lynn Townsend, Jr.

CATEGORY: Activism and advocacy
IDENTIFICATION: American historian and author
BORN: April 29, 1907; San Francisco, California
DIED: March 30, 1987; Los Angeles, California
SIGNIFICANCE: White argued that religion—medieval Christianity, in particular—played a significant role in the environmental crisis that was becoming apparent during the late 1960's. His controversial thesis was influential in spawning several movements in environmentalism, including ecotheology.

Lynn Townsend White, Jr., was a professor of history at Princeton and Stanford universities and at the University of California, Los Angeles. A graduate of Stanford, Union Theological Seminary, and Harvard University, he also served as president of Mills College and was a founding member of the Society for the History of Technology.

As a historian, White specialized in the history of medieval technology. His major work, *Medieval Technology and Social Change*, published in 1962, defined the field of medieval technology historiography. The book details several seemingly small changes in agriculture and animal husbandry—the stirrup, the plow, and crop rotation, for example—that profoundly influenced European culture and land use.

In regard to environmental issues, White is best remembered for his seminal 1966 lecture "The Historical Roots of Our Ecological Crisis," which was published in 1967 in the journal *Science*. In this work White argues that ecology is directly related to human beings' beliefs about themselves and about the world. According to White, Western Christianity promotes the belief that nature exists only to serve humanity, and this idea leads to the ruthless exploitation of nature.

The article has frequently been taken as an attack on Christianity, but White, a lifelong Presbyterian, did not intend to attack Christianity as a whole; he simply sought to urge a reexamination of Christianity's doctrine of nature. Against the hierarchical view that humanity should dominate nature, White's essay proposes Saint Francis of Assisi as the "patron saint for ecologists" because Francis's theology regards all natural things as having been made for the glory of their creator and therefore as intrinsically valuable.

White's thesis has been controversial, and it is not universally accepted by historians, theologians, or ecologists. Nevertheless, its importance and its influence—especially on religious movements such as ecotheology—are broadly recognized.

David L. O'Hara

SEE ALSO: Anthropocentrism; Antienvironmentalism; Ecology as a concept; Environmental ethics; Environmentalism.

Whooping cranes

CATEGORY: Animals and endangered species
DEFINITION: Migratory crane species native to North America
SIGNIFICANCE: Efforts to prevent the extinction of the whooping crane have had some success, but the whooping crane population remains very small. This regal bird has become an important symbol for wildlife conservation.

Standing at a height of 1.5 meters (5 feet) or more, with a wingspan of up to 2 meters (7 feet), whooping cranes are the tallest birds in North America.

An ultralight aircraft leads four whooping cranes along a migration route from a wildlife refuge in Wisconsin to one in Florida. (AP/Wide World Photos)

Named for their loud, bugling call, they are white with black-tipped wings, black beaks, and bare patches of red skin on the tops of their black-banded heads. Marshes are vital to the survival of whooping cranes, which nest in tall marsh grasses and eat blue crabs and other foods found in wetland areas. Unlike other cranes, whoopers are slow to mature. After reaching the age of four years, a whooping crane chooses a mate for life; in the wild, these birds may live up to twenty-four years. Adult females produce only two eggs in a clutch, and often only one of the two chicks survives. For this reason, the whooping crane population has never been abundant.

In the nineteenth century the total number of whooping cranes was estimated at 1,400. Two flocks were known to migrate to Canada from Louisiana and Texas. Their numbers rapidly dwindled as their wetland habitats were destroyed with the expansion of the human population. The magnificent look of adult whoopers also made them targets for sport hunting. Every death of a whooping crane also ended the re-productive life of the surviving mate. By 1900 only about 100 whooping cranes were left. Despite the fact that the Migratory Bird Treaty, signed by the United States and Canada in 1918, made the hunting of whooping cranes illegal, the number of whooping cranes dwindled to only 15 by the late 1930's because of the continuing loss of wetlands.

In 1937 the wintering grounds of the Texas flock were set aside as the Arkansas National Wildlife Refuge. By 1944 this flock had grown from 15 to 21 in number, but the last of 6 whoopers from the Louisiana flock died in 1948. Conservationists were delighted when it was discovered in 1954 that whooping cranes were breeding in a secluded section of Wood Buffalo National Park in Alberta, Canada. To help ensure the flock's survival, officials arranged for eggs to be taken from the nests so that the chicks could be raised safely in captivity. These birds were later returned to the flock, helping build the population. Another flock that does not migrate was established at Florida's Kissimmee Prairie Sanctuary using captive-bred whoopers.

Conservation efforts have increased the whooping crane population to around 300. Of this number, only about 180 are part of the wild flock in the Arkansas National Wildlife Refuge that migrates between Canada and the United States. The proximity of the refuge to shipping routes with barges carrying chemicals and fertilizers has raised fears among conservationists that an accidental spill could wipe out the whole flock. Since 1999, attempts have been made to establish a second migratory flock in the wild. Some success has been seen in a project run by the nonprofit group known as Operation Migration, which began in 2001; the project uses ultralight planes to teach whooping cranes a migration route from Wisconsin to Florida.

Lisa A. Wroble

FURTHER READING

Gibbons, Whit. "Birds and the Ecovoid." In *Keeping All the Pieces: Perspectives on Natural History and the Environment*. 1993. Reprint. Athens: University of Georgia Press, 2010.

Roman, J. "A Whooping Success: The World's Most Endangered Crane Makes a Comeback." *Wildlife Conservation* 110, no. 3 (2007): 38-45.

Scott, Chris. *Endangered and Threatened Animals of Florida and Their Habitats*. Austin: University of Texas Press, 2004.

SEE ALSO: Captive breeding; Endangered Species Act; Endangered species and species protection policy; Migratory Bird Act; Wetlands; Wildlife management; Wildlife refuges.

Wild and Scenic Rivers Act

CATEGORIES: Treaties, laws, and court cases; preservation and wilderness issues

THE LAW: U.S. federal law establishing a system of free-flowing rivers for the protection of their natural, scenic, and recreational values

DATE: Enacted on October 2, 1968

SIGNIFICANCE: Since the late 1960's the Wild and Scenic Rivers Act has been an important legislative tool for protecting America's pristine and free-flowing rivers from impoundments and unrestricted development.

Wild and Scenic Rivers Act

Congress passed the Wild and Scenic Rivers Act of 1968 to recognize and protect the rivers of the United States. Main provisions of the act follow.

Congressional Declaration of Policy. It is hereby declared to be the policy of the United States that certain selected rivers of the Nation which, with their immediate environments, possess outstandingly remarkable scenic, recreational, geologic, fish and wildlife, historic, cultural, or other similar values, shall be preserved in free-flowing condition, and that they and their immediate environments shall be protected for the benefit and enjoyment of present and future generations. The Congress declares that the established national policy of dam and other construction at appropriate sections of the rivers of the United States needs to be complemented by a policy that would preserve other selected rivers or sections thereof in their free-flowing condition to protect the water quality of such rivers and to fulfill other vital national conservation purposes.

Classification. A wild, scenic or recreational river area eligible to be included in the system is a free-flowing stream and the related adjacent land area.... Every wild, scenic or recreational river in its free-flowing condition, or upon restoration to this condition, shall be considered eligible for inclusion in the national wild and scenic rivers system and, if included, shall be classified, designated, and administered as one of the following:

(1) *Wild river areas*—Those rivers or sections of rivers that are free of impoundments and generally inaccessible except by trail, with watersheds or shorelines essentially primitive and waters unpolluted. These represent vestiges of primitive America.

(2) *Scenic river areas*—Those rivers or sections of rivers that are free of impoundments, with shorelines or watersheds still largely primitive and shorelines largely undeveloped, but accessible in places by roads.

(3) *Recreational river areas*—Those rivers or sections of rivers that are readily accessible by road or railroad, that may have some development along their shorelines, and that may have undergone some impoundment or diversion in the past.

During the 1950's and 1960's federal agencies such as the Bureau of Reclamation and U.S. Army Corps of Engineers were engaged in hundreds of projects for building large and small river impoundments for irrigation, flood control, recreation, and hydroelectric power generation. Seeing the loss of free-flowing rivers, the Outdoor Recreation Resources Review Commission, a federal panel of resource experts, suggested that legislation was needed to create a river preservation system. Under the sponsorship of Senator Frank Church of Idaho, the Wild and Scenic Rivers Act was signed into law by President Lyndon B. Johnson in October, 1968. Segments of eight rivers received protection as part of the original legislation: the Middle Fork of the Clearwater River and the Middle Fork of the Salmon River in Idaho, the Eleven Point in Missouri, the Feather in California, the Rio Grande in New Mexico, the Rogue in Oregon, the St. Croix in Minnesota and Wisconsin, and the Wolf in Wisconsin.

A key provision of the act is the way new rivers can be added to the National Wild and Scenic Rivers System. To qualify for protection a river has to be free-flowing and must possess outstanding scenic, recreational, geological, or other qualities. In lieu of an entire river being designated, the act enables river segments, including tributaries, to be added through congressional action, by nomination by the governor of the state in which the river is located, or through recommendation of the secretary of the interior or the secretary of agriculture. Along with state governments, four federal agencies are responsible for managing rivers in the system: the National Park Service, the Bureau of Land Management, the Fish and Wildlife Service, and the Forest Service.

River segments are designated within one of three categories—wild, scenic, or recreational—corresponding to their level of development. Segments categorized as wild must be inaccessible except by trail and free of impoundments. Scenic segments must also be free of impoundments and must have shorelines that are mostly undeveloped, with accessibility to roads located in a minimum of locations. Segments designated as recreational are the most developed and may have experienced some type of impoundment in the past. The act also includes stipulations concerning how rivers should be managed. For example, river management should not interfere with public use or enjoyment and should be respectful of private property rights.

By 2010 more than 19,300 kilometers (12,000 miles) of 252 rivers located in thirty-eight states had become part of the National Wild and Scenic River System. This included approximately 9,800 kilometers (6,100 miles) of river classified as wild, 4,300 kilometers (2,700 miles) classified as scenic, and 5,800 kilometers (3,600 miles) designated recreational. In 1995 the Interagency Wild and Scenic River Coordinating Council was created to provide oversight for the system. The council is made up of representatives from all the federal agencies involved in managing the system's rivers.

Thomas A. Wikle

FURTHER READING

Cech, Thomas V. "Water, Fish, and Wildlife." In *Principles of Water Resources: History, Development, Management, and Policy.* 3d ed. New York: John Wiley & Sons, 2010.

Echeverria, John D., Pope Barrow, and Richard Roos-Collins. *Rivers at Risk: The Concerned Citizen's Guide to Hydropower.* Washington, D.C.: Island Press, 1989.

Palmer, Tim. *Endangered Rivers and the Conservation Movement.* 2d ed. Lanham, Md.: Rowman & Littlefield, 2004.

River Network and David M. Bolling. *How to Save a River: A Handbook for Citizen Action.* Washington, D.C.: Island Press, 1994.

SEE ALSO: Bureau of Land Management, U.S.; Dams and reservoirs; Department of the Interior, U.S.; Floods; Forest Service, U.S.; National Park Service, U.S.; Nature preservation policy.

Wild horses and burros

CATEGORIES: Animals and endangered species

DEFINITION: Feral horses and burros living in the rangelands of the western United States

SIGNIFICANCE: What should be done with the wild horses and burros of the United States has long been the subject of debates, primarily between ranchers who want to protect rangeland for their livestock and animal activists, environmentalists, historians, and others who want to preserve and protect the animals in their wild state.

Bands of untamed horses and burros can be found on the publicly owned rangelands of the western United States. The largest herds live on the public lands of eleven states: Nevada, Wyoming, Utah, Oregon, California, Colorado, Idaho, Arizona, Montana, North Dakota, and New Mexico. Feral horses also can be found on islands off the eastern coast of the United States, including Assateague and Cumberland. The

Wild horses on Steens Mountain in Oregon. (Bureau of Land Management [BLM])

Bureau of Land Management (BLM) and the U.S. Fisheries and Wildlife Service are the primary custodians of these animals.

HISTORY

Horses are natural denizens of grasslands, but they have migrated beyond their original native territory owing to human migrations. During the nineteenth and early twentieth centuries, equines abandoned after mining collapses, for example, discovered bioniches in arid lands, where the hardiest survived. Such horses formed herds or joined herds that had been created earlier by horses descended from those originally brought to the New World during the Spanish Conquest in the sixteenth century. Other equines originated in the great horse cultures of the Native Americans that the U.S. Army destroyed in the nineteenth century. Thriving populations gradually became competitors with the ranching industry (chiefly cattle and sheep ranchers).

Environmental disruptions that resulted from the wild and domesticated beasts sharing unrestrained access to fragile desert ecosystems included overgrazing, soil compaction, degradation of riparian environments, and overpopulation owing to the lack of natural predators (ironically, natural predators such as wolves and mountain lions had been annihilated because they threatened livestock). Before passage of the Taylor Grazing Act of 1934, ranchers routinely grazed livestock on federal lands and their private holdings as well with no restrictions.

In 2010 it was estimated that some 37,000 feral horses and burros were living on millions of acres of public lands in the United States, half of them in Nevada. Another 34,000 that had been gathered in annual roundups were warehoused in government facilities around the country. At such facilities, the younger horses and burros await adoption or sale and the older and weaker animals receive care on an indefinite basis. Some wild horse advocacy groups have challenged the scientific methodology used in arriving at these figures; the BLM conducts only "rough counts." The formulas that the BLM uses to determine the numbers of animals—both wildlife and livestock—that shared rangeland can sustain with grazing are also a source of contention.

DEBATES

A lengthy legal history accompanies disputes over the fate of wild horses. Animal activists, wildlife biologists, environmental impact analysts, and historians often line up on one side to oppose ranchers, underfunded government agencies, and other vested economic interests. During the 1950's a Nevada woman, Velma Johnston, derisively dubbed "Wild Horse Annie," exposed the inhumane treatment of wild horses that were being corralled, shipped, and slaughtered to make pet foods. Her national campaign contributed to the eventual passage of groundbreaking national legislation in 1971 with the Wild Free-Roaming Horses and Burros Act. In 1978 that act was amended to "protect the range from wild horse overpopulation."

Debates surrounding the treatment of wild horses and burros often include examination of whether these animals can be classified as wildlife. Investigation of the genetic history of wild horses has demonstrated that they are indigenous to North America. Genus *Equus* migrated to Asia via the Bering land bridge in Paleolithic times, expanded into Europe, and then returned in the company of the Spanish conquistadores during the sixteenth century. This finding subverts a legal strategy to declare these equines invasive species, which would allow them to be disposed of as nuisances, as they would no longer be protected with native wildlife status.

Battles over the handling of wild horses and burros continue to be waged in the courts. Severe budget cuts have strained the caretaking capabilities of the BLM, and a sense of urgency surrounds the question of what to do with the horses collected in the annual roundups that are not adopted or sold. Debates continue, as advocates for wild horses and burros argue against euthanasia for healthy animals and many U.S. taxpayers express an increasing unwillingness to continue funding their care.

JoEllen Broome

FURTHER READING

Lindholdt, Paul, and Derrick Knowles, eds. *The Individual and Public Lands in the American West.* Spokane: Eastern Washington University Press, 2005.

Marshall, Julie Hoffman. *Making Burros Fly: Cleveland Amory, Animal Rescue Pioneer.* Boulder, Colo.: Johnson Books, 2006.

Ryden, Hope. *America's Last Wild Horses.* Rev. ed. New York: Lyons Press, 1999.

Stillman, Deanne. *Mustang: The Saga of the Wild Horse in the American West.* Boston: Houghton Mifflin Harcourt, 2008.

SEE ALSO: Amory, Cleveland; Bureau of Land Management, U.S.; Grazing and grasslands; *Kleppe v. New Mexico*; Taylor Grazing Act.

Wilderness Act

CATEGORIES: Treaties, laws, and court cases; preservation and wilderness issues

THE LAW: U.S. federal legislation concerning the preservation of designated lands in their most natural condition

DATE: Enacted on September 3, 1964

SIGNIFICANCE: The U.S. Congress has designated almost 5 percent of the total U.S. land area as wilderness under the provisions of the Wilderness Act, which created the National Wilderness Preservation System. Designated wilderness areas are protected from development and from environmentally disruptive activities.

The 1964 Wilderness Act established the National Wilderness Preservation System (NWPS), gave the U.S. Congress authority to designate wilderness areas, and directed the secretaries of the interior and agriculture to review lands for possible wilderness designation. The act initially set aside 54 areas—a total of 3.6 million hectares (9 million acres) of federal Forest Service land—for wilderness classification. In 1968 Congress began adding wilderness areas. By 1999 there were 631 wilderness areas in forty-four states, totaling nearly 42.1 million hectares (104 million acres). A decade later, that total had grown to 756 wilderness areas (including a tropical rain forest in Puerto Rico, the first wilderness area designated in a U.S. territory) that all together occupied more than 44 million hectares (109 million acres). Wilderness areas are located within national forests, wildlife refuges, and parks and are managed by a host of agencies, including the U.S. Forest Service, the Bureau of Land Management (BLM), the National Park Service, and the Fish and Wildlife Service.

The land area of the United States totals 914 million hectares (2.3 billion acres); by 2010 approximately 4.7 percent of that had been designated wilderness. Just over half of this land—23.2 hectares (57.4 million acres)—is in the state of Alaska and accounts for roughly 16 percent of land in the state. Alaska is also the state in which the largest amount of land has actually been set aside: The Alaska National Interest Lands Conservation Act (ANILCA) of 1980 more than tripled the NWPS by establishing 35 new wilderness areas totaling more than 22.7 million hectares (56 million acres). Excluding Alaska, less than 3 percent of land in the United States is classified as wilderness or has been recommended for such designation.

The Wilderness Act defines wilderness as federal land "where the earth and its community of life are untrammeled by man, where man himself is a visitor who does not remain." Although numerous exceptions are made, the act generally prohibits commercial activities, motorized and mechanical access, permanent roads, and human-made structures and facilities within wilderness areas. An area may be determined to be suitable for wilderness designation if it is

> an area of undeveloped land retaining its primeval character and influence, without permanent improvements or human habitation, which is protected and managed so as to preserve its natural conditions and which (1) generally appears to have been affected primarily by the forces of nature, with the impact of man's works substantially unnoticeable; (2) has outstanding opportunities for solitude or primitive and unconfined type of recreation; (3) has at least five thousand acres of land or is of sufficient size as to make practicable its preservation and use in an unimpaired condition; and (4) may also contain ecological, geological, or other features of scientific, educational, scenic, or historic value.

Congress has the authority to designate areas as wilderness and uses its power to do so under the act. Designation is permanent, and new lands are added as Congress sees fit.

Permitted and Prohibited Uses

Although wilderness areas are protected to preserve their natural conditions, a number of nonmotorized activities—such as horseback riding, hiking, camping, fishing, and hunting—are allowed in them. Preexisting and valid extractive uses are also allowed to continue until the permits granted for such activities expire, are abandoned, or are purchased by the government. Preexisting grazing is allowed to continue as long as it is consistent with sound resource management practices. In addition, the Wilderness Act honors all federal-state relationships with regard to state water laws and state fish and wildlife responsibilities. Activities that are generally not allowed in these areas include mining, timber harvesting, water development, mountain biking, and use of any motorized equipment such as snowmobiles and all-terrain vehicles. Allowances for some of these activities made in the act can be seen as a compromise between preservationists and those resource interests concerned with grazing, mining, timber harvesting, and water development.

Given the definition of wilderness and the amount of federally owned land in the western United States, congressionally classified wilderness is a particularly western phenomenon. Whereas states such as Alaska, Arizona, California, Idaho, and Washington have more than 1.6 million hectares (4 million acres) of wilderness each within their borders, the nonwestern states Connecticut, Delaware, Iowa, Kansas, Maryland, and Rhode Island have none.

Although Congress makes the final decision as to the suitability of additional areas for wilderness designation, land management agencies such as the Forest Service and the BLM are often responsible for making official recommendations. For example, the Federal Land Policy and Management Act of 1976 directed the BLM to review land it administers for possible wilderness designation, and the Forest Service has gone through two Roadless Area Review and Evaluation (RARE) plans and the 2001 Roadless Area Conservation Rule to determine suitable wilderness areas. Such plans have often been criticized for not designating enough land area and for refusing to designate land with great economic potential, thus leaving "rocks and ice" for wilderness classification and those lands with economic value under multiple-use management.

Martin A. Nie
Updated by Karen N. Kähler

Further Reading

Campaign for America's Wilderness. *People Protecting Wilderness for People: Celebrating 40 Years of the Wilderness Act.* Washington, D.C.: Author, 2004.

Dawson, Chad P., and John C. Hendee. *Wilderness Management: Stewardship and Protection of Resources and Values.* 4th ed. Boulder, Colo.: WILD Foundation, 2009.

Frome, Michael. *Battle for the Wilderness.* Rev. ed. Salt Lake City: University of Utah Press, 1997.

Harvey, Mark. "Loving the Wild in Postwar America." In *American Wilderness: A New History,* edited by Michael Lewis. New York: Oxford University Press, 2007.

Hays, Samuel P. *Wars in the Woods: The Rise of Ecological Forestry in America.* Pittsburgh: University of Pittsburgh Press, 2007.

Nash, Roderick. *Wilderness and the American Mind.* 4th ed. New Haven, Conn.: Yale University Press, 2001.

Scott, Doug. *The Enduring Wilderness: Protecting Our Natural Heritage Through the Wilderness Act.* Golden, Colo.: Fulcrum, 2004.

SEE ALSO: Alaska National Interest Lands Conservation Act; Gila Wilderness Area; Nature preservation policy; Nature reserves; Preservation; Roadless Area Conservation Rule; Wilderness areas; Wilderness Society.

Wilderness areas

CATEGORIES: Preservation and wilderness issues; land and land use

DEFINITION: Natural, undeveloped areas in the United States that are protected under the Wilderness Act of 1964

SIGNIFICANCE: The designation of large areas of land as protected wilderness areas is the subject of ongoing debate in the United States, with preservationists asserting that more areas need to be protected and critics arguing that the natural resources found in these areas should be available for use.

Preserving areas of unspoiled nature is a relatively new idea, and in the United States, this idea began to make sense to many Americans only when the seemingly inexhaustible wilderness of North America had been, in fact, nearly exhausted. In 1924, at the urging of Forest Service employee and influential conservationist Aldo Leopold, 305,500 hectares (755,000 acres) of the Gila National Forest in New Mexico were set aside as the first federally protected wilderness, the Gila Primitive Area. As the system of primitive areas grew, environmentalists became concerned about inconsistent management and about the fact that these areas were protected only by agency

policy and not by law. They began lobbying for federal legislation that would designate and protect wilderness areas throughout the United States. The concept of preserving wilderness was strongly opposed, however, by many of those who made their livings by using natural resources; these included people involved in ranching and those in the timber and mining industries. They saw protected wilderness lands, which often had great economic value, as being "locked up" for the pleasure of a few.

LEGISLATION

On September 3, 1964, after eight years of debate and compromise, President Lyndon B. Johnson signed the Wilderness Act, creating the National Wilderness Preservation System (NWPS), which consisted of fifty-four areas totaling 3.6 million hectares (9 million acres). The act states:

> A wilderness, in contrast with those areas where man and his own works dominate the landscape, is hereby recognized as an area where the earth and its community of life are untrammeled by man, where man himself is a visitor who does not remain.

The act defines the mechanism for adding more areas to the system in the future. To be considered, an area must be at least 2,023 hectares (5,000 acres) "or of manageable size." This is a far cry from the early days of wilderness advocacy, when the minimum size was thought to be 202,000 hectares (500,000 acres), or, as Aldo Leopold put it, "large enough to absorb a two-week pack trip." Designated wildernesses become part of the NWPS. All roads, structures, and other installations are prohibited in designated wilderness, as is the use of motorized equipment or any mechanical transport. These areas of wild nature have been, and continue to be, the focus of intense controversy regarding their designation and management.

A significant addition to the NWPS came in 1975 with the passage of the Eastern Wilderness Act. The lack of pure, untouched wilderness in the eastern states led to the loosening of the strict standards of the original act to allow the inclusion of ecologically significant areas that show more impact from human activities than would originally have been permitted. In this way, sixteen areas totaling 83,770 hectares (207,000 acres), from 8,900-hectare (22,000-acre) Bradwell Bay in Florida to the 5,670-hectare (14,000-acre) Lye Brook Wilderness in Vermont, were added to the system. As of the late 1990's, the wilderness system encompassed more than 650 areas, ranging in size from the 2-hectare (5-acre) Oregon Islands Wilderness to the 3.6 million hectares (9 million acres) of the Wrangell-St. Elias Wilderness in Alaska, for a total of more than 40 million hectares (100 million acres). By 2010, in part as a result of the passage of the Omnibus Public Land Management Act of 2009, the number of wilderness areas had grown to 756, with a total of more than 44.1 million hectares (109 million acres). Of this total, 52 percent was in the state of Alaska.

DEBATES AND CONTROVERSIES

Although the amount of protected land may seem quite large, preservationists point out that only about 5 percent of the landscape of the United States is protected in its natural state. Some large areas continue to be fought over, such as the fragile Arctic coastal plain of Alaska, home of vast caribou herds and underlain by large oil deposits. Idaho, which is among the top three states in the lower forty-eight in terms of wilderness land area (exceeded only by California and Arizona), still has millions of hectares of undeveloped roadless land that many believe should be protected. Wilderness advocates also point out that many wilderness areas, as well as national parks and other protected lands, have illogical political boundaries, unrecognized by grizzly bears and other important wildlife species. They argue that areas between and adjacent to designated wilderness areas should often be protected as well, to create units based on natural, ecological boundaries.

After a wilderness area is designated, the focus shifts to the maintenance of its desired qualities, leading to the paradox of "wilderness management." Although recreation is only one of the stated uses of wilderness—the others being scenic, scientific, educational, conservation, and historic—agency efforts and budgets are based primarily on the need to manage the often vast numbers of human visitors. One of the stated purposes of preserving wilderness areas is to provide for "primitive and unconfined recreation," but another consideration is the protection of the resource itself. At what point do the camping and trail restrictions, quotas, and permits needed to protect the resource impinge on the unconfined recreation of the visitor?

Another issue related to wilderness areas is that of wildfire suppression. It is now understood that fire is an important component of most ecosystems, but past policies of fire suppression have left unnatural fuel conditions in many areas. Should managers allow wildfires to burn, even though these fires are likely be larger and more destructive than natural, periodic fires of the past? Other major controversies center on the reintroduction of wildlife species (such as the wolf and the grizzly bear) to wilderness areas, the disposition of long-standing mining and drilling claims, and the flying of aircraft over, or even into, remote wilderness areas.

Joseph W. Hinton

FURTHER READING

Allin, Craig W. *The Politics of Wilderness Preservation.* 1982. Reprint. Fairbanks: University of Alaska Press, 2008.

Dawson, Chad P., and John C. Hendee. *Wilderness Management: Stewardship and Protection of Resources and Values.* 4th ed. Boulder, Colo.: WILD Foundation, 2009.

Frome, Michael. *Battle for the Wilderness.* Rev. ed. Salt Lake City: University of Utah Press, 1997.

Nash, Roderick. *Wilderness and the American Mind.* 4th ed. New Haven, Conn.: Yale University Press, 2001.

Scott, Doug. *The Enduring Wilderness: Protecting Our Natural Heritage Through the Wilderness Act.* Golden, Colo.: Fulcrum, 2004.

SEE ALSO: Alaska National Interest Lands Conservation Act; Ecosystems; Gila Wilderness Area; Leopold, Aldo; Marshall, Robert; National parks; Wilderness Act.

Wilderness Society

CATEGORIES: Organizations and agencies; activism and advocacy; preservation and wilderness issues

IDENTIFICATION: American nonprofit organization dedicated to the protection and preservation of wilderness areas and wildlife

DATE: Established in 1935

SIGNIFICANCE: The Wilderness Society has been instrumental in the passage of major conservation legislation in the United States, including the Wilderness Act of 1964, the Wild and Scenic Rivers and National Trails System Acts of 1968, and the National Forest Management Act of 1976.

In January, 1935, a group of eight dedicated conservationists organized the Wilderness Society in Washington, D.C. Among the participants were Robert S. Yard, publicist for the National Park Service; Benton MacKaye, known as the "father of the Appalachian Trail"; Robert Marshall, head of recreation and lands for the U.S. Forest Service; and Aldo Leopold, a wildlife ecologist at the University of Wisconsin. Leopold believed that the new society would form a cornerstone for efforts to preserve America's vanishing wilderness.

After much dedicated work and pressure by the Wilderness Society, the Wilderness Act was finally signed into law by President Lyndon B. Johnson on September 3, 1964. This act established the National Wilderness Preservation System (NWPS), which enabled the U.S. Congress to set aside selected areas within national forests, national parks, national wildlife refuges, and other federal lands as units to be kept permanently unchanged by humans. There would be no roads, structures, vehicles, or any significant impacts of any kind in these selected areas. The Wilderness Act initially designated approximately 3.6 million hectares (9 million acres) as wilderness.

The Wilderness Society has since had a hand in the passage of several other major public lands bills, including the Wild and Scenic Rivers Act and the National Trails System Act, both passed in 1968, and the National Forest Management Act of 1976. The organization's efforts have helped to contribute a total of 44 million hectares (109 million acres) to the NWPS. In particular, the Alaska National Interest Lands Conservation Act of 1980 designated 22.7 million hectares (56 million acres) of pristine land for protection, and the California Desert Protection Act of 1994 designated 3.2 million hectares (8 million acres) of desert lands. After a ten-year effort, the Wilderness Society was instrumental in the enactment of the National Wildlife Refuge System Improvement Act of 1997, which strengthened protections for wildlife in all national wildlife refuges.

During the 1980's and 1990's, despite the successes of the Wilderness Society, public lands continued to be compromised and degraded by air and water pollution, excessive development, road building, logging, cattle grazing, mining, and recreational activities. By the late 1990's, the Wilderness Society had focused on the preservation of a number of high-priority areas, with the overall goal of creating a nationwide network of wildlands. Key campaigns were launched in Montana, California, Idaho, Nevada, Utah, Colorado, Texas, Vermont, the Pacific Northwest, and the southern Appalachians.

In 1997, with backing from the Wilderness Society, the U.S. Congress blocked a plan to create a massive gold mine just outside Yellowstone National Park by appropriating money from the Land and Water Conservation Fund to purchase the mining claims. Similarly, the society helped create public pressure that led the Du Pont Corporation to defer plans for a titanium mine on the border of the Okefenokee National Wildlife Refuge in Georgia. Working closely with local and national groups during 1996 and 1997, the society convinced the federal government to withdraw a proposed logging plan for nine national forests in the Sierra Nevada. Additionally, in response to a Wilderness Society lawsuit, a federal judge blocked logging in four national forests in Texas, pointing out that it would be detrimental to wildlife habitats.

Despite the loosening of some environmental protections under the presidential administration of George W. Bush, the Wilderness Society enjoyed several notable victories during this period. Among them were the 2001 adoption by the U.S. Forest Service of the Roadless Area Conservation Rule, which protects roughly 24 million hectares (60 million acres) within national forests, and the 2005 addition of the first tropical rain forest—4,047 hectares (10,000 acres) in Puerto Rico—to the U.S. national forest system. In late 2008 Wilderness Society staff and policy experts met with U.S. president-elect Barack Obama's transition team and urged that the new administration take action to address climate change and protect roadless forests, the Arctic National Wildlife Refuge, and other fragile lands threatened by oil and gas drilling. The 2009 Omnibus Public Land Management Act, passed during the first year of Obama's presidency, protects 0.85 million hectares (2.1 million acres) of new wilderness areas in nine states.

Alvin K. Benson
Updated by Karen N. Kähler

FURTHER READING

Kline, Benjamin. *First Along the River: A Brief History of the United States Environmental Movement.* 3d ed. Lanham, Md.: Rowman & Littlefield, 2007.

Maher, Neil M. "The Great Conservation Debate." In *Nature's New Deal: The Civilian Conservation Corps and the Roots of the American Environmental Movement.* New York: Oxford University Press, 2008.

Sutter, Paul. "New Deal Conservation: A View from the Wilderness." In *FDR and the Environment*, edited by Henry L. Henderson and David B. Woolner. New York: Palgrave Macmillan, 2005.

SEE ALSO: Conservation Reserve Program; Leopold, Aldo; Marshall, Robert; National Forest Management Act; National Trails System Act; Nature preservation policy; Preservation; Wetlands; Wild and Scenic Rivers Act; Wilderness Act; Wilderness areas.

Wildfires

CATEGORIES: Preservation and wilderness issues; resources and resource management

DEFINITION: Fires occurring in wilderness or open country in various vegetation types, generally characterized by large size, rapid flame spread, intensity of heat and smoke, and difficulty of prediction or control of behavior

SIGNIFICANCE: Growing human control of fire has resulted in intensification of the human relationship with, and responsibility for, fire in the environment. This growing control has resulted in ongoing alterations in the definition of "wild" fire. These changes in definition are themselves reflective of fire-mediated alteration and modification of the preexisting environment by humans.

For more than 450 million years, all fires in plant matter on the earth were both "natural" and "wild." Over the past half million to a million years, and particularly during the past 20,000 years, the "control" (meaning both use and suppression) of fire by humans has changed not only our understanding of "natural fire" and "wildfire" but also the character of myriad environments and the human cultures sustained by them. From the effects of slash-and-burn agriculture on species diversity to the effects of global climate change on future human urbanization patterns, human control of fire has had tremendous consequences—even, paradoxically, increasing the number, intensity, and duration of wildfire events.

USE AND SUPPRESSION OF FIRE

Although wildfires occur on every continent except Antarctica, the consequences of the use and the suppression of fire are best seen in four regions: Amazonia, equatorial Africa, Australia, and western North America. An understanding of the dynamics of use and suppression in regard to wildfire requires expansion of the traditional "fire triangle" of ignition source, combustible material, and oxygen into a "wildfire hexagon" through the addition of three key factors: topography (the "shape" of the landscape), climate or weather (particularly in regard to the effects of drought on fuel moisture and strong winds on fire behavior), and human interaction (both in shorter-term responses to individual fire events and in longer-term transformation of fire-prone environments).

Weather, topography, and the presence of atmospheric oxygen are not easily altered on large scales within short time frames, so only three elements of the wildfire hexagon—ignition source, combustible material, and human interaction—can be readily affected by short-term human activity. Because fire has long been a tool useful in clearing the land for subsequent agricultural production, in many areas of Amazonia and equatorial Africa people intervene by enhancing natural fire cycles through purposely introducing ignition sources and combustible material into the environment. Because fire has also been perceived as a threat to valuable timber, homes, and other property, however, in many areas of Australia and western North America people intervene by suppressing natural fire cycles through reducing ignition sources where possible or, once a fire has been ignited, attempting to reduce the combustibility or availability of fuels through firefighting activities.

Problems with these strategies of use and suppression arise when the use becomes too careless or the suppression too careful. Fire used to clear land in Amazonia and equatorial Africa often ends up making its way into surrounding rain forest, there becoming wildfire. Years and decades of too-careful fire suppression in western North America and Australia have resulted in thickety forests so overburdened with ladder fuels—leading from flashy fuels such as grass and duff along the ground through a midlevel of shrubs and younger trees, finally to the topmost canopy of mature trees—that once a fire gets started in such forest it is much more prone to become a catastrophically destructive burn. Although "natural" and "controlled" are usually seen as opposites, human control of fire in both the "use" and "suppression" situations outlined above has separated "natural" from "wild" such that the resulting wildfires are both uncontrolled and unnatural.

Patchwork use of broadcast burning in wildland—by hunters to stampede game, by pastoralists to open up grazing lands, and by agriculturalists to prepare land for planting—arguably originated many tens of thousands of years ago. Systematic, widespread suppression of fire in wildland arguably began only in response to the Great Fire of 1910 in the United States. With the continued unprecedented growth in human population over the past three centuries, however, the increasing number of wildfires stemming from both fire use and fire suppression has resulted in many of the same negative effects on the environment, from the physical (increased erosion, landslides, mudflows, flash flooding, and altered water quality) to the biological (loss of species habitat, introduction of invasive species, and declines in biodiversity).

Increases in population and demographic changes have meant that more and more land that was formerly wild has become increasingly bordered and penetrated by housing tracts along the wildland-urban interface (WUI; also known as the wildland-urban intermix). The upshot has been still more property at risk, with consequent pressures to suppress fire more thoroughly in naturally fire-prone but increasingly populated and economically valuable areas—with the result that more fuels build up year upon year until, when ignition finally comes, the result is too often a devastating wildfire.

Wildfires affect not only physical, biological, and economic aspects of landscapes and watersheds but the atmosphere as well. Smoke, soot, ash, ozone, greenhouse gases (such as carbon dioxide), and other fire by-products are lofted by wildfire not only throughout the troposphere (the lowest layer of the

Camp crew members of the California Department of Forestry watch as a wildfire approaches in Simi Valley in September, 2005. (AP/Wide World Photos)

atmosphere) but also as high as the lower strato-sphere, where their influence on human health and on climate can be global.

Striking a New Balance

Growing awareness that too-careless use and too-careful suppression of fire in wildlands have resulted in increases in the number, duration, and intensity of wildfires—with cascading local, regional, and global ecological and economic effects—has led fire experts to reevaluate approaches to wildfires. This reevaluation is based in an increasingly nuanced ecological understanding of the roles of natural fire, fire cycles, weather patterns, and fire-dependent, fire-tolerant, and fire-intolerant plant adaptations.

The recognition that many natural landscapes contain plants that are fire-dependent or fire-tolerant (not only in western North American and Australian wildlands but also in Southeast Asia and South Africa) has called into question policies of too-careful or complete fire suppression, particularly when such policies have resulted in the increased presence of both flashy, fire-intolerant vegetation and higher fuel loads generally. Because lower-intensity fires can reduce fuel loads and reduce or eliminate often invasive fire-intolerant vegetation, fire has been selectively allowed back into these wildlands. Carefully monitored against escape onto higher-value locations, fires that have been naturally caused (most often by lightning, but also at times by volcanic eruptions or meteor strikes) are allowed to burn so that they can fulfill their ecological role.

Controlled burning (generally referred to as pre-scribed burning in the United States) is a fire management strategy in which wildland fires are purposely ignited under favorable weather conditions, with the goal of creating lower-intensity fires that help clean out accumulated fuels, foster higher levels of species diversity, and reduce the future risks of intense, long-lived wildfires. The goal of such "allowed fire" strategies is to emulate natural fire through the use of controlled fire of the right type, in the right place, and at the right time.

Just as too-careful fire suppression has had to give way to some controlled use of fire in an attempt to emulate natural fire, so too-careless fire use has had to be curbed through prevention of ignition—again in an attempt to emulate natural fire. Because slash-and-burn agricultural practices damage fire-resistant rain forests and encourage the growth of flammable brush (and the more frequent occurrence of future wildfires), those wishing to curtail careless fire use in rain-forest areas—whether neighboring landowners with flammable tree crops and orchards or environmentalists wishing to preserve rain-forest species diversity—have had to exert pressure on slash-and-burn agriculturalists to be more circumspect in igniting, more careful in monitoring, and more thorough in suppressing fires lit to clear land.

Because human interaction can change both the fuel and the ignition factors in the wildfire hexagon, the fastest and most powerful way to affect the risk of future wildfire is to change the way humans interact with their environments. Those who in the past suppressed fire in order to further their interests now find they must at times incorporate fire use in their approach, and those who in the past used fire to further their own interests now find they must also at times incorporate fire suppression in their approach.

Howard V. Hendrix

Further Reading

Carle, David. *Introduction to Fire in California.* Berkeley: University of California Press, 2008.

Egan, Timothy. *The Big Burn: Teddy Roosevelt and the Fire That Saved America.* Boston: Houghton Mifflin Harcourt, 2009.

Holbrook, Stewart H. *Burning an Empire: The Study of American Forest Fires.* 1943. Reprint. New York: Macmillan, 1960.

United Nations Food and Agriculture Organization. *FAO Meeting on Public Policies Affecting Forest Fires.* Rome: Author, 1999.

Wilson, Bill, et al., eds. *Forest Policy: International Case Studies.* Wallingford, England: CABI, 1998.

See also: Air pollution; Amazon River basin; Coniferous forests; Controlled burning; Deciduous forests; Forest and range policy; Forest management; Forest Service, U.S.; National Park Service, U.S.; Rain forests; Slash-and-burn agriculture; Sustainable forestry; Watershed management.

Wildlife management

CATEGORY: Animals and endangered species
DEFINITION: Control of the population levels of wild
  species and maintenance and improvement of
  the species' habitats
SIGNIFICANCE: Where humans and wildlife coexist,
  wildlife management is necessary for maintaining
  a balance between what the human population
  needs and desires and what the wildlife populations require for survival. Management may be manipulative, acting on wildlife population densities,
  or custodial, minimizing the human factors that
  have impacts on wildlife and wild habitats.

Wildlife has been compromised by human activity
throughout history. Animals have been hunted
for food and skins, trees have been cut down for shelter, and plants have been collected for food, medicine, and clothing. As weapons improved, more animals were hunted. As humans learned to cultivate
their own plants and herd animals, they required
more space, so areas of wild habitat were cleared.
With diminishing habitats, wildlife numbers also diminished. Many species became extinct without human notice.

The rulers of ancient civilizations set aside areas as
game reserves for hunting and protected certain forests for religious reasons. Medieval European kings
continued the practice of creating game reserves for
sport hunting by the privileged classes. These reserves
created artificially high populations of sport species,
such as grouse, mallards, and pheasants, while reducing the numbers of predator species, such as weasels,
otters, wildcats, and badgers.

The amount of land and wildlife in North America
seemed limitless to the early European settlers. As
land was cleared to make way for westward expansion,
forests were lost, marshes were drained, coastlines
were changed, and breeding grounds were destroyed.
While some wild species failed to breed or produced
fewer numbers, others, such as wolves, coyotes, and
grizzly bears, came into constant conflict with humans. Early settlers were so intent on battling nature
to ensure their survival that many never noticed the
demise of numerous animal species.

Hunters in the United States lobbied for the protection of wildlife as early as 1888. During the early
twentieth century, wildlife and habitats began to be
conserved for their beauty and their scientific value.
Most of this conservation came about because of the
work of conservation clubs or through the establishment of reserves and parks through the Forest Reserve Act of 1891. It was not until the environmental
movement of the 1960's and 1970's that wildlife management began to focus on ecology and the balance of
nature.

Modern wildlife management in the United States
and elsewhere has become a complex coordinated effort among many interest groups, including conservation officers, biologists, ecologists, fishers, hunters,
park employees, wildlife enthusiasts, environmentalists, and local, regional, and national governments.

CHALLENGES

The balance of nature is important for the environmental health of the earth. Natural cycles generally
ensure that animals, plants, birds, fish, and insects are
supported by the habitats in which they live. If one element is removed or disrupted, the balance is affected;
if a predator is removed, for example, overbreeding
may result. Wildlife management includes maintaining a balance between the number of animals and the
amount of food and shelter available.

To ensure that their management efforts help keep
a proper balance, wildlife managers study animals,
fish, and waterfowl. They tag animals and track their
migration patterns to gain helpful information. They
also catalog the plants, insects, soil types, and varieties
of species in different areas to learn how habitats support their wildlife. Wildlife management approaches
may be manipulative or custodial; manipulative approaches involve acting directly on species or their
habitats to change species population levels, whereas
in custodial management nature is allowed to run its
course without human interference.

Overbreeding may lead to disease or lack of food
for an animal species. Hunting and fishing regulations are designed to help keep populations under
control. To keep the number of certain species from
becoming too low, legal hunting and fishing seasons
are established. During years when wildlife populations are high, seasons may be extended. During years
when wildlife populations are low, limits are set on the
number of animals each hunter or fisher can take. Another method for maintaining population levels of
certain species is to reintroduce natural predators to
areas where their species had previously been eliminated.

Wildlife management also involves enforcing wildlife laws. State and federal law-enforcement and conservation officers make sure that people enjoy protected habitat and park areas without littering, picking protected plants, or destroying the habitat in any other way. They also enforce laws prohibiting poaching (the killing of species out of season or taking more than the legal limit) and the selling or harming of endangered, threatened, or rare species.

Fines collected from wildlife law-enforcement actions and fees for hunting and fishing licenses help support a variety of wildlife programs, including the protection of breeding grounds. Many animal and bird species need specific habitats in order to breed. Wildlife management includes intervention to ensure that these breeding grounds remain undisturbed. Maintaining the plants, soils, and insects within habitats is also part of wildlife management. Maintenance of plants and insects may involve preventing or cleaning up pollution.

When populations are low, moratoriums on hunting may be announced until repopulation occurs. Most often, depleted animal species need help to repopulate. Wildlife managers may create protected habitat areas or restore breeding grounds, such as marshes. Such restoration efforts are costly, however.

EXAMPLES

Species such as the black-footed ferret and the whooping crane have been rescued from extinction through captive-breeding and reintroduction programs. In such wildlife management programs, wild species are safely captured and moved to zoos or other specialized facilities. Ideally, they will breed while in captivity. Once a captive population has grown sufficiently, individuals in the population are preconditioned for survival in the wild and then released into their natural habitat.

Such programs are costly and fraught with challenges. One experimental program, for example, involved transferring whooping crane eggs to the nests of greater sandhill cranes in an effort to establish a self-sustaining migratory flock of whooping cranes. The young whooping cranes were successfully reared by their sandhill foster parents and learned from them how to migrate. The whooping cranes, however, imprinted on their foster parents to such an extent that when the fosterlings reached sexual maturity they ignored one another and tried unsuccessfully to mate with sandhill cranes. The fostered whooping cranes never reproduced, and the experimental population eventually died out.

In the case of wood ducks, a hunting moratorium and nesting boxes helped the species recover from endangered status. As the result of a program in which artificial nesting boxes were placed for the ducks in areas where predators and humans would not harm the eggs, populations recovered enough that some hunting of wood ducks has been permitted. This method was also used with the snow goose population, which during the early twentieth century was down to a few thousand birds. The program was so successful that by the 1960's snow goose numbers had increased almost twentyfold. The ready availability of waste grain and other farm crops helped the population to continue its expansion. By 2009, the snow goose population numbered in the millions, and the federal government had extended the hunting season and allowed more hunting methods in an effort to bring the population under control. The staggering number of the birds has stressed and degraded their fragile Arctic breeding ground, affecting the health of the snow geese as well as other birds that share the habitat. The extreme rebound of the snow goose population over the course of a century is a prime example of how problematic and complex wildlife management can be.

Lisa A. Wroble
Updated by Karen N. Kähler

FURTHER READING

Adams, Clark E., and Kieran J. Lindsey. *Urban Wildlife Management.* 2d ed. Boca Raton, Fla.: Taylor & Francis, 2009.

Bolen, Eric G., and William Laughlin Robinson. *Wildlife Ecology and Management.* 5th ed. San Francisco: Pearson, 2003.

Dawson, Chad P., and John C. Hendee. *Wilderness Management: Stewardship and Protection of Resources and Values.* 4th ed. Boulder, Colo.: WILD Foundation, 2009.

Krausman, Paul R. *Introduction to Wildlife Management: The Basics.* Upper Saddle River, N.J.: Prentice Hall, 2002.

Sinclair, Anthony R. E., John M. Fryxell, and Graeme Caughley. *Wildlife Ecology, Conservation, and Management.* 2d ed. Malden, Mass.: Blackwell, 2006.

Williams, Byron K., James D. Nichols, and Michael J. Conroy. *Analysis and Management of Animal Populations.* San Diego: Academic Press, 2002.

SEE ALSO: Balance of nature; Captive breeding; Extinctions and species loss; Kaibab Plateau deer disaster; Nature reserves; Poaching; Predator management; Wildlife refuges; Zoos.

Wildlife refuges

CATEGORIES: Animals and endangered species; land and land use

DEFINITION: Regions of land or water set aside by governments or private organizations to protect and preserve one or more species of wildlife

SIGNIFICANCE: The U.S. National Wildlife Refuge System has endured congressional debate and public scrutiny involving environmental issues related to the societal, governmental, and commercial use of designated sanctuaries, culminating in the 1997 National Wildlife Refuge System Improvement Act and the transformation of America's refuges into multiple-use systems.

Prior to 1900, the U.S. federal government aggressively raised much-needed revenue and rewarded growing commerce by selling or giving away nearly 405 million hectares (1 billion acres) of land to states, homesteaders, veterans, railroads, and businesses. President Theodore Roosevelt initiated the protection of habitat for wildlife in 1903 when he set aside Pelican Island, a 1.2-hectare (3-acre) ecosystem of barren sand and scrub in Florida's Indian River, as a federal reservation to protect birds from hunters supplying plumes to the fashion industry. Inspired while camping in California's Yosemite Valley with naturalist John Muir, Roosevelt established more than fifty wildlife refuges, five national parks, and eighteen national monuments, such as the Grand Canyon. He also greatly increased the area of lands designated as national forests before leaving his second term in office.

To preserve additional lands "for our children and their children's children forever, with their majestic beauty all unmarred," Roosevelt guaranteed land for future refuges by separating other federal public domain regions such as national forests and rangelands from the control of commercial interests. More than 90 percent of the refuge land area in existence in the United States in the early years of the twenty-first century resulted from Roosevelt's foresight, which enabled the National Wildlife Refuge System to grow larger than the national park system and entail nearly 4 percent of the surface area of the United States.

With vital assistance by private individuals and organizations such as the Nature Conservancy and the National Audubon Society, wildlife refuges have been established for waterfowl, big game, small resident game, and colonial nongame birds. Wildfowl refuges, easily the most plentiful, are geographically patterned to supply breeding, wintering, resting, and feeding areas along the four major North American migration flyways. The sportsmen who were essential in establishing many national refuges ensured that hunting would be permitted on most sanctuary lands, with trapping allowed on many. Although the entire National Wildlife Refuge System logs substantial numbers of hunting visits annually, visitors who are interested in wildlife education and photography outnumber hunters and anglers by more than four to one.

The National Wildlife Refuge System—overseen by the U.S. Fish and Wildlife Service, which is part of the Department of the Interior—is the most comprehensive nature protection network in the world. The entire system includes more than 60.7 million hectares (150 million acres) inside refuge boundaries, encompassing 553 national wildlife refuges and other units as well as 38 wetland management districts. Nearly all of the refuges in the system are open to the public. Endangered and threatened species are supported on almost 60 refuges created specifically for that purpose, and many urban refuges have been established near large cities. At least one wildlife refuge has been established in each of the fifty states, and several are found in overseas possessions from Puerto Rico to American Samoa in the South Pacific. The two largest refuges are in Alaska: The Arctic National Wildlife Refuge and the Yukon Delta National Wildlife Refuge are both more than 7.7 million hectares (19 million acres) in size. The smallest refuge in the system is the Mille Lacs National Wildlife Refuge in Minnesota, which is just 0.24 hectare (0.60 acre) in size.

ENVIRONMENTAL MANAGEMENT ISSUES

Creating and maintaining a refuge to provide food, cover, and protection from human development for wildlife is considerably more difficult than simply sequestering an area and allowing nature to run its course; continual management of resources is imperative to keep delicate ecosystems in balance.

(continued on page 1327)

Milestones in Wildlife Protection

Year	Event
1870	The first state wildlife refuge is established in California.
1898	Kruger National Park is established in South Africa for the preservation of big game.
1900	The Lacey Act regulates the interstate commerce of birds and mammals; the act is supplemented by a similar act for black bass in 1926.
1903	The first federal bird sanctuary is established by President Theodore Roosevelt at Florida's Pelican Island.
1908	Theodore Roosevelt calls a conference of state governors and related officials to inventory natural resources in the United States.
1916	The National Park Service is established and forbids hunting within its jurisdiction.
1929	The Migratory Bird Conservation Act provides for a system of refuges along major flyways.
1934	The Duck Stamp Act requires hunters of migratory fowl to purchase duck stamps with their waterfowl licenses; proceeds are used to establish wildlife refuges.
1937	Taxes on arms and ammunition are used for wildlife preservation.
1940	The National Wildlife Refuge System is established by a consolidation of the Bureau of Biological Survey and the Bureau of Fisheries; its mission includes biological research and administration as well as enforcement of federal legislation.
1966	The National Wildlife Refuge System Administration Act mandates that all refuge uses be compatible with the primary purpose for which the refuge was established.
1970	The National Environmental Policy Act, as well as other legislation designed to combat pollution, is passed.
1973	The Endangered Species Act, which updates prior acts in 1966 and 1969, requires refuge managers to protect certain species of flora and fauna.
1980	The Alaska National Interest Lands Conservation Act doubles the amount of land in the U.S. refuge and park systems.
1987	A federal court rules that the U.S. Fish and Wildlife Service is responsible for policing the existing ban on spring hunting by native groups in Alaska.
1992	Research commissioned by Defenders of Wildlife finds that the National Wildlife Refuge System is grossly inadequate.
1997	The National Wildlife Refuge System Improvement Act establishes a revamped multiple-use mission statement for refuge habitat conservation.
2001	The Roadless Area Conservation Rule, a federal policy initiative designed to protect national forests from commercial development, is issued.
2005	More than 4,000 hectares (approximately 10,000 acres) of Puerto Rican rain-forest land are added to the U.S. national forest system.
2008	Leaders of the Wilderness Society urge U.S. president-elect Barack Obama to address climate change and increase protection of national refuge lands.
2009	The Omnibus Public Land Management Act adds 850,000 hectares (2.1 million acres) of new wilderness areas in nine U.S. states.

The many tasks that humans must perform on refuge lands include fixing broken floodgates and cleaning clogged ditches, seeding wildlife foods, plowing and burning areas that have been overrun with unwanted vegetation, and closing off areas from the public during sensitive periods for animals, such as mating and birthing seasons. In addition, refuge managers must often battle for their lands' shares of rapidly declining water supplies.

Although many refuges include areas just as spectacular as those within the national park system, the National Wildlife Refuge System as a whole was for many years not well utilized by the public. The U.S. Congress, observing this lack of public use, leased some of these public lands for commercial purposes such as grazing, farming, oil drilling, mining, logging of timber, military maneuvers, and motorized recreation. However, as public use of refuges increased during the 1980's and 1990's, Congress added eighty new refuges to the system, creating such a backlog of environmental preservation issues that the legislators then began to contemplate selling some areas to pay for maintenance. With minimal budgets, refuge managers are charged with making certain that all activities that take place within their refuges are compatible with wildlife while still allowing potentially destructive activities such as off-road driving and motorcycling, powerboating, and commercial fishing. The maintenance of biodiversity is also an important goal of managers; the wildlife refuge system provides homes for about 700 species of birds, 220 species of mammals, 250 species of reptiles and amphibians, and 200 species of fish, in addition to innumerable species of plants.

Refuge areas owned by private corporations have often sacrificed key habitat for short-term economic gain with little regard for long-term environmental and social consequences. However, business executives have realized that environmental issues are of genuine concern to most Americans. In response to increased environmental awareness and pressure from consumers, employees, and stockholders, many large businesses have implemented stewardship strategies that protect natural resources, enhance wildlife habitat, and provide for public enjoyment of their underdeveloped land.

Key Legislative Actions

Following Theodore Roosevelt's initial work, new additions to the refuge system came slowly until the Dust Bowl years of the 1930's, when migratory bird populations became depleted. Congress then passed the 1934 Migratory Bird Hunting and Conservation Stamp Act, widely known as the Duck Stamp Act, which added a conservation fee to the price paid for every waterfowl license purchased by a hunter; the revenue collected in this way enabled the Fish and Wildlife Service to acquire wetlands along major bird migration flyways. Additional moneys to purchase ref-

Cranes and snow geese share space at Bosque del Apache Wildlife Refuge in New Mexico. Such refuges provide essential stopovers for migrating birds and habitat for many plant and animal species. (AP/Wide World Photos)

uge lands came from the Land and Water Conservation Fund set up in the 1960's to increase public space for outdoor recreation, which generated considerable revenues from offshore oil drilling leases. Passage of the 1980 Alaska National Interest Lands Conservation Act (ANILCA) enabled both the refuge system and the national park system to double in size. Although 96 percent of all refuge units are outside Alaska, Alaska contains about 83 percent of National Wildlife Refuge System lands.

The 1973 Endangered Species Act spurred managers of refuges to make more concessions for certain species of flora and fauna. At the beginning of the twenty-first century, refuges in the U.S. system harbored about 280 threatened or endangered species. Surveys in the late 1980's by the Fish and Wildlife Service and the General Accounting Office revealed that more than 60 percent of refuges were permitting activities known to be harmful to wildlife. The most harmful practices, such as military activities and drilling, were not under the control of the Fish and Wildlife Service. The high-profile activist group Defenders of Wildlife organized a citizens' commission in 1992 that confirmed that the National Wildlife Refuge System was "falling far short of meeting the urgent habitat needs of the nation's wildlife" and was suffering from "chronic fiscal starvation and administrative neglect."

The National Wildlife Refuge System Improvement Act, signed into law by President Bill Clinton on October 9, 1997, dramatically shifted the priorities of the refuge system from its original sole purpose of protecting wildlife to the formation of a multiple-use system. The legislation redefined the system's mission regarding the conservation of habitat for fish, wildlife, and plants; designated priority public uses such as hunting, fishing, wildlife observation and photography, and environmental education and interpretation; and required that the environmental health of the refuge system be maintained. This monumental bill gave hunting, fishing, commercial trapping, and recreation equal status in the refuges with the conservation of plants, birds, and animals. It also limited new or secondary refuge use to activities compatible with wildlife protection and made legislative changes more difficult for future congressional cycles. The principle of multiple use, however, continues to allow and may result in increased mining, drilling, grazing, logging, and motorized recreation on refuge lands, in addition to increased military training, including bombing and tank and troop exercises. Upon signing the bill, Clinton stated that he "hoped and trusted that the process by which this bill was enacted will serve as a model for future congressional action on other environmental issues," with the future of the National Wildlife Refuge System to be shaped by the future of other bills such as the Clean Water Act, the Wetlands Protection Act, and the Endangered Species Act.

Daniel G. Graetzer

FURTHER READING

Butcher, Russell D. *America's National Wildlife Refuges: A Complete Guide.* 2d ed. Lanham, Md.: Taylor Trade, 2008.

Dolin, Eric Jay. *Smithsonian Book of National Wildlife Refuges.* Washington, D.C.: Smithsonian Institution Press, 2003.

Nelson, Lisa. "Wildlife Policy." In *Western Public Lands and Environmental Politics,* edited by Charles Davis. 2d ed. Boulder, Colo.: Westview Press, 2001.

Patent, Dorothy Hinshaw. *Places of Refuge: Our National Wildlife Refuge System.* Boston: Houghton Mifflin, 1992.

Riley, Laura, and William Riley. *Guide to the National Wildlife Refuges.* Rev. ed. New York: Macmillan, 1992.

SEE ALSO: Biodiversity; Darling, Jay; Ecosystems; Endangered species and species protection policy; Forest and range policy; Land-use policy; Multiple-use management; Roosevelt, Theodore; Wildlife management.

Wilmut, Ian

CATEGORY: Biotechnology and genetic engineering
IDENTIFICATION: English reproductive biologist
BORN: July 7, 1944; Hampton Lucey, England
SIGNIFICANCE: Wilmut, one of the world's foremost authorities on biotechnology and genetic engineering, conducted a landmark cloning experiment in 1996 that produced Dolly the sheep, the first mammal clone ever produced from adult cells.

Ian Wilmut earned his Ph.D. in animal genetic engineering from Darwin College at the University of Cambridge in 1971. One of his first projects, conducted in 1973, involved the birth of the first calf ever reproduced from a frozen embryo. In 1974 Wilmut took a job with the Animal Breeding Research Station, later known as the Roslin Institute, a nonprofit organization affiliated with the University of Edinburgh in Scotland.

In 1990 Wilmut hired cell cycle biologist Keith Campbell to assist him with cloning studies at the Roslin Institute. Their first success came in 1995 with the birth of Megan and Morag, two Welsh mountain sheep cloned from differentiated embryo cells. The achievement came as the result of a new technique pioneered by Wilmut

Reproductive biologist Ian Wilmut in his laboratory in February, 1997, shortly after his announcement of the successful cloning of Dolly the sheep. (AP/Wide World Photos)

and Campbell that involved starving embryo cells before transferring their nuclei to fertilized egg cells. The technique synchronized the cell cycles of both cells and led Wilmut and Campbell to believe that any type of cell could be used to produce a clone.

In the landmark cloning experiment that produced the Finn-Dorset lamb Dolly, Wilmut removed udder cells from a six-year-old adult ewe and isolated the cells to starve them of nutrients, thereby arresting their growth and division. Next, he extracted an egg cell from another ewe and removed its nucleus, which contained the second ewe's genetic material. Following the joining of the egg cell with one of the udder cells, Wilmut implanted the resultant embryo into a surrogate mother. Dolly was born on July 5, 1996, but the success of the experiment was kept secret until February, 1997, to give Wilmut and his scientific team time to tabulate the results fully, be assured of Dolly's survival, and secure a patent for their technique. Wilmut's breakthrough ended decades of skepticism among scientists, many of whom had believed that cloning would never be possible for any animal.

On July 25, 1997, Wilmut and Campbell shocked the world again when they announced the birth of Polly, a lamb with a human gene in every cell of its body. Using a method similar to the one they had used for cloning Dolly, the scientists had cloned Polly from a fetal skin cell that had a human gene. Wilmut and his research team had opened up a new frontier of science.

The announcements of the births of Dolly and Polly stirred the scientific community and the public, kicking off a large-scale debate regarding the ethics and direction of cloning research. In particular, people feared the possibility that human clones would be produced. Wilmut, however, stated that he saw no reason for scientists to pursue the first cloning of a human; he pointed instead to the new possibilities that the cloning of animals could hold for the treatment of human disease. Cloned animals could act as manufacturing plants for valuable human proteins, which are costly and difficult to produce in large amounts elsewhere. In addition, agriculture could benefit from the cloning of high-quality animals, such as the best milk-producing cows and the best wool-producing sheep.

Alvin K. Benson

FURTHER READING

Avise, John C. *The Hope, Hype, and Reality of Genetic Engineering: Remarkable Stories from Agriculture, Industry, Medicine, and the Environment.* New York: Oxford University Press, 2004.

Wilmut, Ian, Keith Campbell, and Colin Tudge. *The Second Creation: Dolly and the Age of Biological Control.* New York: Farrar, Straus and Giroux, 2000.

SEE ALSO: Biotechnology and genetic engineering; Cloning; Dolly the sheep; Genetically modified organisms.

Wilson, Edward O.

CATEGORIES: Activism and advocacy; ecology and ecosystems

IDENTIFICATION: American evolutionary biologist and author

BORN: June 10, 1929; Birmingham, Alabama

SIGNIFICANCE: An evolutionary biologist with extensive field experience, especially in studying ants, Wilson became a political target during the 1970's because of his application of sociobiology to humans. More recently, he has championed biological diversity and has worked to save species from extinction.

Born in Birmingham, Alabama, to a couple who divorced within a few years, Edward O. Wilson became a dedicated naturalist who moved often and attended numerous schools during his youth, before graduating from high school in 1946 in Decatur, Alabama. From there he went to the University of Alabama, where he earned his bachelor's and master's degrees in biology; then, after a year at the University of Tennessee, he entered Harvard University, where he received his doctorate in 1955 and joined the faculty in 1956.

A childhood injury to his right eye, combined with an adolescent loss of hearing for high-pitched sounds, had pointed Wilson toward entomology because he could pick up insects and hold them close to his left eye, whereas he could not, for example, easily see or hear birds in the forest. Having been, as a thirteen-year-old, the first person to identify imported fire ants in the United States, Wilson chose myrmecology, the study of ants, as a subspecialty. Among his books on the subject are *The Ants* (1990), which won for Wilson and his collaborator Bert Hölldobler a Pulitzer Prize, and *Pheidole in the New World: A Dominant, Hyperdiverse Ant Genus* (2003).

Highly successful in his Harvard career until his re-

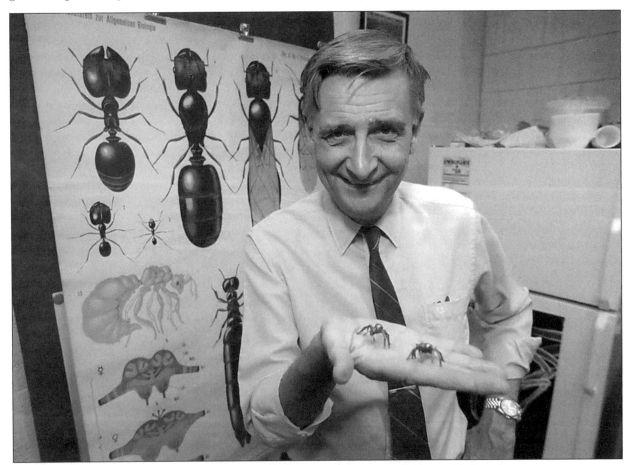

Evolutionary biologist and author Edward O. Wilson. (AP/Wide World Photos)

tirement in 2002, Wilson nevertheless faced two challenges from his colleagues there in biology. The first arose during the late 1950's from molecular biologists such as James D. Watson, who disdained evolutionary biologists such as Wilson as old-fashioned collectors instead of modern scientists. As years went by, however, the two sides began to cooperate in research. The second challenge, which drew more public attention, came from Wilson's effort in one chapter of his book *Sociobiology: The New Synthesis* (1975) to show that the evolution of *Homo sapiens* has influenced human behavior. Although Wilson has stated that he believes his ideas have triumphed, the verbal attacks on him by the Marxists Richard Lewontin and Stephen Jay Gould, both of whom had offices near Wilson's own at Harvard, proved troubling and led indirectly to a physical attack on Wilson by the International Committee Against Racism at a scientific meeting in 1978.

After his book *On Human Nature* (1978), which also won for him a Pulitzer Prize, Wilson began to focus increasing attention on environmentalism, starting with a 1980 article published in *Harvard Magazine* in which he stated that the gravest problem facing humanity is the loss of biodiversity: Species are becoming extinct at an alarming rate, and the descendants of humans currently alive will find it hard to pardon their ancestors' impoverishment of the earth. In later work, Wilson has agreed with other environmentalists in seeing human involvement in all of what he considers to be, in ascending destructive order, the main causes of decline in biodiversity: excessive gathering, fishing, and hunting; too many people; polluted air, water, and soil; biological invaders, microscopic and macroscopic; and the loss of natural habitat.

A secularist who once was a Baptist, Wilson has appealed to Christians to recognize their biophilia, their bond with nature, and to join in an effort to save the living creation, which even technologically primitive humans ravaged. He has proposed measures to save earth's biosphere, with his basic idea being that humans should cease spending finite natural capital and start acting as stewards of the earth. Noting the record of socialism in harming nature, Wilson advocates environmentally farsighted capitalism, nongovernmental organizations such as the Nature Conservancy and the World Wide Fund for Nature, governments that consider a healthy natural environment essential, and the wise application of science.

Victor Lindsey

FURTHER READING

Wilson, Edward O. *The Creation: An Appeal to Save Life on Earth.* New York: W. W. Norton, 2006.

_____. *The Future of Life.* New York: Alfred A. Knopf, 2002.

_____. *Naturalist.* 1994. Reprint. Washington, D.C.: Island Press, 2006.

SEE ALSO: Accounting for nature; Animal rights; Biodiversity; Biophilia; Biosphere; Conservation biology; Darwin, Charles; Environmentalism; Extinctions and species loss; Habitat destruction; Introduced and exotic species; Nature writing.

Wind energy

CATEGORY: Energy and energy use

DEFINITION: Energy harnessed from moving air to produce mechanical or electrical power

SIGNIFICANCE: Wind energy is one of several nonpolluting, renewable types of energy that are considered potential candidates to replace fossil fuels, which are finite in supply and produce by-products that are harmful to the environment. It has been predicted that wind and solar industries will be significant sources of new manufacturing jobs in the twenty-first century.

Human harnessing of wind energy goes back thousands of years. Historically, people used sails to harness wind energy to propel ships long before the invention of the steam engine. Wind energy has also long been used to drive windmills to grind grain, pump water for irrigation, and keep lands from being flooded with seawater. At the dawn of the twentieth century, however, as fossil fuels became cheap and widely accessible and as the usage and applications of electricity became widespread, windmills began to be neglected except by a few interested researchers and users.

Rapid increases in the prices of fossil fuels during the 1970's brought a resurgence of interest in wind energy as an alternative source of power. This led to the progressive evolution of windmills into wind turbines—wind-driven machines connected to electrical generators to produce electricity. By the beginning of the twenty-first century, the combination of practical experience, advances in technology, and scientific research had led to the sophisticated wind turbines that dot the landscapes of many countries, including Germany, Denmark, the Netherlands, South Africa, and the United States.

How Wind Turbines Work

A turbine is a machine that converts the energy that is stored in a moving fluid (such as air, water, or steam) into another form of energy (such as electricity or mechanical work). Wind turbines catch energy from the wind by using blades that are shaped like propellers. The blades are attached to a shaft and are tilted in such a way that the force of the wind on them attempts to lift each blade. The lift is only partially complete, because the shaft begins to turn before the blade rises very high above its original station. This lift effect holds true for each blade, and it is repeated over and over again. The net result is that the shaft rotates continuously as long as the speed of the wind remains above a certain threshold. The assembly con-

A wind farm in Helix, Oregon. (AP/Wide World Photos)

sisting of the blades and the shaft to which they are attached is part of what is called the rotor.

During the early days, windmills had six or more blades. It is now known that, by carefully shaping the blades, one can use fewer of them and capture much more energy than windmills did. Thus, in modern times, turbines are equipped with only two or three blades. The wind turbine assembly is mounted onto a tall tower. As a general rule, the taller the tower, the better. This is because the higher above the ground the turbine is located, the less the wind that reaches it is disturbed or reduced by what is on the ground and by surrounding objects such as trees and buildings.

Applications and Systems

Wind energy applications can be divided into three types: stand-alone wind turbines, distributed energy systems, and turbines that are connected to utility power grids. Stand-alone systems are generally used by home owners and by small business owners—such as ranchers, farmers, and owners of small retail stores—seeking to reduce the size of their electric bills. Others use them for communications and for pumping water. Distributed energy systems are various small power generation technologies that can be grouped and combined for the purpose of improving or expanding the operation and delivery of electrical energy.

For wind turbines to be connected to a local power grid, large numbers of them, generating many megawatts of power, are needed to make the required costs of construction, operation, and maintenance worthwhile. Such arrangements are called wind farms or wind plants. Several providers of electrical power, in the United States and other nations, use wind farms to supply power to their customers.

Benefits

Research by the American Wind Energy Association indicates that two main categories of benefits are associated with the adoption of wind energy: First, the production of electricity using wind energy reduces environmental risks while enhancing health benefits; and second, the installation of wind farms spurs economic development in the areas, usually rural, where they are located.

The generation of electricity using wind energy produces little air pollution. It is estimated that extensive use of this technology could reduce total U.S. emissions of carbon dioxide, a greenhouse gas associated with global warming, by one-third. This corresponds to a reduction of 4 percent at the world level. Survey data show that forty-six of fifty states in the United States have wind resources that could be developed; thus the potential for growth in this area is very great. For example, if ten of the windiest states in the United States were to develop 10 percent of their wind energy potential, the result would offset the carbon dioxide emissions from all U.S. power plants that burn coal.

The American Wind Energy Association estimated in 2010 that every megawatt of electrical power produced from wind energy generated $1 million in economic development. The association found that when wind energy is adopted by rural communities, local farmers and other landowners receive steady income through the lease of their land and the payment of royalties. Furthermore, the advent of wind energy operations brings new jobs to the communities where they are located. Some of these jobs are directly related to the installation and maintenance of wind turbines, whereas others come from such activities as road and building construction and from the transportation, hospitality, and food services needed as the local economy changes.

Josué Njock Libii

FURTHER READING

Burton, Tony, et al. *Wind Energy Handbook.* New York: John Wiley & Sons, 2001.

Eggleston, David M., and Forrest S. Stoddard. *Wind Turbine Engineering Design.* New York: Van Nostrand Reinhold, 1987.

Gipe, Paul. *Wind Power: Renewable Energy for Home, Farm, and Business.* 2d ed. White River Junction, Vt.: Charles Green, 2004.

Hansen, Martin O. L. *Aerodynamics of Wind Turbines.* 2d ed. Sterling, Va.: Earthscan, 2008.

McKinney, Michael L., Robert M. Schoch, and Logan Yonavjak. "Renewable and Alternative Energy Sources." In *Environmental Science: Systems and Solutions.* 4th ed. Sudbury, Mass.: Jones and Bartlett, 2007.

SEE ALSO: Air pollution; Alternative energy sources; Carbon dioxide; Energy conservation; Fossil fuels; Renewable energy; Smart grids; Smog; Solar energy.

Windscale radiation release

CATEGORIES: Disasters; nuclear power and radiation

THE EVENT: Release of radioactive material into the atmosphere as the result of a fire in the reactor core of the Windscale nuclear reactor on the west coast of England

DATE: October 10, 1957

SIGNIFICANCE: The fire in the reactor core at the Windscale plant gave rise to one of the world's first serious nuclear accidents. The release of significant amounts of radioactive material into the atmosphere caused short-term contamination of several hundred square miles of the surrounding countryside.

The Windscale nuclear reactor overheated in October, 1957, and a fire resulted in the reactor core. The first indication that radioactive material was escaping into the atmosphere from the reactor came on the evening of October 10, when a nearby weather station detected an increase in background radiation.

Health physicists considered the release of iodine 131 to be the most serious hazard. Iodine 131 falls to the ground, where it may be consumed by cows eating grass and concentrated in their milk. If humans drink this contaminated milk, the iodine 131 concentrates in the human thyroid gland, where its radioactive decay can cause cancer. The British government monitored milk from the region for evidence of iodine-131 contamination, and two days after the first release of radioactive material, milk samples from farms near the Windscale plant showed evidence of contamination. Initially, the government impounded milk supplies from within a 3-kilometer (2-mile) radius around the plant. However, as iodine-131 contamination was detected over a wider region, milk produced over an area of about 500 square kilometers (200 square miles) was impounded. The contaminated milk was dumped into the sea, and milk for the people living near the Windscale reactor was trucked in from outside the contaminated region. Since iodine 131 decays rapidly, the ban on consumption of milk from the affected area lasted only a few weeks.

The fire in the reactor core also released a significant amount of polonium 210 into the atmosphere. This raised concerns because polonium 210 decays by emitting alpha particles, which are dangerous to the lungs. However, in the areas where the concentration of polonium 210 was highest, the additional exposure to radioactive decay was found to be approximately equivalent to the average annual background rate of radioactive decay in the British Isles.

The design of the Windscale nuclear plant minimized the public health hazard posed by the reactor fire, as filters in the stacks of the plant trapped a large fraction of the radioactive material released from the reactor. The quick action of the British government in collecting and destroying contaminated milk from the affected region also reduced the health effects of the release.

The Windscale event released about 0.001 times the amount of iodine 131 into the atmosphere that was released by the fire at the Chernobyl nuclear reactor in 1986. A study conducted in 1997, forty years after the Windscale release, concluded that individuals who received the most serious exposure to the Windscale radiation experienced a slight increase in their likelihood of developing fatal cancers compared to the normal fatal cancer risk. The conclusion was that the long-term health effects of the Windscale release were minimal. In 2007, however, British scientists announced the results of a new study that used computer modeling and environmental monitoring to examine how the radioactive materials released would have spread. They concluded that the environmental contamination caused by the Windscale release was probably greater than originally thought. Whereas previous estimates had put the number of cases of cancer eventually caused by the radiation at 200, the results of the study suggested that a more accurate estimate would be 240 cases.

George J. Flynn

FURTHER READING

Bodansky, David. "Nuclear Reactor Accidents." In *Nuclear Energy: Principles, Practices, and Prospects*. 2d ed. New York: Springer, 2004.

Cooper, John R., Keith Randle, and Ranjeet S. Sokhi. "Nuclear Power." In *Radioactive Releases in the Environment: Impact and Assessment*. Hoboken, N.J.: John Wiley & Sons, 2003.

SEE ALSO: Chelyabinsk nuclear waste explosion; Chernobyl nuclear accident; Nuclear accidents; Nuclear power; Radioactive pollution and fallout.

Wise-use movement

CATEGORIES: Philosophy and ethics; resources and resource management

IDENTIFICATION: Antienvironmentalist movement that promotes the economic exploitation of natural resources, particularly in the American West

SIGNIFICANCE: Wise-use advocates, who typically criticize environmental policies as excessive and elitist, have succeeded somewhat in damaging the image of environmentalists and in hampering environmental policy making, but without gaining the support of the American public for their own view of the appropriate balance between human and natural values.

Wise-use activists maintain that true conservationists should seek a reasonable balance between human and natural values and that human values should come first. Some wise-use advocates challenge environmental "extremists" and restrictive resource policies through obstruction, lobbying, advocacy in the media, litigation, and (rarely) violence. Although the wise-use movement began in the American West, where federal lands are extensive, wise-use organizations, and the concepts associated with the movement, have grown to national scope.

ROOTS OF THE MOVEMENT

The antienvironmentalist movement developed as environmental policies initiated in the 1960's and 1970's threatened historical values and patterns of natural resource development. Until the early 1970's, policies of public land management in the United States promoted local resource development industries, particularly ranching, logging, and mining. Laws such as the Endangered Species Act of 1973 and various wilderness bills restricted access to federal land resources. Noneconomic and nonlocal values, such as wilderness preservation, advanced at the expense of traditional economic resource use. Environmental policies also affected private lands. Some of these new policies hurt the livelihoods of workers and communities throughout the West, as well as the profits of large and small businesses.

The wise-use movement has strong roots in the Sagebrush Rebellion of the late 1970's, which occurred when western rural interests sought to counter national environmental politics and policies by shifting federal resources to state and local governments. However, Ronald Reagan's election as president of the United States in 1980 and the proexploitation tilt of appointees such as Secretary of the Interior James Watt undercut pressures for such change. The contemporary wise-use movement emerged in the late 1980's when conservative westerners felt that their rural interests had again been forgotten. The antienvironmental movement started as a creature of trade groups and large economic interests. Groups such as People for the West, for example, existed solely through the support of mining interests. Still, the movement attracted publicity, developed strategies, and cultivated ideological links with conservative think tanks and politicians.

WISE-USE VICTORIES

Environmentalists initially underestimated the strength of the wise-use movement, perhaps because of its bombastic antienvironmentalist rhetoric and weak popular base. In the early 1990's, however, wise-use groups scored significant legal and political victories. Their powerful symbolic actions and organizational abilities attracted more popular support in the West than in other regions of the United States. Broad public campaigns, such as that of the Yellow Ribbon Coalition to promote old-growth logging in the Pacific Northwest, demonstrated the new force and appeal of wise-use activism. Coalition activists attacked preservationist ethics by labeling the endangered species conflict as a matter of "people versus owls." Strong traditional interest groups such as the American Farm Bureau Federation came to favor wise-use ideology and strategies. By 2010 hundreds of groups were associated with the wise-use label, with some estimates placing the number at one thousand or more. The movement also enhanced its influence in Congress, especially after Republicans took control in 1994. However, continuing public concern over the environment restrained legislative action on wise-use proposals. Although willing to interfere in the implementation of laws such as the Endangered Species Act, Congress did not overturn any major environmental legislation.

In the early 1990's wise-use advocates broadened their national appeal by moving beyond public lands and traditional western issues to ally themselves with advocates in the property rights movement. In simple terms, the property rights movement argues that the Fifth Amendment to the U.S. Constitution requires that the government compensate citizens for any loss of income arising from the federal regulation of property use. Such an interpretation could effectively preclude most significant environmental regulations, since the amount of compensation could be overwhelming. The U.S. Supreme Court has occasionally favored this view, such as in its decision in the case *Lucas v. South Carolina Coastal Council* (1992), but the Court has not recognized all environmental regulatory costs as "takings" requiring compensation.

FUTURE OF THE MOVEMENT

The wise-use movement appears to be a natural countermobilization against the environmentalist successes of the 1970's. The key question is whether the movement can sustain its momentum. Continued public support for environmental regulations has thus far prevented any significant reform of the Endangered Species Act and keeps the most radical wise-use concepts on the periphery of the debate. The American public in general may never accept the wise-use perspective on the appropriate use of natural resources. Furthermore, wise-use proposals appeal primarily to rural interests and extractive industries, while the country continues to urbanize and shift toward a service economy.

On the other hand, the movement has had powerful allies in Congress, such as Congressman Don Young, a Republican from Alaska, and the property rights issue continues to generate significant public sympathy and business interest. The movement may benefit from the Supreme Court's changing view of property rights and its increased deference to government decisions. The fear of violating property rights may cripple environmental regulators, while groups seeking to stop the sale of federal natural resources may be less welcome in court.

Wise-use advocates have done some damage to the image of environmentalists without successfully selling their own view of the appropriate balance between human and natural values. The movement continues to complicate environmental policy making without redefining the public purpose. The wide-ranging diversity among wise-use activists ensures the movement's survival. Even the radical wing of the movement includes people from many different backgrounds and with widely varying priorities. For example, some Nevada wise-use advocates, arguing that the federal government does not have legal jurisdiction over counties, build illegal roads on federal lands. Others in the movement are like Ron Arnold of the Center for the Defense of Free Enterprise, who has stated that the movement intends to "destroy the environmental movement once and for all." Such extremists draw media attention and public fire. They also clear the way for moderate groups to make seemingly reasonable inroads on environmental regulations. The movement may eventually shed the "wise-use" label if it becomes identified with extreme acts, such as the bombings of U.S. Bureau of Land Management offices in Nevada in the 1990's. Whatever the names and the strategies used, organized and sophisticated resistance to environmentalism has become an important part of the political landscape.

Mark Henkels

FURTHER READING

Brick, Philip D., and R. McGreggor Cawley, eds. *A Wolf in the Garden: The Land Rights Movement and the New Environmental Debate.* Lanham, Md.: Rowman & Littlefield, 1996.

Davis, Charles, ed. *Western Public Lands and Environmental Politics.* 2d ed. Boulder, Colo.: Westview Press, 2001.

Gottlieb, Alan M., ed. *The Wise Use Agenda: The Citizen's Policy Guide to Environmental Resource Issues.* Bellevue, Wash.: Free Enterprise Press, 1989.

Helvarg, David. *The War Against the Greens: The "Wise-Use" Movement, the New Right, and the Browning of America.* Rev. ed. Boulder, Colo.: Johnson Books, 2004.

Jacques, Peter J. *Environmental Skepticism: Ecology, Power, and Public Life.* Burlington, Vt.: Ashgate, 2009.

Switzer, Jacqueline Vaughn. *Green Backlash: The History and Politics of Environmental Opposition in the U.S.* Boulder, Colo.: Lynne Rienner, 1997.

SEE ALSO: Antienvironmentalism; Grazing and grasslands; Land-use policy; Privatization movements; Range management; Sagebrush Rebellion; Watt, James.

Wolman, Abel

CATEGORIES: Activism and advocacy; water and water pollution; urban environments
IDENTIFICATION: American sanitary engineer
BORN: June 10, 1892; Baltimore, Maryland
DIED: February 22, 1989; Baltimore, Maryland
SIGNIFICANCE: Wolman was a pioneer in the field of sanitary engineering. His innovations and advocacy influenced the establishment of sound water-resource management strategies by American cities during the twentieth century.

Abel Wolman graduated from Baltimore's well-known high school Baltimore City College in 1909. He earned a bachelor of arts degree from The Johns Hopkins University in 1913 and then completed a bachelor of science degree in engineering at Johns Hopkins two years later. While he was working toward his second degree in 1914, he took a job as an engineer with the Maryland State Department of Health.

While working in the health department, Wolman began to pioneer the use of engineering methods to improve public health. He was specifically concerned with water supplies, wastewater treatment, and sewage disposal, addressing the dangers of such waterborne diseases as typhoid. In 1919 Wolman worked with chemist Linn Enslow to standardize techniques to chlorinate municipal drinking-water supplies. Many people believed that chlorine in any amount was poisonous, but Wolman convinced officials to adopt the procedure by explaining the disinfectant benefits of chlorination. His efforts in this area are considered to be among the most influential improvements in water management for public health, resulting in decreased death rates from waterborne diseases.

Promoting a regional approach to water supply and sewage disposal, Wolman helped consolidate the Baltimore area into one water-supply region. Named chief engineer for the Maryland State Department of Health in 1922, he analyzed municipal water-supply needs and evaluated how best to recycle wastewater. In 1935 U.S. president Franklin D. Roosevelt appointed Wolman as chairman of the Water Resources Committee of the Natural Resources Planning Board, which was in charge of managing the federal government's water resources projects.

During the 1950's Wolman predicted possible environmental problems from unsafe disposal of nuclear wastes when private companies were granted access to nuclear energy through the Atoms for Peace program. During his service on the Reactor Licensing Board of the Atomic Energy Commission, Wolman insisted that concrete containment structures be built for the first commercial nuclear power plants in the United States. During the 1960's Wolman stressed the dangers of nonorganic environmental hazards to public health, noting that new technologically produced chemicals and contaminants had been introduced to water sources. He warned that humans must be held accountable for how they alter the environment and must envision how to protect resources.

Wolman served as editor in chief of the *Journal of the American Water Works Association* (1921-1937) and the journal *Municipal Sanitation* (1929-1935) and as associate editor of the *American Journal of Public Health* (1923-1927). The Johns Hopkins University presented Wolman with an honorary doctorate in 1937 when he established a Department of Sanitary Engineering at the university; he held the position of department chairman until he retired in 1962. Wolman received numerous awards, including the Tyler Prize for Environmental Achievement. He was a prolific author, and his major articles were collected in *Water, Health, and Society: Selected Papers*, edited by Gilbert F. White, in 1969.

Wolman offered his environmental engineering expertise worldwide as a consultant, and he emphasized the responsibility of engineers to protect environmental quality. An advocate for the poor, Wolman testified against landlords who did not provide clean water for tenants. The *Baltimore Evening Sun* promoted environmental engineering during the International Drinking Water Supply and Sanitation Decade in the 1980's by eulogizing Wolman: "[He] envisioned a world in which the most basic of necessities, water to drink, would be safe and plentiful to all peoples of the world."

Elizabeth D. Schafer

FURTHER READING

Melosi, Martin V. *Effluent America: Cities, Industry, Energy, and the Environment.* Pittsburgh: University of Pittsburgh Press, 2001.

Rogers, Jerry R., ed. *Environmental and Water Resources: Milestones in Engineering History.* Reston, Va.: American Society of Civil Engineers, 2007.

SEE ALSO: Chlorination; Drinking water; Environmental engineering; Sewage treatment and disposal; Water conservation; Water pollution; Water quality; Water treatment.

Wolves

CATEGORY: Animals and endangered species

DEFINITION: Predatory, carnivorous mammals of the family Canidae

SIGNIFICANCE: The gray wolf, once the most widely distributed mammal in North America, inhabits a fraction of its former range, and the red wolf is the world's most endangered wild canid. Widespread habitat destruction, human infiltration, and predator-control programs resulted in considerable wolf population decline in the early twentieth century. Since the 1970's, however, recovery programs, combined with legislative changes, have allowed some wolf populations to increase.

Wolves, the largest members of the dog family (Canidae), were once abundant throughout North America, Europe, and Asia. Gray wolves are the most abundant wolves, with two species in North America: the gray wolf (*Canis lupus*) and the subspecies red wolf (*Canis rufus*). Both vary in color, ranging from gray to brown and black, with some gray wolves even being white in color. Although red wolves generally have a characteristic reddish tint to their fur, they are mostly brown in color. As the most dominant large carnivores in North America, wolves play an important role as predators in the ecosystems of which they are a part; they are considered a keystone species. Wolves generally hunt in packs and feed on large animals such as deer and elk; one of their primary functions within ecosystems is to remove sick or injured animals from these prey populations.

Although wolves rarely attack humans, they are widely regarded as being among the world's fiercest animals. A decline in native prey species for wolves occurred concurrently with an increase in domestic livestock animals during the settlement of North America, and over time, because of the tendency of wolves to prey on livestock, government agencies began to initiate intensive predator-control programs aimed at reducing wolf populations. In the lower forty-eight U.S. states, except Minnesota, gray wolves were hunted nearly to extinction and ultimately were listed as endangered. Only several hundred gray wolves in Minnesota, an isolated wolf population on Michigan's Isle Royale, and a few red wolves remained. In 1926 the last gray wolf in Yellowstone National Park was killed during a time when Congress sanctioned predator control. Red wolf populations were decimated through habitat loss and predator control; the U.S. Fish and Wildlife Service declared the subspecies extinct in the wild in 1980.

Shifts in land use and legislative changes eventually slowed declines in wolf populations in the United States. Recovery programs, natural migration of wolves from Canada, protection from unregulated hunting, and improved prey habitats resulted in growth in the gray wolf population in the lower forty-eight states to numbers exceeding five thousand by the early years of the twenty-first century. The National Park Service ended systematic killing of predators in 1933. The Endangered Species Act was passed in 1973, and in 1980 the U.S. Fish and Wildlife Service established the Northern Rocky Mountain Wolf Recovery Plan. In 1995, after an absence of nearly seventy years, wolves were reintroduced into Yellowstone National Park. Because of the success of recovery programs, in 1996 the International Union for Conservation of Nature reduced the status of the gray wolf on the organization's Red List to that of "least concern." In 1970 a captive-breeding program was established to increase the red wolf population and reintroduce the subspecies to the wild, and by 2010 the red wolf population in the wild was estimated to be between one hundred and two hundred. Restoration efforts for red wolves have included programs at the Alligator River National Wildlife Refuge in northeastern North Carolina and at other national wildlife refuges.

A gray wolf near Yellowstone National Park. (©Jim Kruger/iStockphoto.com)

C. J. Walsh

FURTHER READING

Busch, Robert H. *The Wolf Almanac: A Celebration of Wolves and Their World.* New and rev. ed. Guilford, Conn.: Lyons Press, 2007.

Mech, L. David, and Luigi Boitani, eds. *Wolves: Behavior, Ecology, and Conservation.* Chicago: University of Chicago Press, 2007.

Peterson, Rolf. *The Wolves of Isle Royale: A Broken Balance.* 1995. Reprint. Ann Arbor: University of Michigan Press, 2007.

SEE ALSO: Endangered Species Act; Endangered species and species protection policy; Extinctions and species loss; Predator management; Wildlife management; Yellowstone National Park.

World Commission on Environment and Development. *See* Brundtland Commission

World Fertility Survey

CATEGORY: Population issues

IDENTIFICATION: Study that documented population growth and family-planning measures in various regions throughout the world

DATES: 1973-1984

SIGNIFICANCE: The World Fertility Survey represented the first time information on population growth and family-planning practices had been gathered from several of the world's developing regions. Most of the nations that took part in the survey used the information they gained to cope with and plan for population changes.

During the 1940's demographer Frank Notestein and others developed what became known as transition theory to describe how population change occurred following modernization and industrialization. Notestein observed that populations were dramatically increasing around the world, especially in the poorer regions of the world, as mortality was declining in the context of high fertility. Concerns about overpopulation were voiced, but the nature of the changes taking place was poorly understood. In the late 1960's a series of surveys (known as knowledge, attitudes, and practices, or KAP, surveys) were undertaken to measure people's desires to have children, but this approach was soon criticized as inadequate in terms of methodology and regional coverage.

The widely respected British statistician Maurice Kendall, partly in an effort to revitalize the International Statistical Institute (ISI), proposed the World Fertility Survey (WFS) in 1971 and led the planning in 1972. The first wave of the survey followed in 1973. The WFS entailed the collection of data through comparable, high-quality interviews of 341,300 women around the world. The questionnaire was designed to gather information on maternity and marital history, contraceptive knowledge, work history, and husband's background in order to document population growth and family-planning measures. Households rather than families were surveyed, and only women under the age of fifty who had been married at some point were eligible as respondents. The standard questions could be supplemented by "modules" on abortion and on economic or community factors. Hundreds of technical reports and research articles using information gained during the survey were published, particularly in the 1980's. Although funding for the WFS stopped in 1984, similar data-gathering efforts continued through the Demographic and Health Surveys (DHS), a program based in the United States.

The WFS revealed, among other findings, that many women would use family-planning services if such services were available to them, that infant mortality rises when intervals between births are shorter, and that breast-feeding can increase the period of time between pregnancies. The WFS also showed that although fertility had declined significantly in regions of Asia, Latin America, and the Middle East, it remained relatively high in these regions when they were contrasted to the United States and Europe. No decline in fertility was evident for sub-Saharan Africa, and transition theory did not account for these regional differences.

Seventeen of the forty-two countries that participated in the WFS included a "community module" in the survey instrument with the goal of determining whether reproductive behavior was related to increased access to family-planning services or was a function of declining mortality thanks to better primary health care services. Although community effects were clearly reflected in infant and child mortality patterns, they were not strong for reproductive behaviors. In particular, it was not clear that access to family-planning services alone was sufficient to change reproductive behaviors. Part of the problem was that certain concepts, such as "household," had very different meanings in different areas (for example, in some settings husbands and wives had separate residences). Furthermore, the details that respondents provided concerning the health services available to them were often inaccurate. The earliest versions of the less well-funded DHS had fewer questions on these issues, a deficiency that made it difficult for researchers to identify the motivations and mechanisms driving fertility change.

Joan C. Stevenson

FURTHER READING

Halfon, Saul. "Contesting Surveys: Co-producing Demography and Population Policy." In *The Cairo Consensus: Demographic Surveys, Women's Empowerment, and Regime Change in Population Policy.* Lanham, Md.: Lexington Books, 2007.

Livi-Bacci, Massimo. *A Concise History of World Population.* 3d ed. Malden, Mass.: Blackwell, 2001.

Mazur, Laurie, ed. *A Pivotal Moment: Population, Justice, and the Environmental Challenge.* Washington, D.C.: Island Press, 2010.

SEE ALSO: Ehrlich, Paul R.; Population-control and one-child policies; Population-control movement; Population growth; United Nations population conferences.

World Health Organization

CATEGORIES: Organizations and agencies; human health and the environment
IDENTIFICATION: Agency of the United Nations providing leadership in international matters of public health
DATE: Founded on April 7, 1948
SIGNIFICANCE: People, pathogens, and the vectors that connect them are all elements of the environment. The World Health Organization works to control or eradicate environmental pathogens and at least control disease vectors around the world.

During the later nineteenth century a series of International Sanitary Conferences were held in Paris (1851, 1859), Istanbul (1869), Vienna (1874), Washington, D.C. (1881), Rome (1885, 1892), and other centers in the face of epidemics of cholera, yellow fever, influenza, and plague. Little of a cooperative nature emerged until the last of these meetings, and at subsequent conferences nations agreed to work together to quarantine suspect people and vessels, to take measures against cholera, and to control plague-carrying rat populations. As a result of such efforts, the largely European International Office of Public Hygiene (OHIP) was established in Paris in 1908; it was dedicated to the monitoring of leprosy, typhoid, sexually transmitted diseases, tuberculosis, and cholera-tainted water supplies and to undertaking efforts against these problems. In 1902 nations in the Americas formed the International Sanitary Bureau, later the Pan-American Health Organization (PAHO), centered in Washington, D.C.

Epidemic diseases plagued Europe following World War I, and the League of Nations, formed in 1919, created the League of Nations Health Organization (LNHO). This body was charged with battling disease and the environmental factors and conditions that caused and spread it. The fact that the United States and other important nations were not members of the League of Nations limited LNHO's effectiveness, although it did cooperate with OHIP and PAHO. As World War II reached its climax and the United Nations was being designed to replace the League, the World Health Organization (WHO) was conceived. Allied officials feared that the social and environmental devastation in Europe and parts of Asia would again result in epidemics, and WHO was tasked with reducing these risks and handling problems that did occur. WHO was officially established on April 7, 1948; it convened in Geneva for the first time in June, 1948.

From its inception WHO has focused on the maintenance of good health in general populations, the study and removal of threats to that good health, and the provision of aid to those suffering from disease and other impediments to healthy living. One of its great successes was the eradication of smallpox in 1977 through a decades-long campaign of vaccination. Since that time WHO has made unsuccessful but encouraging efforts to wipe out measles, tuberculosis, malaria, and polio.

The United Nations member states appoint delegates to the World Health Assembly, the body that oversees WHO's efforts; the assembly convenes each May. The assembly elects the members of the Executive Board, a group of thirty-four health experts who serve three-year terms. The board is tasked with advising the assembly as well as with translating its concerns and directives into action. WHO receives member-state funding as well as donations, and has increasingly participated with global and regional organizations and initiatives, such as the Bill and Melinda Gates Foundation and the International AIDS Vaccine Initiative.

Environmental health is a specific subset of WHO concern and effort that includes "physical, chemical, and biological factors external to a person, and all the related factors impacting behaviors." WHO emphasizes that health is not just about germs; it is also about environmental factors such as disease-carrying insects, water contamination, air quality, exposure to radiation, occupational hazards, changes in climate and ground cover, nutrition, noise, waste disposal and sanitation, and personal hygiene. WHO operates from six regional offices around the world, which enables this global body to allocate scarce resources for research, monitoring, education, disease prevention, and treatment as needed locally and in ways that are attuned to local social and cultural norms.

Joseph P. Byrne

FURTHER READING

Corvalán, C., D. Briggs, and G. Zielhuis, eds. *Decision-Making in Environmental Health: From Evidence to Action*. New York: World Health Organization/E & FN Spon, 2000.

Lee, Kelley. *Health Policy in a Globalizing World*. New York: Cambridge University Press, 2002.

_____. *The World Health Organization (WHO)*. New York: Routledge, 2009.

SEE ALSO: Environmental health; Environmental illnesses; Food and Agriculture Organization; Globalization; Pandemics; Smallpox eradication; United Nations Commission on Sustainable Development; United Nations Conference on the Human Environment; United Nations Convention to Combat Desertification; United Nations Educational, Scientific, and Cultural Organization; United Nations Environment Programme.

World Heritage Convention

CATEGORIES: Treaties, laws, and court cases; preservation and wilderness issues

THE CONVENTION: International agreement to protect designated sites of great cultural, historic, or natural value

DATE: Opened for signature on November 16, 1972

SIGNIFICANCE: The World Heritage Convention promotes protection of the environment by obligating signatory nations to identify, maintain, and preserve important natural and cultural sites within their territories as part of the universal heritage of humanity.

The United States proposed the World Heritage Convention in 1972 to commemorate the one hundredth anniversary of the establishment of Yellowstone National Park and was the first nation to sign the convention when it was adopted by the United Nations Educational, Scientific, and Cultural Organization (UNESCO). The convention—formally titled the Convention Concerning the Protection of the World Cultural and National Heritage—essentially promotes the U.S. national park concept worldwide. By the time of the convention's twenty-fifth anniversary in 1997, nearly 150 nations had ratified the agreement and had placed more than five hundred sites on the World Heritage List. By the end of the first decade of the twenty-first century, 187 nations had ratified the convention, and more than nine hundred cultural and natural sites had been named.

Nations that are signatories to the World Heritage Convention nominate sites within their own borders for inclusion on the World Heritage List. The nominations are reviewed by the World Heritage Committee, a body consisting of the representatives of twenty-one signatory nations; the representatives are elected by the General Assembly of the signatory nations. The committee also places sites threatened by natural disaster or civil strife on the List of World Heritage in Danger.

In signing the World Heritage Convention, nations pledge to identify, maintain, and preserve important natural and cultural sites within their territories as part of the universal heritage of humanity; they also pledge to promote and publicize these sites for worldwide public enlightenment. In addition, member nations assist one another with studies, advice, training, and equipment necessary to resolve problems, restore damaged areas, and establish programs to protect, preserve, and publicize the sites. The World Heritage Committee offers technical advice and monetary assistance through its World Heritage Fund. Individual nations also offer direct nation-to-nation assistance.

To be included on the World Heritage List, sites must possess outstanding, universally recognized cultural or natural features. Sites include both human-made constructions and natural areas. Selected natural sites include areas that represent major stages in the earth's evolutionary history, ongoing geological processes or biological evolution, or human interaction with the environment; that contain unique, rare, or superlative national phenomena; or that provide habitats for rare or endangered plants and animals.

Each signatory nation maintains sovereignty over its sites and is responsible for site maintenance and protection. Listed sites in a given country include those owned by the national government (such as national parks and national historic landmarks) as well as those owned by state or tribal governments, local

World Heritage Site Distribution, 2010

REGION	NATIONS WITH SITES	TYPES OF SITES				SHARE OF WORLD SITES (%)
		CULTURAL	NATURAL	MIXED	TOTAL	
Africa	30	42	32	4	78	9
Arab States	15	61	4	1	66	7
Asia and Pacific	31	138	51	9	198	22
Europe and North America	50	377	58	10	445	49
Latin America and Caribbean	25	86	35	3	124	14
Total	151	704	180	27	911	100

Source: United Nations Educational, Scientific, and Cultural Organization.
Note: Uvs Nuur Basin, which straddles Europe and Asia in Mongolia and Russia, is counted here as part of Asia.

governments, and private groups or individuals, with the owners pledging to protect their properties in perpetuity.

Among the sites included on the World Heritage List are the Great Wall of China, the Taj Mahal in India, Ecuador's Galápagos Islands, the Tower of London, and the massive Spanish fortifications at San Juan, Puerto Rico. North American sites include twenty in the United States; among these are Grand Canyon National Park in Arizona, Everglades National Park in Florida, Independence Hall in Philadelphia, Cahokia Mounds State Historic Site in Illinois, Pueblo de Taos in New Mexico, and the Statue of Liberty in New York.

Gordon Neal Diem

FURTHER READING

Di Giovine, Michael A. *The Heritage-scape: UNESCO, World Heritage, and Tourism.* Lanham, Md.: Lexington Books, 2009.

Leask, Anna, and Alan Fyall, eds. *Managing World Heritage Sites.* Burlington, Mass.: Butterworth-Heinemann, 2006.

McHugh, Lois. "World Heritage Convention and U.S. National Parks." In *American National Parks: Current Issues and Developments*, edited by Rony Mateo. Hauppauge, N.Y.: Novinka Books, 2004.

SEE ALSO: Grand Canyon; Lake Baikal; National parks; Preservation; Ramsar Convention on Wetlands of International Importance; Wilderness areas.

World Resources Institute

CATEGORIES: Organizations and agencies; activism and advocacy; resources and resource management

IDENTIFICATION: Nonprofit organization devoted to protecting the environment by supporting sustainable management of the world's resources

DATE: Established on June 3, 1982

SIGNIFICANCE: Through its many programs, the World Resources Institute works to develop international agreements to protect the environment and promotes responsible investments in environment-friendly energy and transportation technologies.

The World Resources Institute (WRI) was launched in 1982 by James Gustave Speth with a $15 million grant from the John D. and Catherine T. MacArthur Foundation and $10 million seed money from the Andrew K. Mellon Foundation and the Rockefeller Foundation. WRI was founded to be a center that would address policy research on and analysis of global resources, but it has since shifted its mission and goals. The stated mission of the World Resources Institute is to "move human society to live in ways that protect Earth's environment and its capacity to provide for the needs and aspirations of current and future generations." A good portion of WRI funding comes from U.S. government agencies, including the Environmental Protection Agency (EPA), the National Aeronautics and Space Administration (NASA), the U.S. Agency for International Development (USAID), and the Departments of Agriculture and Energy.

WRI program goals are centered on four primary areas: climate protection, governance, markets and enterprise, and people and ecosystems. The organization seeks to be a catalyst in the development of international agreements and U.S. policies to protect the environment, to promote responsible investments in energy and transportation technologies that are friendly to the environment, and to help reduce greenhouse gas emissions through clean alternatives supported by all stakeholders. More specifically, WRI works to empower people and institutions to do the following: mitigate climate change (particularly by reducing greenhouse gas emissions) and adapt to climate change as needed; make informed, socially equitable decisions to ensure environmental sustainability; expand economic opportunities while protecting the environment and harnessing markets; and stop and reverse land degradation and other environmental problems to ensure productivity for generations to come.

WRI sponsors and supports research, workshops, conferences, and related activities to address the world's environmental problems. In addition, WRI produces a number of publications and offers online resources, some of the most important of which are *World Resources* (launched in 1986), a biennial country-by-country assessment of environmental conditions and trends; Global Forest Watch (launched in 2000), an independent online network for monitoring forests; and Earth Trends (launched in 2001), a comprehensive online database of information on the world's social, economic, and environmental trends. The NextBillion.net, an online initiative launched by WRI in 2004, is the result of the first conference held for the purpose of finding ways in which the business community can help alleviate and reduce global poverty by addressing and meeting the needs of the world's poorest people at the "base of the economic pyramid."

WRI played an important role in the creation of the Global Environment Facility (GEF) and the adoption of the United Nations Convention on Biological Diversity (CBD), as well as many other international environment-related efforts. Over the years, WRI has joined with other nonprofit organizations as well as with international agencies and other institutions to address the protection of the world's natural resources. WRI seeks to ensure that decisions made about the use and management of natural resources and ecosystems reflect the needs and values of the people who are dependent on those resources and ecosystems.

WRI developed the Greenhouse Gas Protocol Initiative in collaboration with the World Business Council for Sustainable Development. This comprehensive tool provides accounting information and information on standards on nearly every greenhouse gas program in place in the world. WRI is also a founding member and supporter of the U.S. Climate Action Partnership (USCAP), an alliance of businesses and environmental groups that promotes the enactment of U.S. legislative action on the reduction of greenhouse emissions. The "blueprint for legislative action" developed by USCAP members and issued on January 15, 2009, supports the implementation of a cap-and-trade system, investment in carbon capture and storage technology, and free allowances for certain businesses.

Lakhdar Boukerrou

FURTHER READING

Irwin, Frances, et al. *Restoring Nature's Capital: An Action Agenda to Sustain Ecosystem Services.* Washington, D.C.: World Resources Institute, 2007.
Speth, James Gustave. *Red Sky at Morning: America and the Crisis of the Global Environment.* New Haven, Conn.: Yale University Press, 2004.

SEE ALSO: Renewable resources; Resource depletion; Sustainable agriculture; Sustainable development; U.S. Climate Action Partnership; Worldwatch Institute.

World Summit on Sustainable Development

CATEGORY: Resources and resource management
THE EVENT: International conference devoted to finding ways to bring together developed and developing nations in the pursuit of sustainable and equitable development
DATES: August 26-September 4, 2002
SIGNIFICANCE: The World Summit on Sustainable Development brought together thousands of participants from all parts of the world and from all levels of society, and in doing so focused worldwide attention on the needs of developing nations and the impacts of development on the environment. The summit produce the Johannesburg Declaration on Sustainable Development, a statement of principles for international action and debate.

The 2002 World Summit on Sustainable Development was held in Johannesburg, South Africa, and is thus sometimes referred to as the Johannesburg Summit; it is also known as Earth Summit 2002. The purpose of the summit was in part to follow up on and discuss the progress made since the 1992 United Nations Conference on Environment and Development, known as the Earth Summit, which was held in Rio de Janeiro, Brazil. The 2002 summit also sought to focus the world's attention on the problematic effects of current methods of development for the planet's environmental health and the well-being of its peoples.

The charge of the summit was to bring into communication and collaboration a wealth of diverse powers, perspectives, and interests so that strategies could be devised and direct action taken toward a concrete plan for sustainable development. Participants addressed the challenges associated with a globalizing economy that increasingly suffers from mounting industrialization, urbanization, ecological degradation, and the depletion of natural resources, problems exacerbated by the world's growing population and ever-increasing demands for food, water, shelter, energy, health and sanitation services, and economic security.

At the Rio meeting in 1992, the international community adopted Agenda 21, an unprecedented global action plan for sustainable development that was founded on the conviction that human beings had reached a defining moment in the history of humankind and of the earth. Agenda 21 calls for a radical departure from the development policies of the past, which have not only increasingly devastated the environment but also contributed to gaping economic divisions between haves and have-nots around the globe, with consequent increases in the levels of poverty, hunger, disease, and illiteracy worldwide. Agenda 21 is intended to chart a new course to improve the living standards of the masses of hopeless impoverished at the base of the economic system and to restore the earth's ecological health in order to bring about a more prosperous future for humankind and for the planet. The 2002 summit provided a fresh opportunity for international leaders and world citizens to return to the plan laid out in Agenda 21—to assess progress toward its implementation, identify shortcomings in that progress, and renew commitment to quantified targets for its fuller implementation.

On the last day of the 2002 World Summit on Sustainable Development, an assembly of six thousand statues stands near the convention center where the summit participants are meeting. According to the Friends of the Earth, which put on the display, the statues represent the voices of people struggling to defend their lives, communities, and environment; they wear labels that reflect the disappointment of many attendees with the summit's results. The large robot in the center represents the power of major corporations. (AP/Wide World Photos)

PLANNING AND PARTICIPANTS

The World Summit on Sustainable Development was planned and organized by a committee of the tenth session of the United Nations Commission on Sustainable Development in four preparatory meetings held in 2001 and 2002. A bureau consisting of ten representatives, two from each of five world regions, steered the preparations for the summit and worked to raise international awareness of and support for the summit among world governments and lay groups. A key element of the summit planning was the United Nations Secretary-General's Advisory Panel, a team charged with exploring the challenges of sustainable development and making recommendations to the secretary-general concerning how those challenges might be addressed at the summit.

Because the summit was to be a mammoth global event and certainly the largest gathering of international delegates ever convened in Africa, its coordinators were committed to carrying out all aspects of the event according to the ecologically sound "best practices" that the summit would be promoting. The Greening the WSSD Initiative was established to oversee the environmental impacts of the summit and to ensure that minimal waste would be generated by the thousands of participants.

During the summit, while international governmental representatives were meeting at the Sandton Convention Centre on the outskirts of Johannesburg, a nongovernmental forum was also under way at the nearby NASREC Expo Centre. Numerous side events coordinated by the United Nations, as well as parallel events coordinated by independent groups, ensured the broad participation and inclusiveness that the summit's planners considered to be key to the successful setting of realistic but high-reaching goals and gaining widespread commitment to those goals.

More than twenty thousand of people from every corner of the world and every level of society took part in various aspects of the summit, from heads of state and other national delegates to leaders of nongovernmental organizations, leaders in business and industry, and a broad variety of workers and trade unionists, farmers, indigenous peoples, local authorities, and members of the scientific and technological communities. The broad diversity of participants was intended to reflect the major groups identified in Agenda 21.

APPROACHES AND OUTCOMES

Recognizing that poverty in the developing world is a complex, multifaceted problem that is integrally related to nations' external indebtedness, internal corruption, global trade policies, lack of development capital, and many other factors, both national and international, participants in the World Summit on Sustainable Development sought to develop both global and country-specific plans for creating sustainable patterns of consumption and production. These included devising new tactics for optimizing global resource use, minimizing waste, transferring environmentally sound technologies across the globe, achieving greater equity in income distribution, and developing human resources in all corners of the planet through the generation of employment, the extension of basic education and professional training opportunities, and the establishment of effective primary and maternal health care systems. Whether the proposed sustainability initiatives were global or national, the summit encouraged a community-driven approach that would empower local and community groups and include women, youth, and children.

For the most part, the summit was deemed to be successful, because its broad participant base and the global media coverage it received focused international attention squarely on the problem of sustainability. Many celebrated the summit's progress regarding aid to developing nations. Critics, however, argued that the summit's success was reduced by the conspicuous absence of U.S. president George W. Bush from among the more than one hundred world leaders who gathered to address such crucial issues as the spread of acquired immunodeficiency syndrome (AIDS) and other diseases, the depletion of fish stocks around the world, and the need to promote environmentally friendly agriculture. Critics asserted that Bush's absence reduced negotiators' ability to address a number of tough issues, such as the problem of rich nations' policies of farm subsidies, which render poor nations incapable of competing with rich nations agriculturally. Many delegates also complained that the United States led the developed nations in resisting the setting of new targets to phase out export subsidies and other trade-distorting domestic supports.

Wendy C. Hamblet

FURTHER READING

Bigg, Tom. "The World Summit on Sustainable Development: Was It Worthwhile?" In *Survival for a Small Planet: The Sustainable Development Agenda*, edited by Tom Bigg. Sterling, Va.: Earthscan, 2004.

Hens, Luc, and Bhaskar Nath. *The World Summit on Sustainable Development: The Johannesburg Conference.* New York: Springer, 2005.

Speth, James Gustave, and Peter M. Haas. "From Stockholm to Johannesburg: First Attempt at Global Environmental Governance." In *Global Environmental Governance.* Washington, D.C.: Island Press, 2006.

Strachan, Janet R., et al. *The Plain Language Guide to the World Summit on Sustainable Development.* Sterling, Va.: Earthscan, 2005.

SEE ALSO: Agenda 21; Earth Summit; Environmental ethics; Environmental health; Environmental security; Globalization; Johannesburg Declaration on Sustainable Development; Population growth; Renewable resources; Rio Declaration on Environment and Development; Sustainable development.

World Trade Organization

CATEGORIES: Organizations and agencies; resources and resource management

IDENTIFICATION: International organization that deals with the rules of trade between nations

DATE: Established on January 1, 1995

SIGNIFICANCE: The World Trade Organization was created to regulate global trade among member states and to promote trade liberalization. Its primary goal has been to remove barriers to trade, and because national policies concerning environmental protections are often seen as obstructing free trade, the organization's rulings have frequently failed to place a high priority on such protections.

The World Trade Organization (WTO) administers trade between nations in accordance with sets of rules known as the WTO agreements, which are the results of negotiations among the bulk of the world's trading nations, ratified by the members' governments. The WTO was created in 1995 to ensure that global trade runs smoothly, predictably, and as free of obstructions as possible. It seeks to accomplish this overriding goal by providing a forum for the negotiation of agreements among producers of goods and services, importers, and exporters, granting to these parties a legal and institutional framework for imple- menting their agreements, monitoring adherence to agreements, and resolving disputes arising from conflicting interpretations and applications of the agreements.

WTO trade agreements are generally reached by consensus among the entire membership (multilateral agreements), but occasionally individual countries use the WTO forum to develop agreements of more limited scope (plurilateral agreements). By 2010 the body of WTO trade agreements consisted of sixteen multilateral agreements and two plurilateral agreements.

By 2010 the WTO had 153 members, 117 of which were developing countries, and was overseeing more than 97 percent of all global trade. The highest body within the organization is the Ministerial Conference, which meets every two years. Between the Ministerial Conference meetings, a General Council conducts WTO affairs. This council meets in Geneva several times each year and also meets as the Trade Policy Review Body and the Dispute Settlement Body. The Goods Council, Services Council, and Intellectual Property Council report to the General Council. Specialized subsidiary bodies administer trade agreements and monitor their implementation and also deal with individual agreements and other areas of negotiation such as the environment, development, membership applications, and regional trade agreements. The WTO is supported by a secretariat, located in Geneva, that is led by a director-general and employs about seven hundred support staff.

ORIGINS

The WTO continues the work of its predecessor, the General Agreement on Tariffs and Trade (GATT), which was created by the victorious Allied Powers in the wake of World War II to regulate international trade to the best advantage of the founding partners. The system was developed over fifty years through a series of trade negotiation rounds. The earliest rounds sought mainly to reduce tariffs, but later negotiations included other nontariff measures, such as antidumping rules. The round of negotiations that took place from 1986 to 1994, known as the Uruguay Round, led to the creation of the WTO.

Further rounds reached agreement on telecommunications services and broadened liberalization measures well beyond those established in the Uruguay Round. By 1994 forty governments had concluded negotiations for tariff-free trade in information technology products, and seventy members reached a financial services deal covering more than 95 percent of trade in banking and other financial services. In 2000 talks began to address the liberalization of trade in agriculture and related services. The fourth Ministerial Conference, held in Doha, Qatar, in November, 2001, incorporated these changes into the WTO agenda. WTO rules have since come to cover antidumping and subsidies, investment, competition policy, trade facilitation, transparency in government procurement, intellectual property, and a range of broader issues raised by developing countries as they struggle to implement existing WTO agreements.

Activists wear masks during a 2009 demonstration in Manila, Philippines, against the trade liberalization policies promoted by the World Trade Organization. (AP/Wide World Photos)

Mission and Activities

For the most part, the WTO continues the GATT mission of reducing or eliminating obstacles to trade, particularly import tariffs, and governing the rules and conduct of international trade (for example, setting common product standards, monitoring the use of subsidies, and regulating trade-related intellectual property rights). The WTO monitors member adherence to regional and bilateral trade agreements, ensuring transparency in trade activities and intervening to settle member disputes about agreements as they arise. The WTO's mission includes helping developing member countries to build trade capacity and encouraging nonmember countries in the process of becoming members. Other WTO activities include economic research and educating the public about international trade.

Controversies

The WTO asserts that its open-border approach ensures a level playing field among international trading parties, which promotes economic growth in wealthy nations as much as in the developing world. Certainly, GATT and the WTO have helped to build a strong and prosperous trading system that can boast unprecedented economic growth. However, critics argue that the founding and guiding principles of free trade guarantee the most-favored-nation principle, which serves best the interests of the already economically successful while undermining sustainable devel-

opment and social progress in developing countries, as well as global peace and stability.

Critics note how much global trade rules under the WTO fall short of the institutional framework of the proposed International Trade Organization that was developed following World War II; this framework was designed to balance two paramount goals—the economic goal of global trade liberalization and the social goal of stimulating full global employment. These two goals were to be served simultaneously, not through the enforcement of blanket open-border regulations across the globe but through the granting of special treatment to developing nations, many of them newly emerging from colonial histories. The policy of giving poorer nations a leg up into the global economy was abandoned, however, when the U.S. Senate rejected the broad mandate of the proposed framework. The United States instead led the world's foremost nations in approving the much more limited GATT proposal, which focused primarily on reducing barriers to trade and investment and opening all borders as wide as possible while neglecting the goal of proactive development for poorer nations. During the 1980's, the U.S. government led the charge to replace GATT with a larger, more powerful organization better equipped to carry out that primary task of breaking down barriers to trade. Negotiations concluded in 1994, and the World Trade Organization came into being in 1995.

Wendy C. Hamblet

FURTHER READING

Anderson, Sarah, and John Cavanagh, with Thea Lee. *Field Guide to the Global Economy.* Rev. ed. New York: New Press, 2005.

Narlikar, Amrita. *The World Trade Organization: A Very Short Introduction.* New York: Oxford University Press, 2005.

Speth, James Gustave, and Peter M. Haas. "Key Actors, Expanding Roles: The United Nations, International Organizations, and Civil Society." In *Global Environmental Governance.* Washington, D.C.: Island Press, 2006.

Wallach, Lori, and Patrick Woodall. *Whose Trade Organization? A Comprehensive Guide to the WTO.* Rev. ed. New York: New Press, 2004.

SEE ALSO: Commercial fishing; Development gap; Dolphin-safe tuna; Globalization; North American Free Trade Agreement; Race-to-the-bottom hypothesis.

World Wide Fund for Nature

CATEGORIES: Organizations and agencies; activism and advocacy; animals and endangered species; preservation and wilderness issues

IDENTIFICATION: International nonprofit organization dedicated to the preservation and restoration of the natural environment and biodiversity

DATE: Founded on September 11, 1961

SIGNIFICANCE: The World Wide Fund for Nature is one of the largest organizations in the world dedicated to protecting biological diversity by addressing issues such as pollution and habitat destruction. It plays an active role in nature conservation through its funding of research and conservation projects and through joint efforts with governments and business interests.

On a fact-finding trip to East Africa for the United Nations Educational, Scientific, and Cultural Organization (UNESCO) in 1960, Julian Huxley, a British biologist, found what he believed to be the destruction of both wildlife and wildlife habitats at such

Activists with the World Wide Fund for Nature bring attention to the issue of the need for safe drinking water around the world in a demonstration in Mexico City in 2006. The WWF's widely known panda logo is seen on a giant mock-up of a water bucket with a spigot on top. (Henry Romero/Reuters/LANDOV)

a rate that some species were in danger of disappearing within twenty years. Huxley returned to England and published several articles about what he had seen, and the public reaction to his writings—in particular a letter he received from businessman Victor Stolan, who suggested that an organization was needed to fund conservation worldwide—encouraged Huxley to found an organization for wildlife preservation.

Working with ornithologists Max Nicholson, Peter Scott, and Guy Mountfort, Huxley created the World Wildlife Fund (WWF) in 1961. From WWF's inception, Huxley and the other founders planned for the organization to work in cooperation with the International Union for Conservation of Nature and Natural

Resources (IUCN), which Huxley had also helped to found. WWF set up offices in Morges, Switzerland, where IUCN had its offices, and also established offices in various countries to raise funds to provide grants for conservation projects. The organization, which had chosen the panda as its symbol, focused on saving endangered species and preserving their habitats. The first WWF grants went to IUCN, the International Waterfowl Research Bureau, the International Council for Bird Preservation, and the International Youth Council for the Study and Preservation of Nature.

In 1970 the efforts of Prince Bernhard of the Netherlands resulted in the establishment of a WWF trust fund totaling ten million dollars. With its operating costs assured, WWF increased its project funding, and during the 1970's it funded many campaigns to save individual animal species, including tigers, seals, marine turtles, whales, and rhinoceroses. WWF also began to fund campaigns to conserve rain forests and undertook joint efforts with governments to use trade regulation to save both wildlife and plant species from extinction. In 1979 WWF moved its headquarters to Gland, Switzerland.

In 1986, to reflect the expansion of its mission beyond wildlife preservation to include the conservation and preservation of nature in general, the international organization changed its name to the World Wide Fund for Nature (the branches established in the United States and Canada kept the original name); it retained both the acronym, WWF, by which it had been known since its founding and its symbol, the panda. During the 1990's, WWF continued to work with governments and local groups and also began working with businesses to achieve its goals of nature conservation and preservation. WWF interacts globally with the political and economic sectors both as a collaborator and as a lobbyist. In the twenty-first century its concerns include the sustainable use of natural resources, the reduction of deforestation and pollution, and the maintenance of natural wild areas as habitat for endangered species.

Shawncey Webb

FURTHER READING

Carter, Neil. *The Politics of the Environment: Ideas, Activism, Policy.* 2d ed. New York: Cambridge University Press, 2007.

Earle, Sylvia A. *The World Is Blue: How Our Fate and the Ocean's Are One.* Washington, D.C.: National Geographic Society, 2009.

Train, Russell E. *Politics, Pollution, and Pandas: An Environmental Memoir.* Washington, D.C.: Island Press, 2003.

SEE ALSO: Convention on International Trade in Endangered Species; Deforestation; Habitat destruction; Indigenous peoples and nature preservation; International Union for Conservation of Nature; International whaling ban; Lovejoy, Thomas E.; Rain forests; Rhinoceroses; Sea turtles.

World Wilderness Congresses

CATEGORIES: Activism and advocacy; preservation and wilderness issues

IDENTIFICATION: International assemblies devoted to the preservation of wilderness on a global scale

DATE: Inaugurated in October, 1977

SIGNIFICANCE: The World Wilderness Congresses have played a major role in wilderness preservation by providing a forum where international participants from diverse cultures and disciplines can address the implementation of global initiatives aimed at preserving the world's biodiversity and its remaining wilderness areas.

The first World Wilderness Congress began as one of the initiatives undertaken by Ian Player, a South African conservationist, to promote wilderness conservation on a global scale and to ensure wilderness preservation. In 1974, convinced that wilderness conservation must be a concern of every nation and every individual, not just a preoccupation of Western culture, Player resigned from the South African Wildlife Service to pursue efforts to make wilderness conservation a global issue. He established the Wilderness Leadership School to introduce individuals from all walks of life to the wilderness by taking them on foot into the African wilderness. Shortly thereafter he founded the WILD Foundation, the Wilderness Trust, and the Magqubu Ntombela Foundation, which honors his friend and colleague Qumbu Magqubu Ntombela, a Zulu chief.

In October, 1977, working with colleagues in conservation, Player and the WILD Foundation convened the first World Wilderness Congress in Johannesburg, South Africa. The congress provided a forum for discussion and implementation of programs and projects targeting wilderness preservation. The focus of the congress was wilderness preservation on an international, multicultural, multiprofessional, and multi-interest scale. Participants in the congress addressed wilderness preservation from many different viewpoints, including artistic, cultural, governmental, environmental, economic, and academic. The organizers emphasized the congress's importance as an opportunity for the exchange of information—not only among the various groups engaged in wilderness preservation but also among those involved in activities affecting wilderness. The congress introduced programs for involving native peoples in wilderness conservation and addressed banking and economic development as wilderness preservation issues. It also recognized the importance of art and creativity in the fight to preserve wilderness by presenting an extensive exhibition of conservation art.

In keeping with its concept of wilderness as a global issue, the WILD Foundation cohosts a World Wilderness Congress every three to five years to provide opportunities for face-to-face discussion and networking among conservation groups from all over the world. The congresses have been held in many different locations, including in Australia, Scotland, Norway, Mexico, India, and Alaska. Significant contributions to wilderness preservation have come out of the congresses, among them a strengthening of the concept that preservation of wilderness is a task for the global community.

Projects and Initiatives

At the second World Wilderness Congress, the issue of the impact of hydroelectric dams on wilderness conservation in Tasmania received international attention for the first time, and participants formulated a global overview of definitions relating to wilderness. At the third congress participants witnessed the effectiveness of gaining global attention for wilderness preservation as it was announced that the government of Tasmania had opted to protect the Southwest Tasmanian Wilderness rather than to build dams. In addition, two new wilderness preservation organizations, Wilderness Association Italiana and the South African Wilderness Action Group, were formed as a result of the third congress.

Two proposals for programs affecting conservation and preservation were made at the fourth congress in 1987; they called for a world conservation bank and a world conservation corps. In 1991 the Global Environment Facility was jointly established by the United Nations Development Programme, the United Nations Environment Programme, and the World Bank as an independent organization that would give financial assistance in the form of grants to developing countries for projects beneficial to the global environment.

The fifth World Wilderness Congress, held in Norway in 1993, focused on preservation of the Arctic in harmony with sustainable use by the indigenous peoples of the region. The sixth congress established the Asian Wilderness Initiative and gathered support for a joint plan of India and Namibia to return cheetahs to India. The seventh congress concentrated on projects and initiatives to increase protected areas of wilderness on both public and private lands and to implement additional legislation and training programs to guarantee wilderness preservation.

The eighth congress, held in Alaska in 2005, gave priority to issues of global warming and to the topic of the petroleum industry's efforts to gain permission to drill for oil and gas in the Arctic National Wildlife Refuge. The congress also saw the creation of the WILD Planet Fund to maintain wilderness and two new organizations: the Native Lands and Wilderness Council, focusing on the participation of indigenous tribes in wilderness preservation through sound land use and management practices; and the International League of Conservation Photographers, an organization of photographers dedicated to increasing appreciation of wilderness and its preservation through photography.

The ninth World Wilderness Congress was held in Mérida, Mexico, in 2009. Participants in the congress addressed the interrelatedness of human activity, wilderness, and climate change. One outcome of the congress was the Message of Mérida, a plan for making the preservation of wilderness and biodiversity a part of the global strategies used to address climate change and its effects. During the congress the WILD Foundation instigated an international agreement for wilderness preservation that was signed by the United States, Canada, and Mexico. In addition, the first Corporate Commitment to Wilderness was formulated and signed by fifteen corporations, with the expectation of more corporations signing in the near future. The ninth congress also produced the Marine Wilderness Collaborative and established a special program for California's marine wilderness.

At each congress, exhibitions of local art, dance, and music are integrated into the programs and recognized as an essential part of wilderness preservation. From the exhibition of conservation art at the first congress to the nature-related aboriginal art presented in Australia, to the twenty life-size jaguar sculptures displayed and the body painting offered at Mérida, the World Wilderness Congresses use art to reiterate the interrelatedness of all peoples and the necessity of wilderness preservation as part of protecting the earth that they share.

Shawncey Webb

FURTHER READING

Chester, Charles C. *Conservation Across Borders: Biodiversity in an Interdependent World.* Washington, D.C.: Island Press, 2006.

Martin, Vance, and Andrew Muir, eds. *Wilderness and Human Communities: The Spirit of the Twenty-first Century.* Golden, Colo.: Fulcrum, 2004.

Martin, Vance, and Partha Sarathy, eds. *Wilderness and Humanity: The Global Issue.* Golden, Colo.: Fulcrum, 2001.

Player, Ian. *Zulu Wilderness: Shadow and Soul.* Golden, Colo.: Fulcrum, 1998.

SEE ALSO: Arctic National Wildlife Refuge; Global Environment Facility; Indigenous peoples and nature preservation; International Union for Conservation of Nature; Nature preservation policy; Serengeti National Park; Wilderness areas; Wilderness Society; Wildlife refuges; World Wide Fund for Nature.

World Wildlife Fund. *See* World Wide Fund for Nature

Worldwatch Institute

CATEGORIES: Organizations and agencies; activism and advocacy; resources and resource management

IDENTIFICATION: Independent research organization that focuses on critical global issues

DATE: Established in 1974

SIGNIFICANCE: The Worldwatch Institute provides important information about the environment to world leaders, policy makers, and the public in general. Through its fact-based analyses and the ideas it offers, the institute helps to shape world opinion regarding environmental protection, development, and sustainable use of resources.

The Worldwatch Institute, the first research institute devoted to the analysis of global environmental issues, was founded in 1974 by Lester Brown, one of the world's most influential thinkers. Based in Washington, D.C., the institute is an interdisciplinary research organization that aims to help create an environmentally sustainable society that is capable of adequately meeting human needs. For this purpose, it focuses on some of the twenty-first century's most pressing global challenges: climate change, resource degradation, population growth, and poverty.

The Worldwatch Institute uses the best available scientific evidence to perform analyses that help shape the views and positions of decision makers and leaders around the world. It aims to promote the development of innovative solutions to global problems by bringing together the public and private sectors as well as concerned citizens. The institute works with a global network of 150 partners and affiliates in forty countries and produces publications, including most notably the influential yearly report titled *State of the World*, that are translated into thirty-six languages.

The main programs of the Worldwatch Institute are the Climate and Energy, Food and Agriculture, Green Economy, China, India, and Transforming Cultures programs. These different programs are described as follows on the institute's dedicated Web site:

• The Climate and Energy Program is dedicated to accelerating the transition to a low-carbon energy system based on sustainable use of renewable energy resources in concert with major energy-efficient gains.

- The Food and Agriculture Program highlights the benefits to farmers, consumers, and ecosystems that can flow from food systems that are flexible enough to deal with shifting weather patterns, productive enough to meet the needs of expanding populations, and accessible enough to support rural communities.
- The Green Economy Program recognizes that the global environmental and economic crises have common origins and must be tackled together. The program seeks to offer solutions that enhance human well-being and reduce inequities while protecting the planet.
- The China Program seeks to help decision makers within China and around the globe better understand environmental challenges and opportunities.
- The India Program tracks key developments in India and seeks to engage today's decision makers and tomorrow's leaders on all national and global issues.
- The Transforming Cultures Program seeks to transform today's consumerist culture into a culture of sustainability.

In addition to these program areas, the Worldwatch Institute monitors human health, water resources, biodiversity, governance, and environmental security around the world.

The Worldwatch Institute's numerous publications are intended to enable decision makers in government, civil society, business, and academia to keep track of the latest developments concerning the environment and issues of sustainability. The institute's most significant publication, the annual *State of the World*, essentially provides an assessment of global environmental problems and of the innovative ideas proposed and applied across the globe to address them. Every year this report has a particular focus; for example, the 2010 volume is subtitled *Transforming Cultures: From Consumerism to Sustainability*. The 2009 volume focuses on global warming, 2008's examines "innovations for a sustainable economy," and 2007's addresses the "urban future." The institute offers a "*State of the World* at a Glance" feature on its Web site that provides lists and brief explanations of the key facts and innovations noted in each *State of the World* volume.

Among its many other publications, the institute produces *Vital Signs Online*, which provides the latest data and analyses necessary for an understanding of critical global trends, including population growth, biodiversity loss, growth in energy consumption, and rising carbon emissions. *World Watch Magazine*, published bimonthly, offers cutting-edge analysis of social and environmental issues.

Nader N. Chokr

FURTHER READING

Brown, Lester. "Worldwatch." In *Life Stories: World-Renowned Scientists Reflect on Their Lives and on the Future of Life on Earth*, edited by Heather Newbold. Berkeley: University of California Press, 2000.

Nelson, David E. "In Praise of Lester Brown." *Futurist* 42, no. 6 (2008).

Wallis, Victor. "Lester Brown, the Worldwatch Institute, and the Dilemmas of Technocratic Revolution." *Organization and Environment* 10, no. 2 (1997): 109-125.

Worldwatch Institute. *State of the World*. New York: W. W. Norton, 2010.

SEE ALSO: Brown, Lester; Population growth; Renewable resources; Resource depletion; Sustainable agriculture; Sustainable development; Urban sprawl.

Yellowstone National Park

CATEGORIES: Places; preservation and wilderness issues

IDENTIFICATION: U.S. national park located in Wyoming and parts of Montana and Idaho

SIGNIFICANCE: Yellowstone National Park is among the areas of the continental United States containing the widest biodiversity of both plant and animal species. Like all national parks, it faces the constant challenge of balancing protection of the ecosystems it encompasses with enabling access by people who want to take advantage of the land's scenic beauty and the recreational opportunities it offers.

Yellowstone National Park was the first national park, both in the United States and in the world. In 1806 John Colter became the first non-Native American to explore the region in which Yellowstone is located. In 1870 Henry Washburn, Nathaniel Langford, and Gustavus Doane spent more than a month investigating the area and documenting their expedition. Many of those who traveled through the region were deeply impressed by its beauty and spoke out in support of its being declared a public park, and in 1872 President Ulysses S. Grant signed the legislation creating Yellowstone National Park. Langford was appointed the first superintendent of the park; he served for five years and then was replaced by Philetus Norris, who secured the funding necessary to build the first rudimentary roads in the park.

Yellowstone National Park is approximately 8,990 square kilometers (3,470 square miles) in size and expands across the northwest corner of Wyoming as well as parts of Montana and Idaho. The park is home to black bears, grizzly bears, bison, coyotes, deer, elk, and moose that wander freely through the meadows; bobcats, mountain lions, and big-horn sheep live in the mountainous areas of the park. The gray wolf population that originally lived in the park was decimated by predator-control measures in the early twentieth century, but the species was reintroduced in 1995, and by 2010 more than seventeen hundred gray wolves were living within the park's protected habitat. In addition to the wide variety of wildlife, more than seventeen hundred species of plants and trees are found within the park's boundaries.

Yellowstone also contains numerous geological natural wonders. These include almost three hundred waterfalls, one of the world's largest volcanic calderas, and more than ten thousand geothermal features. Perhaps most famous among the geothermal features are the park's hot springs and geysers. Old Faithful Geyser erupts roughly every ninety minutes, much to the delight of the park's visitors. It reaches more than 93 degrees Celsius (200 degrees Fahrenheit) in temperature, releases thousands of gallons of water, and shoots water more than 55 meters (180 feet) into the sky. Nearby are many natural prismatic springs that are known for their vivid display of colors. Because of both its biodiversity and its unique land-

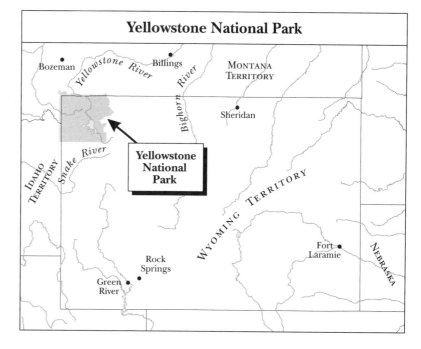

scapes, Yellowstone has been declared a World Heritage Site by the United Nations Educational, Scientific, and Cultural Organization (UNESCO).

The numbers of people visiting Yellowstone each year were relatively small until the early 1880's, when the Northern Pacific Railroad built a station in Livingston, Montana, near the northern entrance to the park. Tourism increased steadily after that, and by 1915 roughly one thousand automobiles were entering the park per year. Tourism declined during both World War I and World War II, but efforts to add visitor facilities and modernize the existing ones throughout the 1950's and 1960's were followed by significant increases in visitors. Throughout the 1960's and 1970's yearly attendance at Yellowstone continued to climb, and by the early twenty-first century some two million people were visiting the park every year. An ongoing challenge at Yellowstone, as at all national parks, is that of protecting the park's ecosystems while allowing human access to the park's lands and recreational opportunities.

In the late 1980's tragedy struck the park in the form of uncontrolled forest fires; an estimated 35 percent of the park's total land area was burned. The damage was so great that the entire park was temporarily closed in September, 1988. Prior to the disastrous Yellowstone fires, the National Park Service had taken a suppression approach to the management of natural fires (such as those caused by lightning) in the national parks. Policy makers began to reevaluate this approach, however, as they realized that the suppression of small fires for decades had allowed the growth of large amounts of fuel to sustain larger, devastating fires. In addition, it became clear that prescribed burning in controlled doses is beneficial for the life of a forest. Fire management policies were subsequently changed to reflect this new knowledge.

Kathryn A. Cochran

FURTHER READING

Garrott, Robert A., P. J. White, and Fred G. R. Watson, eds. *The Ecology of Large Mammals in Central Yellowstone: Sixteen Years of Field Studies*. Burlington, Mass.: Academic Press, 2009.

Heacox, Kim. *An American Idea: The Making of the National Parks*. Washington, D.C.: National Geographic Society, 2001.

Reinhart, Karen Wildung. *Yellowstone's Rebirth by Fire: Rising from the Ashes of the 1988 Wildfires*. Helena, Mont.: Farcountry Press, 2008.

Sellars, Richard. *Preserving Nature in the National Parks: A History*. New ed. New Haven, Conn.: Yale University Press, 2009.

SEE ALSO: Forest Service, U.S.; Kings Canyon and Sequoia national parks; National Park Service, U.S.; National parks; Wildfires; Wildlife management; Wolves; Yosemite Valley.

Yokkaichi, Japan, emissions

CATEGORIES: Disasters; atmosphere and air pollution

THE EVENT: Heavy pollution of the air in the Yokkaichi area, caused by petrochemical plants

DATES: 1950's-1970's

SIGNIFICANCE: The industrial pollution of the air in and around Yokkaichi, Japan, caused widespread health problems for residents and led to landmark court cases and legislation aimed at preventing and reducing air pollution.

Yokkaichi, a port city on Japan's Ise Bay, developed as a major industrial and petrochemical center in the early twentieth century. The demands of World War II and Japan's postwar recovery led to further industrial expansion in the area, and an oil refinery complex known as the Yokkaichi Kombinato was created in the 1950's. Although the complex was an economic success, the pollution it generated was soon linked to breathing difficulties and a variety of other health problems in area residents. Researchers found a high correlation between airborne sulfur dioxide and the incidence of bronchial asthma in children and of bronchitis in older people. Nevertheless, in 1963, a second industrial complex was opened in the region, and a third was added in 1973. In one district of Yokkaichi, airborne sulfur dioxide levels were found to be 800 percent above normal. In the early 1960's, nearly one-half of the area's young children, nearly one-third of its elderly, and approximately one-fifth of its young adults had developed respiratory abnormalities.

In 1967 a group of Yokkaichi residents filed a suit against the Shiohama Kombinato, which ran one of the petrochemical complexes, and in 1972 the plaintiffs were awarded nearly $300,000 in damages. The award marked the first time that a group of Japanese companies had been held liable for damages, setting a precedent that made other companies vulnerable to such litigation. As a result of the case and ensuing controversy, in 1967 Japan's government enacted a basic antipollution law. Within the next several years, additional laws spelled out redress rights for victims from the Yokkaichi area and for residents of polluted areas near Kawasaki and Ōsaka. Regulations requiring refineries to adhere to pollution-abatement policies were also strengthened.

As a result of such measures, by the mid-1970's airborne sulfur dioxide levels in the Yokkaichi region had decreased more than 60 percent, and the rate of respiratory complaints among area residents had declined sharply. By the 1990's nearly 100,000 Japanese citizens had been declared eligible for compensation under the new laws.

Alexander Scott

See also: Air pollution; Ashio, Japan, copper mine; Environmental illnesses; Minamata Bay mercury poisoning; Sulfur oxides.

Yosemite Valley

Categories: Places; preservation and wilderness issues

Identification: Scenic valley of the Merced River located on the western slope of the Sierra Nevada in east-central California

Significance: The Yosemite Valley was part of the first reserve set aside by the United States as an area to be preserved for future generations, a concept that soon thereafter led to the establishment of a national park system.

Yosemite Valley, with Half Dome in the distance. (Courtesy, PDPhoto.org)

Caucasians first entered Yosemite Valley in 1851 during conflicts between California gold miners and local Native Americans. In 1864, after lobbying by early conservationists, U.S. president Abraham Lincoln signed the Yosemite Land Grant, which gave Yosemite Valley and an area a few miles to the south called Mariposa Big Tree Grove—15,900 hectares (39,200 acres) of federal land in all—to the state of California as a reserve to be used for public enjoyment and recreation. The state-supervised reserve was the first area specifically set aside by the United States to be preserved for all future generations. Its inception planted the seed for the national park system, although Yellowstone, not Yosemite, was the first site officially designated a national park, in 1872.

Despite the protection provided by the 1864 land-grant legislation, the floor of Yosemite Valley was used for commercial purposes, including plowing and orchard planting, timber cutting, and grazing. The unprotected, high-mountain country surrounding the valley was also logged and grazed. American preservationist John Muir, while exploring the area in the late 1860's, became concerned about these disturbances to the natural landscape. For the next two decades, he publicized his concerns and worked to preserve the high country.

Muir's efforts paid off in 1890 when some 377,000 hectares (932,000 acres) of the high country surrounding Yosemite Valley gained federal protection through the establishment of Yosemite National Park. However, the valley itself and Mariposa Grove remained under California's jurisdiction. Muir and others then worked to get Yosemite Valley transferred from state to federal government jurisdiction in order to protect it and consolidate the public holdings into a single, unified national park. To help rally public support for Yosemite Valley and other land in the Sierra Nevada, Muir and others founded the Sierra Club in 1892.

In 1903 Muir persuaded President Theodore Roosevelt that the valley needed federal protection, and in 1906 California ceded the area back to the federal government. Yosemite Valley and Mariposa Grove thus became part of Yosemite National Park. There was a cost, however: The overall size of the park was reduced to about 308,000 hectares (761,000 acres), and private mining and timber holdings in the park were excluded from restrictions. In 1913, despite the opposition of Muir and other conservationists, the U.S. Congress approved a project to dam and flood Hetch Hetchy Valley in the northwest corner of Yosemite National Park to create a reservoir to supply water to the city of San Francisco.

Since its inception as a state park in 1864, Yosemite has been a magnet for tourists. By the early years of the twenty-first century, the national park was attracting millions of visitors—and their automobiles—each year. Minimizing the environmental damage that can be caused by such high numbers of visitors is an ongoing problem for Yosemite and for other national parks.

Jane F. Hill

FURTHER READING

Perrottet, Tony. "John Muir's Yosemite: The Father of the Conservation Movement Found His Calling on a Visit to the California Wilderness." *Smithsonian* 39, no. 4 (July, 2008): 48-55.

Radanovich, Leroy. *Yosemite Valley.* Charleston, S.C.: Arcadia, 2004.

SEE ALSO: Grazing and grasslands; Hetch Hetchy Dam; Kings Canyon and Sequoia national parks; Muir, John; National parks; Sierra Club; Yellowstone National Park.

Yucca Mountain nuclear waste repository

CATEGORIES: Nuclear power and radiation; waste and waste management

IDENTIFICATION: Proposed and partially developed radioactive waste storage site located in southern Nevada

SIGNIFICANCE: The selection of Yucca Mountain in Nevada as a repository for the permanent disposal of highly radioactive waste has been controversial because of a number of environmental concerns, including questions about the site's geological suitability and the challenges of safely transporting nuclear waste across the country. In 2010 federal funding for the site was cut and its license application was withdrawn.

The 1982 Nuclear Waste Policy Act began the process of establishing a permanent underground storage repository for America's high-level radioactive waste. In 1987 Congress passed the Nuclear Waste Policy Amendments Act, which singled out Yucca Mountain as the only site to be studied for the national repository. The Department of Energy (DOE) had begun studying the area as a possible repository site almost a decade earlier. Yucca Mountain is located roughly 160 kilometers (100 miles) northwest of Las Vegas, Nevada. Plans were for specially designed waste canisters to be stored in tunnels bored into Yucca Mountain at a depth of approximately 305 meters (1,000 feet) below ground surface and 305 meters above the water table.

Any proposal to construct a high-level nuclear waste repository somewhere in the United States is bound to meet with controversy. In Yucca Mountain's case, many questions have been raised concerning its geologic suitability. One concern is the site's location within a seismically active region. There are more than thirty known earthquake faults in and around Yucca Mountain, some of which have been active since work began on the repository. In 1992 an earthquake with a magnitude of 5.7 caused considerable damage to the aboveground Yucca Mountain project field operations center. The DOE has maintained that the threat of earthquakes to the underground repository is not a significant problem because deep subsurface structures can withstand the ground motion generated by earthquakes.

Another possible problem that has been cited is Yucca Mountain's volcanic history. The mountain was formed millions of years ago by a series of explosive volcanic eruptions. Scientists have studied twelve small, dormant volcanoes located within 20 kilometers (12.4 miles) of the mountain. All have been active within the past four million years, and half of them have erupted within the past one million years. By some estimates, the probability of future volcanic activity disrupting the Yucca Mountain repository during the next ten thousand years is about one in seven thousand. According to the DOE, the volcanic threat to the repository is negligible.

There is also considerable concern about the

Demonstrators at the U.S. Capitol in 2002 protest the planned creation of a nuclear waste repository at Yucca Mountain in Nevada. (AP/Wide World Photos)

movement of subsurface waters that could transport any radioactive material that might leak from waste containers. During the mid-1990's new findings indicated that there were more fractures and more water in Yucca Mountain rock than previously believed. The DOE asserted that corrosion-resistant waste canisters would be sufficient to prevent groundwater from carrying radioactive material off-site.

Finally, opponents of a centralized national repository at Yucca Mountain or elsewhere have pointed out that an accident during nuclear waste transport could result in the release of radiation in a populated area. Project proponents have countered that the likelihood of radioactive contamination is slight, citing the existing successful safety record for the transport of nuclear material worldwide.

In 1993 the DOE scheduled interim waste storage to begin at the facility in 1998. The repository was to be completed and accepting wastes for permanent storage in 2010. However, by 1998 suitability studies were still being conducted and the site was not ready for interim storage. In 2001 the Environmental Protection Agency (EPA) proposed radiation standards for the repository; the state of Nevada, seeking more stringent standards, responded with a lawsuit.

In 2002 Congress authorized the DOE to apply for a license from the Nuclear Regulatory Commission (NRC) to construct the Yucca Mountain repository. Legal, political, technical, and budgetary issues resulted in more postponements. In 2004 the U.S. Court of Appeals ruled that the EPA's standards for the site, which would protect public health and the environment from radioactive exposure risk for ten thousand years, were inadequate in light of recommendations from the National Academy of Sciences. A year later, the EPA issued proposed changes to the standards that extended protection to one million years.

In 2008, by which time the repository's proposed opening date had slid to 2020, the DOE submitted its license application to the NRC. With a Democratic majority in Congress and Senator Harry Reid, a Nevada Democrat and longtime opponent of the repository, serving as Senate majority leader, congressional support for the costly project was flagging. In early 2010 the federal budget for fiscal year 2011 eliminated Yucca Mountain repository funding altogether.

On March 3, 2010, the DOE filed a motion with the NRC to withdraw the license application for the repository. The facility never received any waste for storage.

A special commission was formed in March, 2010, to assess the country's nuclear future and provide recommendations concerning long-term strategies for radioactive waste disposal. The NRC and federal courts have continued to review various legal challenges that have been filed in opposition to Yucca Mountain's closure. Until a solution to America's radioactive waste disposal problem is found, high-level nuclear wastes will remain distributed among various temporary storage locations at facilities across the country.

Kenneth A. Rogers and Donna L. Rogers
Updated by Karen N. Kähler

FURTHER READING

Macfarlane, Allison, and Rodney C. Ewing, eds. *Uncertainty Underground: Yucca Mountain and the Nation's High-Level Nuclear Waste.* Cambridge, Mass.: MIT Press, 2006.

Vandenbosch, Robert, and Susanne E. Vandenbosch. *Nuclear Waste Stalemate: Political and Scientific Controversies.* Salt Lake City: University of Utah Press, 2007.

Walker, J. Samuel. *The Road to Yucca Mountain: The Development of Radioactive Waste Policy in the United States.* Berkeley: University of California Press, 2009.

SEE ALSO: Hanford Nuclear Reservation; Intergenerational justice; NIMBYism; Nuclear and radioactive waste; Nuclear Regulatory Commission; Nuclear regulatory policy.

Z

Zahniser, Howard Clinton

CATEGORIES: Activism and advocacy; preservation and wilderness issues

IDENTIFICATION: American conservationist and nature writer

BORN: February 25, 1906; Franklin, Pennsylvania

DIED: May 5, 1964; Hyattsville, Maryland

SIGNIFICANCE: Zahniser was an influential figure in the wilderness preservation movement of the mid-twentieth century. In addition to serving as executive secretary of the Wilderness Society for more than twenty years, he authored the landmark Wilderness Act of 1964.

Although born in the town of Franklin in northwestern Pennsylvania, Howard Clinton Zahniser grew up to the east in Forest County, in a remote village nestled against the Great Allegheny Forest. After a childhood spent happily roaming the woods and an adolescence during which he absorbed the manifestos of the American Transcendentalists, such as Ralph Waldo Emerson and Henry David Thoreau, Zahniser was convinced that the truest manifestation of the spiritual dimension of the material universe is the untrammeled wilderness. Not educated in the hard sciences (Zahniser completed an English degree in 1928 at tiny Greenville College in Illinois), he responded to the sheer majesty of nature and was certain that such contact is a necessary boon for a humanity bound to technology and cities.

After a stint as a teacher and then working as a journeyman journalist, Zahniser became a kind of media director for the U.S. Department of Agriculture's Bureau of Biological Survey, which later became the U.S. Fish and Wildlife Service, from 1931 to 1943. His heart was in the wilderness, however; he was an avid camper and hiker and a frequent contributor of articles to a variety of nature magazines. In 1945 he went to work for the Wilderness Society, taking a considerable cut in salary to do so. He served as the organization's executive secretary for more than twenty years; during most of that time he also edited its quarterly magazine, *Living Wilderness*.

During the mid-1950's Zahniser spearheaded efforts to stop the U.S. Department of the Interior's proposal to build dams in Dinosaur National Monument in Colorado. Encouraged by the success of these efforts but wary of future projects that might damage the delicate ecostructures of undeveloped federally owned land, Zahniser in 1956 drafted a visionary bill designed to safeguard the American wilderness permanently by setting up a system to coordinate the management of the more than 3.6 million hectares (9 million acres) of forests, national parks, and wildlife sanctuaries under congressional jurisdiction (rather than under the jurisdiction of individual federal agencies). The draft was poetic, even lyrical, in its descriptions of the necessity of the wilderness. It was introduced in the U.S. House of Representatives by John Saylor, a Republican representing Pennsylvania, and in the Senate by Hubert Humphrey, a democrat from Minnesota.

An eight-year battle for the bill's passage ensued, with objections raised by the National Park Service and the Forest Service, the authority of which would be greatly diminished by the law, as well as by entrenched interests of mining, lumber, and farming. Throughout the frustrations of the process—including sixty-six rewrites of the bill and eighteen congressional hearings—Zahniser emerged as the bill's most formidable proponent. In passionate congressional testimony and eloquent magazine columns, Zahniser argued that civilization draws its spiritual strength from interaction with pristine nature. With his folksy charisma, Zahniser forged a national coalition of grassroots supporters, politicians, journalists, conservationists, and scientists to ensure the bill's eventual passage.

Zahniser's health began to fail during the arduous campaign, and he died of heart failure on May 5, 1964, just months before President Lyndon B. Johnson signed the Wilderness Act into law. The members of the generation of environmental activists that emerged during the 1970's revered Zahniser as a folk hero; for many, his resilient determination to protect the wilderness remains a model of humane and effective activism.

Joseph Dewey

FURTHER READING

Harvey, Mark W. *Wilderness Forever: Howard Zahniser and the Path to the Wilderness Act.* Seattle: University of Washington Press, 2005.

Nash, Roderick. *Wilderness and the American Mind.* 4th ed. New Haven, Conn.: Yale University Press, 2001.

Scott, Doug. *The Enduring Wilderness: Protecting Our Natural Heritage Through the Wilderness Act.* Golden, Colo.: Fulcrum, 2004.

Zahniser, Ed, ed. *Where Wilderness Preservation Began: Adirondack Writings of Howard Zahniser.* Utica, N.Y.: North Country Books, 1992.

SEE ALSO: Echo Park Dam opposition; Fish and Wildlife Service, U.S.; Leopold, Aldo; Marshall, Robert; Nature preservation policy; Nature writing; Sierra Club; Thoreau, Henry David; Wilderness Act; Wilderness Society; Wildlife refuges.

Zapovednik system

CATEGORY: Preservation and wilderness issues

IDENTIFICATION: System of nature reserves established in the former Soviet Union

SIGNIFICANCE: The reserves in the *zapovednik* system served two different purposes under the former Soviet Union, with some used as sites of preservation and research and others used mainly for agricultural experimentation. Since the breakup of the Soviet Union in 1991, the reserves have faced difficult times, but some have gained protection and funding after being designated World Heritage Sites or biosphere reserves.

The establishment and administration of nature reserves has generally mirrored the development of biology, ecology, and related sciences. The systematic development of nature reserves with a scientific basis accompanied the developments in biology of the nineteenth century. During that time, Russian biologists related English evolutionist Charles Darwin's evolutionary theories to the environment. From such early scientific studies came a recognition of the importance of natural areas and their preservation.

In the first decade of the twentieth century, Russian zoologists developed the idea of nature reserves, called *zapovedniki*, that would be dedicated to the protection of entire ecosystems. Over the course of the century the popularity of and support for nature reserves in general waxed and waned in the Soviet Union, often influenced as much or more by the political and economic climate as by scientific advances. As was also true in other countries, environmental concerns were seldom of prime importance in the Soviet Union, a giant state that struggled for its own survival for more than seven decades before disintegrating into separate republics in 1991.

In January of 1919, while the new Soviet government was struggling for its existence, agronomist Nikolai N. Pod''iapol'skii proposed the establishment of the regime's first *zapovednik* at Astrakhan. The first five years of the Soviet Union also saw the organization of the All-Russian Society for the Protection of Nature, a volunteer conservation organization that had the effect of enhancing environmental awareness among citizens.

In 1921 Soviet leader Vladimir Lenin signed legislation titled On the Protection of Monuments of Nature, Gardens, and Parks. This empowered the Ministry of Education to declare parcels of nature having special scientific, cultural, or historical value as *zapovedniki*. Between 1919 and 1932 a total of 128 *zapovedniki* were created, with a total area of 12.6 million hectares (31 million acres). The reserves represented 0.56 percent of the total area of the Soviet Union. They varied in size from parcels smaller than 100 hectares (247 acres) to a few of more than 1 million hectares (2.47 million acres) each. Most were in western Russia, Ukraine, or the Caucasus, with a smaller number in Siberia and the far east on the Pacific coast.

The *zapovedniki* were administered by two separate government agencies with different philosophies and sets of goals. The Ministry of Education maintained relatively pristine reserves for their aesthetic properties and as sites for preservation and scientific research. During the 1920's these reserves were utilized by several important Russian ecologists who pioneered studies in such areas as productivity, trophic relationships, and predator-prey interactions in ecosystems. Unlike the national park system in the United States, the *zapovedniki* run by the Ministry of Education did not give great consideration to tourism.

Nature reserves under the management of the Ministry of Agriculture, in contrast, were maintained primarily as centers of agricultural production and experimentation. The Soviets sought to discover scientific management policies that would maximize yields of timber, fur, and other products of value to the economy. These goals compromised conservation efforts.

The emergence of Joseph Stalin as dictator of the Soviet Union in 1929 and the outbreak of World War II in 1939 had disastrous effects on conservation efforts in general and on the *zapovedniki* in particular. Stalin disbanded the Society for the Protection of Nature and introduced a vigorous program of industrialization. Lip service was paid to conservation, but polluting industries were allowed to operate with few regulations.

During this same period, the conservation movement in the Soviet Union was subjected to a fate similar to that of the field of genetics: Scientific principles were abandoned in favor of incorrect, unsupported ideas favored by Marxist theorists. As a result, many unspoiled nature reserves that had been established during the 1920's were dismantled or converted into agricultural enterprises. By 1952 only 40 reserves with a total of 1.5 million hectares (3.7 million acres) were left; this represented just 12 percent of what had existed before the war.

Under Leonid Brezhnev, first secretary of the Communist Party during the 1960's, a new interest in conservation emerged. An improved economy and a reduction in restrictions on individual freedom led to an increase in tourism among Soviet citizens, and appreciation of natural areas grew as international tourists also sought these places out. By 1981 nature reserves had become popular with the Soviet public, and the number had grown to 129, surpassing the number that had existed before World War II.

After the breakup of the Soviet Union in 1991, new challenges emerged. Faced with a legacy of widespread pollution and near bankruptcy, the former states of the Soviet Union struggled to maintain themselves. Many conservationists feared that the environment in general, including the *zapovednik* system, would suffer. The biodiversity of the reserves was threatened both by a lack of funds and by impoverished local people who destroyed the flora and fauna in order to survive. Most observers believed that the assistance of outside agencies would be required to prevent deterioration of the system. By 2010 the number of *zapovedniki* had shrunk to 101. Some of the remaining sites, however, had begun to receive protections and aid with funding through their designation by the United Nations Educational, Scientific, and Cultural Organization (UNESCO) as World Heritage Sites or biosphere reserves.

Thomas E. Hemmerly

FURTHER READING

Oldfield, Jonathan D. *Russian Nature: Exploring the Environmental Consequences of Societal Change.* Burlington, Vt.: Ashgate, 2005.

Schwartz, Katrina Z. S. *Nature and National Identity After Communism: Globalizing the Ethnoscape.* Pittsburgh: University of Pittsburgh Press, 2006.

Weiner, Douglas R. *A Little Corner of Freedom: Russian Nature Protection from Stalin to Gorbachev.* Berkeley: University of California Press, 1999.

_____. *Models of Nature: Ecology, Conservation, and Cultural Revolution in Soviet Russia.* 1988. Reprint. Pittsburgh: University of Pittsburgh Press, 2000.

SEE ALSO: Biodiversity; Biosphere reserves; Ecology as a concept; Ecosystems; Lake Baikal; Nature reserves; Soviet Plan for the Transformation of Nature; Wilderness areas; World Heritage Convention.

Zebra mussels

CATEGORIES: Animals and endangered species; water and water pollution

DEFINITION: Endemic European freshwater bivalves that occur in North America as an exotic species

SIGNIFICANCE: Zebra mussels are a source of concern in North America, where they are an invasive or exotic species, because of the mussels' potential influence on aquatic systems. When zebra mussels colonize a body of water, they reduce the abundance and species diversity of native mussels, and they also reproduce in such density that they restrict water flow in pipes.

The zebra mussel (*Dreissena polymorpha*) is native to southern Russia and is thought to have been introduced into Lake Saint Clair—which lies between Michigan and Ontario, Canada—in 1986 via discharged ballast water. Since its introduction, the zebra mussel has become widely dispersed, occurring in all of the Great Lakes by 1990; by 1994 it had appeared in or adjacent to nineteen U.S. states. This rapid dispersal is largely the result of the mussels' ability to attach to boats that navigate these waters, as well as their ability at all life stages to survive overland transport (for example, living on the hulls of boats as the boats are transported between lakes). It is expected that the species will continue to disperse and increasingly colonize North American inland lakes.

The shell sizes of zebra mussels average 25 to 35 millimeters (1 to 1.4 inches). Zebra mussels typically live three to five years, and females usually reproduce during their second year; each female can produce more than forty thousand eggs in one reproductive cycle. After fertilization, veliger larvae emerge within three to five days and are free-swimming for up to one month. Dispersal during this time is primarily caused by water currents. Larvae then settle to the bottom, where they crawl via a foot, looking for suitable substrate (preferred to be hard or rocky). To attach themselves to the substrate, they secrete proteinaceous byssal threads from a byssal gland located in the foot.

From an ecological and environmental perspective, one of the most important concerns related to ze-

bra mussel colonization of a body of water is that it reduces the abundance and species diversity of native unionid mussels. Because native unionid mussel beds provide the type of hard substrate that zebra mussels prefer, zebra mussels readily colonize such areas, negatively influencing feeding, growth, locomotion, respiration, and reproduction of native unionids.

In addition, zebra mussels have an important influence on the environment because of their role as biofoulers. When they colonize pipes they can restrict water flow, negatively affecting water supply to hydroelectric facilities, nuclear power plants, and public water supply plants. Zebra mussels can also attach to the hulls of boats, leading to increased drag, and their weight can sink navigational buoys. Because zebra mussel densities have been measured as high as 700,000 mussels per 1 square meter (10.8 square feet), they clearly can cause serious problems.

Despite their negative effects on aquatic systems, zebra mussels can have positive effects on water qual-

A native fatmucket mussel is encrusted by invasive zebra mussels. (USFWS)

ity through their role as biofilters. An adult zebra mussel can filter the phytoplankton from as much as 1 liter of water per day and can significantly alter water quality. However, even this influence has potential negative consequences in that it reduces the amount of food available for zooplankton and eventually for recruiting fishes. This can also lead to a change in food webs from phytoplankton-dominated systems to macrophyte-dominated systems. Given their filtering ability, zebra mussels also tend to bioaccumulate substances, possibly increasing the concentration of toxic substances that are passed up the food web.

Dennis R. DeVries

FURTHER READING

Pimentel, David, ed. *Biological Invasions: Economic and Environmental Costs of Alien Plant, Animal, and Microbe Species.* Boca Raton, Fla.: CRC Press, 2002.

Van Driesche, Jason, and Roy Van Driesche. "Refuge for the Mussels: Biotic Integrity and Zebra Mussel Invasion in the Ohio River Basin." In *Nature Out of Place: Biological Invasions in the Global Age.* Washington, D.C.: Island Press, 2000.

SEE ALSO: Biodiversity; Biomagnification; Introduced and exotic species; Lake Erie.

Zero Population Growth. *See* Population Connection

Zoning

CATEGORIES: Land and land use; urban environments

DEFINITION: Government-imposed regulations on the use and dimensions of land and structures

SIGNIFICANCE: Since the adoption of early zoning laws that simply regulated uses and dimensions, governments have begun to innovate with zoning provisions that protect not only the character of communities but also their natural resources. Creative zoning applications include overlay conservation and wetlands districts, performance and incentive zoning, transfer of development rights, and complete bans on land uses that are detrimental to the environment.

Zoning is a police power that permits local governing bodies to infringe on private property rights in order to protect the general public health and safety and to preserve property values. Zoning regulations are found throughout the world; they generally commence with authorization, usually through a state, federal, or national enabling law, that permits local governments to regulate the uses and dimensions of land, buildings, and other structures for the purpose of achieving general and specific purposes. Relief from onerous zoning regulations is available through various uniform permitting processes, such as variances, that are set forth in the authorizing legislation and in local laws.

In 1926 the U.S. Supreme Court decided the seminal case of *Euclid v. Ambler,* which upheld the right of the village of Euclid, Ohio, to regulate an industrial use in order to achieve the legitimate governmental purpose of protecting a residential neighborhood. After the *Euclid* decision urban planning and zoning, often dubbed Euclidean zoning, spread throughout the United States, and local governments began to adopt zoning ordinances and bylaws. Over time zoning laws have become increasingly comprehensive and have placed greater limitations on private property as courts have continued to uphold more complex regulations. However, in the United States zoning regulations may be found invalid and unconstitutional if they go too far in infringing on private property rights without the government's providing just compensation as required by the Fifth Amendment of the U.S. Constitution.

USE AND DIMENSIONAL REGULATIONS

Cities, towns, counties, and other governmental bodies usually commence the zoning process by adopting zoning maps that depict various zoning districts within geographic boundaries. The districts are related to land use and are usually derived from municipal land-use plans. Types of zoning districts include residential, multifamily, commercial, industrial, and open space. Within each zoning district some uses are allowed by right without further governmental review. Other uses may be prohibited within a district, while some are permitted after discretionary review by a local land-use board or official established to conduct such review. Most discretionary reviews require due process, including notice and holding of public hearings, before a regulatory body or official makes a decision. In addition, due process

In the United States and many other nations, urban areas are subject to zoning regulations that specify allowable land uses, such as whether certain areas may be developed for housing, business, recreation, or other uses. (©Gary Blakeley/Dreamstime.com)

provides the property owner with the right to appeal a zoning decision, usually to a local quasi-judicial body and eventually—after all other available remedies are exhausted—to a municipal or state court.

In tandem with the adoption of zoning maps, governments adopt regulations that apply to uses within each zoning district, including dimensional regulations. Dimensional regulations might include regulation of lot sizes through frontage, which is a measurement along the way or road providing access to a lot; depth regulations, concerning the lengths of side lot lines; and rear lot line lengths. In addition, minimal lot sizes based on square footage or acreage are usually stated, and lot coverage regulations may be imposed, requiring that a specific percentage of a lot remain open space, without buildings or other development. Density is controlled through setback regulations, which require buildings and structures to be set back specific distances from established lot lines. Other dimensional regulations include height regulations of buildings and other structures, such as com-

munications towers and church steeples, and regulation of parking through the requirement of specific numbers of parking spaces for particular uses.

COMPLEXITY OF REGULATIONS

Over time, with increases in population and urban growth, as well as the emergence of new technologies, the regulation of land use has become more complex. Under performance zoning regulations, for example, developers whose developments meet specific community goals receive bonus points that allow them more flexibility in design, usually with respect to density controls. In a similar vein, some zoning laws allow a private property owner to protect important natural or historic features on a piece of property by exchanging or transferring development rights on that property for greater development rights on another piece of property. Incentive zoning regulations may involve impact fees, which require monetary payments to the government when specific land uses overburden existing infrastructure, such as roads or schools. Moreover, state and federal laws may infringe on local governments' authority to regulate and prohibit land uses to which residents may have objections, such as transmission towers, wind farms, and hazardous waste facilities.

Zoning has become a popular regulatory vehicle among groups seeking to protect wetlands, floodplains, and other natural resources from any development. Although such regulations were often found to be unconstitutional takings in the United States when first adopted, courts are no longer reluctant to preserve the natural environment for future generations by upholding such provisions and finding them constitutional.

Carol A. Rolf

FURTHER READING

Arnold, Craig Anthony. *Fair and Healthy Land Use: Environmental Justice and Planning.* Chicago: American Planning Association, 2007.

Babcock, Richard F., and Charles L. Siemon. *The Zoning Game Revisited.* 1985. Reprint. Cambridge, Mass.: Lincoln Institute of Land Policy, 1990.

Boone, Christopher G., and Ali Modarres. *City and Environment.* Philadelphia: Temple University Press, 2006.

Elliott, Donald L. *A Better Way to Zone: Ten Principles to Create More Livable Cities.* Washington, D.C.: Island Press, 2008.

Ferrey, Steven. "Local Environmental Controls." In *Environmental Law: Examples and Explanations.* 5th ed. New York: Aspen, 2010.

Merriam, Dwight. *The Complete Guide to Zoning: How to Navigate the Complex and Expensive Maze of Zoning, Planning, Environmental, and Land-Use Law.* New York: McGraw-Hill, 2004.

Nolon, John R. *Open Ground: Effective Local Strategies for Protecting Natural Resources.* Washington, D.C.: Environmental Law Institute, 2003.

Wolf, Michael A. *The Zoning of America: Euclid v. Ambler.* Lawrence: University Press of Kansas, 2008.

SEE ALSO: Land-use planning; Land-use policy; Open spaces; Planned communities; Property rights movement; Sustainable development; Urban planning; Urban sprawl; Wise-use movement.

Zoos

CATEGORY: Animals and endangered species

DEFINITION: Parks where live animals are displayed for public viewing

SIGNIFICANCE: For many years zoos were operated primarily as entertainment venues, but in the mid-twentieth century most zoos began to shift their focus to education and active conservation of endangered and threatened species.

Each year, millions of people visit zoos around the world. Many modern zoos, often called wildlife conservation parks or natural wildlife parks, have replaced cages of concrete and steel with simulated natural environments, and typically modern zoos obtain new animals through selective breeding programs instead of capturing them from the wild.

HISTORY

Early zoos were the sole province of the wealthy; the first recorded zoo in history belonged a Chinese emperor in 1100 B.C.E. It was not until the nineteenth century that zoos became open to the public. The word "zoo" derives from the term "zoological park" or "zoological garden," which reflects what the first zoos were designed to be: afternoon diversions along the same lines as amusement parks or circuses. The companies that ran zoos captured exotic beasts from newly charted regions of the world and displayed them with little regard for the animals' physical health or emotional well-being. Mortality was high, and display animals were constantly replaced with others captured from the wild, where there seemed to be an inexhaustible supply.

The first zoo to use moats to separate animals from visitors was established in Germany by Carl Hagenbeck in 1907. With moats replacing the bars of cages, visitors had unobstructed views of the animals, and, depending on the placement of the moats, visitors could have the impression that the animals were free. The habitats in which the animals lived, however, were nothing like what they were accustomed to in the wild. In their strange surroundings, many animals spent their days sleeping; others often displayed near-psychotic behavior patterns, such as pacing, head butting, and even self-mutilation.

Two developments are credited with influencing the changes that took place in zoo management during the twentieth century. First, with the advent of motion pictures and then television, potential zoo visitors were able to see many animals in their natural habitats, and suddenly giraffes, lions, and zebras no longer seemed quite so exotic. Second, wild animals were becoming increasingly scarce, and words such as "conservation" and "endangered" entered the collective vocabulary. Acquiring specimens from the wilderness became more costly, and zoos began to look at internal breeding programs to replenish their stocks. However, they found that animals kept in unnatural and, in some cases, inhumane conditions would not breed.

New zoo enclosures were designed to encourage natural behavior in animals by replicating their natural environment as much as possible while still ensuring the safety of both the animals and zoo visitors. The animals held in zoos began receiving healthier diets; when possible, they were allowed to feed in much the same way they would in the wild—by digging, foraging, or grazing. Zoos began to keep human contact with orphaned and injured animals to an absolute minimum, and some zoos took the additional step of instituting the policy of not naming their animals in an effort to discourage anthropomorphism. By the end of the twentieth century, 80 percent of the mammals on display in zoos were born in captivity.

DEBATES AND CRITICISMS

Zoo managers continue to struggle to balance science, conservation biology, scarce resource allocation, and ethics. Among the decisions that must be made are whether predators should be offered the chance to exercise natural hunting behaviors by being offered live prey, and whether zoos should maintain potentially deadly animals that are necessary for breeding programs but are dangerous and difficult to control, such as macaques, many of which harbor the

A two-toed sloth, native to the tropical forests of Central and South America, is on display during an educational class held within the Bronx Zoo in New York. Most modern zoos seek to educate the public about animal species and habitats. (AP/Wide World Photos)

deadly hepatitis B virus, and adult male elephants. Another dilemma is the question of what zoos should do with "surplus" animals that are inbred, unable to reproduce, or otherwise genetically inferior.

Municipal bureaucracies can sometimes hamper zoos' conservation efforts. Zoo managers must often combat local governments and public opinion when dealing with unpopular issues, such as the selling or other disposal of surplus animals and the allocation of monetary resources. In addition, budget cuts have forced zoo managers to turn to the private sector for financial assistance. Among the fund-raising activities pursued by some zoos are "adopt an animal" programs and commercial ventures that involve selling "exotic compost."

Some people question whether zoos should exist at all—whether it is cruel to take animals from their natural habitats and place them on display, even if the purpose is education rather than entertainment. The People's Republic of China "rents" giant pandas to zoos around the world; some people assert that these pandas would better serve their species by remaining in the wild or by being part of a captive-breeding program aimed at replenishing the species' numbers. Critics argue that the money devoted to zoos and captive-breeding programs would be better spent on preserving animals' natural habitats. To combat this type of criticism, some zoos have changed their focus from "collecting" wildlife to "protecting" wildlife, also known as field conservation. In these zoos, visitors view exhibits linked with protection and conservation programs in natural habitats; this allows the visitors to connect what they are seeing in captivity to what is worth saving in the wilderness. Some zoos have taken the additional step of "adopting" wildlife refuges.

Despite the criticisms that zoos face, the fact remains that many animal species simply could not survive without the existence of zoos and captive-breeding programs. Ironically, where historical zoos replenished their stock from the wilderness, some zoos are now replenishing the wilderness with captive-bred animals.

P. S. Ramsey

FURTHER READING

Hancocks, David. *A Different Nature: The Paradoxical World of Zoos and Their Uncertain Future.* Berkeley: University of California Press, 2001.

Hanson, Elizabeth. *Animal Attractions: Nature on Display in American Zoos.* Princeton, N.J.: Princeton University Press, 2002.

Kisling, Vernon N., Jr., ed. *Zoo and Aquarium History: Ancient Animal Collections to Zoological Gardens.* Boca Raton, Fla.: CRC Press, 2001.

Norton, Bryan G., et al., eds. *Ethics on the Ark: Zoos, Animal Welfare, and Wildlife Conservation.* Washington, D.C.: Smithsonian Institution Press, 1995.

SEE ALSO: Animal rights; Animal rights movement; Captive breeding; Convention on International Trade in Endangered Species; Poaching.

APPENDIXES

Environmental Organizations

From the late twentieth century onward, especially since the 1970's, worldwide awareness of environmental issues has been heightened by the work of hundreds of nongovernmental organizations established to promote research, influence public policy, and encourage citizen engagement in solving problems associated with the earth's environment and ecosystems. Some of the most important of these organizations are briefly described below; asterisks on entries indicate organizations that are profiled in more depth in individual essays in this encyclopedia.

African Conservation Foundation (ACF)

Year founded: 1999
Web site: http://www.africanconservation.org

ACF focuses on protecting and conserving African wildlife by seeking to find workable approaches to managing natural resources that offer a balance between needs for development and conservation initiatives. The organization provides training, support, and assistance to groups that share similar aims. It also sponsors research and conservation projects in Africa and engages in fund-raising to promote awareness and support of these efforts. ACF has offices in several African countries and in the United Kingdom.

American Farmland Trust (AFT)

Year founded: 1980
Web site: http://www.farmland.org

AFT was founded by farmers and ranchers in the United States to help preserve farms and ranches, promote a healthy and sustainable environment, and build communities. It engages in lobbying activities and produces publications outlining problems associated with the loss of farm and ranch lands and suggesting solutions for retaining these resources. AFT has been influential in the passage of several state and federal laws governing farm and ranch preservation, including special provisions in the 1996, 2002, and 2008 farm bills passed by the U.S. Congress.

Antinea Foundation

Year founded: 2007
Web site: http://www.antinea-foundation.org

The Antinea Foundation, based in Switzerland, supports research, education, and public awareness programs aimed at promoting conservation of the earth's oceans and marine ecosystems. The current organization was formed from a merger of two Swiss groups, Association Pacifique and Association Antinea. The centerpiece of the foundation's initiatives is a ten-year voyage of exploration and scientific research conducted aboard the Antinea Foundation's ship *Fleur de Passion*, a converted German warship.

Association for Environment Conscious Building (AECB)

Year founded: 1989
Web site: http://www.aecb.net

Operating in the United Kingdom, AECB promotes building projects that respect and help preserve the environment. Its membership includes builders, designers, housing professionals, and government officials. The organization conducts seminars on green building practices, develops building standards and codes that enhance conservation and reduce harmful construction practices, and lobbies for implementation of rigorous requirements to reduce the impacts of new construction and renovation on the environment, especially standards that reduce harmful carbon emissions.

Association of Environmental and Resource Economists (AERE)

Year founded: 1979
Web site: http://www.aere.org

AERE is an association of academics, professionals from government agencies and private research organizations, and representatives of consulting firms who are committed to promoting the study of environmental and natural resource economics. The group sponsors workshops, conferences, and symposia as a means of sharing information and stimulating further investigation of the economic ramifications of environmental problems. A sister organization, the European Association of Environmental and Resource Economists, was established in 1990 and works toward the same ends.

Association of Environmental Professionals (AEP)

Year founded: 1975
Web site: http://www.califaep.org

AEP draws its members from the fields of environmental planning, natural resources management, and environmental science. It promotes awareness of environmental issues, serves as a watchdog on governmental policies relating to the improvement or degradation of the environment, monitors the impacts of current legislation, and lobbies for changes to laws and policies to promote sustainability. The organization is also actively involved in helping members improve their skills as environmental activists and natural resource managers.

Australian Conservation Foundation
Year founded: 1966
Web site: http://www.acfonline.org.au

The Australian Conservation Foundation is a nonprofit organization focused on protecting the natural resources of Australia. The group's chief interests are research, policy development, education, and advocacy. Over four decades it has made significant contributions to conservation efforts targeted at the Great Barrier Reef as well as several endangered rivers and rain forests. It has campaigned against hazardous mining practices and encouraged environmentally responsible farming and land management. Since 1990 it has sought more active involvement with businesses to promote environmentally responsible development.

Canadian Parks and Wilderness Society (CPAWS)
Year founded: 1963
Web site: http://www.cpaws.org

CPAWS (also known as Société pour la nature et les parcs du Canada, or SNAP) is actively engaged in efforts to preserve the Canadian wilderness. The group monitors government and private activity in national and province parks, advocates for restricted development in natural environments, and promotes maintenance of healthy natural ecosystems. Annually it honors Canadians who have made notable contributions to conservation.

Center for Health, Environment, and Justice* (CHEJ)
Year founded: 1981
Web site: http://www.chej.org

Formerly known as Citizens Clearinghouse for Hazardous Waste, CHEJ is a grassroots organization that engages citizens in campaigns to promote healthy communities through elimination of pollutants and creation of sustainable urban ecosystems.

Center for International Environmental Law (CIEL)
Year founded: 1989
Web site: http://www.ciel.org

CIEL is a nonprofit organization with offices in the United States and Switzerland. The organization uses international courts to regulate activities that might prove harmful to the environment. CIEL offers legal counsel, conducts policy research, engages in advocacy initiatives, and assists in building other groups' capacities to make meaningful contributions to environmental causes. It also sponsors educational activities, most notably a curriculum in international environmental law at the American University in Washington, D.C.

Ceres*
Year founded: 1989
Web site: http://www.ceres.org

Ceres is a network of investors, environmental organizations, and other public interest groups working to address a variety of sustainability issues, including matters such as global climate change.

Climate Project
Year founded: 2006
Web site: http://www.theclimateprojectus.org

Founded by former U.S. vice president and 2007 Nobel Peace Prize recipient Al Gore, the Climate Project is designed to bring worldwide awareness to the growing problems associated with global warming and climate change. The organization sponsors lectures and seminars at which carefully trained presenters stress the urgency for action to reverse the damages of global warming. By 2009 the group had trained more than two thousand people to help deliver its message. The Climate Project has been the target of criticism from some observers who have expressed concerns that the organization presents an alarmist message that is not always rooted in sound science.

Comité de Liaison Energies Renouvelables (CLER)
Year founded: 1984
Web site: http://www.cler.org/info

CLER is an advocacy and educational association made up of professionals from industry, commerce, architecture, building trades groups, and the academic community working collaboratively to promote sustainable energy for France. Members of CLER engage in educational programs to promote awareness of energy issues and garner support for projects that enhance the development of renewable energy resources.

Conservation International

Year founded: 1987
Web site: http://www.conservation.org

Based in the United States, Conservation International has been active in promoting projects to preserve the world's biodiversity both on land and in marine environments. Working in nearly four dozen countries, the group has been active in sponsoring explorations of remote regions, where dozens of previously unknown species of animals and plants have been identified. The organization has engaged in lobbying efforts aimed at restricting activities harmful to natural ecosystems, but it has also worked to find solutions for allowing humans to live harmoniously within the natural environment.

Conservation Law Foundation (CLF)

Year founded: 1966
Web site: http://www.clf.org

CLF is a regional environmental advocacy group focused on issues affecting the northeastern United States. It has lobbied successfully to halt development that would have degraded several traditionally important natural areas in the region, such as New Hampshire's White Mountains; has become involved in activities to clean up Atlantic coastal areas, several rivers, and Boston Harbor; has advocated for mass transportation as a means of reducing energy pollution; has fought against the proliferation of nuclear power; and has initiated collaborative efforts to combat global warming by rewarding companies that reduce emissions.

Earth First!*

Year founded: 1980
Web site: http://www.earthfirst.org

Earth First! is a radical advocacy group with branches in nearly twenty countries. It promotes direct action, including civil disobedience, to protect the environment from commercial efforts at development.

Earth Policy Institute (EPI)

Year founded: 2001
Web site: http://www.earth-policy.org

EPI, an advocacy group based in the United States, is primarily interested in raising public awareness about dangers facing the world's population, among them environmental problems such as global warming and the loss of plant and animal species. Estab-

lished to promote the ideas of environmentalist and activist Lester Brown, EPI publishes books and reports that lay out a vision of a sustainable environment, document current problems, and track efforts by governments and private groups in meeting attainable systemic changes.

Earthwatch Institute

Year founded: 1971
Web site: http://www.earthwatch.org

The Earthwatch Institute is an international organization that engages people in activities that help them understand what is required to create a sustainable natural environment. Operating from offices in the United States, the United Kingdom, Australia, and Japan, the Institute sponsors research projects around the globe in which volunteers are encouraged to join professionals on expeditions where they can gather data on rain-forest and marine ecology, wildlife conservation, and archaeology.

Energy Action Coalition

Year founded: 2004
Web site: http://energyactioncoalition.org

Energy Action Coalition is a network of more than fifty U.S. and Canadian environmental groups committed to raising awareness among college students regarding environmental problems facing the planet. Headquartered in the United States, the organization lobbies for changes to environmental law and policies and engages communities in efforts to deal with challenges posed by the deteriorating condition of the earth's natural resources. The coalition encourages student involvement in numerous conferences, seminars, and lobbying efforts to get federal governments to reverse long-standing economic and environmental policies thought to favor development over conservation and sustainability of ecosystems.

Environmental and Energy Study Institute (EESI)

Year founded: 1984
Web site: http://www.eesi.org

EESI grew out of a program established by the U.S. Congress in 1975 to gather and disseminate information about environment and energy issues as a means of assisting lawmakers in developing sound policies. The group became an independent nonprofit in 1984 but continued to keep its focus on education and data collection. EESI sponsors congressional briefings, meetings, and seminars and issues publications that address the topics of global warming and air pollution. It endorses policies that promote energy security and rural economic development and encourages increased use of renewable energy sources and improved energy efficiency. The organization also lobbies for the protection of areas such as the nation's Arctic and coastal regions.

Environmental Defense Fund (EDF)

Year founded: 1967
Web site: http://www.edf.org

EDF has been one of the most effective and also one of the most controversial environmental advocacy organizations in the United States. Growing from a grassroots movement to save endangered raptors in New York, EDF emerged as a national force lobbying for laws and policies that promote species conservation, clean water, and sustainable ecosystems. It claims to have been the driving force behind the U.S. ban on the use of the pesticide dichloro-diphenyl-trichloroethane (DDT) and dangerous chemical compounds including chlorofluorocarbons (CFCs), passage of the Safe Drinking Water Act of 1974 and the 1990 amendments to the Clean Air Act, establishment of the Northwestern Hawaiian Islands Coral Reef Ecosystem Reserve, and creation of the U.S. Climate Action Partnership. Critics have charged, however, that EDF has often exercised little concern for human communities in its zeal to enforce tighter restrictions on the use of chemicals or improve natural ecosystems.

Environmental Design Research Association (EDRA)

Year founded: 1968
Web site: http://www.edra.org

EDRA was founded in the United States to bring together design professionals, social scientists, facilities managers, and others interested in creating buildable spaces compatible with and respectful of the natural environment. Its members meet regularly to share scholarship and best practices, and EDRA publishes annual proceedings of the group's meetings. EDRA has also established a number of awards to recognize individuals, organizations, and projects that demonstrate commitment to best practices in environmental management.

Environmental Foundation for Africa (EFA)

Year founded: 1992
Web site: http://www.efasl.org.uk

Although founded in the United Kingdom, EFA is based in West Africa and has as its principal focus the protection and improvement of that region. EFA works locally to sponsor community involvement in projects to improve environmental quality. The organization promotes awareness of environmental problems, provides technical assistance in creating solutions to those problems, serves as an advocacy group in lobbying for funding and policy changes that will help preserve or improve the environment of the region, and organizes networks of environmental organizations, government agencies, and private funding groups to bring about change that will improve the lives of the people of West Africa while respecting the natural world in which they live.

Environmental Investigation Agency (EIA)

Year founded: 1984
Web site: http://www.eia-international.org

Based in the United Kingdom and the United States, EIA is an activist organization that funds undercover investigations to identify violations of environmental law and expose animal cruelty. EIA has been successful in focusing a spotlight on illegal logging activities in Southeast Asia, securing a ban on the illegal trade in tiger parts (particularly in China and India), lobbying for a cessation of commercial whaling, and revealing the extent of illegal trading in hazardous chemicals. Using evidence gathered in its investigations, EIA has lobbied effectively for stricter laws governing environmental issues in numerous countries.

Environmental Protection UK (EPUK)

Year founded: 1898
Web site: http://www.environmental-protection.org.uk

Originally the Coal Smoke Abatement Society and known for years as the National Society for Clean Air and Environmental Protection, EPUK is a nongovernmental organization dedicated to improving air quality in the United Kingdom. It has also been active in efforts to reduce noise pollution and has campaigned to protect and rehabilitate land areas. EPUK lobbies for stronger environmental standards, particularly in the area of air quality, and provides assistance to local governments in achieving those standards.

European Environmental Bureau (EEB)
Year founded: 1974
Web site: http://www.eeb.org

EEB is a federation made up of more than 140 environmental groups from all member countries of the European Union (EU). Organized principally to provide a unified and strong voice for representing the interests of environmentalists, EEB provides information on environmental issues to its members and to officials of individual European governments and the EU Parliament and its subsidiaries. It also represents member organizations in lobbying before these government bodies on pending legislation or on issues that EEB believes require government action to protect or rehabilitate the region's natural resources.

Friends of the Earth International* (FOEI)
Year founded: 1969
Web site: http://www.foei.org

FOEI is an international network of environmental organizations with affiliates in seventy-seven countries. The organization is dedicated to challenging environmentally unsound government policies and to promoting initiatives that will sustain and improve ecosystems worldwide.

Global Witness
Year founded: 1993
Web site: http://www.globalwitness.org

Global Witness is an international nonprofit organization whose aim is to investigate and expose criminal activities and irresponsible behaviors that damage natural resources and have negative impacts on the quality of life for people in impoverished areas of the globe. Operating out of offices in the United States and the United Kingdom, Global Witness uses information gathered in its investigations to lobby governments for changes to policies that harm the environment and perpetuate poverty.

Green Belt Movement
Year founded: 1977
Web site: http://www.greenbeltmovement.org

The Green Belt Movement, established in Kenya by activist Wangari Maathai, is a grassroots organization aimed at getting people, especially women, directly involved in conservation efforts. It has sponsored the planting of trees across the country as a means of replenishing resources used for subsistence by Kenya's rural population. During the 1980's the program expanded to other countries in Africa, and Green Belt officials began actively engaging in efforts to protest irresponsible development in the region.

Greenpeace*
Year founded: 1971
Web site: http://www.greenpeace.org/international

Greenpeace is an international organization with affiliates in more than forty countries. It supports research and encourages lobbying efforts to identify and promote sustainable ecosystems, but it has also engaged in direct action to stop activities considered harmful to the environment.

Groupe Energies Renouvelables, Environnement, et Solidarités (GERES)
Year founded: 1976
Web site: http://www.geres.eu

GERES is a French nongovernmental organization working in France, Africa, and Asia to provide technical expertise to improve environmental conservation efforts, mitigate the effects of climate change, and improve lives through sustainable development activities. It promotes access to and efficient use of energy, develops plans for environmentally friendly waste management, and works to combat climate change. Its members work through local partnerships to facilitate environmental management and development of resources for indigenous populations.

International Network for Sustainable Energy (INFORSE)
Year founded: 1992
Web site: http://www.inforse.org

Headquartered in Denmark, INFORSE is a network of environmental organizations working to promote sustainable energy, protect the environment, and decrease poverty. An outgrowth of the 1992 Earth Summit, INFORSE works through regional offices in Asia, Africa, Europe, and the Americas to help develop programs aimed at transitioning the world's economies to 100 percent renewable energy sources by the year 2050. INFORSE has been active in creating public awareness of energy issues and in collaborative efforts with other organizations to establish and enforce standards relating to energy production and use.

International Union for Conservation of Nature* (IUCN)

Year founded: 1948
Web site: http://www.iucn.org

IUCN is dedicated to promoting the conservation of natural resources; its principal focus has been on identifying endangered species and promoting efforts to prevent extinctions.

Izaak Walton League

Year founded: 1922
Web site: http://www.iwla.org

The Izaak Walton League was founded by American sportsmen to promote protection of natural resources, especially rivers and lakes. The group has lobbied for programs to preserve and rejuvenate America's rivers, lakes, wetlands, and wilderness areas. It was instrumental in the creation of the Upper Mississippi River National Wildlife and Fish Refuge in 1924 and the passage of the Clean Water Act of 1972. Over the years the organization has also lobbied against illegal logging and for legislation to protect endangered species.

League of Conservation Voters* (LCV)

Year founded: 1969
Web site: http://www.lcv.org

LCV is an educational and advocacy group in the United States that lobbies to elect candidates to office who are likely to support environmental issues.

National Audubon Society*

Year founded: 1905
Web site: http://www.audubon.org

The National Audubon Society is a conservancy group noted for its advocacy of programs aimed at the preservation of bird species and habitats. It has been active in supporting bans on harmful chemicals and in creating wildlife sanctuaries.

National Council for Science and the Environment (NCSE)

Year founded: 1990
Web site: http://ncseonline.org

Originally known as the Committee for the National Institute for the Environment, NCSE is a U.S.-based nonprofit organization that works to assist policy makers responsible for making decisions about the environment by providing scientific data that they can use in forming judgments. NCSE supports research, disseminates information, and operates public education and outreach programs to communicate to the public accurate, scientifically based information about the environment.

National Wildlife Federation (NWF)

Year founded: 1936
Web site: http://www.nwf.org

Originally known as the General Wildlife Federation, NWF is the largest environmental educational and advocacy group in the United States. The federation's principal aims are to connect individuals with the natural world, protect and restore critical wildlife habitats, and work toward reversing trends in global warming. NWF members come from a wide variety of interests—sports enthusiasts, nature lovers, environmentalists, and others—but work collaboratively to find ways to balance the needs of human communities with those of the natural world. The group sponsors a number of educational activities, information programs, and conferences and has also partnered with other conservation organizations on a number of important environmental projects.

Natural Resources Defense Council* (NRDC)

Year founded: 1970
Web site: http://www.nrdc.org

NRDC is an advocacy group that has operated principally in the United States but has also branched out to other countries; it engages in lobbying activities and occasionally takes legal action to promote sound environmental policy or seek the prohibition of activities it considers detrimental to the environment.

Naturfreunde International (NFI)

Year founded: 1895
Web site: http://www.nfi.at

NFI (also known as Friends of Nature) is an international organization founded as an offshoot of the Social Democratic movement in Europe to promote appreciation for and responsible use of the region's natural resources. NFI initially encouraged recreation and tourism as a means of stimulating people to become familiar with the natural world; over time the organization has become effective and insistent in lobbying for responsible conservation and sustainable development, especially for regions that cross political boundaries.

Naturschutzbund Deutschland

Year founded: 1899
Web site: http://www.nabu.de

Naturschutzbund Deutschland (also known as the Nature and Biodiversity Conservation Union) is one of Germany's oldest and most widely respected conservation groups, a private nonprofit that works on conservation projects both within Germany and outside the country's borders. The group publishes periodic reports to inform the public about environmental issues, conducts education programs, and works closely with governmental agencies in crafting laws and policies affecting the environment.

Ocean Conservancy

Year founded: 1972
Web site: http://www.oceanconservancy.org

Known as the Center for Marine Conservation until 2001, the Ocean Conservancy is a nonprofit organization based in the United States that is interested in preserving and improving the world's marine resources. The group works to promote healthy and diverse ocean ecosystems and opposes practices that it considers to be threats to both marine and human life. The Ocean Conservancy has been active in efforts to restore sustainable American fisheries, protect wildlife from human activities, and encourage government reforms that can bring about improved stewardship for the oceans.

Organisationen for Vedvarende Energi (OVE)

Year founded: 1975
Web site: http://www.ove.org

OVE (also known as the Danish Organization for Renewable Energy) is an association of individuals, business groups, and educational institutions working to promote renewable energy for Denmark. Often associated with the antinuclear movement, OVE has initiated grassroots campaigns to lobby the Danish government to restrict the use of nonrenewable energy sources and support the development of energy sources that can be replenished. The organization conducts informational campaigns and has become involved in cooperative efforts with other environmental groups to tackle issues such as climate change and global warming.

People for the Ethical Treatment of Animals* (PETA)

Year founded: 1980
Web site: http://www.peta.org

PETA is the largest animal rights group in the world; it campaigns against various forms of animal cruelty, including exploitation of animals in medical research, entertainment, and the use of animals as food for humans.

Rainforest Action Network* (RAN)

Year founded: 1985
Web site: http://ran.org

RAN is an environmental group concerned principally with issues surrounding the sustainability of the world's forests; it has engaged in campaigns to pressure corporations to refrain from activities that would deplete habitat or permanently alter the condition of forest terrains worldwide.

Sierra Club*

Year founded: 1892
Web site: http://www.sierraclub.org

The Sierra Club promotes responsible use of the earth's ecosystems and engages in educational initiatives and lobbying activities to promote conservation and responsible use of natural resources.

Stockholm Environment Institute (SEI)

Year founded: 1989
Web site: http://www.sei-international.org

Although established by the government of Sweden, SEI is an independent organization performing research and developing policies that promote sustainable environments. It has offices in six countries in addition to Sweden, giving it an international reach. SEI researchers focus on overarching issues such as climate change, energy systems, ecosystem vulnerability, and governance issues, as well as specific matters such as water resources and air pollution. The group has influenced governmental policies through its work on sustainability modeling and vulnerability assessments.

Tellus Institute

Year founded: 1976
Web site: http://www.tellus.org

The Tellus Institute, headquartered in the United States, has focused on scientific investigation to advance what it sees as a necessary transition to a sustainable, equitable, and humane global civilization. With funding from a variety of sources, including governments, nongovernmental agencies, and corporations, Tellus Institute researchers have produced reports on topics such as water quality, energy issues, requirements for sustainable communities, corporate social responsibility, and climate change. The institute has consistently sought to promote what it calls the "Great Transition," a paradigm shift in cultural values away from materialism and consumerism to a greater sense of global citizenship in which individual fulfillment can be achieved in societies that provide sufficiently for all their members.

Union of Concerned Scientists* (UCS)

Year founded: 1969
Web site: http://www.ucsusa.org

UCS, which was founded by researchers to investigate the scientific ramifications of government policies regarding the environment, promotes research and lobbying on a wide range of issues, including energy policy, climate change, and developments in technology, especially military technology, that may potentially affect the environment.

U.S. Climate Action Partnership* (USCAP)

Year founded: 2007
Web site: http://www.us-cap.org

USCAP is an umbrella organization uniting efforts of businesses and environmental groups to encourage government action to reduce greenhouse gas emissions.

Wetlands International

Year founded: 1954
Web site: http://www.wetlands.org

Wetlands International, a conservation group headquartered in the Netherlands, began as the International Wildfowl Inquiry, dedicated to the protection of waterfowl. It gradually broadened its focus to include protection of wetlands and changed its name to International Waterfowl and Wetlands Research Bureau (IWRB). In 1991 this group merged with the Asian Wetland Bureau and Wetlands for the Americas and assumed its current name. Wetlands International works to protect and restore the world's wetlands through scientific investigation, educational programs, and advocacy initiatives to influence government policy regarding these areas.

Wilderness Society*

Year founded: 1935
Web site: http://wilderness.org

The Wilderness Society is an American group devoted to the preservation and responsible use of the nation's wilderness areas; it monitors policies and activities related to government-owned areas such as national forests and areas controlled by the Bureau of Land Management.

World Business Council for Sustainable Development (WBCSD)

Year founded: 1995
Web site: http://www.wbcsd.org

The WBCSD, formed by a merger of the Business Council for Sustainable Development and the World Industry Council for the Environment, is an association of more than two hundred companies interested in the relationship between business and the environment. Headquartered in Switzerland with offices in the United States and Belgium, the WBCSD promotes ecofriendly business practices and lobbies governments for policy changes that will help bring about both sustainable environments and economies. Group members operate on the philosophy that responsible policies regarding the environment are ultimately good for business as well.

World Resources Institute* (WRI)

Year founded: 1982
Web site: http://www.wri.org

WRI is a think tank based in Washington, D.C., whose members are concerned with protecting the earth's ecosystems and simultaneously enhancing people's lives. The group unites scientists and business leaders to conduct research, provide information to the public, and promote environmentally responsible development that improves communities worldwide.

Worldwatch Institute*

Year founded: 1974
Web site: http://www.worldwatch.org

The Worldwatch Institute is a research organization that collects and disseminates data on issues such as climate change, degradation of natural resources, and population growth in order to promote sustainable development.

World Wide Fund for Nature* (WWF)

Year founded: 1961

Web site: http://wwf.panda.org

Originally known as the World Wildlife Fund, WWF is an international organization that encourages conservation of natural resources (both animal and plant) and fosters sustainable development; it has actively promoted the harmonious relationship of humans with natural ecosystems worldwide.

Laurence W. Mazzeno

Biographical Dictionary of Key Figures in Environmentalism

The figures described below are important in the history of environmentalism. Asterisked individuals are covered in greater depth in separate essays within these volumes.

Abbey, Edward* (1927-1989): American environmental activist and author. The originality of Abbey's ideas regarding the preservation of nature, expressed with great eloquence in his writings, helped to increase awareness of environmental issues and inspired a radical environmental movement.

Abzug, Bella (1920-1998): American attorney, politician, and social activist. Abzug founded the Women's Environment and Development Organization (WEDO). She argued that women's perspectives are important to policy making concerning the environment and peace efforts, and in 1972 she was elected to the U.S. House of Representatives on a women's rights and peace platform. In 1997 Abzug launched a major initiative to link environmental contaminants to various health issues and other problems.

Adams, Ansel* (1902-1984): American photographer and environmental activist. Through his spectacular photographs and his advocacy, Adams helped to increase Americans' awareness of the beauty of the nation's wilderness areas and the importance of preserving that beauty.

Amory, Cleveland* (1917-1998): American author and animal rights activist. Amory's decades of activism for animal rights and animal protection saved thousands of animals from extermination and helped bring the issue of cruelty to animals into the public spotlight.

Attenborough, Sir David (b. 1926): English naturalist and television broadcaster. Attenborough is an esteemed presenter of nature documentaries in Great Britain. His first success was with the series *Life on Earth* in 1979. *The Living Planet* followed in 1984, and in 1990 *The Trials of Life* looked at animal behavior. His 2006 two-part program *Are We Changing Planet Earth?* addressed the issue of global warming.

Audubon, John James* (1785-1851): French American naturalist and wildlife artist. Through his unique paintings and his writings, Audubon demonstrated ecological relationships among organisms and set new standards for field observation. By illustrating the beauty of birds and animals, he helped to lay the foundation for a national environmental consciousness in the United States.

Bahro, Rudolf (1935-1997): German ecosocialist philosopher and author. Bahro, born in East Germany, joined the East German Socialist Unity Party in 1954. He argued that both the communist and the capitalist powers of the world engage in policies that promote unlimited economic development on a planet with limited resources, and that such policies exploit developing nations as well as nature. In the early 1990's Bahro taught classes in social ecology at Humboldt University in Berlin.

Bailey, Michael (b. 1954): Canadian conservationist and film and video producer. Bailey was an early member of Greenpeace and is especially interested in saving dolphins and whales. Through his involvement with the Climate Summit, a consortium begun by former U.S. vice president Al Gore, he uses video technology to educate the public on climate change and global warming. In 1997 Bailey was instrumental in pressuring the Japanese government to release dolphins in connection with the Japan-based ELSA Nature Conservancy.

Bancroft, Tom (b. 1951): American ecologist. Bancroft was named chief scientist at the National Audubon Society in 2007. In 2008 he testified before the U.S. Congress regarding the decline in bird populations around the world, citing environmental causes for the decline that include global warming, habitat destruction as the result of agriculture and encroachment of human settlements, and introduction of nonnative species.

Bari, Judi (1949-1997): American environmental activist. As a member of the radical environmental organization Earth First!, Bari worked as an organizer in a 1990 campaign to protect a redwood forest in Northern California and raise national awareness of the issue of logging in old-growth forests. During the campaign, she was seriously injured by a bomb that had been placed in her automobile by an unknown assailant.

Berry, Thomas (1914-2009): American Catholic priest, ecotheologian, and cultural historian. Berry taught that science can promote the religious sense of mystery in the universe and that a relationship with the earth is necessary for human spirituality as well as for human life itself. He is best known for his 1988 book *The Dream of the Earth*, which addresses the place of human beings in the cosmos.

Berry, Wendell* (b. 1934): American author of books on conservation and agrarianism. Berry's integrated professions of farmer, writer, and critic of industrial development have placed him among the major figures of the twentieth century in both conservation and literature.

Bookchin, Murray* (1921-2006): American ecological activist, author, and anarchist thinker. Bookchin, the creator of the concept of social ecology, suggested in the 1960's that the prosperity of the post-World War II United States had been bought at the price of serious harm to the environment.

Borlaug, Norman* (1914-2009): American plant pathologist and environmental activist. Borlaug, who became known as the father of the Green Revolution, pioneered efforts to develop high-yield crops to increase food production throughout the world.

Brockovich, Erin* (b. 1960): Legal clerk and environmental activist. Brockovich helped construct a legal case against the Pacific Gas and Electric Company for its role in polluting the drinking water of Hinkley, California, with chromium 6. The clients in the case received the largest settlement ever made in the United States in a direct-action lawsuit.

Brower, David* (1912-2000): American environmental activist and writer. Brower, who was vigorously involved in battles concerning environmental issues for more than fifty years, was one of the twentieth century's most influential and controversial environmental activists and writers.

Brown, Lester* (b. 1934): American agricultural scientist and author. Brown founded the Worldwatch Institute, an environmental think tank the mission of which is to analyze the state of the earth and to act as "a global early warning system."

Browner, Carol M. (b. 1955): American attorney and environmentalist. In 2009 President Barack Obama named Browner to be the first director of the White House Office of Energy and Climate Change Policy. Previously she had held the post of administrator of the U.S. Environmental Protection Agency (the first woman to do so), after serving as head of the Florida Department of Environmental Regulation, where she established herself as an aggressive

advocate for the environment in general and for wetlands in particular.

Brundtland, Gro Harlem* (b. 1939): Norwegian politician, physician, and environmental advocate. Brundtland has been called the "Green Goddess" because of the innovative environmental programs she initiated during her career as prime minister of Norway.

Burroughs, John* (1837-1921): American nature writer. Through his best-selling books, Burroughs raised Americans' awareness of the beauty of nature and the importance of preserving it.

Callicott, J. Baird (b. 1941): American environmental philosopher. Callicott is a leading interpreter of the work of American wilderness conservationist and environmental philosopher Aldo Leopold. Callicott maintains that there should be no sharp division between humans and nature; he asserts that the world should not be thought of as atomistic and mechanical, but rather as organic and systematic.

Carson, Rachel* (1907-1964): American author and environmentalist. As the author of *Silent Spring* (1962) and other best-selling books, Carson helped to spark the modern environmental movement.

Carter, Jimmy* (b. 1924): American politician who served as governor of Georgia and as president of the United States. During his political career, Carter made many decisions that demonstrated an environmentalist agenda.

Commoner, Barry* (b. 1917): American biologist and antinuclear activist. Commoner has raised public awareness of a number of important environmental issues, particularly regarding the use of energy resources, organic farming and pesticides, waste management, and toxic chemicals.

Comstock, Anna Botsford (1854-1930): American artist and conservationist. At a time when many Americans were moving from farms to cities, Comstock developed the idea that nature study is a necessary part of a full human existence; she saw such study as a nurse for human health, an elixir of youth, and even a cure for problems of school discipline. She wrote and illustrated several books on the topic of nature study and helped to create the nature study movement in the United States.

Cousteau, Jacques* (1910-1997): French explorer, conservationist, and filmmaker. Cousteau, one of the twentieth century's best-known explorers and conservationists, gained widespread attention for environmental issues, particularly those concerning the world's oceans.

Darling, Jay* (**1876-1962**): American cartoonist and wildlife conservationist. Darling was instrumental in starting the Federal Duck Stamp Program, which generated revenue to buy new lands to serve as waterfowl refuges, as authorized by the Migratory Bird Conservation Act.

Darwin, Charles* (**1809-1882**): English naturalist. Darwin's theory of evolution through natural selection, the dominant paradigm of the biological sciences, underlies the study of ecosystems.

Davis, Naomi (**b. 1956**): American environmental activist. Davis is the founder of the Chicago-based group Blacks in Green, which focuses on rallying African Americans to link environmental technology with job creation. The long-term goal of Blacks in Green is to convert one thousand blighted acres in Chicago into a mixed-income ecodevelopment that will use both solar power and wind power to generate electricity and will be a site of substantial horticultural activity.

Dubos, René* (**1901-1982**): French-born American bacteriologist and environmental writer. Through his writings, Dubos encouraged exploration of the manner in which humans interact with the environment.

Eaubonne, Françoise d'* (**1920-2005**): French novelist, poet, essayist, and journalist. D'Eaubonne, who coined the term "ecofeminism," influenced thinking about the relationship between ecology and feminist theory, both of which are built on the concept of the interconnectedness of living and nonliving beings.

Ehrlich, Paul R.* (**b. 1932**): American biologist and environmental philosopher. Ehrlich has published several books that have been influential in raising awareness and promoting action concerning such problems as the dangers of overpopulation and the possible effects of nuclear war.

Foreman, Dave* (**b. 1946**): American environmental activist and author. As one of the cofounders of the radical environmental group Earth First! and through his continued leadership in less radical organizations, Foreman has had a great deal of influence on the environmental movement.

Fossey, Dian* (**1932-1985**): American zoologist and author. Fossey influenced views of animal behavior and the need for animal protection through her writings about the mountain gorillas of Central Africa and her passionate attempts to save the gorillas from poachers.

Francis of Assisi, Saint (**c. 1181-1226**): Italian monk. In modern times Saint Francis of Assisi has become the patron saint of environmentalists because of his love of nature. Among his works admired by environmentalists is his "Canticle of the Creatures," in which he uses the expressions "Brother Sun" and "Sister Moon."

Freudenburg, William R. (**b. 1951**): American environmental researcher and educator. Freudenburg, who teaches environmental studies at the University of California, Santa Barbara, is known for his work on the relationship between environment and society. His areas of research include resource-dependent communities, the social impacts of environmental and technological change, and risk analysis. He has examined such topics as the social impacts of U.S. oil dependence and the polarizing nature of debates over protection of the northern spotted owl from logging in old-growth forest habitat.

Garnaut, Ross (**b. 1946**): Australian economist. Garnaut is best known for his influential report the *Garnaut Climate Change Review*, which examines the likely effects of climate change on the Australian economy. From 1981 to1983 he directed the Australia Economic Relations Research Project. He has also written widely on the industrializing economy of China, including the 2004 volume *China: Is Rapid Growth Sustainable?*

Gibbons, Euell* (**1911-1975**): American ethnobotanist and nature writer. Gibbons improved the public image of wild food foraging and thus of environmentalism in general, as his staid, avuncular image made environmentalism acceptable to Americans who had tended to perceive environmental activism as subversive.

Gibbs, Lois* (**b. 1951**): American environmental activist. Gibbs united her community by forming the Love Canal Homeowners Association and leading efforts to compel state and federal officials to relocate residents in her neighborhood whose homes were compromised by exposure to toxic waste.

Gibran, Kahlil (**1883-1931**): Lebanese American artist, philosopher, and writer. Gibran taught that humans are one with nature and that what human-kind does to the earth it does to itself. He is chiefly known for his 1923 book *The Prophet*, a series of philosophical poems written in English; many of the images in this work are taken from nature.

Gill, Frank B.: American ornithologist. From 1996 to 2004 Gill was senior vice president and director of science for the National Audubon Society. He is especially known as the author of the book *Ornithology*, which is considered the leading textbook in the field; several editions of the work have appeared since the first was published in 1990.

Goethe, Johann Wolfgang von (1749-1832): German philosopher, scientist, dramatist, poet, and novelist. Goethe held a holistic view of nature—that is, he believed that although humans are the crowning achievement of nature, they are a part of nature like any other part. He had a passionate respect for and even veneration of the natural world, and this holistic view penetrated every aspect of his literary as well as his scientific work.

Gore, Al* (b. 1948): American environmental activist and politician who served in both houses of Congress and as vice president of the United States. Through his activism and particularly his participation in the documentary film *An Inconvenient Truth*, Gore has brought worldwide attention to the problem of global warming.

Griffin, Susan (b. 1943): American ecofeminist author. Griffin writes poetry and nonfiction prose that makes connections between the mistreatment of women and the destruction of nature. Among her works are the book *Woman and Nature: The Roaring Inside Her* (1978), which explores how female anger can be a transforming force.

Hansen, James E.* (b. 1941): American climate change scientist. As a prominent climate scientist and activist, Hansen has been an important contributor to increased public awareness of global warming.

Hardin, Garrett* (1915-2003): American ecologist. Through his writings—in particular his widely read 1968 article "The Tragedy of the Commons"—Hardin raised awareness of the environmental problems caused by human overpopulation and overexploitation of resources.

Hemenway, Harriet (1858-1960): American socialite and activist for protection of birds. Hemenway began the Massachusetts Audubon Society in 1896 af-ter she realized that many thousands of birds were being slaughtered to provide feathers for women's hats. She convinced women to give up wearing feathers and campaigned for milliners to design featherless hats. By 1905 fifteen other states had formed Audubon Societies, which eventually combined to form the National Audubon Society.

Jeffers, Robinson (1887-1962): American poet. Jeffers is known for his philosophy of "inhumanism"—that is, his high regard for the nonhuman and his low regard for the human. He asserted that the universe is one being, a single organism, and that whatever human beings do to a landscape damages it.

Kress, Stephen W. (b. 1945): American ornithologist. Kress is an expert in seabird conservation and, as vice president for bird conservation for the National Audubon Society, has had extraordinary success leading that organization's Project Puffin seabird restoration program in Maine. He has also published several books that provide expert tips for creating bird-friendly habitats.

Leopold, Aldo* (1887-1948): American wilderness conservationist and environmental philosopher. Leopold, who has been called the father of modern wildlife management and ecology, applied his insightful concepts of ethics and philosophy to conservation strategies and thus helped raise awareness of environmental issues.

Lovejoy, Thomas E.* (b. 1941): American tropical biologist. Lovejoy is recognized for his contributions to conservation policy making. He is best known for developing creative solutions to issues of scientific concern, such as debt-for-nature swaps.

Lovelock, James* (b. 1919): English environmentalist and inventor. Lovelock is best known for his Gaia hypothesis, which suggests that the earth itself is the source of life and that all living things on the planet have coevolved and therefore are inextricably intertwined.

Lovins, Amory* (b. 1947): American physicist. Lovins, cofounder of the Rocky Mountain Institute, has worked to promote the use of sustainable and clean energy, particularly as a means to attain global stability and security.

Lubchenco, Jane (b. 1947): American marine ecologist. Lubchenco is the first woman to serve as the administrator of the National Oceanic and Atmospheric Administration. Previously, she was a member of a team of researchers who conducted the first National Academy of Sciences study on the policy implications of global warming and provided advice on that topic to President George H. W. Bush and the U.S. Congress.

Maathai, Wangari* (b. 1940): Kenyan environmentalist and social activist. A visionary and activist in the fight against deforestation in Africa and beyond, Maathai has spearheaded various initiatives that have resulted in the planting of billions of trees and have brought global attention to this critical environmental issue.

McHarg, Ian L. (1920-2001): Scottish American landscape architect. McHarg was a pioneer in the area of ecologically based landscape architecture. His book *Design with Nature,* first published in 1969, helped to create the field of ecological planning and showed how development can be integrated with aesthetics and environmental concern.

Malthus, Thomas Robert* (1766-1834): English political economist. The author of *An Essay on the Principle of Population, as It Affects the Future Improvement of Society* (1798), Malthus provided the direst explanations of the causes and consequences of population growth.

Manabe, Syukuro* (b. 1931): Japanese meteorological scientist. Manabe's research using computer modeling has improved humankind's understanding of the role that the oceans play in the global climate.

Marsh, George Perkins* (1801-1882): American statesman, diplomat, and author. Marsh's widely read book *Man and Nature: Or, Physical Geography as Modified by Human Action* (1864), a treatise on environmental history, became one foundation for the conservation and environmental movements of the twentieth century.

Marshall, Robert* (1901-1939): American forester and plant physiologist. Marshall influenced both government policy and public opinion through his numerous writings on the need for wilderness conservation and through his participation in the Wilderness Society, an organization he cofounded.

Mather, Stephen T.* (1867-1930): American conservationist. As the first director of the U.S. National Park Service, Mather personified the national parks movement during the early decades of the twentieth century.

Mendes, Chico* (1944-1988): Brazilian rubber tapper and trade union leader. Mendes spent his entire life working against the forces of environmental destruction in the Amazon forest in order to sustain a way of life for his fellow rubber tappers and other indigenous peoples of western Brazil. He earned international recognition as a defender of the Amazon ecosystem.

Molina, Mario* (b. 1943): Mexican chemist. Molina's pioneering work concerning the formation and catalytic decomposition of ozone in the stratosphere led to greater scientific attention to the issue of climate change.

Muir, John* (1838-1914): Scottish American naturalist, preservationist, and writer. Muir, one of America's most notable preservationists and a founder of the Sierra Club, introduced Americans to California's Sierra Nevada and worked hard to protect much of the region's wilderness, including Yosemite, against development.

Mumford, Lewis* (1895-1990): American historian and social philosopher. Although Mumford could not be called an environmental activist in the usual sense, his advocacy of garden cities, his criticism of urban sprawl and the directions taken by modern technology, and his opposition to the use of nuclear energy show his deep-seated concern for the human environment.

Nader, Ralph* (b. 1934): American political activist and consumer advocate. As a consumer advocate and as a Green Party political candidate, Nader has been a champion of the underdog—including the poor, the elderly, and members of minority groups—against corporate and political power structures in the United States.

Naess, Arne* (1912-2009): Norwegian philosopher. Naess's ideas, in particular his introduction of the concept of deep ecology, have had a great deal of influence on environmental philosophy and activism.

Olmsted, Frederick Law* (1822-1903): American landscape architect. Olmsted left a distinct mark on the American environment from New York City to the wilds of California. He synthesized a variety of experiences in his youth and young adulthood to become one of the greatest landscape designers in the history of the United States.

Osborn, Henry Fairfield, Jr.* (1887-1969): American naturalist and conservationist. Through his work with the New York Zoological Society and his writings, Osborn promoted the preservation of endangered species and their habitats and also raised public awareness of the dangers of human overpopulation.

Ostrom, Elinor* (b. 1933): American political scientist. Through her extensive empirical research, Ostrom has shown that it is not inevitable that shared resources will be depleted by overuse.

Passmore, John Arthur (1914-2004): Australian philosopher. Passmore argued that humans cannot continue to exploit the environment, but he also believed that those who say that nature has intrinsic value or that nature has rights of its own are irrational. According to Passmore, the value in nature lies in what it contributes to living things, including humans.

Pinchot, Gifford* (1865-1946): American conservationist and forester. As the first head of the U.S. Forest Service, Pinchot influenced national policy making concerning the conservation of natural resources as well as their management for human use.

Plumwood, Val (b. 1939): Australian ecofeminist author. Plumwood coined the term "human chauvinism" to point to the insensitivity that comes from the human tendency to dominate nature. In her 1993 book *Feminism and the Mastery of Nature*, Plumwood laments the Western world's lack of respect for nature and the idea that only humans can have any direct ethical significance.

Powell, John Wesley* (1834-1902): American geologist and explorer. Powell contributed significantly to scientific knowledge of the American West in the mid-nineteenth century, and his ideas regarding environmental policy are recognized as being ahead of their time.

Revkin, Andrew (b. 1956): American journalist. Revkin reported on the environment for *The New York Times* from 1995 to 2009, when he joined Pace University's Academy for Applied Environmental Studies. He has written on a wide range of environment-related subjects, including the destruction of the Amazon rain forest, the Asian tsunami of 2004, and the North Pole. In 2008 Revkin received Columbia University's John Chancellor Award for Excellence in Journalism, which is presented to reporters committed to writing about difficult stories.

Richards, Ellen Swallow (1842-1911): American environmental chemist. Richards was the first woman to receive a degree in science from the Massachusetts Institute of Technology. During the 1870's she supervised a major project for the newly established Massachusetts Board of Health that involved a sewage and water supply survey. Her research led to the creation of the first water-quality standards in the United States and the establishment of the world's first modern sewage treatment systems.

Roddick, Anita (1942-2007): English businesswoman and environmental activist. In 1976 Roddick founded the Body Shop, a company that sells cosmetics, soaps, and lotions made from natural ingredients and not tested on animals. Roddick designed the Body Shop stores to do more than sell merchandise—they would also serve as centers for education about social justice and environmental issues.

Rolston, Holmes, III (b. 1932): American environmental ethicist. Rolston is known for his writings on environmental ethics and on the relationship between science and religion. In 2003 he was awarded the Templeton Prize for his affirmation of life's spiritual dimension, and his Gifford lectures in 1997-1998 were published under the title *Genes, Genesis, and God: Values and Their Origins in Natural and Human History* (1999).

Roosevelt, Theodore* (1858-1919): American politician and conservationist who served as governor of New York and president of the United States. During his years as U.S. president, 1901-1909, Roosevelt did more to boost conservation efforts in the United States than any other president before him.

Rowell, Galen* (1940-2002): American nature photographer. One of the best-known American nature photographers, Rowell was also a serious mountaineer and helped to develop the field of participatory photography.

Rowland, Frank Sherwood* (b. 1927): American chemist. Rowland was the first person to discover that chlorofluorocarbons released into the atmosphere were destroying the protective ozone layer, and he was influential in the eventual move to ban production of these compounds.

Sale, Kirkpatrick* (b. 1937): American journalist, historian, and environmental writer. In a long career as journalist and activist, Sale has helped shape the modern environmental movement through his writings emphasizing human scale, bioregionalism, decentralization, and a thoroughgoing critique of technology and the idea of progress.

Schumacher, E. F.* (1911-1977): German British economist. Schumacher's promotion of nonmaterialist values and his writings emphasizing the importance of protecting resources while attending to the needs of humans had a great influence on the environmental movement in the 1970's and 1980's.

Schweitzer, Albert (1875-1965): German philosopher and physician. In 1913 Schweitzer established a hospital in west central Africa, in what is now Gabon; in doing so, he took care to preserve the surrounding forest and rejected the use of any technology that would degrade the environment. He believed that all life is precious and that although killing may at times be necessary, it is never ethical. Schweitzer won the Nobel Peace Prize for his work in equatorial Africa, and Rachel Carson dedicated *Silent Spring* (1962) to him.

Seattle (c. 1780-1866): Native American chief. Chief Seattle is known for the peaceful negotiations he carried out with white settlers in what is now the state of Washington. His words have become a source of inspiration for many environmentalists, particularly a speech in favor of ecological responsibility that has been attributed to him. In this speech he asked, "How can you buy or sell the sky—the warmth of the land? The idea is strange to us."

Shiva, Vandana (b. 1952): Indian ecofeminist and environmental activist. Shiva, a leading theoretical physicist in the ecology movement, has devoted herself to environmental activism. She has written numerous influential volumes on topics such as biodiversity, biopiracy, biopolitics, ecofeminism, globalization, and food security.

Silkwood, Karen* (1946-1974): American nuclear industry worker and union activist. Until her death in an automobile crash Silkwood was not widely known, but after the accident, which occurred under mysterious circumstances, many antinuclear activists saw her as a martyr to their cause.

Singer, Peter* (b. 1946): Australian philosopher and bioethicist. Singer, author of the 1975 book *Animal Liberation* and numerous other works in applied ethics, is considered by many to have launched the modern animal liberation movement.

Snyder, Gary* (b. 1930): American poet, essayist, and environmental activist. Snyder was one of the first writers to base his poetry, ethics, and spirituality in environmental ideas and values. He is one of the most influential figures in American nature writing.

Suzuki, David (b. 1936): Canadian genetics scientist and environmentalist. Suzuki is known for his activism regarding climate change and for his television and radio programs that have addressed various issues related to science and the environment. He became the host of the Canadian Broadcasting Corporation television series *The Nature of Things* in 1979; this widely viewed program has focused on the topics of nature, wildlife, and sustainable human societies. Suzuki was awarded the United Nations Environment Programme Medal for his 1985 series *A Planet for the Taking*, in which he called for a change in the way human beings relate to nature.

Swimme, Brian (b. 1950): American mathematical cosmologist. Swimme, who specializes in the study of the evolutionary dynamics of the universe, is director of the Center for the Story of the Universe at the California Institute of Integral Studies. His best-known work is his 1992 book *The Universe Story: From the Primordial Flaring Forth to the Ecozoic Era—A Celebration of the Unfolding of the Cosmos*, which he coauthored with Thomas Berry.

Tansley, Arthur G.* (1871-1955): English botanist. Tansley, who coined the term "ecosystem," published scholarly articles and books on natural processes that have become central to ecological theory.

Thoreau, Henry David* (1817-1862): American naturalist and philosopher. Best remembered as a persuasive advocate of nonviolent civil disobedience to protest unjust laws, Thoreau was also an early advocate of environmentalism.

Udall, Stewart L. (1920-2010): American politician who served as U.S. secretary of the interior. During his term as secretary of the interior (1961-1969), Udall acquired for the federal government 1.56 million hectares (3.85 million acres) of new lands, including four national parks and six national monuments. He was an early supporter of Rachel Carson and was instrumental in the signing of the Radiation Exposure Compensation Act of 1990, which provided benefits for persons sickened by radiation as the result of American nuclear weapons testing.

Vernadsky, Vladimir* (1863-1945): Russian geochemist and mineralogist. Vernadsky developed the concepts of the biosphere and the noosphere, and his 1926 book *The Biosphere* inspired a new vision of humankind's role in shaping the earth's environment.

Watson, Paul* (b. 1950): Canadian animal rights and environmental activist. A dissident Greenpeace member and experienced sailor, Watson founded the Sea Shepherd Conservation Society, one of the world's most aggressive environmental organizations. He and his organization have mounted vigilante (but deliberately nonlethal) attacks against the efforts of seal hunters, whalers, and drift-net fishers.

Watt, James* (b. 1938): American attorney who served as U.S. secretary of the interior. Watt was labeled a major antienvironmentalist during his tenure as secretary of the interior; he was frequently accused of using his office to weaken environmental policies that fell under his domain of authority.

Wayburn, Edgar (1906-2010): American environmentalist and physician. Wayburn joined the Sierra Club in 1939 and was elected to that organization's board of directors in 1961. During the 1960's he was elected president of the Sierra Club five times. His activism has been credited as being an important element in the passage of the Alaska National Interest Lands Conservation Act in 1980. In 1999, when Wayburn received the Presidential Medal of Freedom, President Bill Clinton stated that "over the course of more than a half century, both as president of the Sierra Club and as a private citizen, [Wayburn] has saved more of our wilderness than any person alive."

Weyler, Rex (b. 1947): American Canadian author, journalist, and ecologist. During the 1970's Weyler served as director of the original Greenpeace Foundation. He was a cofounder of Greenpeace International in 1979 and is the author of *Blood of the Land: The Government and Corporate War Against the American Indian Movement* (1982), a book about Native Americans' rights, and *Greenpeace: How a Group of Ecologists, Journalists, and Visionaries Changed the World* (2004).

White, Lynn Townsend, Jr.* (1907-1987): American historian and author. White argued that religion—medieval Christianity, in particular—played a significant role in the environmental crisis that was becoming apparent during the late 1960's. His controversial thesis was influential in spawning several movements in environmentalism, including ecotheology.

Wilmut, Ian* (b. 1944): English reproductive biologist. Wilmut, one of the world's foremost authorities on biotechnology and genetic engineering, conducted a landmark cloning experiment in 1996 that produced Dolly the sheep, the first mammal clone ever produced from adult cells.

Wilson, Edward O.* (b. 1929): American evolutionary biologist and author. An evolutionary biologist with extensive field experience, especially in studying ants, Wilson became a political target during the 1970's because of his application of sociobiology to humans. More recently, he has championed biological diversity and has worked to save species from extinction.

Wolman, Abel* (1892-1989): American sanitary engineer. Wolman was a pioneer in the field of sanitary engineering. His innovations and advocacy influenced the establishment of sound water-resource management strategies by American cities during the twentieth century.

Wright, Frank Lloyd (1867-1959): American architect. Wright believed that daily communion with nature is essential, and in his architectural designs he made the landscape a central element. One of his most famous buildings is the Guggenheim Museum in New York City, which brings light from the sky down into its depths. Wright's "prairie-style" houses are built at the edges of their lots to allow room for gardens in their centers.

Zahniser, Howard Clinton* (1906-1964): American conservationist and nature writer. Zahniser was an influential figure in the wilderness preservation movement of the mid-twentieth century. In addition to serving as executive secretary of the Wilderness Society for more than twenty years, he authored the landmark Wilderness Act of 1964.

Winifred O. Whelan

U.S. Federal Laws Concerning the Environment

Asterisks on entries below indicate that these laws are discussed in more depth in individual essays within these volumes.

Alaska National Interest Lands Conservation Act* (1980): Designated certain public lands in Alaska as units of the national park and national forest systems, the national wildlife refuge system, the National Wild and Scenic Rivers System, and the National Wilderness Preservation System. Provided comprehensive management guidance for all public lands in Alaska, including provisions regarding wilderness; subsistence; transportation and utility corridors; oil and gas leasing; mining; public access; hunting, trapping, and fishing; and implementation of the Alaska Native Claims Settlement Act (1971).

Antiquities Act* (Act for the Preservation of American Antiquities; 1906): Authorized permits for legitimate archaeological investigations and penalties for taking or destroying antiquities. Authorized presidents to protect all forms of American historical sites (natural, scientific, and archaeological) by proclaiming them to be national monuments.

Clean Air Act* (1963): Regulated air emissions from area, stationary, and mobile sources. Amendments in 1970 authorized the Environmental Protection Agency to create national air-quality standards to protect health and the environment, and required states to prepare and submit plans to implement clean air standards. Amendments in 1977 extended the deadline for areas that had not reached compliance levels by 1975. Amendments in 1990 addressed such issues as acid rain, ozone depletion, and air toxins.

Clean Water Act* (1977): An amendment to the Federal Water Pollution Control Act of 1972; prohibited discharge of any pollutant from a source point into navigable waters of the United States unless a special permit had been obtained from the Environmental Protection Agency. Amendment in 1987 (the Water Quality Act) included provisions for toxic pollutants, citizen suits, and funding of sewage treatment plants.

Coastal Zone Management Act* (1972): Provided for management of the nation's coastal resources, including the Great Lakes, and balanced economic developments with environmental conservation. Encouraged states and Native American tribal governments to preserve, protect, develop, and restore or enhance valuable national coastal resources. Amendments in 1990 called on states and tribes to develop and implement coastal nonpoint pollution control programs.

Emergency Planning and Community Right-to-Know Act (1986): Provided assistance to local communities in protecting the environment and public health and safety from chemical hazards. Required each state to create a State Response Commission, to divide itself into districts, and to appoint an Emergency Planning Committee for each district. Required both commissions to provide the community with information on chemical hazards that might affect the public, and required the dissemination of procedures to be followed in the event of emergency hazardous situations.

Endangered Species Act* (1973): Repealed the Endangered Species Conservation Act (1969), which had amended the Endangered Species Preservation Act (1966). Implemented the Convention on International Trade in Endangered Species of Wild Fauna and Flora in 1973 and the 1940 Convention on Nature Protection and Wildlife Preservation in the Western Hemisphere. Provided for the conservation of ecosystems on which threatened and endangered species of fish, wildlife, and plants depend and required the U.S. Fish and Wildlife Service to designate which plants and animals were threatened or endangered. Prohibited activities that could have adverse effects on endangered or threatened species and their habitats.

Energy Policy and Conservation Act* (1975): Enacted to help cut the amount of energy consumed by various industrial and consumer products. Introduced corporate average fuel economy (CAFE) standards for automobile manufacturers, extended oil price controls to 1979, and created the Strategic Petroleum Reserve. Amended in part by the Alternative Fuels Act (1988), which encouraged the development, production, and demonstration of alternative motor fuels and vehicles that could run on such fuels.

Federal Insecticide, Fungicide, and Rodenticide Act* (1947): Amendments in 1972 prohibited the sale, distribution, or use of pesticides that might adversely affect threatened or endangered species. Required users of pesticides to register when they purchase pesticides and to take and pass a certification examination in order to apply pesticides. Required that all pesticides used in the United States be approved and licensed by the Environmental Protection Agency.

Federal Land Policy and Management Act* (1976): Guided the Bureau of Land Management in the management, protection, development, and enhancement of public lands. Required the agency to manage lands for multiple uses and for sustained yield of resources for both present and future generations.

Fish and Wildlife Act* (1956): Directed the secretary of the interior to develop the policies and procedures necessary for carrying out fish and wildlife laws and to research and report on fish and wildlife matters. Authorized the administrator of the Environmental Protection Agency to undertake studies on the effects of insecticides, herbicides, fungicides, and other pesticides on fish and wildlife resources to determine the amounts, percentages, and formulations of chemicals injurious to fish and wildlife and thus the amounts, percentages, and formulations that could be used without losses of fish and wildlife from spraying, dusting, or other treatments using these chemicals.

Flood Control Act* (1944): Limited the authorization and construction of navigation, flood control, and other water projects to those having significant benefits for navigation and that could be operated consistent with other river uses.

General Mining Act* (1872): Authorized and governed prospecting and mining for economic minerals on public lands. Amendments and new acts that superseded the 1872 provisions established regulations on the removal and use of resources such as oil and natural gas and also provided protection for national parks and other historic sites.

Marine Mammal Protection Act* (1972): Enacted in partial response to growing concerns that certain marine mammals were in danger of extinction or depletion as a result of human activities. Prohibited, with certain specified exceptions, the act of hunting, killing, capture, and harassment of mammals in U.S. waters and by American citizens on the high seas, and the importation of marine mammals and their products into the United States.

Marine Protection, Research, and Sanctuaries Act (Ocean Dumping Act; 1972): Prohibited all municipal sewage sludge and industrial waste dumping into the ocean after December 31, 1991.

Migratory Bird Treaty Act (1918): Implemented the 1916 convention between the United States and Great Britain and incorporated the provisions in the 1913 Migratory Bird Act (also known as the Weeks-McLean Act). Made it unlawful to pursue, hunt, take, capture, kill, or sell more than eight hundred species of birds that migrate between the United States and Canada. Scope of the act's protection expanded after similar conventions were signed between the United States and Mexico, and between Japan and the Soviet Union.

National Environmental Policy Act* (1970): Required federal agencies to take environmental factors into consideration before undertaking any major action such as construction of new highways, airports, or military complexes. Required the government to disclose the probable environmental effects of all projects by completing environmental assessments and environmental impact statements. Involved federal courts in environmental questions, expanded judicial review into agency decisions, and gave Congress additional power over matters concerning the environment.

National Forest Management Act* (1976): An amendment to the Forest and Rangeland and Renewable Resources Planning Act (1974). Required the secretary of agriculture to assess forestlands, develop a management program based on multiple-use, sustained-yield principles, and implement a resource management plan for each unit of the national forest system.

National Landscape Conservation System Act (2009): Unified individual units as a public lands system to ensure that the conservation system was appropriately managed, funded, and protected for future generations. Included in the Omnibus Public Land Management Act (2009), which also added newly designated sites to the system, including a national monument and three national conservation areas.

National Trails System Act* (1968): Authorized a national system of trails and defined four categories of national trails. Provided for outdoor recreation opportunities and promoted the preservation of access to the wilderness areas and historic resources of the nation.

National Wildlife Refuge System Administration Act* (1966): Provided guidelines and directives for administration and management of all areas in the system, including wildlife refuges, areas for the protection and conservation of fish and wildlife threatened with extinction, game ranges, wildlife management areas, and waterfowl production areas. Amended by the National Wildlife Refuge System Improvement Act (1997).

Occupational Safety and Health Act (1970): Required employers to provide workers with safe workplaces, addressing such issues as the workplace use of toxic and hazardous substances. Required that state workplace safety and health acts meet or exceed federal requirements.

Oil Pollution Act (1990): Regulated oil storage facilities and required oil-carrying vessels to submit plans for response in the case of large discharges. Required the development of area contingency plans to prepare and plan for oil spill response on a regional scale.

Organic Foods Production Act (1990): Required the U.S. Department of Agriculture to develop national standards for organically produced agricultural products. Required producers to be in full compliance with the resulting standards by October 20, 2002, to be allowed to use the word "organic" in marketing.

Pollution Prevention Act (1990): Focused on source reduction of pollution through requiring cost-effective changes in production, operation, and raw-material use by both private industry and the government. Included provisions regarding recycling and sustainable agricultural practices that increase efficiency in the use of energy, water, and other natural resources.

Reclamation Act* (Newlands Act; 1902): Established the Reclamation Fund and provided for the construction of irrigation projects in the arid lands of the American West. The newly irrigated land would be sold and money put into a revolving fund for similar projects.

Resource Conservation and Recovery Act* (1976): Contained provisions for the control of the generation, transportation, treatment, storage, and disposal of hazardous waste and the management of nonhazardous solid wastes. Amendments in 1984 established the national hazardous waste management program and required the Environmental Protection Agency to identify hazardous waste characteristics and list specific substances as hazardous wastes. Amendments in 1986 addressed problems related to underground storage of petroleum and other hazardous substances.

Safe Drinking Water Act* (1974): Addressed issues relating to the quality and safety of drinking water. Authorized the Environmental Protection Agency to establish purity standards for both underground and surface sources of water for human use.

Small Business Liability Relief and Brownfields Revitalization Act (2002): Provided funds to assess and clean up brownfields (property affected by hazardous substances) and provided funds to enhance brownfield response programs of states and Native American tribal governments.

Superfund* (Comprehensive Environmental Response, Compensation, and Liability Act; 1980): Addressed the handling of hazardous waste sites, accidents, spills, and other emergency releases of pollutants or contaminants. Authorized the Environmental Protection Agency (EPA) to locate the parties responsible for any release and enforce their cooperation in the cleanup. Required the EPA to do cleanup if the releasing parties could not be found or refused to cooperate, but allowed the EPA to recover the costs of the action from those involved. Directed the EPA to revise its Hazard Ranking System and take into account degree of risk, human health, and the environment when placing uncontrolled waste sites on the National Priorities List. Reauthorized by the Superfund Amendments and Reauthorization Act (1986).

Surface Mining Control and Reclamation Act* (1977): Established mandatory uniform standards for surface mining and required minimized adverse impacts on fish, wildlife, and related environmental values. Created a fund for reclaiming and restoring land and water resources adversely affected by coal-mining practices.

Taylor Grazing Act* (1934): Enacted to stop injury to rangelands caused by overgrazing; to provide for the lands' orderly use, improvement, and development; and to stabilize the livestock industry dependent on the public rangelands. Authorized the secretary of the interior to establish grazing districts on public lands and to develop regulations necessary to administer the districts.

Toxic Substances Control Act (1976): Provided for the testing, regulation, and screening of all chemicals produced in or imported into the United

States before they reach the consumer market-place. Required the tracking of chemicals that pose health or environmental hazards and provided for the implementation of cleanup procedures in the case of contamination by toxic materials.

Uranium Mill Tailings Radiation Control Act* (1978): Gave the U.S. Department of Energy the responsibility of stabilizing, disposing, and controlling uranium mill tailings and other radiation-contaminated material at twenty-four uranium mill processing sites located across ten states and at more than five thousand associated properties.

Wild and Scenic Rivers Act* (1968): Created the National Wild and Scenic Rivers System to preserve select rivers with outstanding scenic, recreational, geologic, fish and wildlife, historic, cultural, or other important values in free-flowing conditions for the benefit of present and future generations.

Wilderness Act* (1964): Established the National Wilderness Preservation System and specified criteria for inclusion in the system. Made eligible every roadless area of 2,023 hectares (5,000 acres) or more, every roadless island within the national wildlife refuge and national park systems, and national forestlands.

Anh Tran

Directory of U.S. National Parks

ALASKA

Denali National Park and Preserve
P.O. Box 9
Denali Park, AK 99755
http://www.nps.gov/dena

Gates of the Arctic National Park and Preserve
P.O. Box 30
Bettles, AK 99726
http://www.nps.gov/gaar

Glacier Bay National Park and Preserve
P.O. Box 140
Gustavus, AK 99826
http://www.nps.gov/glba

Katmai National Park and Preserve
P.O. Box 7
King Salmon, AK 99613
http://www.nps.gov/katm

Kenai Fjords National Park
P.O. Box 1727
Seward, AK 99664
http://www.nps.gov/kefj

Kobuk Valley National Park
P.O. Box 1029
Kotzebue, AK 99752
http://www.nps.gov/kova

Lake Clark National Park and Preserve
240 West Fifth Avenue, Suite 236
Anchorage, AK 99501
http://www.nps.gov/lacl

Wrangell-St. Elias National Park and Preserve
P.O. Box 439
Copper Center, AK 99588
http://www.nps.gov/wrst

ARIZONA

Grand Canyon National Park
P.O. Box 129
Grand Canyon, AZ 86023
http://www.nps.gov/grca

Petrified Forest National Park
P.O. Box 2217
Petrified Forest, AZ 86028
http://www.nps.gov/pefo

Saguaro National Park
3693 South Old Spanish Trail
Tucson, AZ 85730
http://www.nps.gov/sagu

ARKANSAS

Hot Springs National Park
101 Reserve Street
Hot Springs, AR 71901
http://www.nps.gov/hosp

CALIFORNIA

Channel Islands National Park
1901 Spinnaker Drive
Ventura, CA 93001
http://www.nps.gov/chis

Death Valley National Park
P.O. Box 579
Death Valley, CA 92328
http://www.nps.gov/deva

Joshua Tree National Park
74485 National Park Drive
Twentynine Palms, CA 92277
http://www.nps.gov/jotr

Kings Canyon National Park
47050 Generals Highway
Three Rivers, CA 93271
http://www.nps.gov/seki

Lassen Volcanic National Park
P.O. Box 100
Mineral, CA 96063
http://www.nps.gov/lavo

Redwood National and State Parks
1111 Second Street
Crescent City, CA 95531
http://www.nps.gov/redw

Sequoia National Park
47050 Generals Highway
Three Rivers, CA 93271
http://www.nps.gov/seki

Yosemite National Park
P.O. Box 577
Yosemite National Park, CA 95389
http://www.nps.gov/yose

Colorado

Black Canyon of the Gunnison National Park
102 Elk Creek
Gunnison, CO 81230
http://www.nps.gov/blca

Great Sand Dunes National Park and Preserve
11500 Highway 150
Mosca, CO 81146
http://www.nps.gov/grsa

Mesa Verde National Park
P.O. Box 8
Mesa Verde, CO 81330
http://www.nps.gov/meve

Rocky Mountain National Park
1000 Highway 36
Estes Park, CO 80517
http://www.nps.gov/romo

Florida

Biscayne National Park
9700 SW 328 Street
Homestead, FL 33033
http://www.nps.gov/bisc

Dry Tortugas National Park
P.O. Box 6208
Key West, FL 33041
http://www.nps.gov/drto

Everglades National Park
40001 State Road 9336
Homestead, FL 33034
http://www.nps.gov/ever

Hawaii

Haleakalā National Park
P.O. Box 369
Makawao, HI 96768
http://www.nps.gov/hale

Hawaii Volcanoes National Park
P.O. Box 52
Hawaii National Park, HI 96718
http://www.nps.gov/havo

Kentucky

Mammoth Cave National Park
1 Mammoth Cave Parkway
P.O. Box 7
Mammoth Cave, KY 42259
http://www.nps.gov/maca

Maine

Acadia National Park
P.O. Box 177
Bar Harbor, ME 04609
http://www.nps.gov/acad

Michigan

Isle Royale National Park
800 East Lakeshore Drive
Houghton, MI 49931
http://www.nps.gov/isro

Minnesota

Voyageurs National Park
3131 Highway 53
International Falls, MN 56649
http://www.nps.gov/voya

Montana

Glacier National Park
P.O. Box 128
West Glacier, MT 59936
http://www.nps.gov/glac

NEVADA

Great Basin National Park
100 Great Basin National Park
Baker, NV 89311
http://www.nps.gov/grba

NEW MEXICO

Carlsbad Caverns National Park
3225 National Parks Highway
Carlsbad, NM 88220
http://www.nps.gov/cave

NORTH DAKOTA

Theodore Roosevelt National Park
P.O. Box 7
Medora, ND 58645
http://www.nps.gov/thro

OHIO

Cuyahoga Valley National Park
15610 Vaughn Road
Brecksville, OH 44141
http://www.nps.gov/cuva

OREGON

Crater Lake National Park
P.O. Box 7
Crater Lake, OR 97604
http://www.nps.gov/crla

SOUTH CAROLINA

Congaree National Park
100 National Park Road
Hopkins, SC 29061
http://www.nps.gov/cong

SOUTH DAKOTA

Badlands National Park
25216 Ben Reifel Road
P.O. Box 6
Interior, SD 57750
http://www.nps.gov/badl

Wind Cave National Park
26611 U.S. Highway 385
Hot Springs, SD 57747
http://www.nps.gov/wica

TENNESSEE

Great Smoky Mountains National Park
107 Park Headquarters Road
Gatlinburg, TN 37738
http://www.nps.gov/grsm

TEXAS

Big Bend National Park
P.O. Box 129
Big Bend National Park, TX 79834
http://www.nps.gov/bibe

Guadalupe Mountains National Park
400 Pine Canyon Road
Salt Flat, TX 79847
http://www.nps.gov/gumo

UTAH

Arches National Park
P.O. Box 907
Moab, UT 84532
http://www.nps.gov/arch

Bryce Canyon National Park
P.O. Box 640201
Bryce Canyon, UT 84717
http://www.nps.gov/brca

Canyonlands National Park
2282 SW Resource Boulevard
Moab, UT 84532
http://www.nps.gov/cany

Capitol Reef National Park
HC 70 Box 15
Torrey, UT 84775
http://www.nps.gov/care

Zion National Park
Springdale, UT 84767
http://www.nps.gov/zion

VIRGINIA

Shenandoah National Park
3655 U.S. Highway 211 East
Luray, VA 22835
http://www.nps.gov/shen

WASHINGTON

Mount Rainier National Park
55210 238th Avenue East
Ashford, WA 98304
http://www.nps.gov/mora

North Cascades National Park
810 State Route 20
Sedro-Woolley, WA 98284
http://www.nps.gov/noca

Olympic National Park
600 East Park Avenue
Port Angeles, WA 98362
http://www.nps.gov/olym

WYOMING

Grand Teton National Park
P.O. Drawer 170
Moose, WY 83012
http://www.nps.gov/grte

Yellowstone National Park
P.O. Box 168
Yellowstone National Park, WY 82190
http://www.nps.gov/yell

NATIONAL PARK SERVICE REGIONAL OFFICES

Alaska Area Region
Regional Director, National Park Service
240 West Fifth Avenue
Anchorage, AK 99501
(907) 644-3510

Intermountain Region
Regional Director, National Park Service
12795 Alameda Parkway
Denver, CO 80225
(303) 969-2500

Midwest Region
Regional Director, National Park Service
601 Riverfront Drive
Omaha, NE 68102
(402) 661-1736

National Capital Region
Regional Director, National Park Service
1100 Ohio Drive SW
Washington, DC 20242
(202) 619-7222

Northeast Region
Regional Director, National Park Service
U.S. Custom House
200 Chestnut Street, Fifth Floor
Philadelphia, PA 19106
(215) 597-7013

Pacific West Region
Regional Director, National Park Service
One Jackson Center
1111 Jackson Street, Suite 700
Oakland, CA 94607
(510) 817-1300

Southeast Region
Regional Director, National Park Service
100 Alabama Street SW
1924 Building
Atlanta, GA 30303
(404) 507-5600

Major World National Parks and Protected Areas

This appendix lists a selection of the world's most important protected areas and national parks, noting the year of establishment and approximate area of each. The list is organized alphabetically by continent and by country within continents. Many of the sites mentioned here are designated by the United Nations Educational, Scientific, and Cultural Organization (UNESCO) as World Heritage Sites; more information on these protected areas is available at the official UNESCO Web site devoted to the World Heritage List: http://whc.unesco.org/en/list. Web site addresses for individual U.S. parks are available in the Directory of U.S. National Parks appendix that appears in this volume.

AFRICA

CAMEROON

Dja Faunal Reserve
Year established: 1950
Area: 5,260 square kilometers (2,031 square miles)

CENTRAL AFRICAN REPUBLIC

Manovo-Gounda St. Floris National Park (World Heritage Site)
Year established: 1988
Area: 17,400 square kilometers (6,718 square miles)

CÔTE D'IVOIRE

Comoé National Park (World Heritage Site)
Year established: 1983
Area: 11,493 square kilometers (4,437 square miles)

DEMOCRATIC REPUBLIC OF THE CONGO

Garamba National Park
Year established: 1938
Area: 4,920 square kilometers (1,900 square miles)

Kahuzi-Biéga National Park
Year established: 1970
Area: 6,000 square kilometers (2,317 square miles)

Okapi Wildlife Reserve
Year established: 1992
Area: 13,726 square kilometers (5,300 square miles)

Salonga National Park
Year established: 1970
Area: 36,000 square kilometers (13,900 square miles)

Virunga National Park
Year established: 1925
Area: 7,800 square kilometers (3,012 square miles)

ETHIOPIA

Simien National Park
Year established: 1969
Area: 220 square kilometers (85 square miles)

KENYA

Lake Turkana National Parks (World Heritage Site)
Year established: 1997
Area: 1,615 square kilometers (624 square miles)

Mount Kenya National Park/Natural Forest (World Heritage Site)
Year established: 1997
Area: 1,420 square kilometers (548 square miles)

MADAGASCAR

Rainforests of the Atsinanana (World Heritage Site)
Year established: 2007
Area: 4,797 square kilometers (1,852 square miles)

NAMIBIA

Etosha National Park
Year established: 1975
Area: 22,270 square kilometers (8,598 square miles)

NIGER

Aïr and Ténéré Natural Reserves (World Heritage Site)
Year established: 1991
Area: 77,360 square kilometers (29,869 square miles)

SENEGAL

Niokolo-Koba National Park
Year established: 1954
Area: 9,130 square kilometers (3,525 square miles)

SOUTH AFRICA

Cape Floral Region Protected Areas (World Heritage Site)
Year established: 2004
Area: 5,530 square kilometers (2,135 square miles)

Golden Gate Highlands National Park
Year established: 1963
Area: 340 square kilometers (131 square miles)

iSimangaliso Wetland Park (World Heritage Site)
Year established: 1999
Area: 3,280 square kilometers (1,266 square miles)

Kruger National Park
Year established: 1926
Area: 18,989 square kilometers (7,332 square miles)

Vredefort Dome (World Heritage Site)
Year established: 2005
Area: 300 square kilometers (116 square miles)

TANZANIA

Kilimanjaro National Park
Year established: 1973
Area: 753 square kilometers (291 square miles)

Ngorongoro Conservation Area
Year established: 1959
Area: 8,288 square kilometers (3,200 square miles)

Selous Game Reserve
Year established: 1922
Area: 44,800 square kilometers (17,297 square miles)

Serengeti National Park
Year established: 1951
Area: 14,760 square kilometers (5,700 square miles)

UGANDA

Bwindi Impenetrable National Park
Year established: 1991
Area: 331 square kilometers (128 square miles)

Rwenzori Mountains National Park
Year established: 1991
Area: 998 square kilometers (385 square miles)

ZAMBIA

Mosi-oa-Tunya/Victoria Falls (World Heritage Site)
Year established: 1989
Area: 88 square kilometers (34 square miles)

ZIMBABWE

Mana Pools National Park, Sapi and Chewore Safari Areas (World Heritage Site)
Year established: 1984
Area: 6,766 square kilometers (2,612 square miles)

ASIA

CHINA

Huanglong Scenic and Historic Interest Area (World Heritage Site)
Year established: 1992
Area: 600 square kilometers (232 square miles)

Sichuan Giant Panda Sanctuaries-Wolong, Mt. Siguniang and Jiajin Mountains (World Heritage Site)
Year established: 2006
Area: 9,245 square kilometers (3,570 square miles)

South China Karst (World Heritage Site)
Year established: 2007
Area: 47,600 square kilometers (18,378 square miles)

Three Parallel Rivers of Yunnan Protected Areas (World Heritage Site)
Year established: 2003
Area: 17,000 square kilometers (6,564 square miles)

Wulingyuan Scenic and Historic Interest Area (World Heritage Site)
Year established: 1992
Area: 264 square kilometers (102 square miles)

INDIA

Great Nicobar Biosphere Reserve
Year established: 1989
Area: 885 square kilometers (342 square miles)

Kaziranga National Park
Year established: 1974
Area: 430 square kilometers (166 square miles)

Manas Wildlife Sanctuary (World Heritage Site)
Year established: 1985
Area: 500 square kilometers (193 square miles)

Nanda Devi and Valley of Flowers National Parks (World Heritage Site)
Year established: 1988
Area: 718 square kilometers (277 square miles)

Sundarbans National Park
Year established: 1984
Area: 1,330 square kilometers (514 square miles)

INDONESIA

Komodo National Park
Year established: 1980
Area: 1,733 square kilometers (669 square miles)

Lorentz National Park
Year established: 1997
Area: 25,056 square kilometers (9,674 square miles)

Tropical Rainforest Heritage of Sumatra (World Heritage Site)
Year established: 2004
Area: 25,951 square kilometers (10,020 square miles)

Ujung Kulon National Park
Year established: 1980
Area: 1,206 square kilometers (466 square miles)

JAPAN

Shirakami-Sanchi (World Heritage Site)
Year established: 1993
Area: 169 square kilometers (65 square miles)

KAZAKHSTAN

Saryarka-Steppe and Lakes of Northern Kazakhstan (World Heritage Site)
Year established: 2008
Area: 4,503 square kilometers (1,739 square miles)

MALAYSIA

Gunung Mulu National Park
Year established: 1974
Area: 529 square kilometers (204 square miles)

Kinabalu National Park
Year established: 1964
Area: 754 square kilometers (291 square miles)

MONGOLIA

Uvs Nuur Basin (World Heritage Site)
Year established: 2003
Area: 8,981 square kilometers (3,468 square miles)

NEPAL

Chitwan National Park
Year established: 1973
Area: 932 square kilometers (360 square miles)

Sagarmatha National Park
Year established: 1976
Area: 1,148 square kilometers (443 square miles)

PHILIPPINES

Tubbataha Reefs Natural Park (World Heritage Site)
Year established: 1993
Area: 1,300 square kilometers (502 square miles)

REPUBLIC OF KOREA (SOUTH KOREA)

Jeju Volcanic Island and Lava Tubes (World Heritage Site)
Year established: 2007
Area: 95 square kilometers (37 square miles)

RUSSIA

Central Sikhote-Alin (World Heritage Site)
Year established: 2001
Area: 15,539 square kilometers (6,000 square miles)

Golden Mountains of Altai (World Heritage Site)
Year established: 1998
Area: 16,115 square kilometers (6,222 square miles)

Lake Baikal (World Heritage Site)
Year established: 1996
Area: 88,000 square kilometers (33,977 square miles)

Natural System of Wrangel Island Reserve (World Heritage Site)
Year established: 2004
Area: 9,163 square kilometers (3,538 square miles)

Volcanoes of Kamchatka (World Heritage Site)
Year established: 1996
Area: 38,302 square kilometers (14,788 square miles)

THAILAND

**Thungyai-Huai Kha Khaeng Wildlife Sanctuaries
(World Heritage Site)**
Year established: 1991
Area: 5,775) square kilometers (2,230 square miles)

VIETNAM

Phong Nha-Ke Bang National Park
Year established: 2001
Area: 858 square kilometers (331 square miles)

AUSTRALIA

Gondwana Rainforests (World Heritage Site)
Year established: 1986
Area: 3,700 square kilometers (1,429 square miles)

Great Barrier Reef (World Heritage Site)
Year established: 1981
Area: 348,700 square kilometers (134,634 square
miles)

Greater Blue Mountains Area (World Heritage Site)
Year established: 2000
Area: 10,326 square kilometers (3,987 square miles)

Heard and McDonald Islands (World Heritage Site)
Year established: 1997
Area: 386 square kilometers (149 square miles)

Purnululu National Park
Year established: 1987
Area: 2,397 square kilometers (925 square miles)

Shark Bay, Western Australia (World Heritage Site)
Year established: 1991
Area: 21,973 square kilometers (8,484 square
miles)

Tasmanian Wilderness (World Heritage Site)
Year established: 1982
Area: 14,075 square kilometers (5,434 square miles)

Willandra Lakes Region (World Heritage Site)
Year established: 1981
Area: 2,400 square kilometers (927 square miles)

EUROPE

BELARUS

**Belovezhskaya Pushcha/Białowieża Forest (World
Heritage Site)**
Year established: 1979
Area: 927 square kilometers (358 square miles)

FINLAND

**High Coast/Kvarken Archipelago (World Heritage
Site)**
Year established: 2000
Area: 1,944 square kilometers (751 square miles)

GERMANY

Wadden Sea (World Heritage Site)
Year established: 2009
Area: 10,000 square kilometers (3,861 square miles)

GREENLAND

Ilulissat Icefjord (World Heritage Site)
Year established: 2004
Area: 4,024 square kilometers (1,554 square
miles)

ICELAND

Surtsey (World Heritage Site)
Year established: 2008
Area: 34 square kilometers (13 square miles)

ITALY

Aeolian Islands (World Heritage Site)
Year established: 2000
Area: 12 square kilometers (5 square miles)

Dolomites mountains range (World Heritage Site)
Year established: 2009
Area: 1,419 square kilometers (548 square miles)

NORWAY

West Norwegian Fjords-Geirangerfjord and Nærøyfjord (World Heritage Site)
Year established: 2005
Area: 1,227 square kilometers (474 square miles)

ROMANIA

Danube Delta (World Heritage Site)
Year established: 1991
Area: 3,124 square kilometers (1,206 square miles)

SPAIN

Doñana National Park
Year established: 1969
Area: 543 square kilometers (210 square miles)

Garajonay National Park
Year established: 1981
Area: 40 square kilometers (15 square miles)

Teide National Park
Year established: 1954
Area: 190 square kilometers (73 square miles)

SWEDEN

Laponian Area (World Heritage Site)
Year established: 1996
Area: 9,400 square kilometers (3,629 square miles)

UNITED KINGDOM

Dorset and East Devon Coast (World Heritage Site)
Year established: 2001
Area: 25 square kilometers (10 square miles)

Giant's Causeway and Causeway Coast (World Heritage Site)
Year established: 1986
Area: 0.7 square kilometer (0.27 square mile)

NORTH AMERICA

CANADA

ALBERTA

Wood Buffalo National Park
Year established: 1922
Area: 44,807 square kilometers (17,300 square miles)

MANITOBA

Wapusk National Park
Year established: 1996
Area: 11,475 square kilometers (4,431 square miles)
Northwest Territories

Tuktut Nogait National Park
Year established: 1996
Area: 16,340 square kilometers (6,309 square miles)

NUNAVUT

Auyuittuq National Park
Year established: 1976
Area: 19,089 square kilometers (7,370 square miles)

Quttinirpaaq National Park
Year established: 1986
Area: 37,775 square kilometers (14,585 square miles)

YUKON TERRITORY

Ivvavik National Park
Year established: 1984
Area: 10,168 square kilometers (3,926 square miles)

Kluane National Park and Reserve
Year established: 1976
Area: 22,013 square kilometers (8,499 square miles)

MEXICO

Calakmul Biosphere Reserve
Year established: 1989
Area: 7,284 square kilometers (2,812 square miles)

El Vizcaíno Biosphere Reserve
Year established: 1988
Area: 143,600 square kilometers (55,444 square miles)

Isla del Golfo Special Biosphere Reserve
Year established: 1978
Area: 1,517 square kilometers (586 square miles)

Islands and Protected Areas of the Gulf of California (World Heritage Site)
Year established: 2005
Area: 7,046 square kilometers (2,720 square miles)

Montes Azules Biosphere Reserve
Year established: 1971
Area: 3,351 square kilometers (1,294 square miles)

Sierra de San Pedro Mártir National Park
Year established: 1947
Area: 757 square kilometers (292 square miles)

UNITED STATES

ALASKA

Arctic National Wildlife Refuge
Year established: 1960
Area: 78,000 square kilometers (30,116 square miles)

Denali National Park and Preserve
Year established: 1917
Area: 24,585 square kilometers (9,492 square miles)

Gates of the Arctic National Park and Preserve
Year established: 1980
Area: 39,460 square kilometers (15,236 square miles)

Glacier Bay National Park and Preserve
Year established: 1980
Area: 13,287 square kilometers (5,130 square miles)

Katmai National Park and Preserve
Year established: 1980
Area: 19,120 square kilometers (7,382 square miles)

Kenai Fjords National Park
Year established: 1980
Area: 4,600 square kilometers (1,776 square miles)

Kobuk Valley National Park
Year established: 1980
Area: 6,757 square kilometers (2,609 square miles)

Lake Clark National Park and Preserve
Year established: 1980
Area: 16,308 square kilometers (6,297 square miles)

Wrangell-St. Elias National Park and Preserve
Year established: 1980
Area: 53,321 square kilometers (20,587 square miles)

Yukon Delta National Wildlife Refuge
Year established: 1909
Area: 77,500 square kilometers (29,923 square miles)

ARIZONA

Coronado National Forest
Year established: 1941
Area: 7,203 square kilometers (2,781 square miles)

Glen Canyon National Recreation Area
Year established: 1972
Area: 4,856 square kilometers (1,875 square miles)

Grand Canyon National Park
Year established: 1919
Area: 4,950 square kilometers (1,911 square miles)

Petrified Forest National Park
Year established: 1962
Area: 380 square kilometers (147 square miles)

Saguaro National Park
Year established: 1994
Area: 370 square kilometers (143 square miles)

ARKANSAS

Hot Springs National Park
Year established: 1921
Area: 23 square kilometers (9 square miles)

Ozark-St. Francis National Forest, Arkansas
Year established: 1908
Area: 4,694 square kilometers (1,812 square miles)

CALIFORNIA

Channel Islands National Park
Year established: 1980
Area: 1,010 square kilometers (390 square miles)

Death Valley National Park
Year established: 1933
Area: 13,638 square kilometers (5,266 square miles)

Joshua Tree National Park
Year established: 1994
Area: 3,196 square kilometers (1,234 square miles)

Kings Canyon National Park
Year established: 1940
Area: 1,873 square kilometers (723 square miles)

Lassen National Forest
Year established: 1905
Area: 4,300 square kilometers (1,700 square miles)

Lassen Volcanic National Park
Year established: 1916
Area: 429 square kilometers (166 square miles)

Mojave National Preserve
Year established: 1994
Area: 6,475 square kilometers (2,500 square miles)

Point Reyes National Seashore
Year established: 1962
Area: 288 square kilometers (111 square miles)

Redwood National and State Parks
Year established: 1968
Area: 456 square kilometers (178 square miles)

Santa Monica National Recreation Area
Year established: 1978
Area: 620 square kilometers (239 square miles)

Sequoia National Park
Year established: 1890
Area: 1,635 square kilometers (631 square miles)

Yosemite National Park
Year established: 1890
Area: 3,081 square kilometers (1,190 square miles)

COLORADO

Black Canyon of the Gunnison National Park
Year established: 1999
Area: 124 square kilometers (48 square miles)

Dinosaur National Monument
Year established: 1915
Area: 853 square kilometers (329 square miles)

Great Sand Dunes National Park and Preserve
Year established: 1932
Area: 170 square kilometers (66 square miles)

Mesa Verde National Park
Year established: 1906
Area: 211 square kilometers (81 square miles)

Rocky Mountain National Park
Year established: 1915
Area: 1,076 square kilometers (415 square miles)

FLORIDA

Big Cypress National Preserve
Year established: 1974
Area: 2,916 square kilometers (1,126 square miles)

Biscayne National Park
Year established: 1980
Area: 700 square kilometers (270 square miles)

Dry Tortugas National Park
Year established: 1935
Area: 249 square kilometers (96 square miles)

Everglades National Park
Year established: 1947
Area: 6,050 square kilometers (2,336 square miles)

GEORGIA

Chattahoochee-Oconee National Forest
Years established: 1936 (Chattahoochee), 1959
    (Oconee)
Area: 3,504 square kilometers (1,353 square miles)

Okefenokee National Wildlife Refuge
Year established: 1937
Area: 1,627 square kilometers (628 square miles)

HAWAII

Haleakalā National Park
Year established: 1961
Area: 116 square kilometers (45 square miles)

Hawaii Volcanoes National Park
Year established: 1916
Area: 931 square kilometers (359 square miles)

IDAHO

**Craters of the Moon National Monument and
    Preserve**
Year established: 1924
Area: 2,893 square kilometers (1,117 square miles)

KENTUCKY

Big South Fork National River and Recreation Area
Year established: 1974
Area: 507 square kilometers (196 square miles)

Mammoth Cave National Park
Year established: 1941
Area: 214 square kilometers (83 square miles)

MAINE

Acadia National Park
Year established: 1929
Area: 186 square kilometers (72 square miles)

MASSACHUSETTS

Cape Cod National Seashore
Year established: 1961
Area: 176 square kilometers (68 square miles)

MICHIGAN

Isle Royale National Park
Year established: 1940
Area: 2,314 square kilometers (893 square miles)

MINNESOTA

Mississippi National River and Recreation Area
Year established: 1988
Area: 218 square kilometers (84 square miles)

Voyageurs National Park
Year established: 1975
Area: 882 square kilometers (341 square miles)

MISSISSIPPI

Gulf Islands National Seashore
Year established: 1971
Area: 549 square kilometers (212 square miles)

MONTANA

Bighorn Canyon National Recreation Area
Year established: 1966
Area: 487 square kilometers (188 square miles)

Glacier National Park
Year established: 1910
Area: 4,101 square kilometers (1,583 square miles)

NEVADA

Great Basin National Park
Year established: 1986
Area: 310 square kilometers (120 square miles)

NEW MEXICO

Carlsbad Caverns National Park
Year established: 1930
Area: 186 square kilometers (72 square miles)

White Sands National Monument
Year established: 1933
Area: 581 square kilometers (224 square miles)

NORTH DAKOTA

Theodore Roosevelt National Park
Year established: 1978
Area: 285 square kilometers (110 square miles)

OHIO

Cuyahoga Valley National Park
Year established: 2000
Area: 136 square kilometers (52 square miles)

OREGON

Crater Lake National Park
Year established: 1902
Area: 732 square kilometers (283 square miles

SOUTH CAROLINA

Congaree National Park
Year established: 2003
Area: 85 square kilometers (33 square miles)

SOUTH DAKOTA

Badlands National Park
Year established: 1978
Area: 970 square kilometers (375 square miles)

Wind Cave National Park
Year established: 1903
Area: 115 square kilometers (44 square miles)

TENNESSEE

Great Smoky Mountains National Park
Year established: 1934
Area: 2,109 square kilometers (814 square miles)

TEXAS

Big Bend National Park
Year established: 1944
Area: 3,242 square kilometers (1,252 square miles)

Guadalupe Mountains National Park
Year established: 1972
Area: 350 square kilometers (135 square miles)

UTAH

Arches National Park
Year established: 1971
Area: 310 square kilometers (120 square miles)

Bryce Canyon National Park
Year established: 1928
Area: 145 square kilometers (56 square kilometers)

Canyonlands National Park
Year established: 1964
Area: 1,366 square kilometers (527 square miles)

Capitol Reef National Park
Year established: 1971
Area: 979 square kilometers (378 square miles)

Zion National Park
Year established: 1909
Area: 579 square kilometers (224 square miles)

VIRGINIA

Shenandoah National Park
Year established: 1935
Area: 796 square kilometers (307 square miles)

WASHINGTON

Mount Rainier National Park
Year established: 1899
Area: 942 square kilometers (364 square miles)

North Cascades National Park
Year established: 1968
Area: 2,769 square kilometers (1,069 square miles)

Olympic National Park
Year established: 1938
Area: 3,584 square kilometers (1,384 square miles)

WYOMING

Grand Teton National Park
Year established: 1929
Area: 1,242 square kilometers (480 square miles)

Yellowstone National Park
Year established: 1872
Area: 8,990 square kilometers (3,470 square miles)

UNITED STATES TERRITORIES

AMERICAN SAMOA

National Park of American Samoa
Year established: 1988
Area: 42 square kilometers (16 square miles)

VIRGIN ISLANDS

Virgin Islands National Park
Year established: 1956
Area: 59 square kilometers (23 square miles)

SOUTH AMERICA AND CENTRAL AMERICA

ARGENTINA

Ischigualasto/Talampaya Natural Parks (World Heritage Site)
Year established: 2000
Area: 2,750 square kilometers (1,062 square miles)

Los Glaciares National Park
Year established: 1981
Area: 4,459 square miles (1,722 square miles)

Valdes Peninsula (World Heritage Site)
Year established: 1999
Area: 3,600 square kilometers (1,390 square miles)

BOLIVIA

Noel Kempff Mercado National Park
Year established: 2000
Area: 15,230 square kilometers (5,880 square miles)

BRAZIL

Central Amazon Conservation Complex (World Heritage Site)
Year established: 2000
Area: 60,000 square kilometers (23,166 square miles)

Cerrado Protected Areas: Chapada dos Veadeiros and Emas National Parks (World Heritage Site)
Year established: 2001
Area: 65,500 square kilometers (25,290 square miles)

Iguazu National Park
Year established: 1939
Area: 1,700 square kilometers (656 square miles)

Pantanal Conservation Area
Year established: 2000
Area: 769 square kilometers (297 square miles)

CHILE

Torres del Paine National Park
Year established: 1959
Area: 2,400 square kilometers (927 square miles)

COSTA RICA

Cocos Island National Park
Year established: 1997
Area: 65,500 square kilometers (25,290 square miles)

Talamanca Range-La Amistad Reserves/La Amistad National Park (World Heritage Site)
Year established: 1983
Area: 5,678 square kilometers (2,192 square miles)

ECUADOR

Cotopaxi National Park
Year established: 1975
Area: 334 square kilometers (129 square miles)

Galápagos Islands (World Heritage Site)
Year established: 1978
Area: 140,665 square kilometers (54,311 square miles)

PERU

Manú National Park (World Heritage Site)
Year established: 1987
Area: 15,328 square kilometers (5,918 square miles)

SURINAME

Central Suriname Nature Reserve
Year established: 2000
Area: 16,000 square kilometers (6,178 square miles)

VENEZUELA

Canaima National Park (World Heritage Site)
Year established: 1994
Area: 30,000 square kilometers (11,583 square miles)

Narayanan M. Komerath and Padma P. Komerath

Time Line

Year	Event
1824	Royal Society for the Prevention of Cruelty to Animals is founded in Great Britain.
1849	U.S. Department of the Interior is established.
1859	British naturalist Charles Darwin publishes *On the Origin of Species by Means of Natural Selection.*
1864	U.S. government cedes Yosemite to the state of California to create a park.
1866	American Society for the Prevention of Cruelty to Animals is incorporated.
1869	Scottish-born naturalist John Muir begins his explorations of California's Sierra Nevada range.
1870	First state wildlife refuge is established in California.
1872	U.S. government makes Yellowstone the world's first national park.
1872	General Mining Act authorizes individual persons and companies to stake mining claims on federal land without paying royalties to the government.
1874	First synthesis of the compound dichloro-diphenyl-trichloroethane (DDT) is reported.
1879	U.S. Geological Survey is established to collect scientific information on natural resources.
1882	First commercial hydroelectric power plant opens in Appleton, Wisconsin.
1883	American Anti-Vivisection Society is founded.
1890	Sequoia National Park is created in California.
1890's-1910's	Science of ecology emerges.
1891	Forest Reserve Act authorizes the president of United States to create national forests.
1892	Sierra Club is founded.
1894	Yellowstone Game Protection Act closes U.S. national parks to hunting and commercial fishing.
1897	Forest Management Act mandates that U.S. national forests be managed to perpetuate water supplies and forests.
1898	Kruger National Park is established in South Africa to protect big game.
1900	Killing of the last wild passenger pigeon prompts public awareness of the dangers of allowing species to become extinct and helps prompt U.S. Congress's passage of the Lacey Act, which regulates commerce in birds and mammals, prohibiting shipment of illegally killed wildlife.
1901	First microbial insecticides are used.
1902	Reclamation Act creates the federal agency that will later become the U.S. Bureau of Reclamation and establishes a fund for constructing irrigation projects to make agriculture possible in marginal western desert lands.
1903	First U.S. national wildlife refuge, a bird sanctuary, is created at Florida's Pelican Island.
1905	National Audubon Society is founded to promote the conservation of birds and other wildlife.
1905	American Bison Society is founded to advocate for protection of remaining bison populations.
1905	U.S. Forest Service is established within the Department of Agriculture to manage national forests.
1906	Antiquities Act authorizes U.S. presidents to create national monuments.
1908	President Theodore Roosevelt calls a conference of state governors and other officials to inventory the nation's natural resources.
1909	Great Lakes International Joint Commission is established as an independent body to resolve transboundary water disputes between the United States and Canada.
1910	Pickett Act authorizes U.S. presidents to set aside land for any public purpose.
1911	Weeks Act authorizes government purchases of national forestlands.
1913	Construction of Hetch Hetchy Dam is authorized within California's Yosemite National Park.
1913	Migratory Bird Act is the first U.S. legislation designed to protect migratory birds.
1913	Los Angeles Aqueduct is completed.

Year	Event
1914	Last captive passenger pigeon dies, completing the extinction of the species.
1916	National Park Service is created within the U.S. Department of the Interior.
1918	Migratory Bird Treaty Act restricts the hunting of migratory birds in the United States and Canada.
1919	Grand Canyon National Park is established.
1923	Hetch Hetchy Dam is completed within Yosemite National Park.
1924	Gila Wilderness Area, the world's first designated wilderness area, is established within Gila National Forest.
1926	Lacey Act (1900) is amended to protect the black bass.
1927	Ansel Adams begins publishing photographs of the American West.
1929	Migratory Bird Conservation Act creates a system of refuges along bird flyways.
1930's	Du Pont Corporation develops the first chlorofluorocarbons as a safe alternative to toxic refrigerants.
1930-1931	Hundreds of workers die of acute silicosis after working on West Virginia's Hawk's Nest Tunnel.
1933	Civilian Conservation Corps is established.
1933	Tennessee Valley Authority is created.
1933	Convention Relative to the Preservation of Fauna and Flora in Their Natural State, an international agreement establishing preservation policies for European colonies in Africa, opens for signature.
1933-1936	Worst droughts in modern U.S. history create conditions that lead to the disaster of the Dust Bowl in the southern Great Plains.
1934	Migratory Bird Hunting and Conservation Stamp Act creates the duck stamp program, which requires hunters to purchase stamps along with their licenses to hunt waterfowl and uses the proceeds to establish wildlife refuges.
1934	Taylor Grazing Act regulates livestock grazing on federal lands.
1935	Wilderness Society is established to help protect wilderness areas and wildlife.
1935	Natural Resources Conservation Service is established within the U.S. Department of Agriculture to coordinate the conservation of soil and water.
1936	Boulder Dam (renamed Hoover Dam in 1947) is completed on the lower Colorado River.
1937	President Franklin D. Roosevelt introduces a tree-planting campaign to alleviate Dust Bowl conditions.
1937	Federal Aid in Wildlife Restoration Act provides federal aid to states for wildlife management.
1937	Migratory Bird Treaty Act of 1918 is amended to include Mexico in a comprehensive plan to protect all migratory birds in North America.
1938	John Muir Trail in California's Sierra Nevada range is completed.
1939	Paul Müller discovers DDT's insecticidal properties.
1940's	American cities begin fluoridating public water supplies.
1940	U.S. Fish and Wildlife Service is established to protect fish and animal habitats.
1940	Kings Canyon National Park is created in California.
1941	Grand Coulee Dam is completed on the Columbia River in Washington State.
1941	City of Los Angeles begins diverting water from streams feeding Mono Lake.
1942	Rocky Mountain Arsenal is established near Denver, Colorado, to produce, store, and decommission highly toxic and hazardous substances.
1942	First commercial DDT formulations are introduced.
1943	Alaska Highway is completed.
1943	Severe episode of photochemical smog in Los Angeles on September 8 is dubbed "Black Wednesday."
1943-1945	DDT is used on civilians and military troops in Europe to control lice and typhus.
1944	Flood Control Act is enacted to coordinate water development projects in the Missouri River basin.

YEAR	EVENT
1945	Atomic bombs dropped by the United States on the cities of Hiroshima and Nagasaki in Japan during World War II result in widespread devastation and hundreds of thousands of deaths caused by the blasts and by the radiation emitted immediately on detonation.
1945	United Nations establishes the Food and Agriculture Organization.
1946	U.S. Bureau of Land Management is created.
1946	U.S. Atomic Energy Commission is established to promote and monitor the use of nuclear energy.
1946	International Whaling Commission is established to regulate global whaling.
1946	United Nations Educational, Scientific, and Cultural Organization (UNESCO) is established.
1946	U.S. Department of Agriculture authorizes the limited use of DDT on crops.
1946	Communicable Disease Center (later renamed the Centers for Disease Control and Prevention) is established by the U.S. government to oversee all areas of public health.
1947	Everglades National Park is created in Florida.
1947	Asbestos mining wastes are dumped into Lake Superior at Silver Bay, Minnesota.
1947	Federal Insecticide, Fungicide, and Rodenticide Act establishes guidelines for U.S. government regulation of pesticides.
1948	International Union for Conservation of Nature is founded.
1948	World Health Organization is established by the United Nations to provide leadership in international public health matters.
1948	Temperature inversion over Donora, Pennsylvania, creates "killer smog" conditions in Donora's industrialized river valley.
1948	Paul Müller receives the Nobel Prize in Physiology or Medicine for developing DDT as an insecticide.
1948-1953	Soviet Plan for the Transformation of Nature is instituted but fails to increase agricultural production dramatically.
1950's	Soviet government's diversion of Aral Sea water begins drastic social, economic, and environmental changes in the region that would later become the republics of Kazakhstan and Uzbekistan.
1950's	As DDT is increasingly used for agriculture, public health, and domestic pest control, studies begin to reveal its negative effects.
1950's	Local food chain in Japan's Minamata Bay is contaminated by industrial wastes containing mercury.
1950	Federal Aid in Fish Restoration Act provides states with federal aid for sport fish management.
1951	Animal Welfare Institute is founded.
1951	Nature Conservancy is founded to help preserve threatened ecosystems.
1951	British government establishes the Serengeti as a national park in Tanganyika (now Tanzania).
1952	World's first serious nuclear reactor accident occurs at Chalk River, Ontario.
1952	Killer smog reduces visibility in London, England, to zero, and causes thousands of deaths from respiratory and other illnesses.
1953	U.S. Department of Agriculture declares kudzu an invasive weed and bans its transport across state lines without permission.
1953	Burning oil slicks on Cuyahoga River near Cleveland, Ohio, reveal poor environmental condition of the Great Lakes, particularly Lake Erie. The river will again burn in 1969.
1954	United Nations holds its first population conference.
1954	First U.S. hydrogen bomb test on the South Pacific's Bikini Atoll spreads radioactive fallout over populated regions of the Marshall Islands.
1955	Air Pollution Control Act is the first U.S. federal law addressing air pollution.
1956	U.S. Fish and Wildlife Service is established within the Department of the Interior.
1956	Proposal to build Echo Park Dam in Dinosaur National Monument is defeated.
1956	Construction begins on Glen Canyon Dam on the Colorado River.

Year	Event
1956	Stringfellow Acid Pits in Southern California begins operating as a hazardous waste disposal facility that will later require decades of cleanup work.
1956	Malathion is first used to counteract fruit fly infestations.
1957	SANE, an organization of environmental activists, scientists, and pacifists, is formed to protest nuclear weapons testing.
1957	Nuclear waste explodes at a weapons production facility in the Soviet Union's Chelyabinsk province.
1957	Fire in the reactor core of the Windscale nuclear reactor in western England releases radioactive material into the atmosphere.
1957	Nuclear plant in Colorado's Rocky Flats releases low-level radioactive contaminants into the region; similar incidents will also occur in 1969 and 1973.
1957	International Atomic Energy Agency is established to promote peaceful uses of nuclear energy.
1957	Price-Anderson Act limits the liability of nonmilitary nuclear facilities in the United States.
1957	Africanized honeybees are introduced to Brazil.
1958	Robert Barker publishes a study linking DDT to declines in robin populations.
1959	Twelve nations with claims on Antarctic territory sign the Antarctic Treaty, pledging to set aside Antarctica for scientific and peaceful pursuits.
1959	St. Lawrence Seaway, an international waterway connecting Lake Superior to the Atlantic Ocean, opens.
1960	Great Swamp National Wildlife Refuge is established in New Jersey.
1960	Multiple Use-Sustained Yield Act clarifies the purposes of U.S. national forests.
1960	Major oil-producing nations join to form the Organization of Petroleum Exporting Countries (OPEC) to control world oil prices.
1961	Idaho National Engineering Laboratory's SL-1 nuclear test reactor has the first fatal nuclear reactor accident in United States.
1961	World Wildlife Fund (later the World Wide Fund for Nature) is founded to help preserve the global environment and biodiversity.
1961	U.S. production of DDT peaks at 160 million pounds.
1962	Rachel Carson's *Silent Spring* calls attention to the dangers of pesticides in the environment.
1962	Euell Gibbons publishes *Stalking the Wild Asparagus*, in which he advocates collecting and eating wild foods.
1962-1971	U.S. military sprays millions of gallons of herbicides over South Vietnam and Laos.
1963	Limited Test Ban Treaty, an international agreement to halt nuclear weapons testing, opens for signature.
1963	Clean Air Act is first U.S. law providing for the monitoring and control of air pollution.
1963	President's Science Advisory Committee releases a report on the dangers of pesticides.
1964	Wilderness Act establishes the National Wilderness Preservation System.
1964	Land and Water Conservation Fund Act provides a trust fund for parkland acquisition.
1964	U.S. Federal Commission on Pest Control is established.
1964-1974	International Biological Program, a coalition of scientists, studies world ecosystems and evaluates the impacts of natural and human-made changes.
1965	In *Scenic Hudson Preservation Conference v. Federal Power Commission*, a federal appeals court grants environmental groups with noneconomic interests the right to intervene in federal regulatory agency licensing decisions.
1965	Council of Europe establishes the European Diploma of Protected Areas as an award to selected European regions that satisfy certain scientific, aesthetic, or cultural criteria.
1966	National Wildlife Refuge System Administration Act mandates that all refuge uses be compatible with the refuges' primary missions.

YEAR	EVENT
1966	Avalanche of coal sludge falls from the side of Merthyr Mountain in Wales onto the village of Aberfan below, causing the deaths of 144 people.
1967	Environmental Defense Fund is formed.
1967	Cleveland Amory founds the Fund for Animals.
1967	Grounding of supertanker *Torrey Canyon* off the southern coast of England spills 119,192 tons of crude oil into English Channel.
1967	Air Quality Act establishes enforcement provisions to reduce interstate air-pollution transport.
1968	Club of Rome is established as an international think tank devoted to studying the complex interrelationships among global problems.
1968	Zero Population Growth (later renamed Population Connection) is founded to work toward population stabilization.
1968	National Wild and Scenic Rivers Act establishes a national river conservation system.
1968	National Trails System Act establishes a national system of recreational hiking trails.
1968	Experimental Lakes Area is established to study environmental problems, especially acidification, in eastern Canadian lakes.
1968	Paul R. Ehrlich publishes *The Population Bomb*, which addresses the dangers of unchecked world population growth.
1968	Garrett Harding's essay "The Tragedy of the Commons" raises awareness of the problem of the depletion of shared resources.
1968-1974	Intense droughts devastate West Africa's Sahel region.
1969	Union of Concerned Scientists is established to advocate for science and to oppose misuses of science and technology.
1969	Ralph Nader begins founding consumer-advocate organizations.
1969	Friends of the Earth (later Friends of the Earth International) is founded.
1969	Blowout in an oil well in the Santa Barbara Channel off the California coast results in a massive oil spill.
1969	China reports virtual elimination of malaria, thanks to DDT use.
1969	Residential use of DDT is banned in the United States.
1970	The National Environmental Policy Act is enacted on January 1. Among other provisions, the act requires that environmental impact assessments be conducted for all federal activities that affect the environment.
1970	Earth Day is inaugurated on April 22 to increase public awareness of environmental issues.
1970	League of Conservation Voters is founded to support proenvironmental policies and political candidates.
1970	Natural Resources Defense Council is established to help protect U.S. natural resources.
1970	U.S. Environmental Protection Agency (EPA) is established.
1970	Egypt's Aswan High Dam on the Nile River is completed.
1970	UNESCO initiates its Man and the Biosphere Programme.
1970	Clean Air Act amendments create stricter air-quality standards and establish the first comprehensive emission regulatory structure, including National Ambient Air Quality Standards.
1970	James Lovelock discovers the accumulation of chlorofluorocarbons in the earth's atmosphere.
1970	National Oceanic and Atmospheric Administration is established to conduct scientific investigations of oceans and the earth's atmosphere.
1970	Norman Borlaug receives the Nobel Peace Prize for his work in developing high-yield grains to alleviate world hunger.
1970	Dian Fossey begins field studies of gorillas in Rwanda that lead to her publication of *Gorillas in the Mist* in 1983.

Year	Event
1970	U.S. oil production peaks, beginning an era of increasing dependence on imported oil.
1970	Council on Environmental Quality is established to coordinate federal environmental efforts.
1971	Animal Welfare Institute begins a campaign against whaling.
1971	The international environmental watchdog organization Greenpeace is established.
1971	Save the Whales Campaign is initiated.
1971	Environment Canada is established to oversee Canadian national environmental policies and programs.
1971	UNESCO's Biosphere Reserve Program is established to recognize areas of global environmental significance.
1971	Ramsar Convention on Wetlands of International Importance opens for signature.
1971	International Institute for Environment and Development is established as a research and advocacy organization.
1972	Environmental Action publishes *Ecotage!*, a handbook of techniques for sabotaging environmentally damaging projects.
1972	Marine Mammal Protection Act imposes a moratorium on the hunting and harassing of marine mammals.
1972	In *Sierra Club v. Morton*, the U.S. Supreme Court rules that injuries to noneconomic interests can provide a basis for challenges to federal agency decisions.
1972	United Nations Environmental Conference in Stockholm, Sweden, is attended by representatives of 113 nations.
1972	World Heritage Convention, an international agreement designed to protect sites of cultural, historic, or natural value, opens for signature.
1972	United Nations Environment Programme is established to coordinate worldwide environmental activities.
1972	London Convention on the Prevention of Marine Pollution, an international agreement to halt reckless dumping of wastes in marine waters, opens for signature.
1972	U.S. Congress votes to ban the use of DDT in the United States.
1972	Clean Water Act establishes stricter water-quality standards.
1972	Federal Water Pollution Control Act amendments provide new protections for wetlands.
1972	Federal Environmental Pesticides Control Act requires the registration of pesticides sold in the United States.
1972	Coastal Zone Management Act provides for management of land and water uses in coastal areas and the Great Lakes.
1972	First recombinant DNA molecules are constructed at Stanford University.
1972	*The Limits to Growth*, published by the Club of Rome, reports on a study that used computer modeling to examine global systems, including world ecosystems.
1972	United States launches first earth resources satellite, known as Landsat.
1973	Endangered Species Act becomes law, requiring wildlife refuge managers to protect certain species of flora and fauna and committing the federal government to the preservation of biological diversity.
1973	Villagers in northern India launch the Chipko Andolan movement to stop lumber companies from clear-cutting mountain slopes.
1973	World oil prices rise steeply after Arab members of OPEC declare production cutbacks and an embargo on exports to the United States and other allies of Israel.
1973	Convention on International Trade in Endangered Species, an international agreement designed to conserve endangered animal and plant species, is opened for signature.
1974	Animal rights advocate Cleveland Amory publishes *Man Kind? Our Incredible War on Wildlife*.

YEAR	EVENT
1974	Worldwatch Institute is established as an independent research organization concerned with global environmental issues.
1974	Safe Drinking Water Act sets federal standards for public water supplies.
1974	Scientists call for a moratorium on recombinant DNA research until the National Institutes of Health can study the subject and develop guidelines.
1975	Energy Policy and Conservation Act addresses American demands for energy while promoting the conservation of energy resources.
1975	Nuclear Regulatory Commission begins operating in the United States.
1975	Ernest Callenbach's novel *Ecotopia* describes an ecologically ideal world.
1976	Federal Land Policy and Management Act directs the Bureau of Land Management to retain public lands and manage them for multiple uses.
1976	National Forest Management Act gives statutory protection to national forests and sets management standards.
1976	In *Kleppe v. New Mexico*, the U.S. Supreme Court makes its most definitive statement on use of the U.S. Constitution's property clause to protect wildlife.
1976	Grounding of the *Argo Merchant* off Rhode Island's coast spills about 181,000 barrels of heavy fuel oil into sea.
1976	Explosion at an Italian chemical factory in Seveso releases toxic dioxin that kills thousands of animals and sickens thousands of people.
1976	Earth-fill Teton Dam collapses in Idaho, flooding a nearby town.
1976	Toxic Substances Control Act authorizes the Environmental Protection Agency to ban substances posing threats to human health or the environment.
1976	Resource Conservation and Recovery Act directs the Environmental Protection Agency to regulate waste production, storage, and transportation.
1976-1980	Love Canal, a community in Niagara Falls, New York, is devastated by toxic chemical wastes buried in the area years earlier.
1977	Peter Singer's *Animal Liberation* is published and becomes an influential articulation of the philosophy of animal rights.
1977	Trans-Alaska Pipeline is completed.
1977	Surface Mining Control and Reclamation Act establishes environmental standards for strip mining.
1977	U.S. Department of Energy is established.
1977	Sea Shepherd Conservation Society is founded to promote international conservation of marine wildlife.
1977	First World Wilderness Congress is held.
1978	In *Tennessee Valley Authority v. Hill*, the U.S. Supreme Court rules that the federal Endangered Species Act gives the protection of endangered species priority over other missions of federal agencies.
1978	U.S. government declares May 3 Sun Day to increase public awareness of solar energy.
1978	Grounding of the *Amoco Cadiz* off the coast of Brittany, France, spills about 1.5 million barrels of crude oil into the sea.
1978	Antarctic and Southern Ocean Coalition, a global coalition of nongovernmental organizations committed to preserving Antarctic region lands and southern oceans as a wilderness area, is founded.
1978	U.S. government bans the use of chlorofluorocarbons as aerosol propellants, but worldwide use continues to grow.
1978	National Climate Program Act establishes a nationwide coordinated program to monitor and predict U.S. climate changes.
1979	Tellico Dam, a hydroelectric dam on the Little Tennessee River near Knoxville, is completed.

Year	Event
1979	Nevada's enactment of the Sagebrush Rebellion Act endorses the goals of a movement seeking greater state—rather than federal—control over public lands.
1979	Three Mile Island nuclear power plant near Harrisburg, Pennsylvania, experiences the worst U.S. nuclear accident to date, exposing weaknesses in U.S. nuclear power plant operations.
1979	Blowout in a PEMEX oil drilling well in the Gulf of Mexico spills 3 million barrels of oil into the sea.
1979	Collision of oil tankers off the coast of Tobago kills twenty-nine crewmen and spills 2.14 million barrels of oil into the Caribbean Sea.
1979	World oil prices double when the Iranian Revolution temporarily stops Iranian oil exports.
1979	Convention on Long-Range Transboundary Air Pollution opens for signature.
1979	Bonn Convention on the Conservation of Migratory Species of Wild Animals opens for signature.
1979	Smallpox is declared to have been eliminated throughout the world.
1980	People for the Ethical Treatment of Animals (PETA) is founded.
1980	Radical environmental movement Earth First! is founded.
1980	Alaska National Interest Lands Conservation Act doubles the territory in U.S. refuge and parks systems.
1980	Fish and Wildlife Conservation Act provides federal aid for protection of nongame wildlife.
1980	Mediterranean Blue Plan, a multinational effort to curb pollution in Mediterranean Sea, is initiated.
1980	Comprehensive Environmental Response, Compensation, and Liability Act, or Superfund, establishes a federal regulatory structure for environmental cleanup of contaminated sites threatening human health and the environment.
1980	In *Diamond v. Chakrabarty*, the U.S. Supreme Court rules that genetically engineered microorganisms are patentable products of human ingenuity.
1980	Drawing on computer modeling, *The Global 2000 Report* analyzes environmental impacts of global human activities.
1981	Citizens Clearinghouse for Hazardous Waste (later renamed the Center for Health, Environment, and Justice) is established to assist local communities in addressing environmental issues.
1981	Association of Zoos and Aquariums establishes the Species Survival Plan to increase the numbers of animals in breeding programs.
1981	United Nations Environment Programme forms the Ozone Group and discusses a global treaty to protect the ozone layer.
1982	United Nations Convention on the Law of the Sea opens for signature.
1982	World Resources Institute is established to help protect the environment by supporting sustainable management of world resources.
1982-1988	Solar One, the first solar power tower system, operates near Barstow, California.
1983	Tom Regan's *The Case for Animal Rights* is published, drawing more supporters to the animal rights movement.
1983	Reservoir of the Colorado River's Glen Canyon Dam is filled.
1983	Eruption of Indonesia's Krakatoa volcano affects weather and sea levels around the world.
1983	Grounding of tanker *Castillo de Bellver* off Cape Town, South Africa, dumps 2 million barrels of crude oil into the Atlantic.
1983	Threat posed by dioxin contamination prompts evacuation of the entire population of Times Beach, Missouri.
1983	World Commission on Environment and Development, also known as the Brundtland Commission, is established by the United Nations to formulate long-term environmental strategies for sustainable development through international cooperation.
1984	Sinking of the cargo ship *Mont Louis*, carrying hundreds of tons of radioactive material, in the North Sea calls international attention to the dangers of transporting such material.

YEAR	EVENT
1984	Accidental release of highly toxic gases from a pesticide plant in Bhopal, India, kills thousands of people.
1984	Trans-Siberian Pipeline, built to carry natural gas from Siberia's Urengoy gas field to Western Europe, is completed.
1984	Worldwatch Institute begins issuing its annual report, *State of the World*, which surveys global environmental issues.
1985	French intelligence agents sink the Greenpeace antiwhaling ship *Rainbow Warrior* in New Zealand.
1985	Rainforest Action Network is founded to protect endangered forests and their native residents.
1985	Conservation Reserve Program is established to reduce agricultural surpluses in the United States, help prevent soil erosion, and reduce carbon in the atmosphere by inducing farmers to reduce the amounts of land they devote to crops.
1985	Scientists observe a hole in the ozone layer over Antarctica that is linked to stratospheric chlorine.
1986	A nuclear power reactor explodes at the Soviet Union's Chernobyl power plant.
1986	Mad cow disease is first identified in Great Britain.
1986	International Whaling Commission approves a worldwide ban on commercial whaling.
1986-1988	Droughts devastate farms in the American Midwest.
1986-1988	Cargo ship *Khian Sea* circumnavigates the globe searching for a place to dispose of hazardous waste from Philadelphia that was originally to be dumped in the Bahamas.
1987	Federal court rules that the U.S. Fish and Wildlife Service must police hunting by Alaska Native groups.
1987	Last six wild California condors are captured and placed with condors already in zoos in a controlled-breeding recovery program.
1987	Accidental release of radioactive powder in Goiania, Brazil, contaminates hundreds of people.
1987	Twenty-four nations sign the Montreal Protocol, which requires that they limit the production and use of chlorofluorocarbons.
1987	*Our Common Future*, a report of the World Commission on Environment and Development, focuses on multilateral global cooperation as the most effective way to address growing environmental crises.
1987	The barge *Mobro* cruises the U.S. Atlantic and Gulf coasts for six months in search of a place to unload a cargo of garbage.
1988	Global ReLeaf, a forest conservation and reforestation program, is initiated.
1988	Collapse of an oil storage tank near the Monongahela River in Pennsylvania results in a massive oil spill.
1988	Intergovernmental Panel on Climate Change is established.
1989	The organization Ceres is established to inject environmental considerations into American investment decisions.
1989	Grounding of the supertanker *Exxon Valdez* off Alaska's coast spills about 262,000 barrels of crude oil into Prince William Sound.
1989	Montreal Protocol enters into force, and signatory nations are required to begin phasing out chlorofluorocarbons.
1989	Basel Convention on the Control of Transboundary Movements of Hazardous Wastes opens for signature.
1989	Controversy develops over the use of Alar, a chemical used as a plant growth regulator.
1989-1990	McToxics Campaign prompts McDonald's fast-food restaurant chain to replace its polystyrene food containers with more environmentally friendly packaging.
1990	Canada initiates its Green Plan as a national strategy to create a cleaner, safer, and healthier environment.
1990	More than one hundred nations agree to an international moratorium on trade in ivory.

Year	Event
1990	Dolphin Protection Consumer Information Act is enacted to protect dolphins from tuna fishers. The law is later strengthened by the 1992 U.S. International Dolphin Conservation Act.
1990	Damaged by a storm off New Brunswick, Norwegian tanker *Berge Broker* spills about 95,000 barrels of oil into the North Atlantic.
1990	Accident on the tanker *Mega Borg* causes 121,000 barrels of crude oil to spill into the Gulf of Mexico.
1990	Iraqi seizure of Kuwaiti oil fields causes a spike in world oil prices.
1990	Alliance of Small Island States, an international lobbying group representing the environmental interests of small island and low-lying coastal nations, is founded.
1990	European Union creates the European Environment Agency to collect and disseminate information about the environment.
1990	International Institute for Sustainable Development is founded.
1990	Amendments to the Clean Air Act of 1963 establish programs to control acid precipitation and many toxic pollutants.
1990	Environmental Protection Agency establishes a broad ban on the manufacture, processing, importation, and distribution of asbestos products.
1990	Global Change Research Act requires U.S. research into worldwide climate change.
1991	Global Environment Facility, an independent financial organization, is established to provide grants for projects protecting the global environment.
1991	Train derailment near Dunsmuir, California, spills thousands of gallons of the pesticide metam sodium into the Sacramento River.
1992	Defenders of Wildlife issues a report stating that the National Wildlife Refuge System is grossly inadequate.
1992	The Earth Summit in Rio de Janeiro, Brazil, is attended by representatives of 179 nations. Important outcomes of the summit include the signing of the Convention on Climate Change and the Convention on Biological Diversity, the endorsement of the Rio Declaration and the Statement of Forest Principles, the adoption of Agenda 21, and the creation of the United Nations Commission on Sustainable Development.
1992	United States, Mexico, and Canada sign the North American Free Trade Agreement to remove trade barriers and reduce legal and financial restrictions among the three countries.
1992	U.S. Food and Drug Administration rules that genetically engineered foods do not require premarket approval or special labeling.
1993	U.S. federal and state officials, timber industry representatives, and environmental groups meet at the Forest Summit on April 3 to resolve issues related to the management of federally owned forests in the Pacific Northwest.
1993	Grounding of the tanker *Braer* at the Shetland Islands spills about 619,000 barrels of crude oil into the sea.
1993	*Journal of the National Cancer Institute* reports that exposure to DDT may increase the risk of breast cancer.
1993-1995	Global Biodiversity Assessment, an independent scientific analysis of all issues, theories, and views regarding biodiversity from a global perspective, is conducted.
1994	Construction begins on China's Three Gorges Dam.
1994	United Nations Convention to Combat Desertification is adopted.
1995	Greenpeace protesters occupy the *Brent Spar*, an abandoned oil storage platform, to persuade Royal Dutch Shell to adopt alternative methods of disposing of such equipment.
1995	World Trade Organization is established to regulate trade among nations.
1995	Oil companies are required to sell reformulated gasoline in U.S. metropolitan regions, and gas stations are required to install vapor-retrieval devices on pumps.

YEAR	EVENT
1996	Grounding of the tanker *Sea Empress* off the coast of Wales releases about 381,000 barrels of crude oil into the sea.
1996	Comprehensive Nuclear-Test-Ban Treaty opens for signature.
1996	Ian Wilmut clones Dolly the sheep, the first mammal successfully cloned from adult cells.
1997	Ecology Hall of Fame is founded in Santa Cruz, California.
1997	National Wildlife Refuge System Improvement Act shifts the priorities of nature preservation systems in the United States toward multiple-use management.
1997	Kyoto Protocol commits signatory nations to place legally binding limits on their emissions of six greenhouse gases.
1998	Drought destroys crops in southern Midwest states.
1998	California institutes tougher emission control standards for new cars; other states follow with similar laws.
1998	International negotiations to phase out the production and use of DDT and other persistent organic pollutants begin in Montreal, Canada.
1999	Grounding of the tanker *Erika* off the coast of France spills 83,333 barrels of crude oil into the Bay of Biscay.
1999	Environmental Protection Agency outlaws the use of methyl parathion on food crops most likely to be eaten by children.
1999	World population tops 6 billion.
2000	Chicago Climate Exchange is established to operate an emissions allowance trading system.
2000	Explosion aboard the tanker *Westchester* spills 15,750 barrels of crude oil into the Mississippi River below New Orleans.
2000	South Africa reintroduces the use of DDT to combat the spread of malaria by mosquitoes.
2000	Toyota begins marketing the first commercially produced hybrid automobile, the Prius, in the United States.
2001	Roadless Area Conservation Rule, a federal policy initiative designed to protect national forests from commercial development, is issued.
2001	Stockholm Convention on Persistent Organic Pollutants, an international agreement to limit the manufacture and use of substances linked to neurological, reproductive, and immune system damage in people and animals, opens for signature.
2002	Johannesburg Declaration on Sustainable Development is adopted.
2003	U.S. occupation of Iraq disrupts Middle East oil markets, raising new international concerns about the stability of world supplies.
2003	Proposed Clear Skies Act is designed to amend the Clean Air Act with a cap-and-trade system.
2003	Dolly the cloned sheep is euthanized after contracting lung disease.
2004	Kenyan environmentalist Wangari Maathai is awarded the Nobel Peace Prize for her work in combating African deforestation.
2004-2006	Federal Bureau of Investigation undertakes Operation Backfire to probe acts of ecoterrorism in the western United States during the previous decade.
2005	More than 4,000 hectares (approximately 10,000 acres) of Puerto Rican rain-forest land are added to the U.S. national forest system.
2005	Environmental Protection Agency's Clean Air Interstate Rule (CAIR) begins a cap-and-trade program to keep air pollution generated in one state from rendering other states noncompliant with air-quality standards.
2005	Energy Policy Act supports voluntary reductions in carbon-intensive activities and export to developing countries of technologies that reduce carbon intensity.

YEAR	EVENT
2006	Sinking of the tanker *M/T Solar 1* off the Philippines dumps 14,720 barrels of crude oil into Guimaras Strait.
2006	Toxic petroleum wastes are illegally dumped in the West African nation of Côte d'Ivoire.
2006	Documentary film *An Inconvenient Truth*, featuring former U.S. vice president Al Gore, increases public awareness of problems related to global warming.
2007	U.S. Climate Action Partnership, an alliance of business corporations and environmental organizations, is established to promote the passage of legislation encouraging reductions in greenhouse gas emissions.
2007	U.S. Conference of Mayors Climate Protection Agreement is adopted by mayors of American cities, who agree to take steps locally to combat global warming.
2007	Collision of a barge with the tanker *Hebei Spirit* off the west coast of South Korea spills 77,775 barrels of crude oil into the sea.
2007	In *Massachusetts v. Environmental Protection Agency*, the U.S. Supreme Court enhances the power of states to mount legal challenges to federal agency actions on matters such as greenhouse gas emissions.
2007	Al Gore is awarded the Nobel Peace Prize for his work in increasing public awareness of climate change.
2008	Hydroelectric power plant at China's Three Gorges Dam becomes fully operational.
2008	Collision of a barge with a tanker near New Orleans spills 11,635 barrels of fuel oil into the Mississippi River.
2008	U.S. federal appeals court rules that the Clean Air Interstate Rule exceeds the Environmental Protection Agency's regulatory authority but later orders temporary reinstatement of the rule.
2009	Omnibus Public Land Management Act adds 850,000 hectares (2.1 million acres) of new wilderness areas in nine U.S. states.
2009	European Union begins phasing out animal testing in the cosmetics industry.
2009	Environmental Protection Agency officially finds that the greenhouse gases methane, carbon monoxide, nitrous oxide, hydrofluorocarbons, perfluorocarbons, and sulfur hexafluoride constitute a threat to public health and welfare.
2009	Cash for Clunkers federal program offers financial incentives for Americans to trade in their old cars for cleaner-running and more fuel-efficient vehicles.
2010	People for the Ethical Treatment of Animals claims a worldwide membership of 2 million.
2010	American Lung Association reports that about 58 percent of Americans endure unhealthy air-pollution levels.
2010	Environmental Protection Agency replaces the Clean Air Interstate Rule with the Transport Rule, which requires eastern states to decrease power plant emissions severely by 2014.
2010	World population is estimated at 6.8 billion.

R. Kent Rasmussen

Glossary

Abatement: Elimination of pollution or the reduction of its degree or intensity.

Absorption: Uptake of water, other fluids, or dissolved chemicals by a cell or an organism (as tree roots absorb dissolved nutrients in soil).

Acclimatization: Physiological and behavioral adjustments of an organism to changes in its environment.

Acid deposition: Complex chemical and atmospheric phenomenon that occurs when emissions of sulfur and nitrogen compounds and other substances are transformed by chemical processes in the atmosphere, often far from the original sources, and then deposited on the earth in either wet or dry form. Wet forms, popularly known as acid rain, can fall to earth as rain, snow, or fog; dry forms are acidic gases or particulates.

Acid mine drainage: Drainage of water from areas that have been mined for coal or other mineral ores. When exposed to air and water, metal sulfides such as pyrite react to produce sulfuric acid.

Activator: Chemical added to a pesticide to increase its activity.

Active solar: Type of solar energy system that uses pumps or fans to circulate fluids that have been heated by the sun.

Adaptation: Changes in an organism's physiological structure, function, or habits that allow it to survive in new surroundings.

Adsorption: Attachment of gaseous, liquid, or solid particles to a surface, a process that is used to remove some pollutants from air or water.

Adulterants: Chemical impurities or substances that by law do not belong in a food, pesticide, or other substance.

Aeration: A process by which air is added to a substance. Aeration of water or soil promotes the biological degradation of some pollutants.

Aerobic: Occurring in, thriving on, or able to survive the presence of oxygen.

Aerosol: A suspension of fine solid or liquid particles in a gas.

Afforestation: Planting of new forests on land that has not recently been forested.

Afterburner: In incinerator technology, a burner located so that the combustion gases are made to pass through its flame in order to remove smoke and odors.

Aftermarket converted vehicle: Factory-produced conventionally fueled vehicle to which nonstandard equipment has been added that enables it to run on alternative fuel.

Agent Orange: Toxic herbicide and defoliant, heavily used during the Vietnam War, containing 2,4,5-trichlorophen-oxyacetic acid (2,4,5-T) and 2-4 dichlorophenoxyacetic acid (2,4-D) with trace amounts of dioxin.

Air emissions: Pollution discharged into the atmosphere from smokestacks, other vents, and surface areas of commercial or industrial facilities; from residential chimneys; and from motor vehicle, locomotive, or aircraft exhausts.

Air permeability: In soils, the property of being able to be penetrated by air.

Air pollutant: Any substance in air that could, in high enough concentrations, harm humans, animals, vegetation, or materials. Pollutants may include almost any natural or artificial composition of airborne matter capable of being airborne. They may be in the form of solid particles, liquid droplets, gases, or combinations thereof.

Airborne particulates: Solid particles or liquid droplets suspended in air.

Alachlor: Herbicide, marketed under the trade name Lasso, used mainly to control weeds in corn and soybean fields; banned for use in Canada (1985) and the European Union (2006).

Alar: Trade name for daminozide, a chemical that makes apples redder, firmer, and less likely to drop off trees before growers are ready to pick them.

Algae: Simple rootless plants that grow in sunlit waters in proportion to the amount of available nutrients. Algae can affect water quality adversely by lowering the dissolved oxygen in the water.

Algal blooms: Sudden spurts in the growth of algae, which can affect water quality adversely and indicate potentially hazardous changes in local water chemistry.

Algicide: Substance or chemical used specifically to kill or control algae.

Allergen: Substance that causes an allergic reaction in individuals sensitive to it.

Ambient air: Any unconfined portion of the atmosphere; open air, surrounding air.

Anaerobic: Occurring in, thriving on, or able to survive the absence of oxygen.

Animal dander: Tiny scales of shed animal skin, a common indoor air pollutant.

Animal feeding operation (AFO): Agricultural lot or facility where livestock or poultry is raised and maintained in confinement. The animals do not forage; rather, feed is brought to them. In an AFO, live animals, feed, manure, urine, and dead animals all share a comparatively small land area. See also **Feedlot**.

Anisotropy: In hydrology, the conditions under which one or more hydraulic properties of an aquifer vary from a reference point.

Anthracite: Hard, black, clean-burning coal containing a high percentage of fixed carbon and a low percentage of volatile matter; used mainly for residential and commercial space heating.

Anthropogenic: Resulting from or produced by humans or human activity.

Antibiotic: Drug used in the treatment of bacterial infections.

Aqueous: Containing, dissolved in, or resembling water.

Aqueous solubility: Maximum concentration of a chemical that will dissolve in pure water at a reference temperature.

Aquifer: Permeable underground geological formation, or group of formations, containing groundwater.

Aquitard: Low-permeability underground geological formation, or group of formations, that restricts the flow of groundwater.

Area source: Any source of air pollution that is released over a relatively small area but cannot be classified as a point source. Such sources may include vehicles and other small engines, small businesses and household activities, and biogenic sources such as forests, which release hydrocarbons.

Arsenicals: Pesticides containing arsenic.

Asbestos: Naturally occurring silicate mineral fiber once widely used for its insulating and fire-retardant properties but later restricted in use because of possible health effects; can cause cancer or asbestosis when inhaled.

Asbestos abatement: Procedures to control fiber release from asbestos-containing materials in a building or to remove them entirely; may include removal, encapsulation, repair, enclosure, encasement, and operations and maintenance programs.

Asbestosis: Lung disease associated with inhalation of asbestos fibers. The presence of the fibers causes lung scarring and inflammation, which makes breathing progressively more difficult and can be fatal.

Ash: Mineral content of a product remaining after complete combustion.

Assay: Test for a specific chemical, microbe, or effect.

Attenuation: Process by which a compound is reduced in concentration over time through absorption, adsorption, degradation, dilution, or transformation.

Background level: Concentration of a substance in an environmental medium (air, water, or soil) that occurs naturally or is not the result of human activities. Also, in exposure assessment, the concentration of a substance in a defined control area during a fixed period of time before, during, or after a data-gathering operation.

Backyard composting: Diversion of organic food waste and yard trimmings from the municipal waste stream to a specially structured enclosure, container, or pile, where they are broken down through controlled decomposition of organic matter by bacteria and fungi into a humus-rich product.

Bacteria: Microscopic living organisms, some of which can aid in pollution control by metabolizing organic matter in sewage, oil spills, or other pollutants. Some bacteria in soil, water, or air can also cause human, animal, and plant health problems.

Bed load: Sediment particles resting on or near a stream channel bottom that are pushed or rolled along by the flow of the water.

Beryllium (Be): Heavy metal hazardous to human health when inhaled as an airborne pollutant. Discharged by machine shops, ceramic and propellant plants, and foundries.

Best available control measures (BACM): In pollution control, the most effective measures (according to regulatory guidance) for controlling pollutant emissions.

Best available control technology (BACT): In pollution control, the most stringent technology available for controlling emissions; major pollution sources are required to use BACT, unless they can demonstrate that it is not feasible for energy, environmental, or economic reasons.

Best demonstrated available technology (BDAT): In hazardous waste treatment, the most effective commercially available means of treating specific types of waste. BDATs may change with advances in treatment technologies.

Bioaccumulants: Substances that increase their concentration within living organisms as the organisms take in contaminated air, water, or food because the substances are very slowly metabolized or excreted.

Bioassay: Test to determine the biological activity or relative strength of a substance by comparing its effect on a test organism with that of a standard preparation.

Biochemical: Relating to chemical processes that occur within living organisms.

Bioconcentration: The accumulation of a chemical within the tissues of an organism to levels greater than in the surrounding medium.

Biodegradable: Capable of decomposing under natural conditions.

Biodiversity: Variety and variability among living organisms and the ecological complexes in which they occur.

Biofuels: Solid, liquid, or gasesous fuels derived from biomass. Examples include biodiesel, typically made from vegetable oils, animals, or recycled grease; and bioethanol, created through the fermentation of the sugar components of wheat, corn, sugar beets, sugar cane, or other plant matter.

Biogenic: Made by living organisms or biological processes.

Biological oxygen demand (BOD): In water, the amount of dissolved oxygen consumed by microorganisms as they break down organic matter; a high BOD in stream water indicates that less oxygen is available for higher forms of life.

Biological stressors: Organisms accidently or intentionally introduced into habitats in which they have not evolved naturally.

Biological treatment: Treatment technology that uses bacteria to consume organic waste.

Biomagnification: Process whereby certain substances such as pesticides or heavy metals move up the food chain, work their way into rivers or lakes, and are eaten by aquatic organisms such as fish, which in turn are eaten by large birds, animals, or humans. The substances become concentrated in animal tissues or internal organs as they move up the chain.

Biomass: All of the living material in a given area; often refers to vegetation.

Biome: The entire community of living organisms in a single major ecological area.

Bioremediation: Use of living organisms to remove pollutants from soil, water, or wastewater; also, the use of organisms such as nonharmful insects to remove agricultural pests or counteract diseases of trees, plants, and garden soil.

Biosphere: The portion of the earth and its atmosphere that can support life.

Biostabilizer: Machine that converts solid waste into compost through grinding and aeration.

Biota: The animal and plant life of a given region.

Biotechnology: Techniques that use living organisms or parts of organisms for a variety of purposes; these include producing a variety of products from medicines to industrial enzymes, improving plants and animals, and developing microorganisms that can consume toxic pollutants.

Biotic community: Naturally occurring assemblage of plants and animals that live in the same environment and are mutually sustaining and interdependent.

Bioturbation: Disturbance of soil or sediment by living organisms.

Blister agent: In chemical warfare, a chemical substance such as mustard gas that burns and blisters the skin, lungs, and eyes.

Bottle bill: Proposed or enacted legislation that requires a returnable deposit on beverage containers and provides for redemption through retail stores or other means; also called container deposit legislation.

Brackish water: Mixture of fresh water and salt water.

Breakdown product: New chemicals formed when a chemical is metabolized or when it reacts with water, light, oxygen, or other elements in its environment.

Breeder reactor: Type of nuclear reactor that produces more fuel than it consumes.

British thermal unit (BTU): Unit of heat energy equal to the amount of heat required to raise the temperature of one pound of water by 1 degree Fahrenheit (0.556 degree Celsius) at sea level.

Bromochlorofluorocarbons (BCFCs): Bromine-containing compounds chemically similar to chlorofluorocarbons (CFCs); used as a fire-extinguishing agent. Like CFCs, BCFCs are highly destructive to upper-atmospheric ozone.

Buffer strips: Strips of grass or other erosion-resistant vegetation between or below cultivated strips or fields.

Buyback center: Facility where individuals or groups bring recyclable items in return for payment.

By-product: Material other than the principal product that is generated as a consequence of an industrial process or as a breakdown product in a living system.

Cadmium (Cd): Heavy metal used in battery production and other industries; toxic to humans when ingested or inhaled, and accumulates in the environment.

Cap-and-trade system: Environmental policy tool that combines emissions trading with a mandatory emissions cap.

Capillary action: Movement of water through very small spaces that results when the attraction between the water molecules and the surfaces of the spaces exceeds the attraction between the water molecules themselves.

Carbon cycle: Exchanges of carbon from one carbon sink to another by means of geological, physical, chemical, and biological processes.

Carbon dioxide (CO_2): Colorless, odorless gas that occurs naturally in the atmosphere. A major greenhouse gas, CO_2 is exhaled by animal life and absorbed by plant life during photosynthesis.

Carbon footprint: Total carbon dioxide and other greenhouse gases that something—such as an individual, a population, an industry, or an activity—emits over a specified period of time; hence, the full impact that something has on climate change.

Carbon monoxide (CO): Colorless, odorless, poisonous gas produced by incomplete fossil-fuel combustion.

Carbon sequestration: Fixation of atmospheric carbon dioxide in a carbon sink through physical or biological processes.

Carbon sink: One of four major reservoirs—the atmosphere, the terrestrial biosphere, oceans, and sediments—that accepts carbon released from another part of the carbon cycle.

Carcinogen: Any substance that can cause or aggravate cancer.

Carnivore: Animal that feeds primarily on the flesh of other animals.

Catalyst: Substance that changes the speed or yield of a chemical reaction without being consumed or chemically changed by the chemical reaction.

Catalytic converter: Air-pollution abatement device that removes pollutants from motor vehicle exhaust, either by oxidizing them into carbon dioxide and water or by reducing them to nitrogen.

Chemical compound: A distinct and pure substance formed by the union of two or more elements in definite proportion by weight.

Chemical stressors: Chemicals released into the environment through industrial operations, modes of transportation, pesticide application, and other human activity that can cause illnesses and even death in plants and animals.

Chemical treatment: Any one of a variety of technologies that use chemicals or a variety of chemical processes to treat waste.

Chlorination: Application of chlorine to drinking water, sewage, or industrial waste to disinfect or to oxidize undesirable compounds.

Chlorofluorocarbons (CFCs): Family of inert, nontoxic, and easily liquefied chemicals used in refrigeration, air-conditioning, packaging, and insulation, or as solvents and aerosol spray-can propellants. Because CFCs are not destroyed in the lower atmosphere, they drift into the upper atmosphere, where their chlorine components destroy ozone.

Chlorosis: Discoloration of normally green plant parts caused by disease, lack of nutrients, or various air pollutants.

Clean fuels: Blends or substitutes for gasoline fuels, including compressed natural gas, methanol, ethanol, and liquefied petroleum gas.

Clear-cutting: Harvesting of all the trees in one area at one time, a practice that can destroy habitat and encourage fast rainfall or snowmelt runoff, with resultant erosion, sedimentation of streams and lakes, and flooding.

Climate change: See **Global climate change.**

Coal: Black or brownish-black carbonaceous rock formed from plant remains that heat, pressure, and geologic time have compacted, hardened, chemically altered, and metamorphosed; a major fossil fuel.

Coal cleaning technology: Precombustion process by which coal is physically or chemically treated to remove some of its sulfur so that when burned the coal produces reduced sulfur dioxide emissions.

Coal gasification: Conversion of coal to a gaseous product by one of several available technologies.

Coastal zone: Lands and waters adjacent to a coast that exert an influence on the uses of the sea and its ecology, or the uses and ecology of which are affected by the sea.

Cogeneration: Consecutive generation of useful thermal and electric energy from the same fuel source.

Coke: Hard, gray, porous residue of coal left after destructive distillation; used as a fuel and in iron smelting processes.

Coliform index: A rating of the purity of water based on a count of fecal bacteria.

Coliform organism: Microorganism found in the intestinal tracts of humans and animals. Presence of such organisms in water indicates fecal pollution and potentially adverse contamination by pathogens.

Colloids: Very small, finely divided solids that remain dispersed in a liquid for a long time because of their small size and electrical charge.

Combustion product: Substance produced during the burning or oxidation of a material.

Commercial waste: All solid waste emanating from business establishments, such as stores, markets, office buildings, restaurants, shopping centers, and theaters.

Community: In ecology, an assemblage of populations of different species within a specified location in space and time.

Compact fluorescent lightbulbs: Lightbulbs that combine the convenience of standard incandescent bulbs with the energy efficiency of fluorescent lighting; designed for residential and commercial use.

Compaction: Reduction of the bulk of solid waste through rolling and tamping.

Compost: Relatively stable humus material that is produced from the composting process, in which bacteria in soil mixed with garbage and degradable trash break down the mixture into organic fertilizer.

Comprehensive Environmental Response, Compensation, and Liability Act (CERCLA): U.S. legislation that provides funds for cleaning up uncontrolled or abandoned hazardous waste sites and emergency releases of pollutants; familiarly known as Superfund. CERCLA also gives the U.S. Environmental Protection Agency enforcement power to obtain private-party cleanup where possible.

Concentration: The relative amount of a substance mixed with another substance.

Conductivity: A measure of the ability of a solution to carry an electrical current.

Confined aquifer: Aquifer in which groundwater is confined under pressure between two aquitards.

Contact pesticide: Chemical that kills pests when it touches them.

Contaminant: Any physical, chemical, biological, or radiological substance or matter that has an adverse effect on air, water, soil, or associated organisms.

Contour mining: See **Surface mining**.

Contour plowing: Soil tilling method that follows the shape of the land to discourage erosion.

Contour strip farming: Farming in which row crops are planted in strips between alternating strips of close-growing, erosion-resistant forage crops.

Corporate average fuel economy (CAFE) standards: Regulations, intended to improve energy conservation, that require automobile manufacturers to meet assigned average fuel economy for all their vehicles made in a given model year.

Corrosion: Dissolution and wearing away of metal caused by chemical reactions, such as that which occurs between water and the pipes, chemicals touching a metal surface, or contact between two metals.

Corrosive: Chemical agent that reacts with the surface of a material, causing it to deteriorate or wear away.

Cover crop: Crop that supplies temporary protection for delicate seedlings or provides a cover canopy for seasonal soil protection and improvement between normal crop production periods.

Cradle-to-grave system: System in which hazardous materials are identified and tracked as they are produced, treated, transported, and disposed of. Tracking is conducted by means of a series of permanent, linkable, descriptive documents.

Crop rotation: Planting of a succession of different crops on the same land area, as opposed to planting the same crop time after time.

Cross-contamination: Transfer of a contaminant from an affected individual, object, or area to an unaffected one; in hydrogeology, movement of underground contaminants from one level or area to another as the result of invasive subsurface activities.

Crude oil: Hydrocarbon mixture that occurs naturally in liquid phase in natural underground reservoirs;

refined to produce petroleum products such as heating oils, fuels, lubricants, and asphalt.

Cultural eutrophication: Human-caused increase in the rate at which a lake, estuary, or bay becomes overenriched with nutrients, which stimulates the growth of plants and results in the depletion of oxygen in the water.

Cumulative exposure: The sum of exposures to a pollutant that an organism experiences over a period of time.

Curbside collection: Method of collecting recyclables, electronic waste, hazardous substances, or other materials from homes, community districts, or businesses.

Dam: Physical barrier erected across a river or other waterway to control the flow of or raise the level of water; may be constructed for flood control, hydroelectric power production, irrigation, or recreation purposes.

DDT: See **Dichloro-diphenyl-trichloroethane**.

Decay products: Degraded radioactive materials, often referred to as "daughters" or "progeny."

Dechlorination: Removal of chlorine from a substance.

Decomposition: Breakdown of matter by bacteria and fungi, a process that changes the chemical makeup and physical appearance of materials.

Decontamination: Removal of harmful substances such as noxious chemicals, harmful bacteria or other organisms, or radioactive material from exposed individuals, rooms, equipment, furnishings, structures, or the exterior environment.

Defluoridation: Removal of excess fluoride in drinking water to prevent the staining of teeth or to render the water safe for consumption.

Defoliant: Herbicide that removes leaves from trees and growing plants.

Deforestation: Removal of trees from forested land.

Degasification: Water treatment process that removes dissolved gases from the water.

Degradation product: See **Breakdown product**.

Degree-day: Rough measure used to estimate the amount of heating required in a given area; defined as the difference between the mean daily temperature and 65 degrees Fahrenheit (18.3 degrees Celsius).

Demineralization: Treatment process that removes dissolved minerals or mineral salts from water or other liquids.

Denitrification: Biological reduction of nitrate to nitrogen gas by denitrifying bacteria in soil.

Density: Measure of how heavy a specific volume of a solid, liquid, or gas is; often expressed in comparison to water or another reference substance.

Deregulation: Elimination of some or all government controls from a previously regulated industry or sector.

Desertification: Transformation of productive or habitable land to desert, generally as a result of changing climate or unsustainable land management.

Destination facility: Facility to which a regulated waste is shipped for treatment and destruction, incineration, or disposal.

Destratification: Within a lake or reservoir, vertical mixing that totally or partially eliminates separate layers of temperature, plant, or animal life.

Desulfurization: Removal of sulfur from coal and other fossil fuels to reduce the pollution that will be generated when the fuel is burned.

Detection limit: Lowest concentration of a chemical that analysis can reliably distinguish from a zero concentration.

Deuterium: A hydrogen isotope with one proton and one neutron in its nucleus; possesses twice the mass of ordinary hydrogen. Also called heavy hydrogen.

Dichloro-diphenyl-trichloroethane (DDT): Synthetic organochlorine insecticide once used extensively in agriculture and efforts to control insect-borne diseases; widely banned or restricted for decades because of its environmental persistence, bioaccumulative characteristics, and damaging impact on wildlife.

Diffusion: Movement of suspended or dissolved particles (or molecules) from an area having a higher concentration of that substance to an area of lower concentration.

Digestion: Biochemical decomposition of organic matter, resulting in partial gasification, liquefaction, and mineralization of pollutants.

Dike: Low wall that can act as a barrier to prevent a spill from spreading or to provide flood protection.

Dioxin: Any of a family of compounds known chemically as dibenzo-p-dioxins. Dioxins are among the most toxic anthropogenic compounds; some are known or suspected human carcinogens with no known safe minimum dose.

Discharge: Flow of surface water in a stream or canal

or the outflow of groundwater from a flowing artesian well, ditch, or spring. Can also apply to discharge of liquid effluent from a facility or to chemical emissions into the air through designated venting mechanisms.

Disinfectant: Chemical or physical process that kills pathogenic organisms in water, in air, or on surfaces.

Dispersant: Chemical agent used to break up concentrations of organic material such as spilled oil.

Disposables: Consumer products, other items, and packaging designed to be used once or a few times and then discarded.

Dissolved solids: Disintegrated organic and inorganic material in water.

Downgradient: Direction in which groundwater flows; similar to "downstream" for surface water.

Drainage basin: Area of land that drains water, sediment, and dissolved materials to a common outlet at some point along a stream channel.

Dump: Site used for the disposal of solid waste without environmental controls.

Dystrophic lakes: Acidic, shallow bodies of water that contain much humus or other organic matter but few fish.

Ecological exposure: Exposure of a nonhuman organism to a stressor.

Ecology: The relationship of living things to one another and their environment, or the study of such relationships.

Economic poisons: Chemicals used to control pests and weeds and to defoliate cash crops such as cotton.

Ecosphere: The part of the earth, its waters, and its atmosphere capable of supporting life.

Ecosystem: The interacting system of a biological community and its nonliving environmental surroundings.

Ecotone: Habitat created through the juxtaposition of distinctly different habitats, an edge habitat, or an ecological zone or boundary where two or more ecosystems meet.

Ejector: Device used to disperse a chemical solution into water being treated.

Electric hybrid vehicle: Motor vehicle that operates using both an electric system and an internal combustion system. In a parallel hybrid, the vehicle can operate on either system; in a series hybrid, the vehicle operates solely on electricity, using its internal combustion motor only to generate addition electricity.

Emission: Release of a pollutant into the environment.

Emission offsets: Balancing of pollutant emissions from proposed new or modified stationary sources by reductions from existing sources to stabilize total emissions.

Emission standard: Legally set maximum amount of polluting discharge allowed from a single source, mobile or stationary.

Enclosure: In asbestos abatement, the placement of an airtight, impermeable, permanent barrier around asbestos-containing materials to prevent the release of asbestos fibers into the air.

End-of-the-pipe technologies: Technologies that reduce emissions of pollutants after the pollutants have formed, such as scrubbers on smokestacks and catalytic converters on automobile exhaust systems.

Endangered species: Animals, birds, fish, plants, or other living organisms threatened with extinction by anthropogenic or natural changes in their environment. Requirements for declaring a species endangered in the United States are contained in the Endangered Species Act.

Energy consumption: Use of energy to produce power or heat or as part of a manufacturing process.

Energy reserves: Estimated quantities of energy sources. Proved reserves are those that have been demonstrated to exist with reasonable certainty based on geologic and engineering data. Probable or indicated reserves are those that can reasonably be expected to exist based on geologic evidence and projections.

Enrichment: Addition of nutrients from sewage effluent or agricultural runoff to surface water; greatly increases the growth potential for algae and other aquatic plants.

Environment: The sum of all external conditions affecting the life, development, and survival of an organism.

Environmental impact statement (EIS): Document required of U.S. federal agencies by the National Environmental Policy Act for major projects or legislative proposals that may significantly affect the environment. A tool for decision making, an EIS describes the probable positive and negative effects of a proposed undertaking and notes alternative actions.

Environmental Protection Agency (EPA): U.S. federal agency responsible for the nation's environmental quality. Some U.S. states also have state-level agencies bearing the same name.

Environmental sustainability: Long-term maintenance of ecosystem components and functions for future generations.

Environmetrics: Application of statistics to environmental studies.

Enzyme: Protein that acts as a catalyst within a living organism.

Epiphyte: Plant that grows nonparasitically on another plant or object rather than in the soil; known informally as an air plant.

Erosion: The wearing away of land surface by wind or water; intensified by land-clearing practices related to farming, residential and industrial development, road building, and logging.

Ethanol: Clear, colorless, flammable alcohol typically produced biologically from biomass such as agricultural crops and their cellulose residues.

Eutrophication: Slow aging process during which a lake, estuary, or bay becomes overenriched with nutrients, which stimulates the growth of plants and results in the depletion of oxygen in the water.

Evaporation ponds: Areas where sewage sludge or other aqueous waste is discharged and allowed to dry.

Evapotranspiration: Loss of water from the soil through both evaporation and transpiration from the plants growing in the soil.

Exposure: Amount of radiation or other pollutant present in a given environment that represents a potential health threat to living organisms.

Facultative bacteria: Bacteria that can live under aerobic or anaerobic conditions.

Fallout: Particulate debris, often radioactive, that settles out of the atmosphere after a nuclear explosion.

Fecal coliform bacteria: Bacteria found in the intestinal tracts of mammals. The presence of such bacteria in water or sludge is used as an indicator of pollution and possible contamination by pathogens.

Feedlot: Confined area for the controlled feeding of animals. Feedlots tend to concentrate large amounts of animal waste that cannot be absorbed by the soil and hence may be carried to nearby streams or lakes by rainfall runoff.

Fill: Human-made deposits of natural soils, rock products, waste materials, or a mixture of some or all of these.

Filtration: Treatment process for removing solid (particulate) matter from water by means of porous media such as sand or manufactured filters; often used to remove particles that contain pathogens.

Fissile material: Material (such as uranium 235, plutonium 239, and uranium 233) that will undergo atomic fission when bombarded by neutrons.

Fission: The splitting of an atomic nucleus, which releases two or more neutrons and large amounts of energy.

Floc: In sewage and water treatment, a clump of solids formed by biological or chemical action.

Flocculation: Process by which clumps of solids in water or sewage aggregate through biological or chemical action so they can be separated from the surrounding fluid.

Floodplain: The flat or nearly flat land along a river or stream or in a tidal area that is covered by water during flooding.

Flue gas: Combustion exhaust gas.

Flume: Natural or human-made channel that diverts water.

Fluoridation: Addition of a chemical to increase the concentration of fluoride ions in a drinking-water supply to reduce the incidence of tooth decay.

Fluorides: Gaseous, solid, or dissolved compounds containing fluorine.

Fluorosis: A health condition resulting from excessive fluoride consumption; a common symptom is a characteristic mottling of the teeth.

Fly ash: Noncombustible residual particles expelled by flue gas.

Food chain: Sequence of organisms, each of which uses the next-lower member of the sequence as a food source.

Food web: Feeding relationships by which energy and nutrients are transferred from one species to another.

Fossil fuel: Fuel derived from ancient organic remains; fossil fuels include peat, coal, crude oil, and natural gas.

Fresh water: Water that generally contains less than 1,000 milligrams per liter of dissolved solids.

Fuel cell: Device that generates an electrical current by converting the chemical energy of hydrogen or another fuel directly into electrical energy.

Fuel efficiency: The proportion of energy released by fuel combustion that is converted into useful energy.

Fugitive emissions: Pollutant emissions not caught by a capture system.

Fume: Tiny particles trapped in vapor in a gas stream.

Fungicide: Pesticide used to control, deter, or destroy unwanted fungi.

Fungistat: A chemical that keeps fungi from growing.

Garbage: Animal and vegetable waste resulting from the handling, storage, sale, preparation, cooking, and serving of foods.

Gasification: Conversion of solid material, such as coal, into a gas for use as a fuel.

Gasohol: Mixture of gasoline and ethanol derived from fermented agricultural products containing at least 9 percent ethanol.

Genetic engineering: Process of inserting new genetic information into existing cells in order to modify a specific organism for the purpose of changing one of its characteristics.

Geographic information system (GIS): Computer system capable of storing, manipulating, analyzing, and displaying data in a geographic context.

Geological repository: Facility for the disposal of radioactive waste within rock; waste packaging and natural geology serve as barriers to isolate the waste from the environment.

Geothermal energy: Heat energy produced beneath the earth's surface by the decay of naturally occurring radioactive elements.

Global climate change: The phenomenon of intensification or enhancement of the greenhouse effect. Most scientists give credence to this concept and attribute the change to increased atmospheric concentrations of greenhouse gases resulting from industrialization and other human activity.

Greenhouse effect: Natural process of atmospheric warming in which solar energy absorbed by the earth's surface is reradiated and absorbed by certain atmospheric gases, primarily carbon dioxide and water vapor; term is often used in reference to an acceleration of this process due to human activity.

Greenhouse gas: Gas that contributes to the greenhouse effect. Major greenhouse gases are water vapor, carbon dioxide, methane, nitrous oxide, fluorocarbons, and ozone.

Ground cover: Plants grown to keep soil from eroding.

Groundwater: Fresh water found beneath the earth's surface, usually in aquifers, which supply wells and springs.

Habitat: Place where a population (human, animal, plant, or microorganism) lives and its surroundings, both living and nonliving.

Halons: Compounds used as fire-extinguishing agents. The bromine content of halons makes them potent greenhouse gases.

Hazardous substance: Any material that poses a threat to human health or the environment. Typical hazardous substances are characterized by toxicity, corrosivity, ignitability, or chemical reactivity.

Hazardous waste: By-products of society that can pose substantial or potential hazards to human health or the environment when improperly managed.

Heat island: Dome-shaped area of elevated temperatures over an urban area caused by structural and pavement heat fluxes in combination with pollutant emissions.

Heat pump: Equipment that extracts heat from a warmer medium and transfers it to a cooler one, thereby cooling the first and warming the second.

Heavy metal: Metallic element with high atomic weight; examples include mercury, chromium, cadmium, arsenic, and lead. Heavy metals tend to accumulate in the food chain and, while some are important for plant and animal nutrition in trace concentrations, higher concentrations are toxic to living organisms.

Heavy water: Water containing a significantly greater proportion of deuterium atoms to ordinary hydrogen atoms than occurs in ordinary (light) water; used to moderate the reaction in some nuclear reactors.

Herbicide: Chemical pesticide designed to control or destroy weeds, grasses, or other plants.

Herbivore: Animal that feeds primarily on plants.

Heterotrophic organism: Species that is dependent on organic matter for food.

Hormone: Natural or synthetic chemical substance that controls and regulates the activity of specific cells or organs.

Household waste: Solid waste, both garbage and trash or rubbish, that normally originates in private homes.

Humus: The organic component of soil, comprising partially decomposed plant and animal matter.

Hydrocarbons: Chemical compounds that consist entirely of carbon and hydrogen.

Hydrochlorofluorocarbons (HCFCs): Synthetic compounds used in refrigeration, foam manufacture, air conditioning, and aerosol propellants. While less destructive to stratospheric ozone than chlorofluorocarbons, HCFCs are potent greenhouse gases.

Hydrofluorocarbons (HFCs): Synthetic compounds that can serve as industrial substitutes for hydrochlorofluorocarbons. HFCs are believed not to deplete stratospheric ozone, but they are potent greenhouse gases.

Hydrogeological cycle: Natural process of recycling water from the atmosphere down to and through the earth and back to the atmosphere again.

Hydrologic cycle: Movement or exchange of water between the atmosphere and the earth.

Hydrology: Science that deals with the properties, distribution, and circulation of water.

Hydrolysis: Decomposition of organic compounds by interaction with water.

Ignitable: Capable of burning or causing a fire.

Immiscibility: Inability of two or more substances or liquids—such as soil and water—to dissolve readily into one another.

Impoundment: Confinement of a body of water, sludge, or other aqueous waste by a dam, dike, floodgate, or other barrier.

Incineration: Destruction of waste through controlled burning at high temperatures.

Incompatible waste: Waste unsuitable for mixing with another waste or other material because it may react to produce effects (heat, pressure, fire, explosion, violent chemical reaction, toxic emissions) that are harmful to human health and the environment.

Indicator: In biology, any biological entity, biological process, or community whose characteristics reflect the existence of specific environmental conditions.

Indirect discharge: Introduction of pollutants from a nondomestic source into a publicly owned wastewater treatment system.

Indirect source: In air-pollution control, any facility or building, property, road, or parking area that attracts motor vehicle traffic and, indirectly, causes pollution; in water-pollution control, a pollution source such as agricultural runoff, urban runoff, and airborne contaminants.

Indoor air pollution: Chemical, physical, or biological contamination of the breathable air inside a habitable or commercial structure or a conveyance.

Industrial sludge: Semiliquid residue or slurry remaining after treatment of industrial water and wastewater.

Industrial source reduction: Practices that reduce the amount of any hazardous substance, pollutant, or contaminant entering any waste stream or otherwise released into the environment.

Industrial waste: Unwanted materials resulting from industrial operations; may be liquid, sludge, solid, or hazardous waste.

Inert ingredient: In pesticides, components such as solvents, carriers, dispersants, and surfactants that are not active against target pests.

Infectious agent: Any organism, such as a pathogenic virus, parasite, or bacterium, that is capable of invading body tissues, multiplying, and causing disease.

Infectious waste: Hazardous waste capable of causing infections in humans. Examples include contaminated animal waste, human blood and blood products, pathological waste, and discarded sharps.

Influent: Water, wastewater, or other liquid flowing into a reservoir, basin, or treatment plant.

Inhalable particles: Particulate matter capable of entering the human respiratory tract.

Inorganic chemicals: Chemical substances of mineral origin, not of basically carbon structure.

Insecticide: Pesticide compound specifically used to kill or prevent the growth of unwanted insects.

Insectivore: Animal that feeds primarily on insects.

Institutional waste: Waste generated at institutions such as schools, libraries, hospitals, and prisons.

Integrated pest management (IPM): Use of coordinated tactics aimed at maintaining populations of insect, animal, or plant pests below damaging levels in an economical and environmentally compatible manner.

Ion: Electrically charged atom or group of atoms.

Irradiated food: Food subjected to brief radioactivity, usually gamma rays, to kill insects, bacteria, and mold, and to enable storage without refrigeration.

Irradiation: Exposure to radiation of wavelengths shorter than those of visible light (gamma, X-ray, or ultraviolet) for medical purposes, to sterilize milk or other foodstuffs, or to induce polymerization of monomers or vulcanization of rubber.

Irrigation: Intentional application of water to land areas to supply the water and nutrient needs of plants.

Irritant: Substance that can cause irritation of the skin, eyes, or respiratory system.

Isotope: Variation of an element that has the same atomic number of protons but a different weight because of the number of neutrons. Various isotopes of the same element may have different radioactive behaviors; some are highly unstable.

Isotropy: Condition in which the hydraulic or other properties of an aquifer are the same in all directions.

Jar test: Laboratory procedure that simulates a water treatment plant's coagulation/flocculation units with differing chemical doses, mix speeds, and settling times to estimate the minimum or ideal coagulant dose required to achieve certain water-quality goals.

Joint and several liability: Under the Comprehensive Environmental Response, Compensation, and Liability Act, this legal concept relates to the liability for Superfund site cleanup and other costs on the part of more than one potentially responsible party.

Karst: Geologic formation of irregular limestone deposits with sinks, underground streams, and caverns.

Kinetic energy: Energy possessed by a moving object, body, or substance.

Kinetic rate coefficient: Number that describes the rate at which a water constituent such as a biochemical oxygen demand or dissolved oxygen rises or falls, or at which an air pollutant reacts.

Landfill gas: Gas formed during the decomposition of organic materials in landfills, comprising primarily methane, carbon dioxide, and water vapor.

Latency: Time from the first exposure to a chemical until the appearance of a toxic effect.

LC$_{50}$: Median lethal concentration, or how much of a substance is needed to kill half of a group of experimental organisms in a given time; a standard measure of toxicity.

LD$_{50}$: Median lethal dose, or the dose of a toxicant or microbe that will kill 50 percent of test organisms within a designated period.

Leachate: Water that collects contaminants as it percolates through wastes, pesticides, or fertilizers.

Leaching: Process by which soluble constituents are dissolved and filtered through the soil by a percolating fluid.

Lead (Pb): Heavy metal that is hazardous to health if breathed or swallowed. Its use in gasoline, paints, and plumbing compounds has been sharply restricted or eliminated by laws and regulations in the United States and many other countries.

Legionella: Genus of bacteria, some species of which can cause Legionnaires' disease, a type of pneumonia.

Level of concern: Concentration of a given pollutant at and above which exposed persons may experience immediate health effects.

Lime (CaO): Calcium oxide, a chemical compound derived from limestone; also called quicklime. Widely used in industrial and potable water treatment and a variety of industrial processes.

Limnology: Science concerned with the physical, chemical, hydrological, and biological aspects of freshwater bodies.

Lindane: Environmentally persistent, bioaccumulative organochlorine pesticide; banned for agricultural use in many countries, including the United States (2007).

Liquefaction: Conversion of a solid into a liquid.

Lithology: Mineralogy, grain size, texture, and other physical properties of granular soil, sediment, or rock.

Manhattan Project: World War II-era U.S. government project that produced the world's first nuclear weapons.

Manufacturing use product: In pesticides, any product intended (labeled) for formulation or repackaging into other pesticide products.

Marine sanitation device: Any equipment or process installed aboard a vessel to receive, retain, treat, or discharge sewage.

Marsh: Type of wetland that does not accumulate appreciable peat deposits and is dominated by herbaceous vegetation. Marshes may be either freshwater or saltwater, tidal or nontidal.

Material safety data sheet (MSDS): Form required by U.S. law detailing the properties of a chemical to inform workers about its hazards and how to handle it safely.

Maximum available control technology (MACT): Emission standard for sources of air pollution requiring the maximum reduction of hazardous emissions, taking cost and feasibility into account.

Maximum contaminant level (MCL): Highest legally permissible level of a contaminant in water delivered to any user of a public system.

Mechanical separation: Use of mechanical means to separate waste into various components.

Medical waste: Solid waste generated in the diagnosis, treatment, or immunization of human beings or animals, in research pertaining thereto, or in the production or testing of biologicals, excluding hazardous waste and household waste.

Methane (CH_4): Colorless, nonpoisonous, flammable gas created through the anaerobic decomposition of organic compounds; a major component of natural gas and also a major greenhouse gas.

Methanol: Alcohol that can be used as an alternative fuel or as a gasoline additive; also known as wood alcohol or methyl alcohol.

Microbial growth: Amplification or multiplication of microorganisms such as bacteria, algae, diatoms, plankton, and fungi.

Microbial pesticide: Microorganism that is used to kill pests and is of minimum toxicity to humans.

Microenvironment: Well-defined small area—such as the home, office, or kitchen—that can be treated as uniform in terms of stressor concentration.

Mineral: Naturally occurring inorganic substance; examples include metals, salts, and sand.

Mining waste: Residues resulting from the extraction of raw materials from the earth.

Mitigation: Measures taken to reduce adverse impacts of particular activities on the environment.

Mobile source: Any nonstationary source of air pollution, such as cars, trucks, motorcycles, buses, airplanes, and locomotives.

Molecule: The smallest division of a compound that still retains or exhibits all the properties of the substance.

Monitoring: Periodic or continuous surveillance or testing to determine the level of compliance with statutory requirements or pollutant levels in various media or in humans, plants, and animals.

Monsoon: Seasonal wind in southern Asia and the season of heavy rainfall that occurs in India in association with this wind.

Morbidity: Rate of disease incidence.

Mortality: Death rate.

Muck soils: Soils made from decaying plant materials.

Mulch: Layer of material (such as wood chips, straw, or leaves) placed around plants to hold moisture, prevent weed growth, and enrich or sterilize the soil.

Multiple-use management: Policy under which land and water resources are managed for more than one purpose. For example, rangeland and forestland may accommodate grazing of livestock, watershed and wildlife protection, recreation, and timber production; bodies of water may be used for recreational purposes, fishing, and water supply.

Municipal discharge: Discharge of effluent from wastewater treatment plants that receive wastewater from households, commercial establishments, and industries in the coastal drainage basin.

Municipal sewage: Wastes (mostly liquid) originating from a community; may be composed of domestic wastewaters and industrial discharges. Also known as municipal waste.

Municipal solid waste: Common garbage or trash generated by a community's industries, businesses, institutions, and homes.

Mutagen: Agent that causes a permanent genetic change in a cell other than changes that occur during normal growth.

National Priorities List (NPL): U.S. Environmental Protection Agency's list of the nation's most serious uncontrolled or abandoned hazardous waste sites. These may undergo long-term remedial action under the Comprehensive Environmental Response, Compensation, and Liability Act, or Superfund.

Navigable waters: Traditionally, waters sufficiently deep and wide for navigation by all, or specified, vessels; such waters in the United States come under federal jurisdiction and are protected by certain provisions of the Clean Water Act.

Necrosis: Death of plant or animal cells or tissues. In plants, necrosis can discolor stems or leaves or kill a plant entirely.

Nematocide: Chemical agent that is destructive to nematodes, or parasitic worms.

Nephelometric: Method of measuring turbidity in a water sample by passing light through the sample and measuring the amount of the light that is deflected.

Nerve agent: In chemical warfare, a chemical substance (such as sarin) that interferes with the transmission of nerve impulses.

Netting: Concept in which all emissions sources in the same area that are owned or controlled by a single company are treated as one large source, thereby allowing flexibility in the control of individual sources in order to meet a single emissions standard.

Neutralization: Decrease in the acidity or alkalinity of a substance through the addition of alkaline or acidic materials, respectively.

NIMBYism: Opposition to the construction of useful but hazardous or aesthetically unappealing features (such as landfills or power plants) in one's own neighborhood; from the acronym NIMBY, for "not in my backyard."

Nitrate: Compound containing nitrogen that can exist in the atmosphere or as a dissolved gas in water and can have harmful effects on humans and animals. A plant nutrient and inorganic fertilizer, nitrate is found in septic systems, animal feedlots, agricultural fertilizers, manure, industrial waste waters, sanitary landfills, and garbage dumps.

Nitric oxide (NO): Gas formed by combustion under high temperature and high pressure in an internal combustion engine; converted by sunlight and photochemical processes in ambient air to nitrogen oxide. NO is a precursor of ground-level ozone pollution, or smog.

Nitrogen oxides (NOx): Compounds that result from photochemical reactions of nitric oxide in ambient air; major components of photochemical smog.

Nitrogenous wastes: Animal or vegetable residues that contain significant amounts of nitrogen.

Nitrophenols: Synthetic organopesticides containing carbon, hydrogen, nitrogen, and oxygen.

Nonattainment area: Any area that fails to meet the U.S. Environmental Protection Agency's national primary or secondary ambient air-quality standard for designated pollutants such as carbon monoxide and ozone.

Nonpoint sources: Diffuse pollution sources that have no single point of origin and are not introduced into a receiving stream from a specific outlet.

Nonpotable water: Water that is unsafe or unpalatable to drink because it contains pollutants, contaminants, minerals, or infective agents.

Nuclear fuel: Fissionable materials that have been enriched so that they can support a self-sustaining fission chain reaction within the reactor.

Nuclear reactor: Apparatus within which a nuclear fission chain reaction can be initiated, controlled, and sustained.

Nuclear winter: Post-nuclear war scenario, predicted by some scientists, in which smoke and debris rising from massive fires caused by the detonation of nuclear weapons could block sunlight for weeks or months, with a predicted outcome of the cooling of the earth's surface and consequent climate changes that could negatively affect world agricultural and weather patterns.

Nuclide: Atom characterized by the number of protons, neutrons, and energy in the nucleus.

Nutrient: Any substance assimilated by living things that promotes growth. The term is frequently applied to nitrogen and phosphorus in wastewater but can also refer to other essential and trace elements.

Nutrient pollution: Contamination of water resources by excessive inputs of nutrients, often accompanied by excess algal production.

Odor threshold: Minimum odor of a water or air sample that can be detected after successive dilutions with odorless water or air.

Off-site facility: A hazardous waste treatment, storage, or disposal area that is located away from the generating site.

Offstream water use: Withdrawal of water from surface or groundwater sources for use at other locations.

Oil reservoir: Underground liquid deposit of hydrocarbons, sulfur, oxygen, and nitrogen trapped and protected from evaporation by surrounding geological strata.

Oil spill: Accidental or intentional discharge of oil onto land or into a body of water.

Oligotrophic lake: Deep clear lake with few nutrients, little organic matter, and a high level of dissolved oxygen.

Omnivore: Animal that feeds on both vegetation and animal flesh.

Onboard controls: Devices placed on vehicles to capture gasoline vapor during refueling and route it to the engine when a vehicle is starting so that the fuel can be burned efficiently.

Onconogenicity: The capacity to induce cancer.

On-site facility: Facility for the treatment, storage, or disposal of hazardous waste that is located on the site where the waste is generated.

Opacity: In air-pollution control, the amount of light obscured by particulate pollution in the air; used as an indicator of changes in the performance of particulate control systems.

Open dump: Uncovered site used for disposal of waste without environmental controls.

Open pit mining: See **Surface mining**.

Oral toxicity: Ability of a substance to cause injury when ingested.

Organic: Derived from living organisms; in chemistry, any compound containing carbon.

Organic matter: Matter originating from living or once-living organisms.

Organism: Any form of animal or plant life.

Organophosphate pesticide: Phosphorus-containing pesticide that disrupts nerve function; such pesticides are short-lived, but some are acutely toxic when first applied.

Osmosis: Process in which a liquid is converted from a weak solution to a more concentrated solution through passage across a semipermeable membrane that allows the solvent but not the dissolved solids to go through.

Outfall: Place where effluent is discharged into receiving waters.

Overburden: Rock and soil overlying and surrounding mineral deposits; overburden is cleared away before mining.

Oxidant: In air-pollution control, a collective term for some of the primary constituents of photochemical smog.

Oxidation: Chemical addition of oxygen to break down pollutants or organic waste.

Oxygenated fuel: Gasoline that has been blended with alcohols or ethers that contain oxygen in order to reduce carbon monoxide and other emissions.

Ozone hole: A thinning of the stratospheric ozone layer.

Ozone layer: The protective layer in the upper atmosphere that absorbs some of the sun's ultraviolet rays, thereby reducing the amount of potentially harmful radiation that reaches the earth's surface.

Ozone precursor: Chemical compound that reacts with other compounds in the presence of sunlight to produce ozone. Examples include carbon monoxide, methane, nonmethane hydrocarbons, and nitrogen oxides.

Packed tower: Pollution-control device that forces dirty air through a tower packed with crushed rock or wood chips while liquid is sprayed over the packed material.

Pandemic: Outbreak of infectious disease that affects large numbers of people over a large geographic area.

Parabolic trough: Type of solar thermal concentrator commonly used in commercial solar plants; characterized by a long, curved reflector mounted on a motorized base that allows the concentrator to track the sun.

Paraquat: Fast-acting, nonselective herbicide used to kill various types of plants; used by the United States on Mexican marijuana fields during the 1970's.

Particle count: Count of the suspended particles in a water sample achieved through microscopic examination of treated water with a special device that classifies particles by number and size.

Particulate loading: Mass of particulates per unit volume of air or water.

Particulate matter: Fine liquid or solid particles such as dust, smoke, mist, fumes, or smog, found in air or emissions; also, very small solids suspended in water.

Passive solar heating: Solar energy system that does not use pumps, fans, or other external mechanical power to move the solar heat.

Pathogen: Microorganism that can cause disease in humans, animals, and plants.

Pathway: Physical course a chemical or pollutant takes from its source to the exposed organism.

Peak level: Level of airborne pollutant contaminants much higher than average or occurring for a short period of time in response to sudden releases.

Percolation: Seepage of water through soils or other media.

Permeability: Rate at which liquids pass through soil or other porous materials in a specified direction.

Permissible dose: Amount of a chemical that may be received by an individual without the expectation of a significantly harmful result.

Persistence: Length of time a compound stays in the environment once introduced.

Pest: Unwanted insect, rodent, nematode, fungus, weed, or other form of terrestrial or aquatic plant or animal life; pests are generally injurious to health or the environment.

Pesticide: Substance or mixture intended for preventing, destroying, repelling, or mitigating any pest.

Petrochemical: Organic or inorganic compound or mixture derived from petroleum. Examples include organic chemicals, plastics, resins, synthetic fibers, organic dyes, organic pigments, detergents, surface active agents, carbon black, and ammonia.

Petroleum: Crude oil or any fraction thereof that is liquid under normal conditions of temperature and pressure.

pH: An expression of the intensity of the basic or acid condition of a liquid; may range from 0 to 14, where 0 is the most acid and 7 is neutral. Natural waters usually have a pH between 6.5 and 8.5.

Pharmacokinetics: Science concerned with the way that drugs move through an organism after they are swallowed, injected, or otherwise introduced into the body.

Phenol: Organic compound that is a by-product of petroleum refining, tanning, and textile, dye, and resin manufacturing. Low concentrations of phenols cause taste and odor problems in water; higher concentrations can kill aquatic life and humans.

Phosphate: Chemical compound containing phosphorus. Phosphates are used in foods and beverages, in a wide range of consumer products and industrial processes, and as fertilizer.

Phosphorus: Essential chemical food element that can contribute to the eutrophication of lakes and other water bodies. Increased phosphorus levels result from discharge of phosphorus-containing waters such as farm runoff into surface waters.

Photochemical oxidant: Air pollutant formed by the action of sunlight on oxides of nitrogen and hydrocarbons.

Photochemical smog: Hazy air pollution caused by chemical reactions between sunlight and various pollutants emitted from different sources.

Photosynthesis: The manufacture by plants of carbohydrates and oxygen from carbon dioxide mediated by chlorophyll in the presence of sunlight.

Photovoltaic cell: Electronic device made of layered semiconductor materials that is capable of converting light directly into electricity.

Phytoplankton: That portion of the plankton community comprising tiny plants.

Phytotoxic: Harmful to plants.

Phytotreatment: Cultivation of specialized plants that absorb specific contaminants from soil or water through their roots or foliage.

Plankton: Tiny plants and animals that live in water.

Plasmid: Minute, circular piece of genetic material occurring in bacteria. Plasmids can be augmented with new genetic material and introduced to live bacteria, some of which will take up the altered plasmids and acquire the new genes.

Plutonium (Pu): Radioactive, fissionable metallic element chemically similar to uranium; occurs naturally or as a by-product of the fission reaction that occurs in a uranium-fuel nuclear reactor.

Point source: Stationary location or fixed facility from which pollutants are discharged, or any single identifiable source of pollution, such as a pipe, ditch, ship, ore pit, or factory smokestack.

Pollen: The fertilizing element of flowering plants; a nontoxic but allergenic background air pollutant.

Pollutant: Any substance introduced into the environment that adversely affects the usefulness of a resource or the health of humans, animals, or ecosystems.

Pollution: Presence of a substance in the environment that, because of its chemical composition or quantity, prevents the functioning of natural processes and produces undesirable environmental and health effects.

Polonium (Po): Radioactive element that occurs in pitchblende and other uranium-containing ores. From a public health standpoint, the radon decay products of most concern are polonium 214 and polonium 218.

Polychlorinated biphenyls (PCBs): Group of toxic, environmentally persistent compounds once widely used in the insulation of electrical transformers and capacitors. Production of PCBs was banned in the United States in 1979.

Polyelectrolytes: Synthetic chemicals used to help solids to clump during sewage treatment.

Population: Group of interbreeding organisms occupying a particular space; the number of humans or other living creatures in a designated area.

Porosity: Degree to which soil, gravel, sediment, or rock is permeated with pores or cavities through which water or air can move.

Postconsumer recycling: Use of materials generated from residential and consumer waste for new or similar purposes.

Potable water: Water that is safe for drinking and cooking.

Precipitator: Pollution control device that collects particles from an airstream.

Precursor: In photochemistry, a compound antecedent to a pollutant.

Prion: Abnormal protein particle believed to be the cause of diseases such as bovine spongiform encephalopathy (mad cow disease).

Prior appropriation doctrine: Legal principle that allocates the rights to the use of water resources on a first-come, first-served basis.

Process wastewater: Any water that comes into contact with any raw material, product, by-product, or waste.

Putrefaction: Biological decomposition of organic matter; associated with anaerobic conditions.

Pyrite: Common iron sulfide mineral (FeS_2); the sulfate produced when it reacts with oxygen and water results in acid mine drainage.

Pyrolysis: Decomposition of a chemical by extreme heat.

Quad: Measure of heat energy equaling one quadrillion (1×10^{15}) British thermal units.

Quench tank: Water-filled tank used to cool incinerator residues or hot materials during industrial processes.

Radiation: Transmission of energy through space or any medium; also known as radiant energy.

Radiation standards: Regulations that set maximum exposure limits for protection of the public from radioactive materials.

Radioactivity: Spontaneous emission of energetic particles from the nucleus of an atom.

Radioisotopes: Chemical variants of radioactive elements with potentially oncogenic, teratogenic, and mutagenic effects on the human body.

Radon: Colorless, naturally occurring, radioactive inert gas formed by radioactive decay of radium atoms in soil or rocks.

Raw water: Intake water prior to any treatment or use.

Recarbonization: Process in which carbon dioxide is bubbled into water to lower the water's pH.

Receiving waters: Watercourse—river, lake, ocean, or stream—into which wastewater or treated effluent is discharged.

Receptor: An ecological entity exposed to a stressor.

Reclamation: In land use, restoration of a surface environment to acceptable conditions approaching the state of the site before it was disturbed by development, mining, or other human activities. In recycling, the restoration of materials found in the waste stream to a beneficial use that may be for purposes other than the original use.

Recycling: Minimization of waste generation through the recovery and reprocessing into new products of usable items that might otherwise become waste.

Red tide: Proliferation of a marine plankton toxic and often fatal to fish, perhaps stimulated by the addition of nutrients; such a tide can be red, green, or brown in color, depending on the coloration of the plankton.

Refinery: Facility that produces finished petroleum products from crude oil, natural gas liquids, and related materials.

Reforestation: Replanting of trees on forestlands that have recently been cleared of trees.

Refueling emissions: Pollutants emitted during vehicle refueling.

Release: Any spilling, leaking, pumping, pouring, emitting, emptying, discharging, injecting, escaping, leaching, dumping, or disposing into the environment of a hazardous or toxic chemical or extremely hazardous substance.

Remediation: Removal, containment, or other cleanup of a toxic spill or a site contaminated by hazardous waste.

Remote sensing: Collection and interpretation of information about objects without physical contact with the objects; examples include satellite imaging and aerial photography.

Removal action: Short-term immediate actions taken to address releases of hazardous substances that require expedited response.

Renewable energy resources: Naturally replenishing energy resources. Examples include biomass, hydroelectric, geothermal, solar, wind, ocean thermal, wave action, and tidal action.

Reservoir: Any natural or artificial holding area used to store, regulate, or control water.

Residue: Dry solids remaining after the evaporation of water or sludge.

Resistance: For plants and animals, the ability to withstand poor environmental conditions or attacks by chemicals or disease. Resistance may be inborn or acquired.

Resource recovery: Process of obtaining matter or energy from materials formerly discarded.

Riparian rights: Entitlement of a property owner to certain uses of water on or bordering the owner's property, including the right to prevent diversion or misuse of upstream waters.

Risk: A measure of the probability that damage to life, health, property, or the environment will occur as a result of a given hazard.

River basin: Land area drained by a river and its tributaries.

Rodenticide: Chemical or other agent used to destroy rodent pests, such as rats and mice.

Rubbish: Solid waste, excluding food waste and ashes, from homes, institutions, and workplaces.

Runoff: That part of precipitation, snowmelt, or irrigation water that runs off the land into streams or other surface water. Runoff can carry pollutants from the air and land into receiving waters.

Safe water: Water that contains no harmful bacteria, toxic materials, or chemicals and is considered safe for drinking, even if it may have taste, odor, color, and certain mineral problems.

Salinity: Amount of salt in water.

Saltwater intrusion: Contamination of fresh surface water or groundwater sources by the movement of salt water into those sources.

Salvage: Utilization of waste materials.

Sanitary sewers: Underground pipes that carry off only sewage from residences and businesses, not stormwater.

Sanitation: Control of physical factors in the human environment that could harm development, health, or survival.

Saturation: Condition of a liquid when it has taken into solution the maximum possible quantity of a given substance at a given temperature and pressure; in soil, the degree to which its pore spaces are occupied by water.

Scrap: Materials discarded from manufacturing operations that may be suitable for reprocessing.

Scrubber: Air-pollution control device that uses a spray of water or reactant or a dry process to trap pollutants in emissions and prevent their release.

Secondary effect: Action of a stressor on supporting components of the ecosystem, which in turn affect the ecological component of concern.

Secondary materials: Materials that have been manufactured and used at least once and are to be used again.

Sedimentation: In wastewater treatment, the process of allowing suspended solids to settle out; in general, the increase of suspended solids in water or the buildup of deposited sediments in stream channels, reservoirs, and the like.

Sediments: Soil, sand, and minerals washed from land into water, generally by rain, snowmelt, or irrigation runoff.

Seepage: Percolation of water through the soil from unlined canals, ditches, laterals, watercourses, or water storage facilities.

Selective pesticide: Chemical designed to affect only certain types of pests, leaving other plants and animals unharmed.

Septic system: On-site system designed to treat and dispose of domestic sewage.

Septic tank: Underground storage tank for domestic sewage wastes; within the tank, bacteria disintegrate the solid matter.

Settling tank: Holding area for wastewater in which heavier particles sink to the bottom for removal and disposal.

Sewage: Human waste and wastewater produced by residential and commercial sources and discharged into sewers or septic systems.

Sewer: Channel or conduit that carries wastewater and stormwater runoff from the source to a treatment plant, receiving stream, or other body of water.

Sewerage: The entire system of sewage collection, treatment, and disposal.

Sharps: Hypodermic needles, syringes (with or without attached needles), Pasteur pipettes, scalpel blades, blood vials, needles with attached tubing, and culture dishes used in animal or human patient care or treatment, or in medical, research, or industrial laboratories.

Silt: Sedimentary materials composed of fine or intermediate-sized mineral particles.

Silviculture: Management of forestland for timber.

Sink: Place in the environment where a compound or material collects.

Sinking: Method of controlling a marine oil spill in which an agent is used to trap the oil and sink it to the bottom of the body of water, where the agent and the oil are biodegraded.

Siting: Process of choosing a location for a facility.

Skimming: Use of a machine to remove oil or scum from the surface of water.

Sludge: Semisolid residue from any of a number of industrial or technological processes; some types of sludge are considered hazardous waste.

Slurry: Watery mixture of insoluble matter resulting from some pollution-control techniques.

Smog: See **Photochemical smog**.

Smoke: Particles suspended in air after incomplete combustion.

Soft water: Any water that does not contain a significant amount of dissolved minerals such as salts of calcium or magnesium.

Solar cell: See **Photovoltaic cell**.

Solid waste: Nonliquid, nonsoluble materials ranging from municipal garbage to industrial wastes that contain complex and sometimes hazardous substances.

Soot: Carbon dust formed by incomplete combustion.

Source reduction: Minimization of the amount of materials entering the waste stream from a specific source through the redesign of products or patterns of production or consumption.

Source separation: Segregation of various wastes at the point of generation (such as the household separation of discarded paper, metal, and glass from other wastes to make recycling simpler and more efficient).

Spent fuel: Irradiated fuel that is no longer able to sustain a fission reaction.

Spoil: Dirt or rock removed from its original location—destroying the composition of the soil in the process—in the course of processes such as strip mining, dredging, or construction.

Spongiform encephalopathy: Degenerative brain disease in which the brain develops porous sponge-like lesions; believed to be caused by prions. Examples include scrapie and mad cow disease.

Sprawl: Unplanned development of open land.

Spray tower scrubber: Device that sprays alkaline water into a chamber where acid gases are present to aid in neutralizing the gas.

Stabilization: In wastewater sludge management, conversion of active organic matter in the sludge into inert, harmless material.

Stack: A chimney, smokestack, or vertical pipe that discharges used air.

Stack effect: Upward movement of air, as in a chimney, because the air is warmer than the ambient atmosphere.

Stagnation: Lack of motion in a mass of air or water that holds pollutants in place.

Stakeholder: Any organization, governmental entity, or individual that has a stake in or may be affected by a given approach to environmental regulation, pollution prevention, energy conservation, or other activity.

Storm sewer: System of pipes (separate from a sanitary sewer) that carries stormwater and other runoff from building exteriors, pavement, and land surfaces.

Stratification: Separation into layers.

Stratigraphy: Study of the formation, composition, and sequence of sediments, whether consolidated or not.

Stratosphere: Region of the earth's upper atmosphere that includes the ozone layer.

Stressors: Physical, chemical, or biological entities that can induce adverse effects on ecosystems or human health.

Strip mining: See **surface mining**.

Sulfur dioxide (SO$_2$): Pungent, colorless toxic gas ejected by volcanoes, produced during fossil-fuel combustion, and emitted from various industrial operations; plays a role in the formation of acid deposition and particulate pollution.

Sump: Pit or tank that catches liquid runoff for drainage or disposal.

Superfund: See **Comprehensive Environmental Response, Compensation, and Liability Act**.

Surface mining: Extraction of comparatively shallow mineral deposits using methods in which the overburden is scraped away and the exposed deposits are mined with surface excavation equipment. Forms of surface mining include open-pit, strip, contour, area, and auger mining.

Surface runoff: Precipitation, snowmelt, or irrigation water in excess of what can infiltrate the soil surface or be stored in small surface depressions; a major transporter of nonpoint source pollutants in rivers, streams, and lakes.

Surface water: All water naturally open to the atmosphere (rivers, lakes, reservoirs, ponds, streams, impoundments, seas, estuaries).

Surfactant: Detergent compound that promotes lathering.

Suspended solids: Small solid particles that float on the surface of, or are suspended in, water, sewage, or other liquids.

Systemic pesticide: Chemical absorbed by an organism that interacts with the organism to make it toxic to pests.

Tail water: Runoff from the lower end of an irrigated field.

Tailings: Residues of raw material or waste separated out during the processing of crops or mineral ores.

Temperature inversion: Condition in which a layer of warm air prevents the rise of cooling air and traps pollutants beneath it. Temperature inversions can cause acute air-pollution episodes.

Teratogen: Substance that can cause nonhereditary birth defects in a developing fetus.

Teratogenesis: Introduction of nonhereditary birth defects in a developing fetus by exogenous factors such as physical or chemical agents acting in the womb to interfere with normal embryonic development.

Terracing: Building of dikes along the contour of sloping farmland to hold runoff and sediment to reduce erosion.

Thermal pollution: Discharge of heated water from industrial processes into surface or marine waters; such pollution can kill or injure aquatic organisms.

Threatened species: Any species of living organism that is at risk of becoming endangered in the near future.

Threshold: Lowest dose of a chemical at which a specified measurable effect is observed and below which it is not observed.

Threshold level: Time-weighted average pollutant concentration values, exposure beyond which is likely to adversely affect human health.

Toxic cloud: Airborne plume of gases, vapors, fumes, or aerosols containing toxic materials.

Toxic concentration: Concentration at which a substance produces a toxic effect.

Toxic dose: Dose level at which a substance produces a toxic effect.

Toxic pollutants: Materials that cause death, disease, or birth defects in organisms that ingest, inhale, or absorb them. The quantities and exposures necessary to cause these effects can vary widely.

Toxic substance: Chemical or mixture that may present an unreasonable risk of injury to health or the environment.

Toxic waste: Waste that can produce injury if inhaled, swallowed, or absorbed through the skin.

Toxicant: Harmful substance or agent that may injure an exposed organism.

Toxicity: Degree to which a substance or mixture of substances can harm humans or animals.

Toxicity assessment: Characterization of the toxicological properties and effects of a chemical, with special emphasis on establishing how dosage relates to physiological response.

Toxicity testing: Biological testing (usually with an invertebrate, fish, or small mammal) to determine the adverse effects of a substance or mixture of substances.

Transboundary pollution: Air or water pollution that travels from one jurisdiction to another, often crossing state or international boundaries.

Transpiration: Process by which water vapor is lost to the atmosphere from living plants. The term can also be applied to the quantity of water thus dissipated.

Transuranic: Pertaining to an element such as plutonium that has an atomic number greater than that of uranium (92).

Trash: Material considered worthless or offensive that is thrown away. Generally defined as dry waste material, but in common usage it is a synonym for garbage, rubbish, or refuse.

Treated wastewater: Wastewater that has been subjected to one or more physical, chemical, and biological processes to reduce its potential of being a health hazard.

Treatment, storage, and disposal facility (TSDF): Hazardous waste management facility that treats, stores, and disposes of hazardous wastes; in the United States TSDFs must comply with the provisions of the Resource Conservation and Recovery Act.

Troposphere: Layer of the atmosphere closest to the earth's surface.

Tundra: Type of treeless ecosystem dominated by lichens, mosses, grasses, and woody plants.

Turbidity: In water, cloudiness caused by suspended silt or organic matter; in air, haziness resulting from the presence of particulate matter and pollutants.

Ultraclean coal (UCC): Coal that is washed, ground into fine particles, then chemically treated to remove sulfur, ash, and other impurities; usually briquetted and coated with a sealant made from coal.

Unconfined aquifer: Aquifer containing water that is not under pressure and able to receive water from the surface. The water level in a well drilled into an unconfined aquifer is the same as the water table outside the well.

Unsaturated zone: Area above the water table where soil pores are not fully saturated, although some water may be present.

Upgradient: Direction opposite the one in which groundwater flows; similar to "upstream" for surface water.

Upper detection limit: The largest concentration that an instrument can reliably detect.

Upwelling: Movement of subsurface ocean water, often cold and nutrient-rich, toward the ocean surface.

Urban runoff: Stormwater from city streets and adjacent domestic or commercial properties that carries pollutants of various kinds into sewer systems and receiving waters.

Use cluster: Set of competing chemicals, processes, or technologies that can substitute for one another in performing a particular function.

User fee: Fee collected from only those persons who use a particular service, in contrast to a fee collected from the public in general.

Utility load: The total electricity demand for a utility district.

Vapor dispersion: Movement of vapor clouds in air caused by wind, thermal action, gravity spreading, and mixing.

Vapor plumes: Flue gases that are visible because they contain water droplets.

Vapor pressure: A measure of a substance's propensity to evaporate.

Variance: Government permission for a delay or exception in the application of a given law, ordinance, or regulation.

Vector: Organism, often an insect or rodent, that carries disease.

Vegetative controls: Nonpoint source pollution-control practices that involve vegetative cover to reduce erosion and minimize migration of pollutants.

Ventilation rate: Rate at which air enters and leaves a building.

Venturi scrubbers: Air-pollution-control devices that use water to remove particulate matter from emissions.

Vessicant: See **Blister agent**.

Vinyl chloride: Chemical compound used in producing some plastics; associated with central nervous system effects, liver damage, and cancer in humans.

Virgin materials: Resources extracted from nature in their raw form, such as timber or metal ore.

Virus: Minute infectious agent that can grow and reproduce only within the living cells of another organism.

Viscosity: Molecular friction within a fluid that produces flow resistance.

Volatile organic compounds (VOCs): Large class of organic chemical compounds that evaporate under normal atmospheric temperatures and pressures. VOCs have a wide variety of household and industrial uses and are common pollutants found in water as well as indoor and outdoor air.

Volume reduction: Processing of waste materials to decrease the amount of space they occupy, usually through compacting, shredding, incinerating, or composting.

Vulnerability analysis: Assessment of elements in the community that are susceptible to damage if hazardous materials are released.

Vulnerable zone: Area over which the airborne concentration of a chemical accidentally released could reach the level of concern.

Waste: Unwanted materials left over from manufacturing processes and refuse from places of human or animal habitation.

Waste characterization: Identification of the chemical and microbiological constituents of waste materials.

Waste minimization: Use of measures or techniques that reduce the amount of wastes generated during industrial production processes or other waste-producing activities.

Waste reduction: Use of source reduction, recycling, or composting to prevent or reduce waste generation.

Waste stream: The flow of solid waste from homes, businesses, institutions, and manufacturing plants that is recycled, burned, or disposed of in landfills.

Wastewater: Spent or used water from homes, institutions, communities, farms, and industries that contains dissolved or suspended matter.

Wastewater treatment plant: Facility in which a series of tanks, screens, and filters are used in a series of processes to remove pollutants from wastewater.

Water pollution: Harmful or objectionable material present in water to an extent sufficient to damage the water's quality and affect its ability to support human or animal life or health.

Water table: The upper surface of groundwater.

Water well: Excavation made for the purpose of locating, acquiring, developing, or artificially recharging groundwater.

Watershed: Land area that drains into a water stream. The watershed for a major river may encompass a number of smaller watersheds that ultimately combine at a common point.

Wetlands: Areas with soils saturated by surface or groundwater, with vegetation adapted for life under those soil conditions; examples include swamps, bogs, fens, marshes, and estuaries.

Wildlife refuge: Area designated for the protection of wild animals, within which hunting and fishing are either prohibited or strictly controlled.

Xenobiota: Any biota displaced from its normal habitat; a chemical foreign to a biological system.

Yard waste: The part of solid waste composed of grass clippings, leaves, twigs, branches, and other garden and landscaping refuse.

Zero air: Atmospheric air purified to contain less than 0.1 parts per million total hydrocarbons.

Bibliography

CONTENTS

AGRICULTURE AND FOOD

Conkin, Paul Keith. *A Revolution Down on the Farm: The Transformation of American Agriculture Since 1929.* Lexington: University Press of Kentucky, 2009.

Conway, Gordon, and Jules N. Pretty. *Unwelcome Harvest: Agriculture and Pollution.* Sterling, Va.: Earthscan, 2009.

Falkner, Robert, ed. *The International Politics of Genetically Modified Food: Diplomacy, Trade, and Law.* New York: Palgrave Macmillan, 2007.

Hahlbrock, Klaus. *Feeding the Planet: Environmental Protection Through Sustainable Agriculture.* London: Haus, 2009.

Holthaus, Gary H. *From the Farm to the Table: What All Americans Need to Know About Agriculture.* Lexington: University Press of Kentucky, 2009.

Imhoff, Dan, ed. *The CAFO Reader: The Tragedy of Industrial Animal Factories.* Healdsburg, Calif.: Watershed Media, 2010.

Koul, Opender, and Gerrit W. Cuperus, eds. *Ecologically Based Integrated Pest Management.* Wallingford, England: CABI, 2007.

Lyman, Howard F., and Glen Merzer. *Mad Cowboy: Plain Truth from the Cattle Rancher Who Won't Eat Meat.* New York: Touchstone, 2001.

Nestle, Marion. *Safe Food: Bacteria, Biotechnology, and Bioterrorism.* Berkeley: University of California Press, 2004.

Pretty, Jules N., ed. *Sustainable Agriculture and Food.* 4 vols. Sterling, Va.: Earthscan, 2008.

United Nations Food and Agriculture Organization. *The State of Food and Agriculture 2009: Livestock in the Balance.* Rome: Author, 2010.

_____. *The State of Food Insecurity in the World 2010: Addressing Food Insecurity in Protracted Crises.* Rome: Author, 2009.

ANIMALS AND ENDANGERED SPECIES

Barrow, Mark V., Jr. *Nature's Ghosts: Confronting Extinction from the Age of Jefferson to the Age of Ecology.* Chicago: University of Chicago Press, 2009.

Bräutigam, Amie, and Martin Jenkins. *The Red Book: The Extinction Crisis Face to Face.* Gland, Switzerland: IUCN, 2001.

Clover, Charles. *The End of the Line: How Overfishing Is Changing the World and What We Eat.* Berkeley: University of California Press, 2008.

Fascione, Nina, Aimee Delach, and Martin E. Smith, eds. *People and Predators: From Conflict to Coexistence.* Washington, D.C.: Island Press, 2004.

Groom, Martha J., Gary K. Meffe, and Carl Ronald Carroll. *Principles of Conservation Biology.* 3d ed. Sunderland, Mass.: Sinauer, 2006.

McGavin, George. *Endangered: Wildlife on the Brink of Extinction.* Buffalo, N.Y.: Firefly Books, 2006.

Mackay, Richard. *The Atlas of Endangered Species.* Rev. and updated 3d ed. Sterling, Va.: Earthscan, 2009.

Sinclair, Anthony R. E., John M. Fryxell, and Graeme Caughley. *Wildlife Ecology, Conservation, and Management.* 2d ed. Malden, Mass.: Blackwell, 2006.

Stolzenburg, William. *Where the Wild Things Were: Life, Death, and Ecological Wreckage in a Land of Vanishing Predators.* New York: Bloomsbury, 2009.

Vié, Jean-Christophe, Craig Hilton-Taylor, and Simon N. Stuart, eds. *Wildlife in a Changing World: An Anal-*

ysis of the 2008 IUCN Red List of Threatened Species. Gland, Switzerland: IUCN, 2009.

Williams, Byron K., James D. Nichols, and Michael J. Conroy. *Analysis and Management of Animal Populations.* San Diego, Calif.: Academic Press, 2002.

Wilson, Edward O. *The Diversity of Life.* New ed. New York: W. W. Norton, 1999.

ATMOSPHERE AND AIR POLLUTION

Burroughs, H. E., and Shirley J. Hansen. *Managing Indoor Air Quality.* 4th ed. Lilburn, Ga.: Fairmont Press, 2008.

Finlayson-Pitts, Barbara J., and James N. Pitts. *Chemistry of the Upper and Lower Atmosphere: Theory, Experiments, and Applications.* San Diego, Calif.: Academic Press, 2000.

Griffin, Roger D. *Principles of Air Quality Management.* 2d ed. Boca Raton, Fla.: CRC Press, 2007.

Jacobson, Mark Z. *Atmospheric Pollution: History, Science, and Regulation.* New York: Cambridge University Press, 2002.

Kessel, Anthony. *Air, the Environment, and Public Health.* New York: Cambridge University Press, 2006.

Lane, Carter N., ed. *Acid Rain: Overview and Abstracts.* New York: Nova Science, 2003.

Lipton, James P., ed. *Clean Air Act: Interpretation and Analysis.* New York: Nova Science, 2006.

McCarthy, Tom. *Auto Mania: Cars, Consumers, and the Environment.* New Haven, Conn.: Yale University Press, 2007.

Seinfeld, John H., and Spyros N. Pandis. *Atmospheric Chemistry and Physics: From Air Pollution to Climate Change.* 2d ed. Hoboken, N.J.: John Wiley & Sons, 2006.

Simpson, John A., ed. *Preservation of Near-Earth Space for Future Generations.* New York: Cambridge University Press, 2007.

Sokhi, Ranjeet S., ed. *World Atlas of Atmospheric Pollution.* London: Anthem Press, 2008.

U.S. Environmental Protection Agency. *The Plain English Guide to the Clean Air Act.* Research Triangle Park, N.C.: Author, 2007.

BIOTECHNOLOGY AND GENETIC ENGINEERING

Aldridge, Susan. *The Thread of Life: The Story of Genes and Genetic Engineering.* 1996. Reprint. New York: Cambridge University Press, 2000.

Bauer, Martin W., and George Gaskell, eds. *Biotechnology: The Making of a Global Controversy.* New York: Cambridge University Press, 2002.

Bodiguel, Luc, and Michael Cardwell, eds. *The Regulation of Genetically Modified Organisms: Comparative Approaches.* New York: Oxford University Press, 2010.

Castilho, Leda R., et al., eds. *Animal Cell Technology: From Biopharmaceuticals to Gene Therapy.* New York: Taylor & Francis, 2008.

Crommelin, Daan J. A., Robert D. Sindelar, and Bernd Meibohm, eds. *Pharmaceutical Biotechnology: Fundamentals and Applications.* 3d ed. New York: Informa Healthcare, 2008.

Drlica, Karl. *Understanding DNA and Gene Cloning: A Guide for the Curious.* 4th ed. Hoboken, N.J.: John Wiley & Sons, 2004.

Grace, Eric S. *Biotechnology Unzipped: Promises and Realities.* Rev. 2d ed. Washington, D.C.: National Academies Press, 2006.

International Union for Conservation of Nature. *Genetically Modified Organisms and Biosafety: A Background Paper for Decision-Makers and Others to Assist in Consideration of GMO Issues.* Gland, Switzerland: Author, 2004.

Klotzko, Arlene Judith. *A Clone of Your Own? The Science and Ethics of Cloning.* New York: Cambridge University Press, 2006.

Morgan, Rose M. *The Genetics Revolution: History, Fears, and Future of a Life-Altering Science.* Westport, Conn.: Greenwood Press, 2006.

Stewart, C. Neal. *Genetically Modified Planet: Environmental Impacts of Genetically Engineered Plants.* New York: Oxford University Press, 2004.

Watson, James D., et al. *Recombinant DNA: Genes and Genomes—A Short Course.* 3d ed. New York: W. H. Freeman, 2007.

ECOLOGY AND ECOSYSTEMS

Begon, Michael, Colin R. Townsend, and John L. Harper. *Ecology: From Individuals to Ecosystems.* Malden, Mass.: Blackwell, 2008.

Bryant, Raymond L., and Sinéad Bailey. *Third World Political Ecology.* London: Routledge, 2005.

Carson, Rachel. *Silent Spring.* 40th anniversary ed. Boston: Houghton Mifflin, 2002.

Chapin, F. Stuart, III, Pamela A. Matson, and Harold A. Mooney. *Principles of Terrestrial Ecosystem Ecology.* New York: Springer, 2002.

Chasek, Pamela S., David L. Downie, and Janet Welsh Brown. *Global Environmental Politics.* 5th ed. Boulder, Colo.: Westview Press, 2010.

Levin, Simon A., ed. *The Princeton Guide to Ecology.* Princeton, N.J.: Princeton University Press, 2009.

Molles, Manuel Carl. *Ecology: Concepts and Applications.* 5th ed. Boston: McGraw-Hill Higher Education, 2009.

Novacek, Michael J. *Terra: Our 100-Million-Year-Old Ecosystem—and the Threats That Now Put It at Risk.* New York: Farrar, Straus and Giroux, 2007.

Odum, Eugene Pleasants, and Gary W. Barrett. *Fundamentals of Ecology.* 5th ed. Belmont, Calif.: Thomson Brooks/Cole, 2005.

Real, Leslie A., and James H. Brown, eds. *Foundations of Ecology: Classic Papers with Commentaries.* Chicago: University of Chicago Press, 1991.

Walker, Brian Harrison, and David Andrew Salt. *Resilience Thinking: Sustaining Ecosystems and People in a Changing World.* Washington, D.C.: Island Press, 2006.

Woodward, Susan L. *Biomes of Earth: Terrestrial, Aquatic, and Human-Dominated.* Westport, Conn.: Greenwood Press, 2003.

Energy and Energy Use

Ayres, Robert U., and Ed Ayres. *Crossing the Energy Divide: Moving from Fossil Fuel Dependence to a Clean-Energy Future.* Upper Saddle River, N.J.: Wharton School Publishing, 2010.

Elliott, David. *Energy, Society, and Environment.* 2d ed. New York: Routledge, 2003.

International Energy Agency. *Key World Energy Statistics 2010.* Paris: Author, 2010.

Kutz, Myer, and Ali Elkamel, eds. *Environmentally Conscious Fossil Energy Production.* Hoboken, N.J.: John Wiley & Sons, 2010.

McCully, Patrick. *Silenced Rivers: The Ecology and Politics of Large Dams.* Enlarged and updated ed. London: Zed Books, 2001.

McLean-Conner, Penni. *Energy Efficiency: Principles and Practices.* Tulsa, Okla.: PennWell, 2009.

Murray, Raymond L. *Nuclear Energy: An Introduction to the Concepts, Systems, and Applications of Nuclear Processes.* 6th ed. Burlington, Vt.: Butterworth-Heinemann/Elsevier, 2009.

National Research Council. *Electricity from Renewable Resources: Status, Prospects, and Impediments.* Washington, D.C.: National Academies Press, 2010.

Nivola, Pietro S. *The Long and Winding Road: Automotive Fuel Economy and American Politics.* Washington, D.C.: Brookings Institution Press, 2009.

O'Keefe, Philip, et al. *The Future of Energy Use.* 2d ed. Sterling, Va.: Earthscan, 2010.

Sandalow, David. *Freedom from Oil: How the Next President Can End the United States' Oil Addiction.* Columbus, Ohio: McGraw-Hill Professional, 2008.

Twidell, John, and Anthony D. Weir. *Renewable Energy Resources.* 2d ed. London: Taylor & Francis, 2006.

Environment and Society

Brohé, Arnaud, Nick Eyre, and Nicholas Howarth. *Carbon Markets: An International Business Guide.* Sterling, Va.: Earthscan, 2009.

Bullard, Robert D., ed. *The Quest for Environmental Justice: Human Rights and the Politics of Pollution.* San Francisco: Sierra Club Books, 2005.

Cato, Molly Scott. *Green Economics: An Introduction to Theory, Policy, and Practice.* Sterling, Va.: Earthscan, 2009.

Dryzek, John S., et al. *Green States and Social Movements: Environmentalism in the United States, United Kingdom, Germany, and Norway.* New York: Oxford University Press, 2003.

Field, Barry C., and Martha K. Field. *Environmental Economics: An Introduction.* 5th ed. New York: McGraw-Hill/Irwin, 2009.

Ghimire, K. B., and Michel P. Pimbert, eds. *Social Change and Conservation.* Sterling, Va.: Earthscan, 2009.

Grant, John. *The Green Marketing Manifesto.* Hoboken, N.J.: John Wiley & Sons, 2007.

Guruswamy, Lakshman D. *International Environmental Law in a Nutshell.* 3d ed. St. Paul, Minn.: Thomson/West, 2007.

Hansjürgens, Bernd, ed. *Emissions Trading for Climate Policy: U.S. and European Perspectives.* New York: Cambridge University Press, 2010.

Makower, Joel, and Cara Pike. *Strategies for the Green Economy: Opportunities and Challenges in the New World of Business.* New York: McGraw-Hill, 2009.

Schreurs, Miranda A. *Environmental Politics in Japan, Germany, and the United States.* New York: Cambridge University Press, 2002.

Steinway, Daniel M., et al. *Environmental Law Handbook.* 20th ed. Lanham, Md.: Government Institutes, 2009.

Forests and Plants

Deal, Robert L., Rachel White, and Gary L. Benson, eds. *Sustainable Forestry Management and Wood Production in a Global Economy.* Binghamton, N.Y.: Haworth Food & Agricultural Products Press, 2007.

Hays, Samuel P. *The American People and the National*

Forests: The First Century of the U.S. Forest Service. Pittsburgh: University of Pittsburgh Press, 2009.

Higman, Sophie, et al., eds. *The Sustainable Forestry Handbook.* Sterling, Va.: Earthscan, 2005.

Humphreys, David. *Logjam: Deforestation and the Crisis of Global Governance.* Sterling, Va.: Earthscan, 2009.

Johnson, Edward A., and Kiyoko Miyanishi, eds. *Forest Fires: Behavior and Ecological Effects.* San Diego, Calif.: Academic Press, 2001.

London, Mark, and Brian Kelly. *The Last Forest: The Amazon in the Age of Globalization.* New York: Random House, 2007.

Omi, Philip N. *Forest Fires: A Reference Handbook.* Santa Barbara: ABC-CLIO, 2005.

Palmer, Charles, and Stefanie Engel, eds. *Avoided Deforestation: Prospects for Mitigating Climate Change.* New York: Routledge, 2009.

Perry, David A., Ram Oren, and Stephen C. Hart. *Forest Ecosystems.* 2d ed. Baltimore: The Johns Hopkins University Press, 2008.

Primack, Richard B., and Richard Corlett. *Tropical Rain Forests: An Ecological and Biogeographical Comparison.* Malden, Mass.: Blackwell, 2005.

United Nations Food and Agriculture Organization. *Global Forest Resources Assessment 2010.* Rome: Author, 2010.

_____. *State of the World's Forests 2009.* Rome: Author, 2009.

HUMAN HEALTH AND THE ENVIRONMENT

Brown, Phil. *Toxic Exposures, Contested Illnesses, and the Environmental Health Movement.* New York: Columbia University Press, 2007.

Conant, Jeff, and Pam Fadem. *A Community Guide to Environmental Health.* Berkeley, Calif.: Hesperian Foundation, 2008.

Fong, Ignatius W., and Karl Drlica, eds. *Antimicrobial Resistance and Implications for the Twenty-first Century.* New York: Springer, 2008.

Frumkin, Howard, ed. *Environmental Health: From Global to Local.* San Francisco: Jossey-Bass, 2010.

Hansen, Deborah Kay, and Barbara D. Abbott, eds. *Developmental Toxicology.* 3d ed. New York: Informa Healthcare, 2009.

Hernan, Robert Emmet. *This Borrowed Earth: Lessons from the Fifteen Worst Environmental Disasters Around the World.* New York: Palgrave Macmillan, 2010.

Kerns, Thomas A. *Environmentally Induced Illnesses:*

Ethics, Risk Assessment, and Human Rights. Jefferson, N.C.: McFarland, 2001.

Lashley, Felissa R., and Jerry D. Durham, eds. *Emerging Infectious Diseases: Trends and Issues.* 2d ed. New York: Springer, 2007.

Loomis, Ted A., and A. Wallace Hayes. *Loomis's Essentials of Toxicology.* 4th ed. San Diego, Calif.: Academic Press, 1996.

Maxwell, Nancy Irwin. *Understanding Environmental Health: How We Live in the World.* Sudbury, Mass.: Jones and Bartlett, 2009.

Moore, Gary S. *Living with the Earth: Concepts in Environmental Health Science.* 3d ed. Boca Raton, Fla.: CRC Press, 2007.

Schardein, James L. *Chemically Induced Birth Defects.* 3d ed. New York: Marcel Dekker, 2000.

LAND AND LAND USE

Braimoh, Ademola K., and Paul L. G. Vlek, eds. *Land Use and Soil Resources.* Dordrecht, Netherlands: Springer, 2008.

Chalifour, Nathalie J., et al., eds. *Land Use Law for Sustainable Development.* New York: Cambridge University Press, 2007.

Chiras, Daniel D., and John P. Reganold. *Natural Resource Conservation: Management for a Sustainable Future.* 10th ed. Upper Saddle River, N.J.: Prentice Hall, 2010.

Dieterich, Martin, and Jan van der Straaten, eds. *Cultural Landscapes and Land Use: The Nature Conservation-Society Interface.* Dordrecht, Netherlands: Kluwer Academic, 2004.

Duerksen, Christopher J., and Cara Snyder. *Nature-Friendly Communities: Habitat Protection and Land Use.* Washington, D.C.: Island Press, 2005.

Geist, Helmut. *The Causes and Progression of Desertification.* Burlington, Vt.: Ashgate, 2005.

King, Michael D., et al., eds. *Our Changing Planet: The View from Space.* New York: Cambridge University Press, 2007.

Kleppel, G. S., M. Richard DeVoe, and Mac V. Rawson, eds. *Changing Land Use Patterns in the Coastal Zone: Managing Environmental Quality in Rapidly Developing Regions.* Albany, N.Y.: Springer, 2006.

Mason, Robert J. *Collaborative Land Use Management: The Quieter Revolution in Place-Based Planning.* Lanham, Md.: Rowman & Littlefield, 2008.

Travis, William R. *New Geographies of the American West: Land Use and the Changing Patterns of Place.* Washington, D.C.: Island Press, 2007.

Unger, Paul W. *Soil and Water Conservation Handbook: Policies, Practices, Conditions, and Terms.* Binghamton, N.Y.: Haworth Press, 2006.

United Nations Convention to Combat Desertification Secretariat. *Desertification: Coping with Today's Global Challenges.* Eschborn, Germany: Deutsche Gesellschaft für Technische Zusammenarbeit, 2008.

Nuclear Power and Radiation

Bayliss, Colin R., and Kevin K. Langley. *Nuclear Decommissioning, Waste Management, and Environmental Site Remediation.* Amsterdam: Butterworth-Heinemann, 2003.

Bodansky, David. *Nuclear Energy: Principles, Practices, and Prospects.* 2d ed. New York: Springer, 2004.

Bréchignac, François, and Brenda J. Howard, eds. *Radioactive Pollutants: Impact on the Environment.* Les Ulis, France: EDP Sciences, 2001.

Edelstein, Michael R., Maria Tysiachniouk, and Lyudmila V. Smirnova, eds. *Cultures of Contamination: Legacies of Pollution in Russia and the U.S.* Amsterdam: Elsevier JAI, 2007.

Hore-Lacy, Ian. *Nuclear Energy in the Twenty-first Century: The World Nuclear University Primer.* London: World Nuclear University Press, 2006.

Mahaffey, James A. *Atomic Awakening: A New Look at the History and Future of Nuclear Power.* New York: Pegasus Books, 2009.

Miller, Alexandra C., ed. *Depleted Uranium: Properties, Uses, and Health Consequences.* Boca Raton, Fla.: CRC Press, 2007.

Murray, Raymond L. *Nuclear Energy: An Introduction to the Concepts, Systems, and Applications of Nuclear Processes.* 6th ed. Burlington, Vt.: Butterworth-Heinemann/Elsevier, 2009.

Pusch, Roland. *Geological Storage of Highly Radioactive Waste: Current Concepts and Plans for Radioactive Waste Disposal.* New York: Springer, 2008.

Schneider, Mycle, et al. *The World Nuclear Industry Status Report 2009, with Particular Emphasis on Economic Issues.* Bonn, Germany: German Federal Ministry for Environment, Nature Conservation, and Reactor Safety, 2009.

Vandenbosch, Robert, and Susanne E. Vandenbosch. *Nuclear Waste Stalemate: Political and Scientific Controversies.* Salt Lake City: University of Utah Press, 2007.

Wittner, Lawrence S. *Confronting the Bomb: A Short History of the World Nuclear Disarmament Movement.* Stanford, Calif.: Stanford University Press, 2009.

Philosophy and Ethics

Anderson, Terry L., and Donald Leal. *Free Market Environmentalism.* Rev. ed. New York: Palgrave, 2001.

Barnhill, David Landis, and Roger S. Gottlieb, eds. *Deep Ecology and World Religions: New Essays on Sacred Grounds.* Albany: State University of New York Press, 2001.

DesJardins, Joseph R. *Environmental Ethics: An Introduction to Environmental Philosophy.* Belmont, Calif.: Thomson/Wadsworth, 2006.

Foltz, Bruce V., and Robert Frodeman, eds. *Rethinking Nature: Essays in Environmental Philosophy.* Bloomington: Indiana University Press, 2004.

Ip, King-Tak, ed. *Environmental Ethics: Intercultural Perspectives.* Amsterdam: Rodopi, 2009.

Liddick, Donald R. *Eco-Terrorism: Radical Environmental and Animal Liberation Movements.* Westport, Conn.: Praeger, 2006.

Moore, Kelly. *Disrupting Science: Social Movements, American Scientists, and the Politics of the Military, 1945-1975.* Princeton, N.J.: Princeton University Press, 2008.

Newbold, Heather, ed. *Life Stories: World-Renowned Scientists Reflect on Their Lives and the Future of Life on Earth.* Berkeley: University of California Press, 2000.

Petit, Patrick Uwe. *Earth Capitalism: Creating a New Civilization Through a Responsible Market Economy.* New Brunswick, N.J.: Transaction, 2011.

Pojman, Louis P., and Paul Pojman, eds. *Environmental Ethics: Readings in Theory and Application.* 5th ed. Belmont, Calif.: Thomson/Wadsworth, 2008.

Steiner, Gary. *Anthropocentrism and Its Discontents: The Moral Status of Animals in the History of Western Philosophy.* Pittsburgh: University of Pittsburgh Press, 2005.

Warren, Karen. *Ecofeminist Philosophy: A Western Perspective on What It Is and Why It Matters.* Lanham, Md.: Rowman & Littlefield, 2000.

Pollutants and Toxins

Batty, Lesley C., and Kevin B. Hallberg, eds. *Ecology of Industrial Pollution.* New York: Cambridge University Press, 2010.

Cooper, John R., Keith Randle, and Ranjeet S. Sokhi. *Radioactive Releases in the Environment: Impact and Assessment.* Hoboken, N.J.: John Wiley & Sons, 2003.

Eisler, Ronald. *Eisler's Encyclopedia of Environmentally Hazardous Priority Chemicals.* Amsterdam: Elsevier, 2007.

Harrad, Stuart, ed. *Persistent Organic Pollutants.* Hoboken, N.J.: John Wiley & Sons, 2010.

Hill, Marquita K. *Understanding Environmental Pollution.* 2d ed. New York: Cambridge University Press, 2004.

Johansen, Bruce E. *The Dirty Dozen: Toxic Chemicals and the Earth's Future.* Westport, Conn.: Praeger, 2003.

Matthews, Graham. *Pesticides: Health, Safety, and the Environment.* Ames, Iowa: Blackwell, 2006.

Newman, Michael C. *Fundamentals of Ecotoxicology.* 3d ed. Boca Raton, Fla.: CRC Press, 2009.

Ross, Benjamin, and Steven Amter. *The Polluters: The Making of Our Chemically Altered Environment.* New York: Oxford University Press, 2010.

Walker, Colin Harold. *Organic Pollutants: An Ecotoxicological Perspective.* Boca Raton, Fla.: CRC Press, 2009.

Wang, Lawrence K, et al. *Heavy Metals in the Environment.* Boca Raton, Fla.: CRC Press, 2009.

Williamson, Mark. *Space: The Fragile Frontier.* Reston, Va.: American Institute of Aeronautics and Astronautics, 2006.

POPULATION ISSUES

American Association for the Advancement of Science. *AAAS Atlas of Population and Environment.* Berkeley: University of California Press, 2001.

Brown, Lester R., Gary Gardner, and Brian Halweil. *Beyond Malthus: Nineteen Dimensions of the Population Challenge.* Sterling, Va.: Earthscan, 2000.

Eager, Paige Whaley. *Global Population Policy: From Population Control to Reproductive Rights.* Burlington, Vt.: Ashgate, 2004.

Gordon, Linda. *The Moral Property of Women: A History of Birth Control Politics in America.* 3d ed. Urbana: University of Illinois Press, 2002.

Greenhalgh, Susan. *Just One Child: Science and Policy in Deng's China.* Berkeley: University of California Press, 2008.

Kahl, Colin H. *States, Scarcity, and Civil Strife in the Developing World.* Princeton, N.J.: Princeton University Press, 2008.

McKee, Jeffrey Kevin. *Sparing Nature: The Conflict Between Human Population Growth and Earth's Biodiversity.* New Brunswick, N.J.: Rutgers University Press, 2003.

Mazur, Laurie Ann, ed. *A Pivotal Moment: Population, Justice, and the Environmental Challenge.* Washington, D.C.: Island Press, 2010.

Meadows, Donella H., Jørgen Randers, and Dennis L. Meadows. *Limits to Growth: The Thirty-Year Update.* White River Junction, Vt.: Chelsea Green, 2004.

Newbold, K. Bruce. *Six Billion Plus: World Population in the Twenty-first Century.* 2d ed. Lanham, Md.: Rowman & Littlefield, 2007.

Reichenbach, Laura, and Mindy Jane Roseman. *Reproductive Health and Human Rights: The Way Forward.* Philadelphia: University of Pennsylvania Press, 2009.

Tobin, Kathleen A. *Politics and Population Control: A Documentary History.* Westport, Conn.: Greenwood Press, 2004.

PRESERVATION AND WILDERNESS ISSUES

Beach, Ben, et al., eds. *The Wilderness Act Handbook.* 5th ed. Washington, D.C.: Wilderness Society, 2004.

Campaign for America's Wilderness. *People Protecting Wilderness for People: Celebrating Forty Years of the Wilderness Act.* Washington, D.C.: Author, 2004.

Carey, Christine, Nigel Dudley, and Sue Stolton. *Squandering Paradise? The Importance and Vulnerability of the World's Protected Areas.* Gland, Switzerland: World Wide Fund for Nature International, 2000.

Chape, Stuart, Mark D. Spalding, and Martin D. Jenkins, eds. *The World's Protected Areas: Status, Values, and Prospects in the Twenty-first Century.* Berkeley: University of California Press, 2008.

Dawson, Chad P., and John C. Hendee. *Wilderness Management: Stewardship and Protection of Resources and Values.* Boulder, Colo.: WILD Foundation, 2009.

Duncan, Dayton, and Ken Burns. *The National Parks: America's Best Idea—An Illustrated History.* New York: Alfred A. Knopf, 2009.

Lockwood, Michael, Graeme Worboys, and Ashish Kothari, eds. *Managing Protected Areas: A Global Guide.* Sterling, Va.: Earthscan, 2006.

Lowry, William R. *Dam Politics: Restoring America's Rivers.* Washington, D.C.: Georgetown University Press, 2003.

Riley, Laura, and William Riley. *Nature's Strongholds: The World's Great Wildlife Reserves.* Princeton, N.J.: Princeton University Press, 2005.

Scarce, Rik. *Eco-Warriors: Understanding the Radical Environmental Movement.* Updated ed. Walnut Creek, Calif.: Left Coast Press, 2006.

Scott, Doug. *The Enduring Wilderness: Protecting Our Natural Heritage Through the Wilderness Act.* Golden, Colo.: Fulcrum, 2004.

Sellars, Richard. *Preserving Nature in the National Parks: A History.* New ed. New Haven, Conn.: Yale University Press, 2009.

Resources and Resource Management

Anderson, David A. *Environmental Economics and Natural Resource Management.* 3d ed. London: Routledge, 2010.

Barrow, C. J. *Environmental Management: Principles and Practice.* London: Routledge, 2002.

Bringezu, Stefan, and Raimund Bleischwitz, eds. *Sustainable Resource Management: Global Trends, Visions, and Policies.* Sheffield, England: Greenleaf, 2009.

Ffolliott, Peter F., Luis A. Bojorquez-Tapia, and Mariano Hernandez-Narvaez. *Natural Resources Management Practices: A Primer.* Ames: Iowa State University Press, 2001.

Graziani, Mauro, and Paolo Fornasiero, eds. *Renewable Resources and Renewable Energy: A Global Challenge.* Boca Raton, Fla.: CRC Press, 2007.

Kudrow, Nikolas J., ed. *Conservation of Natural Resources.* New York: Nova Science, 2009.

McPherson, Guy R., and Stephen DeStefano. *Applied Ecology and Natural Resource Management.* New York: Cambridge University Press, 2003.

Menzies, Charles R. *Traditional Ecological Knowledge and Natural Resource Management.* Lincoln: University of Nebraska Press, 2006.

Mihelcic, James R., and Julie Beth Zimmerman, eds. *Environmental Engineering: Fundamentals, Sustainability, Design.* Hoboken, N.J.: John Wiley & Sons, 2010.

Nemetz, Peter N., ed. *Sustainable Resource Management: Reality or Illusion?* Northampton, Mass.: Edward Elgar, 2007.

Steelman, Toddi A. *Implementing Innovation: Fostering Enduring Change in Environmental and Natural Resource Governance.* Washington, D.C.: Georgetown University Press, 2010.

Young, Anthony. *Land Resources: Now and for the Future.* New York: Cambridge University Press, 2000.

Urban Environments

Adams, Clark E., and Kieran J. Lindsey. *Urban Wildlife Management.* 2d ed. Boca Raton, Fla.: Taylor & Francis, 2009.

Benton-Short, Lisa, and John R. Short. *Cities and Nature.* London: Routledge, 2008.

Birch, Eugenie Ladner, and Susan M. Wachter, eds. *Growing Greener Cities: Urban Sustainability in the Twenty-first Century.* Philadelphia: University of Pennsylvania Press, 2008.

Brunn, Stanley D., Maureen Hays-Mitchell, and Donald J. Zeigler, eds. *Cities of the World: World Regional Urban Development.* 4th ed. Lanham, Md.: Rowman & Littlefield, 2008.

Droege, Peter, ed. *Urban Energy Transition: From Fossil Fuels to Renewable Power.* New York: Elsevier, 2008.

Frumkin, Howard, Lawrence D. Frank, and Richard Jackson. *Urban Sprawl and Public Health: Designing, Planning, and Building for Healthy Communities.* Washington, D.C.: Island Press, 2004.

Gillham, Oliver. *The Limitless City: A Primer on the Urban Sprawl Debate.* Washington, D.C.: Island Press, 2002.

Kahn, Matthew E. *Green Cities: Urban Growth and the Environment.* Washington, D.C.: Brookings Institution Press, 2006.

Pearson, C. J., Sarah Pilgrim, and Jules N. Pretty, eds. *Urban Agriculture: Diverse Activities and Benefits for City Society.* Sterling, Va.: Earthscan, 2010.

Pugh, Cedric, ed. *Sustainability, the Environment, and Urbanization.* 1996. Reprint. Sterling, Va.: Earthscan, 2002.

Spence, Michael, Patricia Clarke Annez, and Robert M. Buckley, eds. *Urbanization and Growth.* Washington, D.C.: Commission on Growth and Development, 2009.

Squires, Gregory D., ed. *Urban Sprawl: Causes, Consequences, and Policy Responses.* Washington, D.C.: Urban Institute Press, 2002.

Waste and Waste Management

Benidickson, Jamie. *The Culture of Flushing: A Social and Legal History of Sewage.* Vancouver: University of British Columbia Press, 2007.

George, Rose. *The Big Necessity: The Unmentionable World of Human Waste and Why It Matters.* New York: Metropolitan Books, 2008.

Grossman, Elizabeth. *High Tech Trash: Digital Devices, Hidden Toxics, and Human Health.* Washington, D.C.: Island Press, 2006.

Hamblin, Jacob Darwin. *Poison in the Well: Radioactive Waste in the Oceans at the Dawn of the Nuclear Age.* New Brunswick, N.J.: Rutgers University Press, 2008.

Kreith, Frank, and George Tchobanoglous. *Handbook of Solid Waste Management.* 2d ed. New York: McGraw-Hill, 2002.

LaGrega, Michael D., Phillip L. Buckingham, and Jeffrey C. Evans. *Hazardous Waste Management.* 2d ed. Boston: McGraw-Hill, 2001.

Pichtel, John. *Waste Management Practices: Municipal, Hazardous, and Industrial*. Boca Raton, Fla.: CRC Press, 2005.

Sigman, Hilary, ed. *The Economics of Hazardous Waste and Contaminated Land*. Northampton, Mass.: Edward Elgar, 2008.

Sprankling, John G., and Gregory S. Weber. *The Law of Hazardous Wastes and Toxic Substances in a Nutshell*. 2d ed. St. Paul, Minn.: Thomson/West, 2007.

Switzer, Carole Stern, and Peter L. Gray. *CERCLA: Comprehensive Environmental Response, Compensation, and Liability Act (Superfund)*. 2d ed. Chicago: American Bar Association, 2008.

United Nations Environment Programme. *Marine Litter: A Global Challenge*. Nairobi: Author, 2009.

United Nations Human Settlements Programme. *Solid Waste Management in the World's Cities: Water and Sanitation in the World's Cities 2010*. Sterling, Va.: Earthscan, 2010.

WATER AND WATER POLLUTION

Allsopp, Michelle, et al. *State of the World's Oceans*. New York: Springer, 2009.

Asano, Takashi, et al. *Water Reuse: Issues, Technologies, and Applications*. New York: McGraw-Hill, 2007.

Copeland, Claudia. *Clean Water Act: A Summary of the Law*. Washington, D.C.: Congressional Research Service, 2008.

Field, John G., Gotthilf Hempel, and Colin P. Summerhayes, eds. *Oceans 2020: Science, Trends, and the Challenge of Sustainability*. Washington, D.C.: Island Press, 2002.

Gleick, Peter H., et al. *The World's Water, 2008-2009: The Biennial Report on Freshwater Resources*. Washington, D.C.: Island Press, 2009.

Glennon, Robert Jerome. *Unquenchable: America's Water Crisis and What to Do About It*. Washington, D.C.: Island Press, 2009.

Kenny, Joan F., et al. *Estimated Use of Water in the United States in 2005*. Reston, Va.: U.S. Geological Survey, 2009.

Pearce, Fred. *When the Rivers Run Dry: Water, the Defining Crisis of the Twenty-first Century*. Boston: Beacon Press, 2006.

Ryan, Mark. *The Clean Water Act Handbook*. 2d ed. Chicago: American Bar Association, 2003.

Shiklomanov, I. A., ed. *World Water Resources at the Beginning of the Twenty-first Century*. New York: Cambridge University Press, 2003.

U.S. Environmental Protection Agency. *National Water Quality Inventory: Report to Congress, 2004 Reporting Cycle*. Washington, D.C.: Author, 2009.

Vickers, Amy. *Handbook of Water Use and Conservation: Homes, Landscapes, Businesses, Industries, Farms*. Amherst, Mass.: Waterplow Press, 2001.

WEATHER AND CLIMATE

Bakker, Sem H., ed. *Ozone Depletion, Chemistry, and Impacts*. Hauppauge, N.Y.: Nova Science, 2009.

Bolin, Bert. *A History of the Science and Politics of Climate Change: The Role of the Intergovernmental Panel on Climate Change*. New York: Cambridge University Press, 2007.

Clarke, Allan J. *An Introduction to the Dynamics of El Niño and the Southern Oscillation*. San Diego, Calif.: Academic Press, 2008.

Dessler, Andrew Emory, and Edward Parson. *The Science and Politics of Global Climate Change: A Guide to the Debate*. New York: Cambridge University Press, 2006.

Hewitt, C. Nicholas, and Andrea V. Jackson, eds. *Atmospheric Science for Environmental Scientists*. New York: Wiley-Blackwell, 2009.

Houghton, John Theodore. *Global Warming: The Complete Briefing*. 4th ed. New York: Cambridge University Press, 2010.

Intergovernmental Panel on Climate Change. *Climate Change 2007: Synthesis Report*. Geneva: Author, 2008.

Schneider, Stephen Henry, et al., eds. *Climate Change Science and Policy*. Washington, D.C.: Island Press, 2010.

Shulk, Bernard F., ed. *Greenhouse Gases and Their Impact*. New York: Nova Science, 2007.

Turco, Richard P. *Earth Under Siege: From Air Pollution to Global Change*. 2d ed. New York: Oxford University Press, 2002.

United Nations Environment Programme. *Environmental Effects of Ozone Depletion and the Interaction with Climate Change: 2006 Assessment*. Nairobi: Author, 2006.

Zerefos, Christos, Georgios Contopoulos, and Gregory Skalkeas, eds. *Twenty Years of Ozone Decline: Proceedings of the Symposium for the Twentieth Anniversary of the Montreal Protocol*. Dordrecht, Netherlands: Springer, 2009.

INDEXES

Category Index

List of Categories

ACTIVISM AND ADVOCACY

AGENCIES. *See*
ORGANIZATIONS AND AGENCIES

AGRICULTURE AND FOOD

ANIMALS AND ENDANGERED SPECIES

WILDERNESS ISSUES. *See*
PRESERVATION AND
WILDERNESS ISSUES

Subject Index